Signals and Communication Technology

This series is devoted to fundamentals and applications of modern methods of signal processing and cutting-edge communication technologies. The main topics are information and signal theory, acoustical signal processing, image processing and multimedia systems, mobile and wireless communications, and computer and communication networks. Volumes in the series address researchers in academia and industrial R&D departments. The series is application-oriented. The level of presentation of each individual volume, however, depends on the subject and can range from practical to scientific.

Indexing: All books in "Signals and Communication Technology" are indexed by Scopus and zbMATH

For general information about this book series, comments or suggestions, please contact Mary James at mary.james@springer.com or Ramesh Nath Premnath at ramesh.premnath@springer.com.

Amy Neustein • Parikshit N. Mahalle •
Prachi Joshi • Gitanjali Rahul Shinde
Editors

AI, IoT, Big Data and Cloud Computing for Industry 4.0

Springer

Editors
Amy Neustein
Linguistic Technology Systems
Fort Lee, NJ, USA

Parikshit N. Mahalle
Dept of Artificial Intelligence and Data Science
Vishwakarma Institute of Information Technology
Pune, Maharashtra, India

Prachi Joshi
Vishwakarma Institute of Information Technology
Pune, India

Gitanjali Rahul Shinde
Department of Computer Engineering
Vishwakarma Institute of Information Technology
Pune, India

ISSN 1860-4862 ISSN 1860-4870 (electronic)
Signals and Communication Technology
ISBN 978-3-031-29712-0 ISBN 978-3-031-29713-7 (eBook)
https://doi.org/10.1007/978-3-031-29713-7

This Springer imprint is published by the registered company Springer Nature Switzerland AG
The registered company address is: Gewerbestrasse 11, 6330 Cham, Switzerland

Introduction

AI, IoT, Big Data, and Cloud Computing for Industry 4.0 offers a stimulating discussion of some of the most advanced leading-edge technologies for the fourth Industrial Revolution – known as "Industry 4.0." In composing this 24-chapter anthology, we have hand-selected contributors engaged in some of the most fascinating work at universities, research institutes, and think tanks – spanning the United States, Europe, and India.

Our goal in assembling this collection was to provide the reader with a comprehensive understanding of the interconnections of AI, IoT, Big Data, and Cloud Computing as integral to the technologies that revolutionize the way companies produce and distribute products and the way local governments deliver their services. Therefore, at every phase of the supply chain, manufactures are found to be interweaving AI, Robotics, IoT, Big Data/Machine Learning, and Cloud Computing into their production facilities and throughout their distribution networks. Equally important, the authors show how their research can be applied to computer vision, cyber-security, natural language processing, healthcare, education, and agriculture.

This compendium is divided into five sections:

The first section, which presents a comprehensive exposition of the fundamentals of Industry 4.0, begins with a rich discussion of the challenges and opportunities encountered when transitioning to Industry 4.0, thereby substituting AI and robotics – backed by IoT and cyber-physical systems technologies – for human labor used to perform a variety of tasks. The authors of that chapter point out that in order to achieve full automation – in which the entire manufacturing process is sans human intervention – we avail ourselves of sensors, actuators, and automation.

The chapter that follows provides keen insights into making the Industry 4.0 ecosystem run more smoothly. Namely, the authors explore Human Computer Interaction (HCI) principles within the context of Industry 4.0, emphasizing the seamless interface between humans and machines – particularly in those instances when humans are needed for supervision, maintenance, confirmation of critical decisions made by machines, and/or handling any malfunctioning of machinery. The chapter expounds on how in an era where machines are getting smarter and making human lives easier, Industry 4.0 architecture elevates the power and capability of machines – and how it utilizes the greater interconnectivity and communication between already smart machines. As such, combinatorially, this makes the entire industrial ecosystem smarter and more efficient, and less prone to error.

The next chapter provides a rich and substantive literature review of affective computing technology, looking at among other things the different modalities, techniques, and tools for detecting the affective state of the user. The authors examine the published research on technologies for assessing the affective state of the user during an e-Learning session, no doubt an excellent domain for testing the arsenal of technologies, such as Big Data – and more specifically, data mining – made available for Industry 4.0. They show how, in recent years, web based education has been perceived as a support tool for instructors, as it can be easily used at any time or at any place. However, they astutely point out that most courses emphasize the cognitive area, while wholly ignoring the learner's feelings and learning mentalities due to the intrinsic remoteness of distance learning.

Given that emotions play a noteworthy part in the cognitive procedure of individuals, technology is deficient when it fails to capture the student's emotive state. Thus, with the aid of affective computing technology, it is possible for researchers to understand the learner's affect during online learning, so that they are able to take corrective action. The chapter presents an organized review related to affective computing technology – exploring and enumerating the different modalities, techniques, and emotions identified by researchers, as well as the tools they have developed to realize e-learning practicability and improvement. Finally, several challenges are listed for helping researchers in the course of applying affective computing technology.

After assiduously analyzing a literature base of 61 research papers, the chapter authors conclude that though many researchers have used the facial expression modality for prediction of affective states, few datasets are available which recognize a broader spectrum of modalities. On the whole, there is still a lot of potential for predicting learning emotions like confusion, engagement, frustration, and so forth from a variety of modalities, which gives strong support for exploring modalities beyond facial expression.

The section closes with a fascinating discussion of how enterprises are expanding their services onto the network edge, wherein the edge performs all cloud-related tasks with little or no interaction with the cloud. The authors demonstrate how AI-enabled edge nodes integrated with IoT devices provide enhancements in computation power, accuracy, and low latency in service delivery as part of the broader benefits of intelligent edge-enabled IoT devices. The authors explain how digital transformation is accelerating the development of the Internet of Things (IoT) devices, pointing out that with technological advancements digital transformation is no longer limited to the Internet of Things (IoT) but has extended to the Internet of Everything (IoE).

That is, smart environments require digitally connecting every device and collaborative interaction among them. Such a huge development of IoT devices is promoting voluminous data generation, and these billions of data are being managed by cloud data centers. Nevertheless, as the authors point out, there is a need for data to be managed, processed, and computed outside the cloud, preferably near the data source. There is likewise a necessity to avoid overburdening of the cloud and the same can be achieved by adopting edge computing. They explain how edge computing convention means pushing cloud services onto the "edge" of IoT device networks.

In the second section, which offers a diversified look at the emerging trends in AI, we begin with the authors' presentation of cutting-edge research in the design of a Chatbot to understand Hindi natural language input from users. Given that advances in machine learning have greatly improved the accuracy of NLP (natural language processing), the use of Chatbots across many languages and dialects, and even more importantly in under-resourced languages, opens up new frontiers for this technology. The authors present their research on an emotionally therapeutic Chatbot in the Hindi language. They've built their CBT (cognitive behavioral therapy) Chatbot using the Transformer model architecture, using the Seq-to-Seq transformer based on a deep learning model.

The authors open their chapter with a discussion of how Chatbots, also known as conversational interfaces, provide a new way for users to interact with computer systems. Chatbot software is a feature or an application for communicating directly through natural language with online users, so as to allow their queries to be solved without having to wait in a queue for a call center agent to pick up. The authors focus on Covid-19 as a use case for a CBT-driven Chatbot in development which provides frequently asked questions about COVID-19 and relevant information resources. They emphasize the mental health aspect of the disease because Covid-19 not only impacts physical health but mental health as well.

The authors describe their design of a special Chatbot that offers a cognitive Patient Health Questionnaire to respond to the mental health needs of users during the Covid-19 pandemic. Based on the user's sentiment analysis score, the Chatbot will then guide the user to find helpful resources to address mental health issues such as depression. They candidly discuss the many challenges encountered while building a Chatbot in Indian languages: namely, no appropriate data and information in Hindi language related to COVID-19 and mental health were available. Therefore, maintaining and adding data in the database became a tedious task. Being morphologically rich, free word order languages, Indian dialects carry ambiguities and complexities and are, therefore, complex to handle. Most of the Indian languages fall under the low resource language category as the datasets currently available for most Indian languages are limited when compared to the English language.

The chapter contributors explain that because the World Health Organization had declared Covid-19 as a public health emergency of international concern (PHEIC) making information about Covid-19 available to speakers of many languages became a healthcare priority. Guided by this initiative to help make information available in a public health crisis, they designed the Chatbot by collecting relevant data from the user and then displaying the results in web and mobile devices. This CBT-driven Chatbot is built using the Transformer model architecture, which is based on the attention mechanism.

Previously, the authors had worked on this model for the English language, and now to advance the Chatbot in different languages, the authors are developing the Chatbot in the Hindi Language as well. They have applied a Seq-to-Seq transformer-based deep learning model into proposed work which presents an end-to-end potential in the domain. They found that the development of a Chatbot in Hindi was necessary to help people who know only Hindi language to converse with the Chatbot easily.

The following chapter presents authors' findings of an extensive literature review of machine learning algorithms for predictive maintenance (PdM). The authors explain that maintaining product and machine safety while lowering maintenance costs has recently become a serious concern. Predictive maintenance (PdM) is one important strategy for achieving these goals. They explain that machine learning (ML) is a good candidate for PdM since current machines come with a large quantity of operational data. PdM and ML for rotary systems (bearing failures) have both been discussed extensively in review publications.

They point out that the number of articles in this sector is growing, thereby emphasizing the necessity of a comprehensive study. Thus, they conduct an analysis of recent academic articles that were published in the database of ScienceDirect, Scopus, Institute of Electronics and Electrical Engineers (IEEE), and Google Scholar from 2017 until June 2021. In summing up, they explain that predictive maintenance is a popular issue in the context of Industry 4.0, but it comes with a number of problems that need to be properly researched in the fields of machine learning. Their literature review study presents a detailed overview of machine learning algorithms for bearing failure prediction, including the most often used ML techniques.

In the next chapter, the authors show how soil and fertilizers are an integral part of the yield to be harvested since the right type of soil and fertilizers in accurate amounts can enhance both the growth and health of the crops to a substantial degree. They point out, however, that the present-day system does not comprehend intelligent recommendations, but instead a scarcity (or nescience) of experimentation and legacy knowledge serves to strip farmers of additional money and time. As a result, they recommend the adoption of a system that is both effective and user-friendly. In so doing, their chapter proposes and implements a system to predict suitable crops and fertilizers according to geographic location and soil quality by applying machine learning algorithms.

The authors' proposed system mainly comprises four stages: analyzing and visualizing the data; separating the data into training and testing sets; training the machine learning models; and, finally, comparing the accuracy of each model. XGBoost and Random Forest were the two models that were the most accurate when it came to determining the most efficient recommendation model. They point out that the highest accuracy was found to be 99.31% for crops and 90% for fertilizers, which makes their system extremely valuable for the farmers to amplify their harvest yield.

The next chapter closely examines the use of machine learning in cybersecurity, which has emerged as a natural next step in modern network intrusion detection systems. The authors point out that as the whole world is moving toward a digital first economy, the number of people with malicious intent who wish to exploit weaknesses in any system has also increased proportionally. In response to this growing threat, there has been an emergence of a number of Intrusion Detection Systems (IDS) to perform anomaly/outlier detection to distinguish anomalous traffic from normal traffic.

The authors performed a comparative study of several machine learning techniques on the well-known NSL-KDD Dataset, and analyzed their effectiveness for anomaly detection. In particular, they analyzed the use of OneClass SVM, Isolation Forest, Auto Encoders, and traditional binary classification algorithms. The authors amassed data suggesting that a system of stacked autoencoders is the most efficient and accurate solution for creating a network intrusion detection system.

In the next chapter, the authors present an optimized and efficient facial recognition system using a very advanced face recognition algorithm. They point out that facial recognition is one of the most important and widely studied topics in the field of Computer Vision and Artificial Intelligence. The ability of automation to detect and verify a person's face using photography is a very important factor in a number of domains: monitoring, device access control (laptop and mobiles), security, tracking, law enforcement, biometrics, information security, smart cards, surveillance systems, and so forth. Various applications help to identify specific people in specific areas that help to find intruders. Real-time recognition is necessary for surveillance purposes.

The authors show how their facial recognition algorithm is processed into five main stages: (i) data collection, (ii) data cleaning, (iii) fine-tuning of the VGG16 model, (iv) model training, and (v) performance checking. The first step is collecting face images of different people. In the second step, face images are cropped and saved to remove unnecessary, irrelevant, or meaningless data (noise). In the third step, a very popular image recognition model, VGG16, is fine-tuned for the specific case of facial recognition with the help of a convolutional neural network. The proposed work is implemented using a python module named Keras. To insure robust experimental design, the authors measure performance using a testing dataset that is different from the training dataset.

The section concludes with a fascinating discussion of digitization in teaching and learning. The authors draw the reader's attention to the fact that Covid-19 compelled students to perform their scholastic work from home, working in a digital environment. However, they point out that both students and teachers faced numerous problems, including internet troubles, software requirements for holding classes, and so on. Witnessing these difficulties, the authors decided to undertake a poll that takes into account a variety of factors such as the location of students and professors, Internet connectivity, engagement in college work, online assessments, etc.

The authors conducted a survey with a large sample size of 1060 students from various departments (e.g., Computer Engineering, Electronics and Telecommunications Engineering, Mechanical Engineering, and Civil Engineering) and 180 teachers who likewise participated in this survey. They found that few students were interested in online learning, and there were a variety of reasons for this, such as technical limitations and problems interacting with the instructors. For example, those residing in rural locations have to contend with more connectivity issues than students living in cities – which served as one of the explanations given by the students for their lack of interest and engagement in online classes. They also found that online classes were ripe for cheating notwithstanding the precautions that were taken.

Based on the survey results, the authors were able to ascertain which instructional strategies engage students the most and how many students actually received the necessary work assistance with their course assignments and class materials. Analyzing these data, the authors could identify possible solutions to make online learning more engaging for both students and teachers. Overall, their research can improve online learning both during and after the pandemic, and this represents an important contribution to e-learning.

The third section, which examines AI-based data management, architecture, and frameworks, begins with an exciting presentation of artificial intelligence-based voice recognition for a remotely operated robot using Arduino. Currently, there are a variety of methods for controlling a robot. Using voice commands provides one of the most user-friendly options. For the disabled, presented with mobility challenges, using voice commands to control the robot's operations is a viable solution to the problem.

The authors show how controlling the robot using the user's vocal commands in conjunction with the input-data visual feeds makes it easier and more precise for the robot to operate. Voice-controlled robots powered by artificial intelligence (AI) assist people in saving time and effort in their daily activities and tasks. Following the interpretation of voice instructions given by users, a set of control data for completing a job is created by the robot's processors. The robot can perform different movements: going forward or backward, making turns, starting and stopping, and activating night mode operation.

The next chapter presents the illuminating findings of a group of researchers who have meticulously examined real-time interactive AR (augmented reality). The authors describe multidisciplinary research covering the domains of clinical study, computational engines, run-time interactive user interface, and software integration, with a ready-to-deploy solution that also serves as a proof-of-concept prototype for each stage of the AR pipeline (in effect an end-to-end demo or reference implementation). Among other stages, the implementation encompasses an Automatic Text Visualizer (ATV). Accordingly, the authors point out that this is the only published work on ATV for an Indian language, specifically Hindi. The authors faced a serious challenge given that the Hindi language is morphologically richer and has a free word order that makes it more complex for computational language processing. Here, the considered user domain is for cognitive learning. Visualization may help in decreasing the cognitive load of a person who has difficulties in comprehension, such as dyslexia. To that end, a dynamic simulation of a behavior-rich interactive 3D virtual environment supports cognitive learning.

The authors show that pedagogies such as constructivist learning, situated learning, game-based learning, and inquiry-based learning can be reinforced by Augmented Reality (AR). Based on Piaget's theory of cognitive learning, they propose visual and linguistic analytics to promote cognitive development by effectively using AR. To that end, they hypothesize that visual images may be easy to comprehend for a person having linguistic learning difficulties.

Prior to this work, the authors had studied, designed, and developed Preksha – a Hindi Text Visualizer – which has a user interface to take language input and produce a 3D virtual environment. Concomitant with their research on run-time generation via AR technology, our present work uses cloud porting, computer vision, interactive virtual assistance, and voice support pertaining to the important domain of mental health support. All in all, the advancement in AR technology has led them in the creation of an intelligent live Avatar (3D character) in the immersive real world as an ally of the end-user. This avatar comprehends and performs actions based on linguistic instruction using natural language processing (NLP) in text and automatic speech recognition (ASR) in speech.

The next chapter examines sentiment analysis as a useful tool for social media and customer analysis, allowing one to glean a summary of the views of a large population regarding a particular topic. The authors present the findings of an empirical survey of different techniques for sentiment analysis; they cover the implementation, advantages, and limitations of each of these methods and conduct an experiment to find out which of these methods is best suited for sentiment analysis in today's scenario. Their experiments conducted on benchmark datasets and the results they obtained highlight the fact that supervised learning algorithms like support vector machines (SVMs), multinomial logistic regression (MLR), and deep learning-based algorithms like convolutional neural network (CNN) and bi-directional recursive neural network (RNN) show improved performances over the lexicon-based methods.

In the chapter that follows, the authors point out that their purpose is to cover Sign Language preliminaries and a global survey of Sign Language Machine Translation Systems with particular reference to Indian Sign Language. They explain that for computer technology to make sense, it is imperative to have social relevance, and the ability to serve across sections of human society where there are physical impairments.

The authors make the incisive point that while growth of IT applications in different real-life domains has boosted research in assistive technology for persons with physical disabilities and for those who are visually impaired, there has been significantly less IT-enabled technical aid observed for persons with hearing impairment. They point to the fact that Sign Language has been a prominent source of communication in this community for social connections. The difference in modalities of Spoken and Sign Language has been one of the natural reasons for communication barriers between the hearing impaired and the rest of society. They show that advances in Machine Translation Systems have triggered the idea of translation between Spoken Language and Sign Language, and proceed to analyze in their chapter different issues and aspects of Sign Language Machine Translation

Systems such as translation methodology, grammatical features, domain, output interpretation, and handling of simultaneous morphological features. In addition, they review published work on Sign Language Machine Translation Systems and astutely identify the gaps in reported systems while proposing a tool to handle simultaneous morphological features of sign languages at run time.

The section concludes with a chapter on how to effectively digitize Devanagari, the official Indian language used in preparing documentation. Devanagari is the most commonly used script in India. It is the official writing system in many institutes or organizations for documentation purposes. Due to digitization, these documents need to be stored in digital format. However, automatic transcription of images of the handwritten text is a tedious task. The authors address the problem of handwritten Devanagari character recognition. CNN (convolutional neural networks) was found to be able to classify handwritten text. However, as the authors revealed, CNN fails to recognize the image correctly for rotated samples – due to the changed spatiality of the rotated image – and the pooling layer of CNN likewise failed in cases of disfiguration and/or proportional transformation of any image.

To meet these challenges, the authors introduced the use of capsule networks for Devanagari text classification. That is, a capsule network uses a dynamic routing algorithm that explores the spatial relationship among the features such as size, feature orientation, perspective, etc. The experiment results show that the CapsNet (capsule network) framework improves results over CNNs for handwritten Devanagari character recognition.

In the fourth section, which examines important security issues for Industry 4.0, we begin with a presentation of the authors' novel research of a predictive model that explores the effect of personalization on users' trust and privacy concerns when making purchases from an e-commerce site. The authors begin by elucidating that real-time personalization is adopted by e-commerce websites to leverage business opportunities with the offering of recommendations that meet users' implicit needs. However, as they aptly point out, little is known about its effect on users' trust and privacy concerns in the context of purchasing behavior. In their chapter they present research on a predictive model for exploring the effect of personalization on users' trust and privacy concerns toward personalization, and the interrelation of these concerns with their willingness to purchase on an e-commerce site. SEM (Structural Equation Modeling) is used to build a model on EFA (Exploratory Factor Analysis) and CFA (Confirmatory Factor Analysis) results.

The authors' experimental results show how the model fits to the parameters of personalization and the role of personalized information relevance toward the belief of trust and privacy concerns in e-commerce websites. Their findings suggest that: personalized recommendation is positively related to users' satisfaction and privacy concerns. Results show that users' trust is not positively correlated with privacy concerns; instead users experience a lower degree of trust when they have higher privacy concerns. Users with higher satisfaction with ecommerce websites are likely to develop more trust. Furthermore, they point out that users' purchase intentions in general are not affected by satisfaction, but satisfaction positively affects the users' purchase intention if their privacy concerns are addressed when generating personalized recommendations. This points out the importance of privacy concerns as a single variable which has an impact on users' interaction with ecommerce sites particularly for those returning to that ecommerce site.

In the chapter that follows, the authors focus on a particular kind of ad hoc networking, namely, wireless sensor networks (WSN). They indicate that dynamic WSNs are in high demand due to recent advances in hardware design, rapid growth in wireless network communications and infrastructure, and increased user demands for node mobility and regional delivery processes. Various studies have been conducted to determine how to improve the reliability of routing nodes by using cryptographic applications, confidence management, or central routing solutions among other methods. However, as the authors show, a reliable routing structure is required to confirm the security and efficiency of wireless sensor network routing.

In contrast, the majority of routing patterns are difficult to execute in real-world circumstances due to the difficulty of dynamically identifying untrustworthy routing node behavior. Simultaneously, there is no effective way to defend against attacks from malicious hosts. In light of the aforementioned concerns, the authors propose a secure routing system for WSNs that incorporates blockchain and reinforcement learning to continually enhance routing efficiency and security. Applying blockchain, a feasible routing mechanism for gaining routing evidence from routing nodes is presented, making routing information detectable and unalterable. The trained reinforcement model is used by routing nodes to dynamically select more efficient and reliable routing channels.

In the next chapter, the authors start by explaining that the popularity of digital media is on the rise today: besides multilevel images, video, and audio, binary document images are also digitized in many applications. These include legal documents, digital books, maps, pictures, and architectural drawings. Security and confidentiality of information transmitted through digital media are the primary concerns. Employing cryptography is an imperfect solution because doing so can expose sensitive information to intruders. Information hiding is more promising as it focuses on imperceptibility. However, hiding information in binary media is more difficult than color and grayscale media, as it causes more visual distortion due to two-shade limitations. Apart from this, the hiding technique one uses needs to take into account the requisite hiding capacity and the security of hidden information.

The authors present an approach for hiding information in binary images which tries to address these issues. Thus, the message to be hidden is initially encrypted by a key known to both transmitter and receiver. This encrypted message is then represented as a two-dimensional sparse matrix and then transformed to a one-dimensional matrix where only information about locations of black pixels "1" in the sparse matrix is preserved. An arithmetic compression algorithm is used to compress the sparse matrix. Black and white pixel run length (RL) pairs are utilized for hiding information which satisfies the threshold set for RL pair length. On the receiver side, a reverse process is applied to extract secret messages. The encryption, sparse matrix representation, and arithmetic data compression secure the information and enhance hiding capacity. Utilization of run length pairs minimizes visual distortion. The authors' study results show that high capacity information is hidden with less distortion. As digital multimedia sources become more specialized, the information hidden within them is used for steganography and watermarking.

In the final chapter in this section, the authors examine the Federated Learning (FL) technique that is widely applied in many fields because models are trained locally with participant data instead of the training data being gathered centrally. Applications such as medical imaging, next-word prediction, and speech prediction widely use FL because these use cases involve sensitive and private data. Although FL avoids sharing actual data, it still faces various privacy and security concerns. Adversaries can actively or passively attack the participant's privacy from the shared model. There are mainly two issues to handle in FL: (1) protecting user's privacy; and (2) integrity of the averaged model (gradients).

The authors point out that currently there is no single solution for privacy preservation that applies to all situations. But applications may adopt different Privacy Enhancing Technologies (PETs) to fulfill the additional privacy and security priorities. Many researchers have proposed a solution to tackle FL's security and privacy challenges. Ongoing research addresses the possibility of improving the privacy and security of participants' private data using various techniques such as Homomorphic Encryption (HE), Trusted Execution Environments (TEE), Differential Privacy Preservation (DPP), and Secure Multiparty Computation (SMC). This chapter demonstrates how the gradients can be inverted to get the original image, which reveals limitations in these methodologies. In light of such limitations, the authors present ways to protect the privacy and security of participants' data in FL.

The fifth section of this anthology, which is the last section of this book, is devoted to Software Language Implementation, Linguistics, and Virtual Machines. The section begins with a chapter that examines data-interoperation protocols for independently engineered software components. The authors propose hypergraph-based metamodels as a useful abstraction for designing data-representation formats that mutually autonomous components can collaboratively adopt as a neutral encoding strategy. They consider ways to rank data-formats' expressiveness, and compare hypergraphs to comparatively "less" expressive metamodels, such as those implicit in XML and JSON, suggesting in concrete terms why expressiveness is beneficial to efficient software development. Concrete examples are drawn from Architecture, Engineering, and Construction (AEC), bioimaging, and, in particular, Geographic Information Systems (GIS) and digital cartography. The authors take the discussion to a broader level by considering how to properly frame the "linguistic" nature of computer programming languages and how software applications have their own "semantics," in the sense of representational connections to empirical objects.

In the next chapter, the authors explore Virtual Machines (VMs) pointing out that because they are "virtual" and not restricted to physically realizable operations, they provide a flexible and extensible compilation target that can promote expressive and adaptable high-level programming languages. The authors examine hypergraph representations of source code, and analyze how hypergraph constructions can serve as a point of orientation for designing a VM's instruction set; they sketch, semi-formally, a mathematical overview suggesting how systems similar to the lambda calculus can be represented in the context of hypergraphs, yielding graph-theoretic models of such calculi that might be adopted as alternatives to their original formulations in symbolic logic. They then outline how this approach to Virtual Machine design and code-modeling may benefit software engineering in contexts such as data sharing protocols and GUI front-end programming.

In the chapter that follows, the authors focus on image-processing and Computer Vision. They consider requirements for a hypothetical Virtual Machine whose intent is consolidation of access to image-analysis functionality by exposing sets of Computer Vision algorithms through a common interface. They review potential Virtual Machine features that are not unique to image-processing but may be especially relevant to that domain, such as recognizing different varieties of number-pairs as distinct built-in types.

In their discussion they also consider functionality that becomes consequential insofar as GUI front-ends are engineered alongside image-processing technology, either through user-guided algorithms (such as interactive segmentation) or through interactive visualization wherein users review workflow architectures, intermediate results, color histograms (and other special-purpose GUI components), or make comparisons between distinct processing algorithms—and similar tools that help users select and fine-tune image-analysis techniques for specific image series. The authors provide a case study for algorithms that may be integrated with Virtual Machines or application GUIs and summarize some facets of a novel database-related image format and explain in detail the mathematical background of its color models and in-memory layout for image data.

The next chapter explores animal cognition in the context of linguistics and robotics, looking at possible models for intermediate-level languages which are more lifelike and situationally grounded than formal computer code, but less complex and nuanced than natural (human) language. The authors show how "animal language" can provide a model for intermediate languages in this sense. They focus on observations and data generated in the context of research on human-canine interactions in the specific modality where people train dogs to use "talking buttons," inspired by Augmentative and Alternative Communication (AAC) devices designed to help speech-language pathologists and their patients, such as children on the autism spectrum. They point to studies that show how dogs (albeit a small sample size) have achieved surprising levels of communicative sophistication – which is not to imply that they "talk" in human language, but they (dogs) do reveal situational and intersubjective/collaborative awareness that potentially goes beyond dogs' reasoning abilities even as measured by very recent dog-cognition research. Their point is that such observations about animal cognition elucidates language at an intermediate stage which may have applications for research in robotics.

In the last chapter in this section, as well as the coda for the book, the authors investigate the interconnections between human language, computer languages, and intermediate forms of communication that may be appropriate for Human-Robot Interaction (HRI). They argue that much of the cognitive processing intrinsic to understanding language is actually extra-linguistic, and in particular that the detailed propositional content conveyed by linguistic expressions is, in many contexts, not structurally present in expressions themselves (or at least, not in its entirety).

In making this argument, the authors consider semantic models which accordingly construe language as providing an "interface" to cognitive "procedures" rather than a logical encoding of predicate meaning. They propound that these approaches have merit in the domain of human language, as they offer potential insights in the realm of programming languages as well. The authors reflect on how linguistic models may be adapted to human-robot communications, and outline plausible conventions governing the interface through which humans will examine, guide, or visualize robots' movements and surroundings. Similarly, they consider GUI programming elements and Human-Computer Interaction (HCI) conventions in traditional computing environments as analogs to HRI (Human Robotic Interactions), suggesting that configurations of GUI controls—onscreen in application windows—serve as a proxy structure for developing technologies to model robots' orientation.

In conclusion, it has been the endeavor of the editors to marshal the most advanced technologies in AI, IoT, Big Data, and Cloud Computing for inclusion in a comprehensive volume that openly explores how such technologies facilitate and guide the fourth industrial revolution – no doubt among the most creative and most promising phases of industrial development in the history of technology.

Contents

Part I
Fundamentals of Industry 4.0

Chapter 1
Opting for Industry 4.0: Challenge or Opportunity

Kirti Wanjale, A. V. Chitre, and Ruchi Doshi

1.1 Introduction

Integrating the physical and computational worlds has led to the fourth industrial revolution, also termed as Industry 4.0. Instead of having mass and semi-customized products it has turned to mass and fully customized products is the meaning of Industry 4.0. Industry 4.0 gives products; Internet enabled facilities, innovative services, Internet based diagnostics and maintenance in an efficient manner. Moreover, it helps in understanding new business models, operating concepts and smart controls, and focusing on the user with his individual needs. These are the systems with industrial automation that enables innovative functionality through Internet and their access to the cyber world, thus changing the lives every day. All future factories should operate in accordance with the Industry 4.0 concept [1]. Although this notion was initially proposed in the context of manufacturing systems, it can be utilized in a variety of industries, including oil and gas, chemical plants, and power plants.

The purpose of this chapter is to provide a high-level overview of Industry 4.0 technology. Our purpose is to provide a viewpoint on what Industry 4.0 is, what its issues are in today's setting, and what the future holds. In this chapter, we have discussed how new technologies help in Industry 4.0 transformation, challenges

K. Wanjale
Computer Engineering Department, Vishwakarma Institute of information Technology, Pune, India
e-mail: kirti.wanjale@viit.ac.in

A. V. Chitre (✉)
EnTC Department, Vishwakarma Institute of information Technology, Pune, India
e-mail: abhijit.chitre@viit.ac.in

R. Doshi
University of Azteca, Chalco de Díaz Covarrubias, Mexico
e-mail: ruchi.doshi@univ-azteca.edu.mx

A. Neustein et al. (eds.), *AI, IoT, Big Data and Cloud Computing for Industry 4.0*, Signals and Communication Technology, https://doi.org/10.1007/978-3-031-29713-7_1

faced to implement Industry 4.0, benefits of Industry 4.0, and finally some case studies where Industry 4.0 is already in place. After reading this chapter, you will come to know about various aspects of implementing Industry 4.0 standards in industries. Also, case studies will demonstrate on how Industry 4.0 is already implemented in various sectors.

1.1.1 Role of Technologies in Industry 4.0 Transformation

It is the *digital transformation* of manufacturing processes and also production in the industry. Meaning of Industry 4.0 is with the help of information and communication technology, switching to the intelligent networking of machines and different processes for industry. Certain goals need to understand behind Industry 4.0 and they are as follows (Fig. 1.1):

- Automate the decision-making processes
- Monitor various assets and processes in real time
- Enable equally connected value creation networks by involving stakeholders
- Implement information technology (IT) and operational technology (OT) convergence

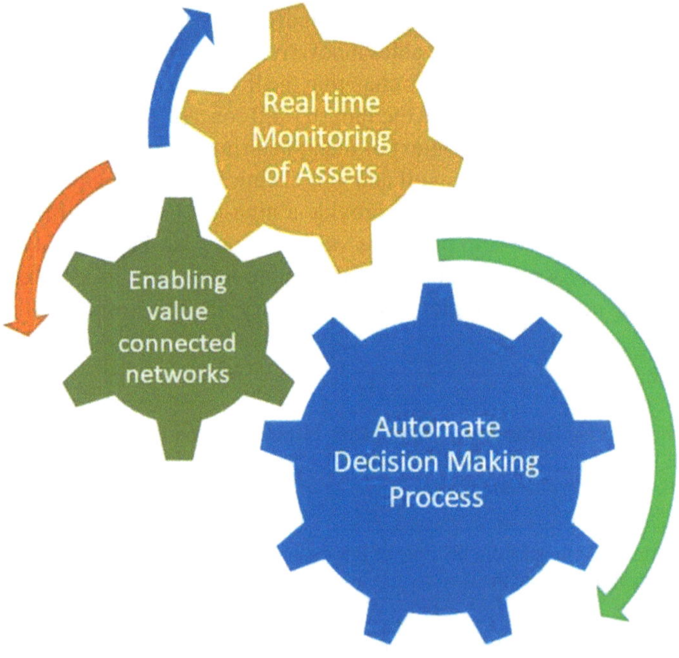

Fig. 1.1 Goals behind Industry 4.0

Automate the decision-making processes: At the very core, Industry 4.0 includes the partial transfer of autonomy and autonomous decisions. It is possible to leverage the information systems with the help of cyber physical systems and machines.

Monitor various assets and processes in real time: In Industry 4.0, it is important to monitor various assets of the industry. Also, the different processes the industry is following since long [2].

Enable equally connected value creation networks by involving stakeholders: When it comes to Industry 4.0, stakeholder involvement is also critical. With the support of stakeholders, we need to enable value generation networks. It is critical to comprehend the entire industry value chain. It contains information on suppliers as well as the sources of the materials. Components required for various types of smart manufacturing are also critical for improved output. Regardless of the number of intermediary procedures and stakeholders, the eventual destination of all output, the end client, must be understood.

Revolutions are disruptive by definition, and the fourth industrial revolution, often known as Industry 4.0, is no exception. Experts believe that this new wave of revolution will bring about just as much change as the previous ones. Unlike previous technological advances such as steam power, electricity, and digital machinery, which were all based entirely on new technology, Industry 4.0 is focused on how new and old tools may be used in novel ways [3].

Robots now work alongside industry workers, and autonomous cars refill production line supplies, according to Industry 4.0. Designers and industrial workers have been connected via sensor networks and communications technologies, with intelligent machines and software interacting autonomously over the cloud, and facilities connected in real time to suppliers and consumers.

Smart technologies, or rather smart technological utilization, offer the manufacturing industry so much potential. Engineers can get instant feedback on costs and performance predictions. Factory machines and logistics equipment can automatically assign factory processes. Cloud-based artificial intelligence (AI) systems can compare parts and processes to optimize performance and computer systems equipped with machine-based learning algorithms enable robotic systems to learn and operate with limited input from human operators.

In the manufacturing industry, it is impossible to go a day without hearing the phrase "Industry 4.0." Manufacturing technologies are growing to fully integrate both automation information and data interchange, which makes sense in an age where most firms' final aim is digital transformation. There is no end to what Industry 4.0 may bring to factory settings and beyond, thanks to the integration of cyber-physical systems, augmented reality, the industrial Internet of things (IIoT), and cloud computing.

Many businesses, however, continue to struggle with successful adoption; according to *IndustryWeek*, two out of every three enterprises piloting digital manufacturing solutions fail when scaling up to large-scale implementation. It begs the following question: why have so many businesses recognized that transformation is required to reach the future of digital manufacturing, but so few have been able to realize its full potential?

When it comes to speeding up factory transformation, edge AI and IoT are crucial. What, on the other hand, is required to accelerate the adoption of these technologies while also assisting businesses in avoiding the pilot stage?

1.1.2 Key Technologies to Transform Production Industry

Industry 4.0 encompasses the latest technologies that bring digital and physical world together [4]. It has been like a new industrial stage in which several emerging technologies are converging to provide digital solutions. Generally speaking, Industry 4.0 supports the growing trend toward automation and data exchange in technology including the following (see Fig. 1.2):

1. *IoT (Internet of things):* It plays a very important role in Industry 4.0. IoT makes possible to connect all physical devices to Internet and bring automations in the work.
2. *Industrial sensors:* The sensors like proximity, temperature, acoustic, and position which sense data from large distance brought revolutionary change in working pattern of industries.
3. *CPS (cyber physical systems):* Cloud computing along with IoT is already opted in Industry 4.0 which has changed most of the processes in industries. Also, it secures all connected assets by applying highly secure policy to protect.
4. *Digital twin:* Industry 4.0 allows to simulate the change in the production line before actual change which is also called as digital twin. It saves huge amount as

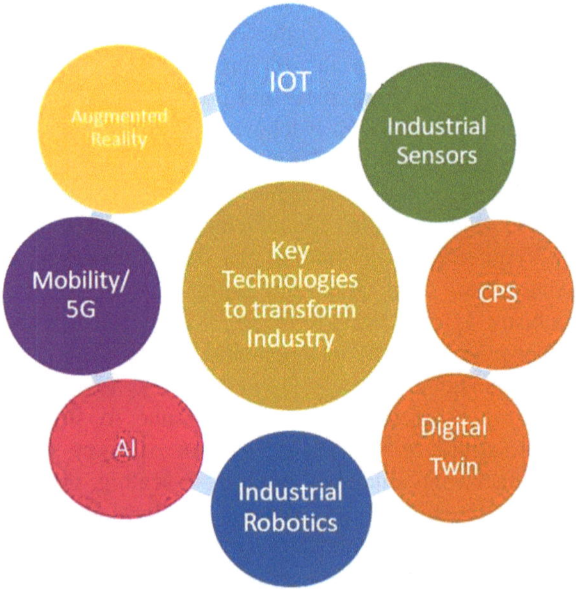

Fig. 1.2 Key technologies to transform production industry

the prototypes are available beforehand. We relate digital twin to be a digital artifact that simulates the behavior and/or appearance of real-world objects. Making use or combination of digital twins can further assist in predictive analysis for real-world problems, for example, in production lines.

5. *Industrial robotics:* This technology has been already opted by many of the manufacturing industries. It automates the action based on machines. It is also useful for heavy lifting.
6. *AI (artificial intelligence):* AI helps in analyzing the huge data generated. Not only analysis it is possible to add intelligence to machines based on certain trainings.
7. *Mobility/5G:* Because of this technology, now it is possible to share the information and intelligence to stakeholder at rapid speed.
8. *Augmented reality:* It is possible to interact with physical things virtually through digital channels.

Implement IT and OT convergence: For better production, we need to implement vertical and horizontal integration among various processes. A *convergence* is between information technology *(IT)* and operational technology *(OT)* [5]. Without this, there is no industrial transformation. The convergence of IT, OT, and their backbones includes networks or communication technology and infrastructure like *cloud infrastructure, server infrastructure, storage infrastructure, and edge infrastructure*. The essence of IT and OT convergence includes the data, processes, and people or teams. In fact, we can say that Industry 4.0 surely gives increased productivity but along with that it helps in increasing quality, flexibility, and efficiency too.

Overall, considering the tremendous impact and usage for improved processes and skilled automation, the discussed technologies are instrumental in driving it. To mention, with AI, machine learning, CPS, and IoT taking a huge leap to determine the analytics, real-time data collection has and will carry a significant transformation toward the manufacturing, thereby staying true to "smart factory" tag [6]. Many companies today are using the techniques for simulation and are in position to determine issues, predict the outcomes, and thereby build better products.

1.2 Challenges When One Wants to Switch to Industry 4.0

While Industry 4.0 has good opportunities in the world, we should not ignore the challenges that the countries will face as they try to ride the Industry 4.0 wave. Whenever any industry wants to adopt some changes in the industry functionalities, there is always a bit oppose to it. In case of adopting Industry 4.0 again many industries are facing various challenges which need to be addressed. Some of them are as follows.

1.2.1 Major Challenges

The pursuit of Business 4.0 presents a variety of technological obstacles, with significant ramifications across numerous facets of today's manufacturing industry. Thus, it is critical to design a strategy for all parties involved in the full value chain and to reach agreement on security concerns and the appropriate architecture prior to execution. Additionally, some authors assert that implementing Industry 4.0 is a difficult task that will likely take 10 or more years to complete. Adopting this new manufacturing process entails a variety of factors and presents a variety of obstacles and challenges, including scientific, technological, and economic difficulties, as well as social and political ones.

Following is the mention of few of these significant challenges.

(a) *Economic:* Whenever an established industry would like to undergo some changes, the cost matters a lot. It requires high economic costs. New business model adaptation becomes necessary which again involves some costing. Before going for some changes with respect to any factor, industry management must be aware of the benefits, but many times it happens that they do not have foresight to clearly understand economic benefits.

(b) *Social:* In case if industry decides to go for Industry 4.0, still there is a big issue of data privacy. The security of data is a critical issue that needs to be addressed first.

It is always observed that whenever some change is proposed, stakeholders are a bit reluctant to the change. The main reason is employer has a threat of redundancy of jobs [7]. It might be possible that people will lose their jobs especially for blue-collar workers. Social advancement is consequently characterized as an extension of the opportunities and open doors for individuals to carry on with an existence that they esteem and have motivation to esteem. This is a development of decision (UNDP, 2012). Social improvement is a significant device for the execution of manageable turn of events, thus the quest for social advancement cultivates economical turn of events. Supportable improvement should be founded on working on individuals' personal satisfaction, and that implies that it ought to be intended to build individuals' capacity to meet their financial requirements without hurting the concerned employees.

(c) *Political:* In case of switching to new ideation, there is problem from regulatory point of view. It is found that there is lack of rules and regulations, standards and forms of certifications, unclear legal issues, and data security [8].

(d) *Organizational/internal:* From organizational perception, reliability and stability are needed for critical machine-to-machine communication. Also, it becomes necessary to maintain the integrity of production processes which the established organizations found bit difficult. Many of the times it is observed that there is less support from top management.

(e) *Ethical challenges:* Past industrial revolutions raised their own moral difficulties at that point, especially the substitution of talented work, such as winding, with additional proficient mechanical weavers, the subsequent abuse of ladies and youngsters in the untalented workforce expected to work the weaving machines. It required some investment before these moral issues were tended to, yet robotization expanded efficiency and made an entirely different circle of gifted positions like bookkeeping and the executives. The advancements of the 4IR, alongside the force of big tech behind their arrangement, raise moral issues that go past the eventual fate of business; they strike at the core of being human.

1.2.2 Some More Challenges

Even after finding solution to some of the above challenging issues, organizations still have some more challenges as follows.

1. *Information management:* An actionable intelligence and connected information are very much necessary in Industry 4.0. Process excellence in a context of relevance, innovation, and timely availability for any desired business is equally important. In all the information either received or generated while following processes should be managed properly is one of the big challenges in Industry 4.0 [9].
2. *Cyber security or data privacy:* Information or data security has also become a major challenge. Increasing number of attacks in the industrial Internet became big issue when industry decides to go for certain advancements.
3. *Awareness gap:* There are many gaps in our comprehension of the technological developments that are occurring in the tech industry. The workforce's requirements are always changing. Is it possible for your employees to keep up? When looking for candidates to fill unfilled roles, seek for people who have "digital dexterity," or the ability to grasp both industrial processes and the digital tools that support them. Business models will only be able to successfully deploy new technology and manage operations if they have the suitable employees.
4. *Data sensitivity:* With the advancement of technology, there has been an increase in worries about data and IP privacy, ownership, and management. Is there a common example? Data is required to train and test an AI algorithm before it can be properly implemented. The data must be provided in order for this to happen. Many businesses, on the other hand, are hesitant to share their data with third-party solution developers. Furthermore, our present data governance regulations for internal usage within enterprises are insufficient to facilitate data exchange across organizations. Data is a valuable asset, so be sure it is safe!
5. *Interoperability:* The lack of separation between protocols, components, products, and systems is another key concern. Interoperability, unfortunately, limits firms' ability to innovate. Interoperability also limits possibilities for upgrading system components because they cannot easily "swap out" one vendor for another or one aspect of the system for another [10].

6. *Security:* Threats in the manufacturing industry in terms of current and emerging vulnerabilities are another major worry. Real-time interoperability is possible because to the physical and digital components that make up smart factories, but it comes with the risk of a larger attack surface. When various machines and gadgets are connected to a single or multiple networks in a smart factory, flaws in any one of them could make the entire system vulnerable to attack. Companies must anticipate both enterprise system vulnerabilities and machine-level operational weaknesses to assist tackle this issue. Many businesses rely on their technology and solution providers to identify vulnerabilities; therefore, they are not completely equipped to deal with these security concerns. A cyber assault can devastate a company's reputation as well as the personal information of its personnel.
7. *Handling data growth:* As more companies become dependent on AI usage, companies will be faced with more data that is being generated at a faster pace and presented in multiple formats. To wade through these vast amounts of data, AI algorithms need to be easier to comprehend. Further, these algorithms need to be able to combine data that might be of different types and time frames [11].
8. *Lack of global visibility:* Many times the changes and revolutions taking place in one region are not known to other regions. One of the possible reason could be lack of communication or significant versatility in working culture as well.
9. *Lack of innovation:* If the organization is quite old, then there is a possibility of lack of innovative ideas from senior employees due to conventional way of thinking.
10. *Insufficient investment plan:* This is a very important issue faced by many of the industries. If any change is required in any industry, then big investments are required. And same has to be proposed in company's budget as well. In case if no provision is done, the situation may go panic resulting in failure in implementation of Industry 4.0.
11. *Lack of visibility the performance:* In case even though company decided to go for Industry 4.0 changes, then there has to be proper vision behind this decision. What could be benefits or losses that company may face that study should be ready beforehand.
12. *Market Plan:* Once the company opts for Industry 4.0, marketing team should have proper plan before sending product to the market. A proper marketing strategy needs to be defined in this case.

In spite of having some challenges or issues, following are the fields where Industry 4.0 can be implemented:

1. Manufacturing operations and quality
2. Production asset monitoring and management
3. Inventory logistics and transportation optimization

1.3 Benefits of Industry 4.0

Higher production and efficiency, improved flexibility and agility, and increased profitability are some of the advantages of Industry 4.0, according to its proponents. In addition, Industry 4.0 enhances the consumer experience. In addition to being engaging and exciting, smart factory technologies should always be placed at the center of any conversation about Industry 4.0. After all, any investment you make in technology, improved production processes, or enhanced systems should generate a return on your initial investment in those areas. Because of the advantages that the technologies provide, the return on investment (ROI) prospects are enormous with Industry 4.0. This comprises technology that improve automation, machine-to-machine communication, production oversight, decision-making processes, and so on.

Following diagram depicts some of the advantages of Industry 4.0 prominently on the basis of productivity, flexibility, quality, operation, and speed improvement perspective (see Fig. 1.3).

1. *Enhanced productivity using optimization and automation*

   In Industry 4.0, optimization of processes and of productivity is the first benefit that everyone can see. Basically, we can say that it is one of the major goals of Industry 4.0 projects. In other words, it saves costs, increases profitability, and reduces waste too. Automation can prevent errors and delays, speed up the production, and in all improve the overall value chain and so on. Industry 4.0 gives various solutions to optimize, from optimized asset utilization and smoother production processes which also results into better logistics and inventory management.

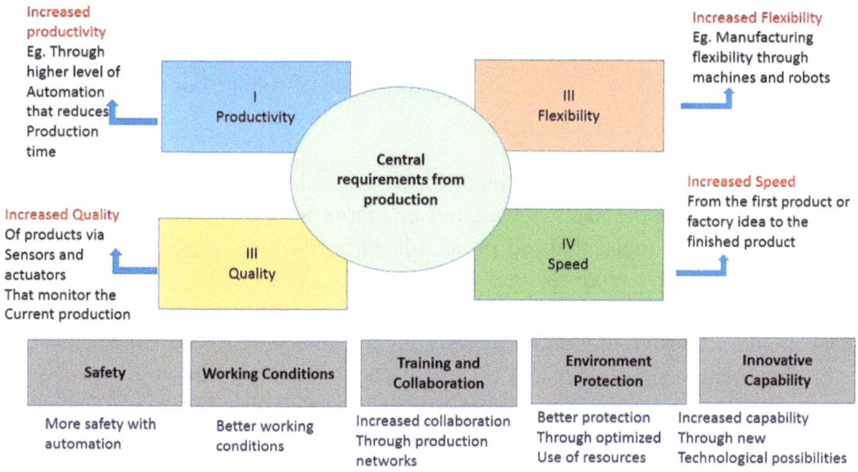

Fig. 1.3 Advantages of Industry 4.0

2. *Real-time supply chain management in a real-time working environment*
   Industry 4.0 also concentrates on enhanced customer centricity. It is about the entire life cycle of products and manufacturing. If we consider the entire value chain and ecosystem, there are many stakeholders involved in the process manufacturing. They are all customers. And a customer always wants enhanced productivity, regardless of where they are in this whole cycle or supply chain. If the final customer wants good products fast and have increased expectations regarding customer experience, quality, and service, then this impacts the whole supply chain, all the way up to manufacturing and beyond.
3. *Increased business continuity through advanced maintenance and automated monitoring possibilities*
   Lets take an example that an industrial robot in a car manufacturing plant gives up, and then it is not just the robot that is broken. Production gets affected which leads to wastage of money and unhappy customers, and sometimes production can be fully disrupted. On top of all the replacement or fixing, the error may damage the reputation. Orders may get cancelled too. If industrial assets are interconnected and can be e-monitored through the Internet of Things, then it is possible to tackle it before they even happen and then the benefits are huge. Alarming or alerting systems can be set up. In this case, the assets can be proactively maintained resulting in real-time monitoring, and diagnosis becomes possible beforehand. No wonder that asset management and maintenance are the second largest area of IoT investments in manufacturing.
4. *Better quality products*
   It is true that customers want speed but that does not mean they will compromise on quality. If your production system has many sensors and IoT techniques, then definitely you can enhance the quality of your products. The typical components of cyber-physical systems and the Internet of Things can be automated so as to monitor the quality aspects. On the other side, we can say that the more the automation, the lesser the errors resulting in better quality. At the same time, it is also true that robots are not going to take all human jobs over soon. Many companies have increased the usage of robots, but at the same time they hired more people.
5. *Better work environment*
   Improving working environment based on real-time temperature, humidity, and other data in the plant or warehouse is more important part of Industry 4.0. Quick detection and enhanced protection in case of certain bad incidents and accidents will give more secured work environment for the workers too. One of the important factors that results in better work environment is the accessibility of the data through cloud and thereby giving flexibility to the employees. The environment will be safer and effective where the key technologies will yield for intelligent assistance in decision-making. So, in a way, Industry 4.0 provides improved work environment where human will be at the epitome to drive it with technologies.

6. *Personalization and customization for the "new" consumer*
   We all know that nowadays digital tools have changed the ways we work, shop, and live. Consumer behavior and preferences are changed. People are now more demanding; they do require fast responses and timely information/deliveries. On top of that consumers also like a degree of personalization, depending on the context. They are more interested in customization as per their needs. Consumers demand to have a direct interaction with a brand and its manufacturing capability. Digital platforms are used to customize products as mentioned, which in turn shortened the routes between production and delivery. In many manufacturing sectors, these things had already happened. Along with the consumer environment, industries have also employed customization in B2b context.
7. *Adopting state-of-the art models to generate more revenue*
   It is possible to transform processes, specific functions, customer service, and experiences, but in the end true value is tapping into new revenue sources and ecosystems. Handling innovative capabilities for customers to give advanced maintenance services is also a prominent factor.

1.4 Applications

Nowadays, CPS has many applications as shown in Fig. 1.4. It can be found in smart health-care wearable devices, smart grid, smart water networks, smart manufacturing, smart factory, gas and oil pipelines monitoring and control, unmanned and autonomous underwater vehicles, hybrid electrical vehicles, and greenhouse control.

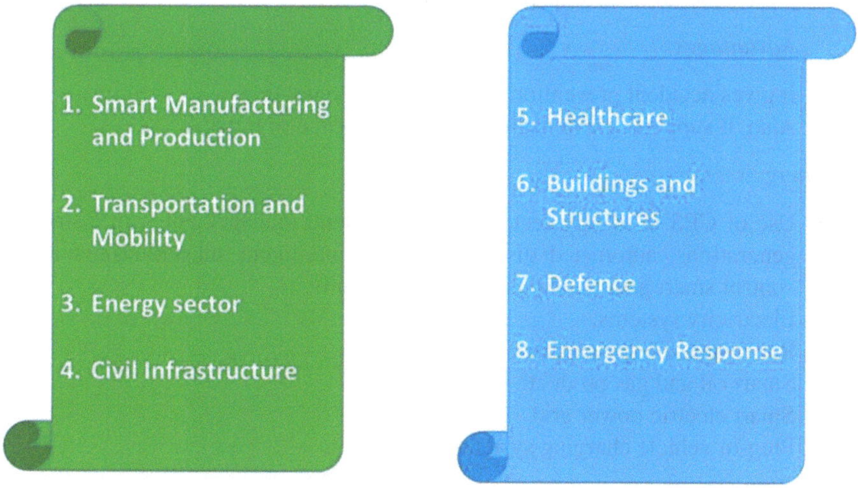

Fig. 1.4 Applications of Industries

CPS is going to generate unique opportunities for economic growth. It will also create skilled jobs. CPS will help ensure the health, safety, and security of the nation. CPS is going to drive the innovation in a broad range of industries and can lead to new products. Manufacturing techniques will increasingly rely upon CPS technologies for advanced and computer-controlled manufacturing processes like automated design tools [12]. It will also bring dynamism in management of production lines, factories, and supply chains.

1. *Smart manufacturing and production*

   - CPS in manufacturing systems is used for logistics integrated with communication abilities. In manufacturing sectors, it will be used in sensors and actuators, robot-operated machines, laid machines, mining machines, and welding machines to improve efficiency of production.
   - Agile manufacturing, supply chain connectivity
   - Intelligent controls, process, and assembly automation

   Advantages

   - It helps to enhance global competitiveness.
   - It has increased the efficiency, agility, and reliability.

2. *Transportation and mobility*

   - It is possible to have vehicle-based transportation CPS. Proximity detection for safety (vehicles coming close).
   - Vehicle health monitoring.
   - Autonomous or smart vehicles (surface, air, water, and space).
   - Vehicle-to-vehicle and vehicle-to-infrastructure communication.
   - Drive-by-wire vehicle systems, plug-ins, and smart cars.
   - Interactive traffic control systems.
   - Next-generation air transport control.

   Advantages

   - It gives accident prevention and congestion reduction.
   - Also, it supports greater safety and convenience of travel.

3. *Energy sector*

   - Use of CPS in energy sector helps in demand management with distributed generations, automated distribution with intelligent substations, wide-area control smart grids, and data aggregation units.
   - Electricity systems.
   - Renewable energy supply.
   - Smart oil and gas production.
   - Smart electric power grid.
   - Plug-in vehicle charging systems.

Advantages

- CPS in this field also helps in greater reliability and security.
- CPS enables diversity of energy supply, which increases energy efficiency.

4. *Civil infrastructure*

- Bridges and dams
- Municipal water and wastewater treatment
- Active monitoring and control system
- Smart grids for water and wastewater
- Early warning systems

Advantages

- It increases assurance of quality.
- CPS gives more safe, secure, and reliable infrastructure.

5. *Health care*

- Highly accurate medical devices and systems
- Image-guided surgery and therapy (robotic surgery, accurately precisely)
- Control of fluid flow for medical purposes and biological findings
- Intelligent operating theaters and hospitals
- Engineered systems based on cognition and neuroscience (e.g., brain–machine interface, orthotics, exoskeletons, and prosthetics)
- Personal care equipment, disease diagnosis, and prevention
- Wireless body area networks
- Assistive health-care systems
- Wearable sensors and implantable devices

Advantages

- In this industry, the use of CPS helps in expanding the life span of every individual.
- Hospital-based to home-based health care is possible because of CPS.
- CPS-based medical instruments enable more individualized health care.
- Advanced CPS can lead to new capabilities to diagnose, treat, and prevent disease.

6. *Buildings and structures*

- High-performance residential and commercial buildings
- Net-zero energy buildings and appliances
- Whole building controls, smart HVAC equipment
- Building automation systems
- Network appliance systems

Advantages

- It has increased building efficiency, comfort, and convenience.
- Improved occupant health and safety.
- Also, we can control indoor air quality using CPS-enabled household devices.
- CPS enables early bomb disposal system and emergency response robotics to sensor networks providing advance warning of catastrophic events.

7. *Defence*

- Intelligent unmanned aircraft and ground vehicles
- Autonomous and smart underwater surface sensors
- Overarching systems that integrate the nation's fighting forces
- Soldier equipment
- Weapons and weapons platforms
- Supply equipment
- Smart (precision-guided) weapons
- Wearable computing/sensing uniforms
- Intelligent supply chain and logistics systems

Advantages

- It has increased war-fighter effectiveness.
- Increased security and agility.
- Greater capability for remote warfare reduces exposure for human war-fighters.

8. *Emergency response*

- First responder equipment
- Communications equipment
- Firefighting equipment
- Detection and surveillance systems
- Resilient communications networks
- Integrated emergency response systems

Advantages

- Increased emergency responder effectiveness, safety, efficiency, and agility
- Rapid ability to respond to natural and other disasters

1.5 Societal Impact

In a Business round table held in August 2019, the fact of benefit and income being the principal factors along with systematic procedures and ventures was underlined by business pioneers. Our review repeats this developing comprehension and obligation to society. A long time back, just 35% of CEOs trusted that the main associations representing things to come expected to spend more time planning for the possible effect that new innovative arrangements might have on society. In this year's study, many CEOs said expanding their organizations' positive effect on society was among their main five wanted results of their Industry 4.0 ventures—a sign that leaders are beginning to figure out that business plays a significant part in forming what these innovations could mean for society.

1.5.1 Relation Between Profit and Purpose

Innovation is taking a front seat: 62% of CEOs demonstrated that creating a gain while emphatically adding to society was an Industry 4.0 venture need for their associations—the second-most-referred to need subsequent to preparing and creating ability. Of those 24% proposed they are gaining a lot of headway against this objective, a similar rate who accept their associations are in front of their rivals in doing as such [13].

It may be advising that of those professing to make extraordinary advancement, 69% have an extensive Industry 4.0 system. Having a conclusive system will likewise assist with dominating contenders: 60% "with methodology" guarantee to be in front of the opposition, while just 13% "without methodology" make this same dispute.

1.5.2 Employee and Customer Advocacy Is Increased

At the point when inquired as to why their organizations decide to concentrate on cultural issues, 42% of CEOs referred to the valuable chance to produce income which recommends that benefit and income proceed to drive associations' procedures and inspirations. Numerous recent college graduates wish to work for organization that have a history of making money. Associations that do not uphold more extensive cultural responsibilities could begin seeing it influence their enrolment, maintenance, and generally speaking main concerns [14]. Issues that appear to have soar in significance for leaders are environmental change and natural manageability [15]. A long time back, as it were 10% of CEOs said their organizations could impact ecological manageability to a huge degree. This year, 48% see handling environmental change as a top liability; 38% put empowering manageability at the same need level.

With a rising number of horrendous, environment-related occasions influencing populaces and geologies, CEOs are starting to feel, or possibly figure out, the business basics of environmental change. Almost 50% of CEOs (48%) totally concur that the impacts of environmental change will adversely influence their associations, and very nearly 90% totally or to some degree concur.

1.6 Case Study: Challenges in Manufacturing Sector

Let us consider the case study of manufacturing sector that if these industries ready to opt for Industry 4.0 then what all issues they may have to face. Also refer to Fig. 1.5.

(a) *Shortening of product life cycle:* The complete product development life cycle gets changed. As some sort of automation is involved in this, it might be possible that product may get ready in less time period. So being an owner of the company one has to focus a lot on this [16].
(b) *Reduction of lot sizes:* It is well known that in industry the scale is always versus cost. If we increase the scale, the manufacturing cost varies accordingly. And same selling cost may be not possible with reduced scale.
(c) *Increased diversity of versions:* Industry also need to opt for diversity in productions versions. If a company is producing 300 mg and 500 mg pouches, then there is a need to produce small pouches too for better selling of the product.
(d) *Shortening of ROI:* We all know that every investor always thinks about return of investments. If company is opting for Industry 4.0, a thought has to be given on ROI.

Fig. 1.5 Challenges in manufacturing sector

(e) *Energy efficient*: One needs to address the issue of power required after opting for Industry 4.0. As many of the processes transition to automation, how much extra electricity would be required? That is also an issue.

1.7 Conclusion

Industry 4.0 is future of industries. If one wants to survive in the competitions companies should opt the changes happening around. As previously stated, the integration of smart products with smart manufacturing, smart logistics, smart networks, and the Internet of Things results in the transformation of existing value chains and the emergence of new and innovative business models, positioning the smart factory as a critical component of future smart infrastructures. Numerous benefits and revenues will accrue as a result of this new infrastructure approach. Indeed, virtual and augmented prototyping facilitates the interactive exploration of all product functionalities by all stakeholders. Industry 4.0 enables a new method of doing business and a new source of value creation, particularly for traditional manufacturing firms. One of Industry 4.0's most significant disruptors is the ever-increasing value and importance of data. Businesses must view data as a rare and valuable raw commodity. As a result, businesses will need to rethink how they approach and manage massive amounts of data and information. This will be one of the most significant issues confronting traditional manufacturing firms. Utilizing dynamically programmable production technology in conjunction with increased machine flexibility (e.g., flexible grip hooks) has a number of advantages, including increased customization, more dynamic resource/capacity allocation, shorter changeover times, and reduced production complexity due to fewer constraints. This enables more rapid, cost-effective, simple, and diversified manufacturing methods. Industry 4.0 provides numerous benefits to businesses across multiple dimensions.

When the fourth generation of the industry arrived, it created an opportunity for reliability engineering to improve system dependability by utilizing big data, the Internet of things, and quick response to changes. In contrast, increasing complexity, dependencies and interconnections between components, dynamic behavior, and newer components like CPS and sensors provide difficulties for reliability engineers. In order to keep up with the times, traditional approaches must be updated and a new framework for reliability, risk, safety, and security should be devised. In this chapter, the concept of Industry 4.0 is introduced, as well as some of the potential difficulties it presents. It is not a pure or faultless review, and it is not focused on specific areas of Industry 4.0 and reliability engineering, but we are there to gain a perspective on these themes and to share our experiences. Some important problems such as system modeling, data, CPS, uncertainty, the interface problem, human–machine interaction, optimization, and other topics are taken into consideration. Throughout each part, the underlying principle of the object is described, and some potential new study possibilities are suggested. Today, multicomponent system modeling, the interdependence of these components, the optimization of the supply chain while also optimizing maintenance and production, and the modeling of resilience are all receiving increased attention.

References

1. "A Roadmap to Industry 4.0: Smart Production, Sharp Business and Sustainable Development", Springer Science and Business, Media LLC, 2020
2. Michael Sony," Design of cyber physical system architecture for industry 4.0 through lean six sigma: conceptual foundations and research issues", Production & Manufacturing Research 2020, VOL. 8, NO. 1, 158–181
3. "Industrial Internet of Things", Springer Science and Business Media LLC, 2017
4. Jeschke, S, Brecher, C, Song, H. Industrial Internet of Things. Cham: Springer, 2017.
5. Zhang, Y, Qiu, M, Tsai, C-W. Health-CPS: healthcare cyber-physical system assisted by cloud and big data. IEEE Syst J 2017; 11: 88–95
6. J Lee, B Bagheri, HA Kao, A cyber-physical systems architecture for industry 4.0-based manufacturing systems Manufacturing letters, 2015 - Elsevier
7. Colombo, AW, Bangemann, T, Karnouskos, S. Industrial cloud-based cyber-physical systems: the IMC-AESOP approach. Cham: Springer, 2014
8. Shah Ahsanul Haque, Syed Mahfuzul Aziz, "False Alarm Detection in Cyber-physical Systems for Healthcare Applications", AASRI Procedia, 2013
9. Zhang, Z., Wang, H., Wang, C., Fang, H. Interference mitigation for cyber-physical wireless body area network system using social networks IEEE Transactions on Emerging Topics in Computing 2013
10. Yilmaz, T., Munoz, M., Foster, R. N., Hao, Y. Wearable wireless sensors for healthcare applications Proceedings of the International Workshop on Antenna Technology (iWAT '13) 2013
11. Haque, S. A., Aziz, S. M. Storage node based routing protocol for wireless sensor networks Proceedings of the 7th International Conference on Sensing Technology (ICST '13) 2013 Wellington, New Zealand
12. Avrunin, G. S., Clarke, L. A., Osterweil, L. J., Goldman, J. M., Rausch, T. Smart checklists for human-intensive medical systems Proceedings of the IEEE/IFIP 42nd International Conference on Dependable Systems and Networks Workshops (DSNW '12) 2012 Boston, Mass, USA
13. Lee, I., Sokolsky, O., Chen, S., Hatcliff, J., Jee, E., Kim, B., King, A., Mullen-Fortino, M., Park, S., Roederer, A., Venkatasubramanian, K. K. Challenges and research directions in medical cyber-physical systems Proceedings of the IEEE, 2012
14. Tang, L. A., Yu, X., Kim, S., Han, J., Peng, W. C., Sun, Y., Leung, A., La, P. T. Multidimensional sensor data analysis in cyber-physical system: an atypical cube approach International Journal of Distributed Sensor Networks 2012.
15. Huang, Q., Ye, L., Yu, M., Wu, F., Liang, R. Medical information integration based cloud computing Proceedings of the International Conference on Network Computing and Information Security (NCIS '11) May 2011
16. Wang, J., Abid, H., Lee, S., Shu, L., Xia, F. A secured health care application architecture for cyber-physical systems Control Engineering and Applied Informatics 2011

Chapter 2
Exploring Human Computer Interaction in Industry 4.0

Varad Vishwarupe, Prachi Joshi, Shrey Maheshwari, Priyanka Kuklani, Prathamesh Shingote, Milind Pande, Vishal Pawar, and Aseem Deshmukh

2.1 Introduction

Affiliation of industrial machines, brought about by the collaboration of automation technology, machine learning, and communication abilities for the independent running of an industry is the concrete objective of the Industry 4.0 framework. Industry 4.0 plays an important role in the manufacturing industry by ensuring security, flexibility, customization, time efficiency, and a dynamic environment as well as increased productivity and quality [1, 2]. Through connecting smart devices and machinery, employing self-learning solutions, and enhancing self-direction capabilities, it is envisioned that the communication cost is reduced while flexibility for manufacturing, mass customization capabilities, production speed, and quality are increased [3–5].

V. Vishwarupe (✉)
Department of Computer Science, University of Oxford, Oxford, UK

Human Inspired AI Research, London, UK
e-mail: varad@huminspaire.com

P. Joshi
Department of AI & DS, Vishwakarma Institute of Information Technology, Pune, India

S. Maheshwari
R&D Division, Tata Elxsi, Pune, India

P. Kuklani
School of Engineering, Northeastern University, Boston, MA, USA

P. Shingote · M. Pande · V. Pawar
Maharashtra Institute of Technology WPU, Pune, India
e-mail: vishal.pawar@mitwpue.edu.in

A. Deshmukh
Business Analytics Division, System Applications and Products (SAP), Pune, India

A. Neustein et al. (eds.), *AI, IoT, Big Data and Cloud Computing for Industry 4.0*, Signals and Communication Technology, https://doi.org/10.1007/978-3-031-29713-7_2

Though the number of benefits of Industry 4.0 is tempting, Industry 4.0 also reveals several challenges or in a more critical way obstacles to improving productivity at the workplace [6]. One such challenge is the work division between manpower and machines, which allows us to explore Industry 4.0-catered human–computer interaction (HCI) to reap the benefits of automated and smart machines, without compromising on the strength of the current manpower. Work division would enable a role change of humans to shift from low-level operations—which can be dangerous, dirty, difficult, and dull tasks—to high expertise and safe tasks [7–10]. Moreover, human intelligence and intervention remain a key role because of the safety, security, social aspects, and uncertainties posed by such autonomous systems [11–14]. Even with the dangerous roles out of the way of industry workers with the fourth industrial revolution, a good design and Human-Computer Interaction (HCI) in place would be required to make humans adept in the newly defined roles, so as to operate complex machines with negligible or no error. Thus, this research work aims to review the current vision of HCI and user experience (UX) design of Industry 4.0 and recommends practices best suited for the fourth industrial revolution.

2.2 Related Work

With the increasing stress on human satisfaction, ease and focus on better experiences, there has been an emerging trend in research work related to human-centered design and development of HCI from an Industry 4.0 perspective. Extracts from such work are mentioned in this section, for obtaining a better picture of the overall context. A good user experience (UX) and intuitive HCI is the cornerstone for the smooth operation of any industry, and with Industry 4.0 mechanisms, these become crucial. A system needs to be designed to generate positive UX for increasing user association and encouragement [15–18] presented that bad user interface design can lower user motivation to use a system. Beard-Gunter [19] worked with HCI design in industries to develop and optimize engagement metrics compared to games. To ensure the effectiveness of a system, a proper balance is required to be maintained between usability and system functionality [20, 21]. Considering design elements in developing the industrial automation system can facilitate the meaningful interactions between the system and the user [22–24].

Recent advances in artificial intelligence (AI) and HCI have seen remarkable contributions that have shaped the world. From being a virtual concept in scientific fiction(sci-fi) a few years ago, AI is literally changing the way we act and interpret the world around us. HCI and AI can be coined as two sides of the same coin according to us, and can be termed as entailing the same story in different ways or entailing different parts of the same story, but ultimately working in unison for a common goal. Ben Shneiderman, one of the pioneers in the field of HCI, has emphasized the use of AI in collaboration with HCI researchers to further develop the field of human-inspired AI [27–35].

There has also been a huge upheaval in the way HCI is used in the manufacturing industry to optimize for quality and not cost. The work of Dudley, Jacques, and

Kristensson is particularly intriguing in this regard wherein they have tweaked a very important facet of Bayesian optimization, the activation function: a particularly important statistical machine learning method to improve user interface design using crowdsourcing [4]. The use of virtual reality (VR) and mixed reality (MR) as a driving force for the domain of HCI has been on the rise too [5, 6]. Using intuitive user interfaces (IUI), VR and MR gadgets, and head-mounted displays (HMDs) have also made it possible to interact with the physical medium in a way that is more immersive. Simulation of touch displays for people with motor disabilities, gesture typing, 3D-based VR models, gaze-based motion tracking gadgets are also some of the advancements that have helped in fusing the two realms of AI and HCI, by developing smart and interactive systems [7–11].

The use of AI-based HCI systems and HCI-oriented AI systems has also been observed, especially in the medicine sector. AI-based diagnosis is difficult to identify diseases and in areas where human intervention seems to have stuck a dead end, enables clinicians to find out ways which are improving health care such as Suckling et al.'s work on gender classification in task functional brain networks [14, 15]. Use of AI–HCI conjunct systems in improving language models for natural languages also depicts the pervasive nature of the said field. Korhonen et al. have done some remarkable work in this regard [16, 17]. Inferring web page relevance, use of HCI models in certain Internet of things (IoT) tasks, generating personalized recommendations for certain users in the browser, and also using few facets of HCI in the development of smart set-top box TV recommendation systems have enabled the development of smart AI based systems [12, 18–22].

While there has been a substantial work at the crossroads of AI and HCI, it is still not extensive enough to be able to use HCI and AI in scenarios wherein it becomes difficult to have experts from both the fields working together such as cyber-physical systems [24–26]. Thus, it is important that the stakeholders from respective fields are shown what, when, why, and how they can contribute, what are the major roles for each discipline and its experts are, and how can AI be used as a catalyst for developing HCI systems, and vice versa. In the context of this chapter, we try to gather insights from the previous works in the aforementioned and present our research that shall help develop this exciting field of knowledge.

When it comes to HCI in Industry 4.0, we need to understand the prior work in the context of a plethora of subfields, namely human–machine interaction (HMI), virtual, augmented, and mixed reality applications in the context of cyber-physical systems, and surveys which have been conducted keeping HCI and HMI under the purview on Industry 4.0 as a research topic. A large number of surveys covering peculiar aspects of HMI or human factors in I4.0 has been identified in the endeavor of collecting relevant literature for this chapter. The vast majority of these studies are specifically oriented toward either VR or AR applications, or both, within I4.0 operations and, thus, covers only a subset of this chapter's scope. Büttner et al. conduct a survey on AR and VR applications in I4.0 manufacturing activities, more precisely on the available platform technologies and application areas, creating a small-scale design space for such Mixed Reality applications in manufacturing [36].

This design space differentiates among four general application scenarios and four types of Mixed Reality technology platforms available for application, respectively.

The application scenarios are manufacturing, logistics, maintenance, or training, while the available platforms comprise mobile devices (AR), projection (AR), and head-mounted displays (HMDs) (AR or VR). Dini and Dalle Mura and Wang et al. both present surveys on AR applications; however, not restricted to I4.0-related application scenarios [39]. Dini and Dalle Mura investigate general commercial AR applications including besides industrial scenarios also, among others, civil engineering. The specific AR application scenarios which they examine based on related scientific literature are maintenance and repairing, inspection and diagnostics, training, safety, and machine setup [8].

Wang et al., Vishwarupe et al, Zahoor et al. and Bedekar et al. [46–51] in turn, examine scientific research on AR applications for assembly purposes from a time span of 26 years starting as early as 1990 and concentrating mainly on the period from 2005 until 2015. Thus, they extend the scope to many years before the advent of I4.0-related initiatives and ideas. The major application purposes of AR in assembly tasks which they investigate are assembly guidance, assembly design and planning, and assembly training.

Lukosch et al. provide a literature review on AR applications with an even less specific focus on industrial deployment by examining the state of the art at the time in research on collaboration in AR considering a wide range of possible application fields, the industrial sector being only one of those [36]. As a result, they identify remaining research challenges relating to collaboration in AR which are the identification of suitable application scenarios and interaction paradigms as well as an enhancement of the perceived presence and situational awareness of remote users.

Palmarini et al., in turn, conduct a structured literature review on different software development platforms and types of data visualization and hardware available for AR applications in various maintenance scenarios [42]. The aim and purpose of their study is to derive a generic guideline facilitating a firm's selection process of the appropriate type and design of AR application, tailored to the specific type of maintenance activity at hand which the firm is planning to enhance utilizing AR technology. Lastly, Choi et al. and Turner et al. provide surveys on VR technology in an industrial environment, the latter group of authors concentrating on a potential combination of VR technology with discrete event simulation for scenario testing in I4.0 activities [37]. Choi et al., on the other hand, present a survey on VR applications in manufacturing, concentrating on potential contributions of VR deployment in the development process for new products and deriving a mapping of different types of VR technology toward the different steps of the product development process. Therefore, Hermann et al. consider applicability of various VR technologies for the phases of concept development, system-level design, design of details, testing and refinement, and launch of production [38].

Besides those surveys on AR and VR applications, a literature review by Hecklau et al. exists on the major challenges as well as skills and competencies needed for future employees under an I4.0 scenario [41]. The authors utilize the insights from the literature analysis to structure the required skills according to different categories, based on which a competence model is created analyzing employees' levels of skills and competencies which will be particularly important in an I4.0 working environment. Those main categories for the competence model are technical, methodological, social, and personal competencies [21].

Uzor et al. [46] also present a very lucid caste study of using Amazon Mechanical Turk, a crowdsourcing platform for facilitating HCI studies in the context of industry 4.0 applications, wherein the importance of having a synergy between the user interface for crowdsourcing-based studies and industrial requirements is enunciated. While we have tried to cover maximum bases for our work related to HCI and HMI in the context of Industry 4.0, it is important to highlight that this list of related work is not all-encompassing. Instead, it is rather supposed to provide an outlook of other existing scientific literature regarding the topic of HCI and Industry 4.0 and thereby reaffirm that, to the best of the author's knowledge, a study matching of this magnitude ceases to exist. This aspect facilitates the justification of this research study signifies the necessary importance as well as the relevance of this research paper.

Though user experience and human-centered design have been explored in different contexts, there is a dearth of studies and research on its consideration, use, and impact on an Industry 4.0 setup. This paper focuses on the exploration of human-centered design principles and customized HCI to tend to the complexities involved in Industry 4.0, and the role of humans in such an ecosystem.

2.3 Research Questions

At the outset, we need to base our study on the fundamental understanding of what HCI is and how it is relevant under the purview of Industry 4.0. In its entirety, HCI is the study of the interaction between humans and computers, particularly as it pertains to the design of technology. HCI is at the crossroads of user-centered design, UI, and UX to create intuitive products and technologies. Researchers who specialize in HCI think about conceptualizing and implementing systems that satisfy human users. HCI also helps to make interfaces that increase productivity, enhance user experience, and reduce risks in safety-critical systems. This is especially relevant in heavy industrial applications of Industry 4.0 such as manufacturing and cyber-physical systems. This is why HCI is on the rise for developing intelligent interactive systems [43].

While defining the research questions for this paper, it is important to identify the key principles of interaction. The gold standard for this is the Norman's model of interaction. Norman's interaction model is a noteworthy and pioneering model for HCI-based studies. It proposes that a user first establishes a goal and then performs actions using the system to achieve that goal. Thereafter the system the end result of the user actions on the UI. A particular user then minutely assesses this result and sees if their objective has been achieved or not. If not, a new goal is established, and

Fig. 2.1 Norman model of interaction

the cycle is repeated [40, 44, 45]. This model of interaction explained is divided into seven primary stages (Fig. 2.1).

This model helps us to understand where things go haywire in our designs. Under the context of Industry 4.0, there are issues when it comes to the sections of machine execution and problem evaluation by the cyber-physical systems. Machine perceive the world in binaries, as opposed to sensory inputs that are so readily perceived by humans. There are also issues with the interpretation of perception which further causes issues with the evaluation of those interpretations. There is a measurable difference between user actions and those that a particular system can perform. A potent and engaging interface allows a user to perform an action without system limitations. This is especially important for Industry 4.0 which is nothing but an interconnected web of system of subsystems consisting of IoT and cyber-physical systems. The second consideration wherein the importance of having a proper interaction mechanism between humans and machine for cyber-physical systems is the difference between the presentation of an output and the user's expectations. An effective interface can be easily evaluated by a user, but it fails to evaluate when only machines are involved in the process. As a preliminary step, research questions were chalked out to define the basis of this research paper.

The focus of these research questions is to primarily study the current visions of an Industry 4.0 ecosystem and especially explore this industrial revolution from the perspective of humans. Answering the formulated research questions would provide an overarching view of the human-centered approach of building the next industrial revolution. For answering the research questions, domain-specific keywords from each research question were used to explore the current research work through Google Scholar. Google Scholar was preferred as the source of all research content, as it gave an expansive search result and encompassed research work published from a wide range of publication organizations, thus covering a broad spectrum of resources. Research material from a diverse set of publication houses—IEEE, Research Gate, Elsevier, Science Direct, and MDPI—were studied and analyzed. Providing better user support whenever there is a tangible difference expected output and the output generated by the user can be a vital factor in deciding the end goal of an efficient cyber-physical system. Thus, we pose the following important three aspects and five questions and try to answer them with our study on the principles of HCI under the context of Industry 4.0:

- The user interface and user experience design paradigm:
  What facets of the UI-UX components of the Human-Computer Interaction be envisioned from an Industry 4.0 perspective?
- The interaction paradigm:
  How should the interaction mechanism between humans and machines be shaped under the context of Industry 4.0?
- The manpower–machinery paradigm:
  How will manpower be trained to interact with, supervise, and maintain complex machinery of Industry 4.0?

1. Why is there a need to involve humans in the process of developing this I4.0-HCI confluence?
   Naturally, it is a human tendency to question things and have a reasoning that is both acceptable as well as viable when it comes to perceiving the world around us. The question of why to involve humans in the I4.0-HCI confluence is vital because autonomous intelligent systems can falter at times, where only human intervention works.
2. Which stakeholders need to be present in the development of the I4.0-HCI confluence?
   Traditionally, AI has been looked upon as a branch of mathematics and computer science which tries to mimic the reasoning abilities of the human brain. However, with the pervasive nature of AI and its encroachment on our day-to-day life, it is essential to broaden the scope of involving different stakeholders in this major confluence.
3. When to trust the AI in I4.0 and when to trust humans, in the decision making at the I4.0-HCI confluence?
   Gradually, with the advent of large amount of data that is available, issues pertaining to trust, ethics, privacy, and security of user's data have been on the rise. It is important to be able to distinguish when to trust the AI and when not to, and rely on human judgment instead. This research question delves into the ethical side of AI and HCI.
4. Where can the users work in tandem with the AI/where they cannot, pertaining to the confluence?
   Eventually, the choice of where to use AI-enabled systems and where to rely on human acumen is an important part of deploying AI and HCI systems. The roles and responsibilities of both the parties should be aptly defined and coherently distinguished. This research question tries to draw a lucid distinction between these two.

5. How much, is too much when it comes to the I4.0-HCI confluence?

   Incidentally, as they say, everything is good only in moderation. So, in this section we try to answer an important prerogative on what should be the extent of the I4.0-HCI confluence and what regularization should be enforced at their crossroads, so that the I4.0-HCI ecosystem works in tandem with humans and does not supersede them.

2.4 Research Solutions

Complex machinery of the fourth industrial revolution can be quite overwhelming to work with, and with massive chunks of information with respect to each machine, and the interconnection of machines, obtaining the right information at the right time can be quite difficult. Information management and control over such information then become critical to avoid chaos. A carefully designed and well-placed HCI would help to steer clear of such a situation. The meticulous design and information display of an HCI system can be achieved by incorporating design process elements such as qualitative and quantitative analysis and information architecture. We further stress that this inclusion of AR and VR would come with a requirement for training to be given to employees working with Industry 4.0 machines, in which case the user interface should be intuitive and the user experience should be a good one. At the same time, employing technologies like AR/VR comes with its very own list of disadvantages. AR/VR headsets are difficult to wear for a long duration of time, and would cause an uncomfortable experience. We recommend for implementing the AR/VR correctly into the Industry 4.0 ecosystem, users need to be considered from the start in the design process of these interactions, to understand them and unearth their concerns and pain points through qualitative analysis such as interviews. These concerns should be further validated with a quantitative study such as surveys to realize a hypothesis. This will enable to gauge the value of introducing AR and VR at the initial level, and will also help in making modifications along the way, according to the best-fit scenario with respect to industry personnel.

To answer the aforementioned questions on what recommendations or solutions can be conceived to have a seamless UI-UX component, interaction mechanism and how the manpower–machinery paradigm should work in the HCI which is specifically developed for and from an Industry 4.0 perspective, we define the following six components of the HCI centered on the Industry 4.0 realm.

As is evident from Fig. 2.2, we envision that the UI-UX that is designed keeping industrial recommendations in mind should involve a subset of six core components. AR-VR-based HMDs are the way to go ahead, when it comes to industrial supervision roles. It would lessen the need to physically visit the site for inspection and avoid unnecessary hazards. The second component should be the use of explainable AI models which not only predict and classify but also explain the rationale behind arriving at a specific decision. This is particularly important in the context of Industry 4.0, since there is a significant level of reliance on machines to

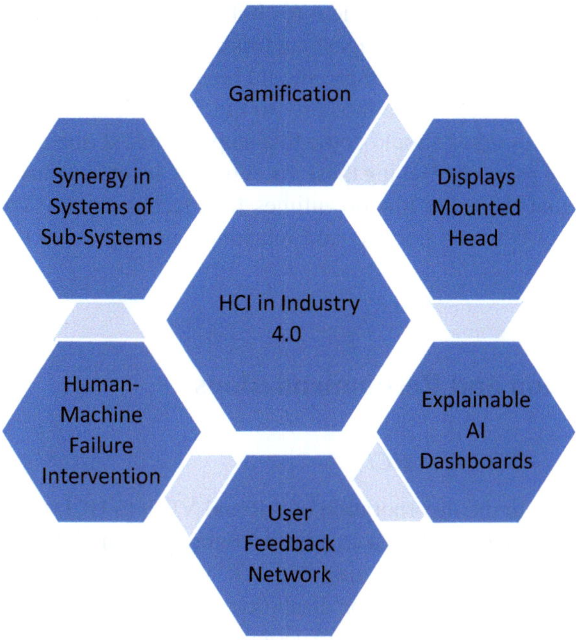

Fig. 2.2 Components of HCI centered on Industry 4.0

perform seamlessly day in and out. Thus, knowing why a particular machine took a particular decision at a particular time helps in reasoning and also accounts for outlier detections, when the mechanism or machine fails.

Third component of this HCI focuses on the need for a user-feedback network which is pivotal to the smooth functioning of a proper HCI mechanism in cyber-physical systems. User should receive both instantaneous and cumulative feedback for examining fault tolerance and industry-specific bottlenecks in the manufacturing processes, which are often the first deterrents on the assembly line, wherein the entire process comes to a standstill.

Fourth consideration includes the ability to have a combination of human–machine intervention when there is critical failure. Oftentimes, there are problems that are only identifiable by human intervention such as system breakdown, actual physical issues with the machinery, and leaks and overflows. During such catastrophic situations, it is important to have a synergy between how much human versus machine intervention is needed for the failure to be resolved in the least possible time, with minimal resources at hand.

The last component of a successful HCI system tailored for the Industry 4.0 framework should include the amalgamation of the system of subsystems which is also known as Internet of Things(IoT). In the IoT framework, device, network, and application are the three most important components which make up the internals in a manufacturing facility. These systems should communicate with each other on a network which is separate from the main hub, so as to accommodate the passage and relay of commands and messages in a brisk and seamless manner. Whenever a certain subsystem malfunctions, there should be immediate relaying of the same to the interconnected nodes with a modality to quickly intimate the nearest operator, wherein the other systems should come to a stop and avoid dire consequences such as fire and loss of property. Having these six components in an industrial HCI system would enable that there are low downtimes for machinery and would establish a cohesive synergy between humans and machines, through an actionable intuitive mechanism.

2.5 Discussion and Recommendations

- *Incorporating AR/VR into HCI*

When it comes to the incorporation of AR and VR into HCI, there are a plethora of considerations that can be taken into account. One modality of research and discussion includes people interacting with AR-VR content. Some of the examples of these include people with motor disabilities using AR-VR content to simplify their lives and get a more immersive experience of the world around them. Another use case includes people with agoraphobia and allied mental disorders who are unable to leave their house due to intense anxiety and can rely on such AR-VR mediums to make their lives more worthwhile, by resorting to alternative means of experiences provided by AR and VR. Head mounted displays(HMDs are also a classic example of another AR-VR implementation that can specifically help in Industry 4.0 environments where it is hazardous for humans to trespass near perilous machinery. Technologies for generating AR-VR can be streamlined by using a human-centered HCI-based approach, by tweaking the design cycle as mentioned above. It includes involving a human-centered approach right from the genesis of the application to the actual deployment and testing, so that AR-VR gets amalgamated with humans in the Industry 4.0 framework seamlessly. One such example includes using digital twins that can be transformed into AR-VR resources. Facilities and technology professionals can also use digital twinning to model the physical environments of systems and appliances. For example, if an organization is building out its own data center, a digital twin can model power, heating and cooling systems against rack layout. Facilities professionals can use this to check for hot spots and make sure that all control panels are easily accessible by facility personnel. Finally, organizations can use the digital twin to optimize human operational processes, such as mapping how technicians walk through the data center.

One more important aspect of incorporating AR-VR into HCI is the use of rigorous mathematical modeling. Some of the common mathematical and computational elements shared by most AR-VR systems include Bayesian optimization, especially parametric estimation and the use of acquisition function and computational elements such as the use of Unity 3D, C#, and VEX CODE VR generator which are some of the state-of-the-art techniques used for developing AR-VR applications. Another of these coding techniques include the use of KONTENT.AI, AR KIT, and UNREAL ENGINE, which are software suites that are very useful in designing elements such as virtual maps, environments and in conjunction with software languages such as C#, enable the development of a wide variety of AR-VR applications.

Some of the next steps involving the implementation of AR-VR in HCI involves ray tracing, mesh-based rendering techniques, and usage of NVIDIA AR WORKS which uses deep learning and CUDA cores to briskly create and mimic the real world using AI-based algorithms such as the widely popular GANs and LSTMs. However, just to keep the scope of this chapter limited and focused on Industry 4.0, further enunciation of these techniques can be included in the future work pertaining to this book chapter.

Current modalities and practices of training can become quite mundane and overwhelming for Industry 4.0-specific training. With high-technology procedures to complete tasks on smart machines, the traditional training method will overburden the manpower, and would not be effective in retaining operating procedures. With AR and VR technologies to be used in Industry 4.0 ecosystem, the training provided will be highly engaging and playful, and due to the visual aspects of these technologies, the training imparted would be retained by industry professionals. It could identify the impact of learning with AR in comparison to traditional vocational training approaches.

Similarly, gamification is another modality that could be used in Industry 4.0-specific HCI to train and assist industrial manpower. Gamification is the method of introducing gamelike elements into a nongame environment, to incentivize users to an otherwise mundane task. However, gamification if not done right can lead to an even more unpleasant user experience than to start with. Therefore, the context of use should be properly defined, and on that front, the concerned manpower should be evaluated in terms of their interest level in game elements. Moreover, stakeholders must be involved in taking an appropriate decision on the type and extent of gamification to be involved in the industry, to benefit the industry as well as personnel mutually (Fig. 2.3).

In our humble opinion, an "ideal" Industry 4.0 HCI system should feature a balanced scale of humans and machines working together, with a positive collaboration between them. To achieve such a synergy, humans should not feel excluded from the interconnected smart machines of the industry. Apart from a rational work division between the two entities, for humans to be inclusive in Industry 4.0, high-technology modalities should be used to foster an engaging and exciting experience of coworking between human and machine. As discussed previously, the intent of using these high-technology modalities, whether they are used to provide information across the industry or as an assistive technology to complete tasks successfully, needs to be defined clearly and a design structure based on the intent is mandatory to be constructed to gain a first-hand view of the space and function of such technological implementation. At the same time, stakeholders

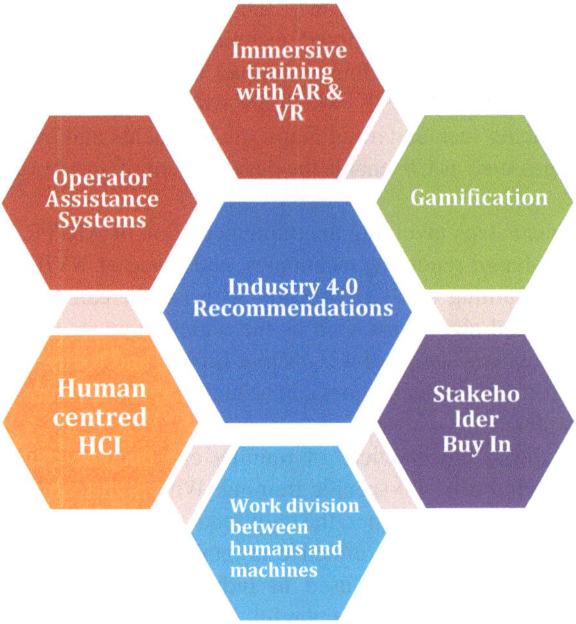

Fig. 2.3 I4.0-HCI confluence framework

and users must be involved in the design process of Industry 4.0 HCI to take care of the business requirements and the perspective of users, respectively.

For an Industry 4.0 to effectively function with the collaboration of humans and machines, a human-centered HCI design process must be established. This would require a customization from the traditional HCI design process to fit the characteristics and operational mechanism of the Industry 4.0. Our recommendation is to personalize the process of gathering and understanding requirements, designing solutions based on requirements, testing, and evaluating, in the context of Industry 4.0. The below personalization will cater to the efficient growth of Industry 4.0, all the while maintaining a strong relationship and codependency between humans and machine. To do this, we suggest involving domain experts and stakeholders in the user requirement part of the cycle. This will enable the stakeholder groups to minutely specify requirements at the outset, thereby honing the process.

Thereafter, using XAI-based modeling shall help eliminate the black box nature of AI based solutions and also help in bringing fairness and accountability to the process. In the next step, we believe that it is important to not only evaluate the solutions with initial user requirements but also perform usability testing involving stakeholder requirements. This shall help in developing a synergy between users and stakeholders, which shall benefit in the longer run by removing the actual constraints faced by users and alleviating them by involving the knowledge of stakeholders. In the last leg of the I4.0-HCI design process cycle, it is important to incorporate user feedback and not only explain them the mechanisms of how the system works but also why it arrived at a certain decision. If there is a large gap between the user-expected output, the stakeholder-predicted output, and the actual output, it is important to reiterate the process by removing critical bottlenecks. This can range from increasing the dataset size by conducting more expansive user surveys and user behavior studies, or it can involve changing the train-validate-test dataset sample sizes and percentages, so as to counter for overfitting, especially when deep learning models are involved, which are inherently more black-boxed in nature.

Thus, there needs to be a consensus between AI and HCI practitioners on when to draw and derive from their respective fields and when to pause and introspect. Iterative nature of both HCI design processes and AI-ML models make it easier to duly bifurcate and then combine certain tasks with each another. Using the I4.0-HCI confluence framework along with the aforementioned modifications in the HCI user-centered design process can definitely help in making the merger of AI and HCI systems transparent, seamless, and cohesive. Thus, after developing the above framework and recommendations for tweaking the user-centered HCI design process, we believe that we are in a position to answer the questions posed above. For the sake of reading convenience, we have included the questions before the respective answers using bullet points. They are as follows:

- Why is there a need to involve humans in the process of developing this I4.0-HCI confluence?

The involvement of users in developing the I4.0-HCI confluence is of utmost importance because without involving domain experts across a wide variety of fields and also without involving a randomized, all-encompassing group of users, the solutions would be black box in nature, similar to traditional AI models and would severe the purpose of developing human-centered intelligent interactive systems, which depict at least some form and type of intelligence.

- Which stakeholders need to be present in the development of the I4.0-HCI confluence?

The type and kind of stakeholders which shall be involved in the merger of these two disciplines shall involve AI researchers, factory workers, data scientists, social scientists, philosophers, mathematicians, and other domain experts as per the application area of the respective domain. This involvement is essential to the conducive growth of the field.

- When to trust the AI and when to trust humans in decision-making at the I4.0-HCI confluence?

While it is imperative that users feel empowered when it comes to decision making, it is equally important for the AI systems to disempower them in a way that they feel less in control of certain processes, which can pose hazards to them. As a very crude and general guideline, if the AI is able to come to a correct decision repeatedly, with excellent fault tolerance, precision-recall values, as well as confusion matrices, the users can learn to trust the AI, especially if explainable

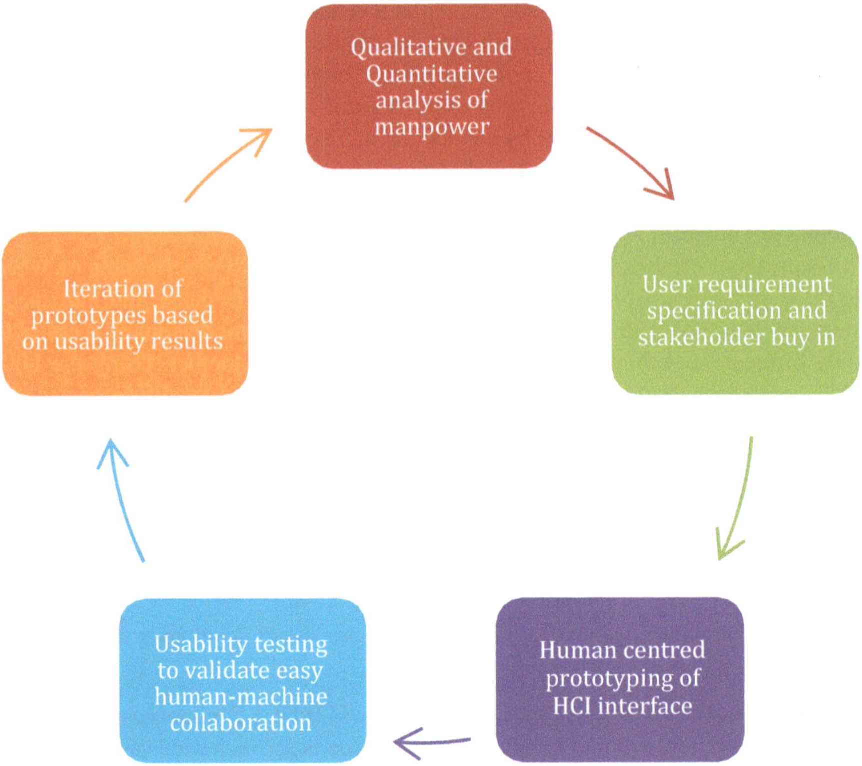

Fig. 2.4 HCI design process for Industry 4.0

AI toolkits are used to justify and interpret the outputs of the system. All in all, the AI systems should be entrusted with tasks and decisions that need huge and fast computation, and for others, humans can be more in control.

- Where can the users work in tandem with the AI/where they cannot, pertaining to the confluence?

Users can work in tandem with AI in the initial design and evaluation phase of the I4.0-HCI confluence and possibly cannot work in certain situations where there is a considerable gap between the respective operational capacities.

- How much, is too much, when it comes to the I4.0-HCI confluence?

AI and HCI systems need to be regularized, made transparent, and trusted with sensitive data that should be kept safe and away from potential attacks. They should also work in tandem with, and for humans, by not overpowering them (Fig. 2.4).

In the five-step process represented above, the first and crucial step of designing and setting up Industry 4.0 is to study and analyze the industrial manpower. This is important so as to keep their experience and concerns in mind while designing intelligent systems and their interfaces around them. The next step involves stating the user requirements analyzed from the qualitative and quantitative study of the concerned manpower, and involving every stakeholder to ensure full acceptance of user requirements. Although the above representation only lists stakeholder buy in the second stage, it is a practice that should be carried out at each stage. Once the requirements are zeroed upon, prototyping with respect to the complexity of Industry 4.0 machinery and ease of use for manpower should be done. In an Industry 4.0 perspective, it becomes important to make use of a good information architecture and easy-to-use interfaces to operate complex machines. Usability testing on such prototypes reveals the inconsistencies and confusion with respect to the usage of interfaces, on the realization of which, iterations of the prototypes are constructed, and the process is repeated until a good HCI is obtained.

2.6 Conclusion and Future Work

The development of a HCI system specifically made for the Industry 4.0 environment could be made a reality focusing on the three paradigms discussed in the paper. It can be a cause of concern for currently working industrial manpower, as it will automate industries extensively. While there are immediate ramifications of this on the numbers of jobs that would be available in the production industry, it does pave the way for developing better industrial practices which are focused on creating a synergy between humans and machines. The first industrial revolution started with the advent of the steam engine, second industrial revolution brought electricity to the world, the third and ongoing revolution brought automation as a boon for the industry, and, imminently, the fourth industrial revolution would usher a new era in the world, where previously imaginable phenomena would take actual shape. Such a revolution is inevitable, and industries would lose out on competition if they steer clear from Industry 4.0 modifications. But, with a well-designed HCI system in place, not only will the industry workers shift to better roles and responsibilities, but also be able to learn new skills, and will experience an engaging job with the help of well-designed assistive technologies to aid them in supervising and maintenance of complex machinery. Future work in this regard would include more case studies which are pertaining to the actual implementation of such cyber-physical systems, especially in the production industry, taking cues from the same, a robust framework for implementing HCI systems in the Industry 4.0 framework would open up unending opportunities at the confluence of humans and machines, thereby changing the world for the better, one industry at a time.

References

1. C. Kruppize et al, “A Survey on Human Machine Interaction in Industry 4.0, Vol. 1, No. 1, Article. Association for Computing Machinery (ACM), February 2020.
2. I. Islam, K. M. Munim, M. N. Islam, and M. M. Karim, “A proposed secure mobile money transfer system for sme in Bangladesh: An industry 4.0 perspective,” in 2019 International Conference on Sustainable Technologies for Industry 4.0 (STI). IEEE, 2019, pp. 1–6.
3. J. Lee, B. Bagheri, and H.-A. Kao, “A cyber-physical systems architecture for industry 4.0-based manufacturing systems,” Manufacturing letters, vol. 3, pp. 18–23, 2015.
4. H. Gruber, Innovation, skills and investment: a digital industrial policy for Europe, Economia e Politica Industriale, Springer, 2017, pp.1–17.
5. C. J. Bartodziej, The Concept Industry 4.0: An Empirical Analysis of Technologies and Applications in Production Logistics, Springer, 2017.
6. J. A. Saucedo-Martínez, et al., Industry 4.0 framework for management and operations: a review, Journal of Ambient Intelligence and Humanized Computing, 2017, pp. 1-13.
7. Fazel Ansari, Selim Erol, Wilfred Sihn, Rethinking Human-Machine Learning in Industry 4.0: How Does the Paradigm Shift Treat the Role of Human Learning?, Science Direct, 2018
8. Nahavandi, S. (2017). Trusted autonomy between humans and robots: Toward human-on-the-loop in robotics and autonomous systems. IEEE Systems, Man, and Cybernetics Magazine, 3(1), 10–17.
9. Bauer, D., Schumacher, S., Gust, A., Seidelmann, J., & Bauernhansl, T. (2019). Characterization of autonomous production by a stage model. Procedia CIRP, 81, 192–197.
10. Campbell, F. J. (2021). Human factors: The impact on industry and the environment. In E. R. Rhodes & H. Naser (Eds.), Natural resources management and biological sciences (pp. 1–14). IntechOpen.
11. Zhang, T., Li, Q., Zhang, C., Liang, H., Li, P., Wang, T., Li, S., Zhu, Y., & Wu, C. (2017). Current trends in the development of intelligent unmanned autonomous systems. Frontiers of Information Technology & Electronic Engineering, 18(1), 68–85.
12. Fosch-Villaronga, E., Lutz, C., & Tamò-Larrieux, A. (2020). Gathering expert opinions for social robots' ethical, legal, and societal concerns: Findings from four international workshops. International Journal of Social Robotics, 12(2), 441–458.
13. Gil, M., Albert, M., Fons, J., & Pelechano, V. (2019). Designing human-in-the-loop autonomous cyber-physical systems. International Journal of Human-Computer Studies, 130, 21–39
14. Santoni de Sio, F., & van den Hoven, J. (2018). Meaningful human control over autonomous systems: A philosophical account. Frontiers in Robotics and AI, 5, 15.
15. Weichhart, G., Ferscha, A., Mutlu, B., Brillinger, M., Diwold, K., Lindstaedt, S., Schreck, T., & Mayr-Dorn, C. (2019). Human/machine/roboter: Technologies for cognitive processes. Elektrotechnik Und Informationstechnik, 136(7), 313–317
16. A. M. Isen, “An influence of positive affect on decision making in complex situations: Theoretical issues with practical implications,” Journal of consumer psychology, vol. 11, no. 2, pp. 75–85, 2001.
17. M. N. Islam, S. J. Oishwee, S. Z. Mayem, A. N. Mokarrom, M. A. Razzak, and A. H. Kabir, “Developing a multi-channel military application using interactive dialogue model (idm),” in 2017 3rd International Conference on Electrical Information and Communication Technology (EICT). IEEE, 2017, pp. 1–6.
18. T. Zaki, Z. Sultana, S. A. Rahman, and M. N. Islam, “Exploring and comparing the performance of design methods used for information intensive websites,” MIST International Journal of Science and Technology, vol. 8, pp. 49–60, 2020.
19. H. Thimbleby and I. H. Witten, “User modeling as machine identification: new design methods for hci,” 1991.
20. A. Beard-Gunter, G. Ellis, C. Dando, and P. Found, “Designing industrial user experiences for industry 4.0,” 2018.

21. M. A. Razzak and M. N. Islam, "Exploring and evaluating the usability factors for military application: A road map for hci in military applications," Human Factors and Mechanical Engineering for Defense and Safety, vol. 4, no. 1, p. 4, 2020.
22. J. Nielsen, Usability engineering. Elsevier, 1994.
23. S. Deterding, R. Khaled, L. E. Nacke, and D. Dixon, "Gamification: Toward a definition," in CHI 2011 gamification workshop proceedings, vol. 12. Vancouver BC, Canada, 2011.
24. D. Gorecky, M. Schmitt, M. Loskyll, and D. Zuhlke, "Human-machine- ¨ interaction in the industry 4.0 era," in 2014 12th IEEE international conference on industrial informatics (INDIN), 2014, pp. 289–294.
25. E. Aranburu, G. S. Mondragon, G. Lasa, J. K. Gerrikagoitia, I. S. Coop, and G. S. Elgoibar, "Evaluating the human machine interface experience in industrial workplaces," in Proceedings of the 32nd International BCS Human Computer Interaction Conference. BCS Learning & Development Ltd., 2018, p. 93.
26. Peter Papcun, Erik Kajáti, Jiſ Koziorek, "Human Machine Interface in Concept of Industry 4.0," IEEE, October 2018.
27. Sebastian Büttner, Henrik Mucha, Markus Funk, Thomas Kosch, Mario Aehnelt, Sebastian Robert, Carsten Röcker, "The Design Space of Augmented and Virtual Reality Applications for Assistive Environments in Manufacturing: A Visual Approach," ResearchGate Conference Paper, June 2017.
28. Markus Funk, Albrecht Schmidt, Tilman Dingler, Jennifer Cooper, "Stop Helping Me - I'm Bored! Why Assembly Assistance needs to be Adaptive," ResearchGate, September 2015.
29. C. D. Fehling, A. Mueller, and M. Aehnelt. Enhancing vocational training with augmented reality. In S. Lindstaedt, T. Ley, and H. Sack, editors, Proceedings of the 16th International Conference on Knowledge Technologies and Data-driven Business. ACM Press, 2016.
30. S. Deterding, D. Dixon, R. Khaled, and L. E. Nacke, Gamification: Toward a Definition. CHI 2011, Vancouver, BC, Canada.
31. R. Tulloch, Reconceptualising gamification: Play and pedagogy. [Online] Available: http://www.digitalcultureandeducation.com/cms/wpcontent/uploads/2014/12/tulloch.pdf.
32. T. Schmidt, I. Schmidt, and P. R. Schmidt, "Digitales Spielen und Lernen – A Perfect Match?: Pädagogische Betrachtungen vom kindlichen Spiel zum digitalen Lernspiel.," in Gesundheit spielend fördern: Potenziale und Herausforderungen von digitalen Spieleanwendungen für die Gesundheitsförderung und Prävention, K. Dadaczynski, S. Schiemann, and P. Paulus, Eds., 2016, pp. 18–49.
33. Jacqueline Schuldt, Susanne Friedemann, 2017. "The Challenges of Gamification in the Age of Industry 4.0." IEEE Global Engineering Education Conference (EDUCON), April 2017.
34. T. Schmidt, MALL meets Gamification: Möglichkeiten und Grenzen neuer (digitaler) Zugänge zum Fremdsprachenlernen. [Online] Available: http://www.uni-potsdam.de/fileadmin01/projects/tefl/
35. Sebastian Büttner, Henrik Mucha, Markus Funk, Thomas Kosch, Mario Aehnelt, Sebastian Robert, and Carsten Röcker. "The Design Space of Augmented and Virtual Reality Applications for Assistive Environments in Manufacturing: A Visual Approach", Proceedings of the 10th International Conference on Pervasive Technologies Related to Assistive Environments - PETRA '17. ACM Press, Island of Rhodes, Greece, 433–440. https://doi.org/10.1145/3056540.3076193
36. SangSu Choi, Kiwook Jung, and Sang Do Noh., 2015, "Virtual reality applications in manufacturing industries: Past research, present findings, and future directions", Concurrent Engineering: Research and Applications 23, 1 (2015), 40–63. https://doi.org/10.1177/1063293X14568814
37. Mario Hermann, Tobias Pentek, and Boris Otto, "Design Principles for Industrie 4.0 Scenarios, 2016", 49th Hawaii International Conference on System Sciences (HICSS). IEEE, 3928–3937. https://doi.org/10.1109/HICSS.2016.488
38. V. V. Vishwarupe and P. M. Joshi, "Intellert: a novel approach for content-priority based message filtering," 2016 IEEE Bombay Section Symposium (IBSS), 2016, pp. 1–6, https://doi.org/10.1109/IBSS.2016.7940206

39. G. Dini and M. Dalle Mura, 2015, "Application of Augmented Reality Techniques in Through-life Engineering Services", Procedia CIRP 38 (jan 2015), 14–23. https://doi.org/10.1016/J.PROCIR.2015.07.044
40. Vishwarupe, V., Bedekar, M., Pande, M., Bhatkar, V. P., Joshi, P., Zahoor, S., & Kuklani, P. (2022). Comparative Analysis of Machine Learning Algorithms for Analyzing NASA Kepler Mission Data. *Procedia Computer Science*, *204*, 945–951. https://doi.org/10.1016/j.procs.2022.08.115
41. Vishwarupe, V., Joshi, P. M., Mathias, N., Maheshwari, S., Mhaisalkar, S., & Pawar, V. (2022). Explainable AI and Interpretable Machine Learning: A Case Study in Perspective. *Procedia Computer Science*, *204*, 869–876. https://doi.org/10.1016/j.procs.2022.08.105
42. Vishwarupe, V., Bedekar, M., Joshi, P. M., Pande, M., Pawar, V., & Shingote, P. (2022). Data Analytics in the Game of Cricket: A Novel Paradigm. *Procedia Computer Science*, *204*, 937–944. https://doi.org/10.1016/j.procs.2022.08.114
43. Vishwarupe, V., Maheshwari, S., Deshmukh, A., Mhaisalkar, S., Joshi, P. M., & Mathias, N. (2022). Bringing Humans at the Epicenter of Artificial Intelligence: A Confluence of AI, HCI, and Human-Centered Computing. *Procedia Computer Science*, *204*, 914–921. https://doi.org/10.1016/j.procs.2022.08.111
44. Vishwarupe V., All Things Policy: Synthetic Content Generation Using Artificial Intelligence, 2022/2/10, https://shows.ivmpodcasts.com/show/all-things-policyRx64RVpQImivrNQ8/episode/synthetic-content-generation-and-chinas-worries-ja9s-I7rfgZE3IhXRg2Fk?startTime=0
45. Vishwarupe V et al. (2021), A zone-specific weather monitoring system, Australian Patent No. AU2021106275, Australian Government, IP Australia, https://patents.google.com/?inventor=Varad+Vishwarupe
46. S Uzor, JT Jacques, JJ Dudley, PO Kristensson, "Investigating the Accessibility of Crowdwork Tasks on Mechanical Turk", Proceedings of the 2021 CHI Conference on Human Factors in Computing 2021, https://doi.org/10.1145/3411764.3445291
47. Wu, Y., Jiang, A.Q., Li, W., Rabe, M., Staats, C., Jamnik, M. and Szegedy, C., 2022. Autoformalization with large language models. *Advances in Neural Information Processing Systems, 35*, pp.32353–32368.
48. Zahoor S., Bedekar M., Vishwarupe V. (2016) A Framework to Infer Webpage Relevancy for a User. In: Satapathy S., Das S. (eds) Proceedings of First International Conference on Information and Communication Technology for Intelligent Systems: Volume 1. Smart Innovation, Systems and Technologies, vol 50. Springer, Cham. https://doi.org/10.1007/978-3-319-30933-0_16
49. Saniya Zahoor, Mangesh Bedekar, Vinod Mane, Varad Vishwarupe (2016), Uniqueness in User Behavior While Using the Web. In: Satapathy, S., Bhatt Y., Joshi A., Mishra D. (eds) Proceedings of the International Congress on Information and Communication Technology. Advances in Intelligent Systems and Computing, vol 438. Springer, Singapore. https://doi.org/10.1007/978-981-10-0767-5_24
50. V. Vishwarupe, M. Bedekar and S. Zahoor, "Zone specific weather monitoring system using crowdsourcing and telecom infrastructure," 2015 International Conference on Information Processing (ICIP), 2015, pp. 823–827, https://doi.org/10.1109/INFOP.2015.7489495.
51. Vishwarupe V., Bedekar M., Pande M., Hiwale A. (2018) Intelligent Twitter Spam Detection: A Hybrid Approach. In: Yang XS., Nagar A., Joshi A. (eds) Smart Trends in Systems, Security and Sustainability. Lecture Notes in Networks and Systems, vol 18. Springer, Singapore. https://doi.org/10.1007/978-981-10-6916-1_17

Chapter 3
Embedding Affect Awareness in e-Learning: A Systematic Outline of the Literature

Snehal R. Rathi and Yogesh D. Deshpande

3.1 Introduction

Education via the web has opened many doors in the present education scheme. E-learning is fundamentally the framework-enabled trade of talent and information by the online instructors to the online students. Web-based learning beats the constraint of reality by presenting extravagance of its client to make utilization of it at, whenever, anyplace. Similar to usual learning, web-based adapting likewise relies upon influential communication of human information, in spite of whether this happens in an eye-to-eye study hall or over the Internet. An educator depends on scientific classification, like Bloom's scientific classification to make learning goals used to build up teaching contents and exercises. Many times, faculty uses these goals to assess student achievement. It has attractive impact on objective-based appraisal [6].

Bloom's taxonomy in e-learning helps the instructors to think and analyze their teaching and student's learning. It is used to state clear objectives which can help the teachers to plan lessons accordingly [5]. Bloom's taxonomy guides the teachers to change the complexity of the questions and helps scholars to acquire greater ranges of hierarchy. Further, it helps to develop critical thinking among teachers. According to Bloom's scientific categorization, lessons' goals can be prepared into three spaces: intellectual area, knowledge; affective area, mentality; and psychomotor area, skills [7]. After the educational strategies are completed, college students must be assessed whether or not they have achieved the presented goals. In the group of students, when the guidance strategy is preparing, the educator always applies several varieties of educating competencies or create inquiries to inspire

S. R. Rathi (✉) · Y. D. Deshpande
Department of Computer Engineering, Vishwakarma University, Pune, India
e-mail: snehal.rathi-023@vupune.ac.in; yogesh.deshpande@vupune.ac.in

A. Neustein et al. (eds.), *AI, IoT, Big Data and Cloud Computing for Industry 4.0*, Signals and Communication Technology, https://doi.org/10.1007/978-3-031-29713-7_3

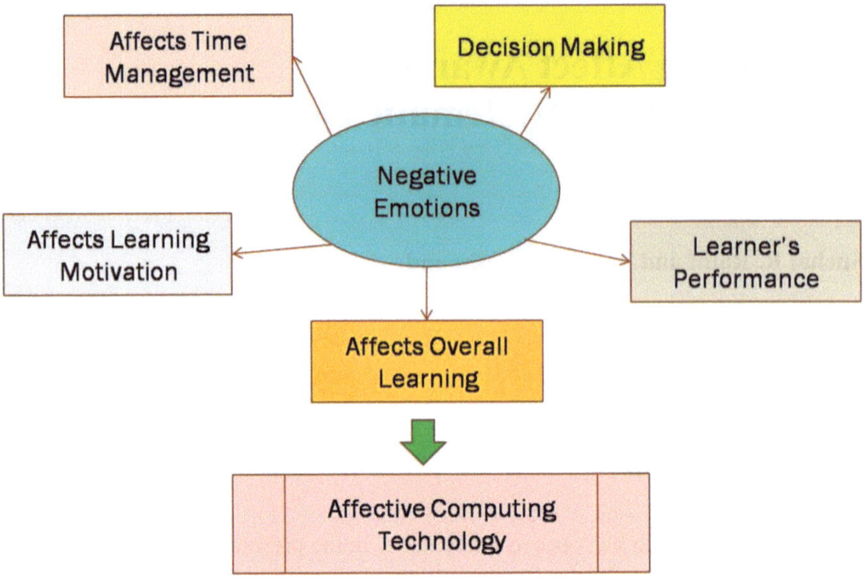

Fig. 3.1 Role of emotions in learning

learners to be thoughtful for accomplishment of affective area instructing goals. The greater part of these aptitudes and approaches are the method of collaboration [3]. E-learning has grown quickly in the last few years, and turned into a pattern of learning [4]. For students' assessment, this framework needs to be planned with the evaluation feature [8]. The widely used evaluation method in distance teaching and learning is online examination [8, 9]. Nonetheless, most instructing courses just put down weight on information broadcast, in particular, the knowledge area, and forget about learner's learning mentalities and the affective area educating goals. Because of the distance, instructors cannot survey the accomplishment of affective area educating goals. But, if instructors can understand the student's affective condition via the mastering framework, they can defeat the trouble.

As shown in Fig. 3.1, emotions have giant influences on learning and have a very noteworthy role in conclusion making and reflecting on the studies. Negative affect comprises an assortment of bad feelings such as outrage, disturb, blame, fear, and nervousness, which can deeply impact training and learning. Bad feelings influence learner's inspiration and learning. It gets hard for a student to manage learning exercises and timings. Answer for this is affective computing technology. Thus, understanding a learner's affect throughout the learning process is crucial for understanding motivation. In ordinary training research, student motivation can be recognized via questionnaires after occasion. With the development of affective computing, creators can fairly perceive and assess a learner's affective state at the time of learning in a continuous way [14], and thereafter they can appreciate the interrelationship between feeling, inspiration, and learning execution.

Therefore, e-courses must place the same importance on the needs of the affective domain along with the cognitive area.

E-learning has been depicted as less emotional and more impersonal due to not recognition of body language, facial expressions, and gestures of the learner when compared to face-to face learning [9]. If online learning systems do not provide real time feedback to the learner, at that point learner does not get the advantage of getting the real comments of learning performances at some stage in the learning procedure. The student's feelings greatly affect the student's inspiration toward better learning.

With this understanding of importance of affect in e-learning, we are presenting here an organized review of 61 selected papers in this area from 2009 to 2021. These papers are included in the survey after intense scrutiny. Here, we focus on different modalities/instruments, techniques, tools developed, and emotions identified by the researchers.

Section 3.2 includes motivation for this research. Section 3.3 represents strategy used in this survey and research questions based on which the survey is designed. Section 3.4 gives the details of related literature included in the survey. Section 3.5 gives survey outcomes and discussion and taken as a whole survey is concluded in Sect. 3.6.

3.2 Motivation

As shown in Fig. 3.2, with traditional teaching methods and high understudy/educator proportions, an instructor faces great obstructions in the study hall. Customarily, educators convey the substance and understudies gain knowledge of it. Using this technique, instructors are normally not ready to react to the individual needs of understudies. Besides, because of the huge quantities of understudies in a class, educators cannot concentrate toward every individual understudy. For reviewing, they consistently direct a test toward finish of the course. Nonetheless, in spite of the high number of understudies in a study hall, experienced educators typically watch, perceive, and tackle the learning way and affective state of the students [31]. The gifted instructors make proper move to emphatically affect knowledge. Be that as it may, the question is the thing that these accomplished instructors "see" and how they appear at a procedure and whether this movement drives the understudy to a gainful way. Student's learning not only relies on factors like the instructor's teaching style, a student's capacity, and a prerequisite but also on the compatibility of emotions.

The degree to which enthusiastic miracles can meddle with mental existence is no news to educators, understudies who are irate, on edge, or discouraged, experience issues in learning. Individuals trapped in these states do not receive data effectively or manage it very well [1]. Nonetheless, the connection among learning and feelings is a long way from being that straightforward and direct. Good and bad affective states produce distinctive sort of reasoning and this may hold noteworthy

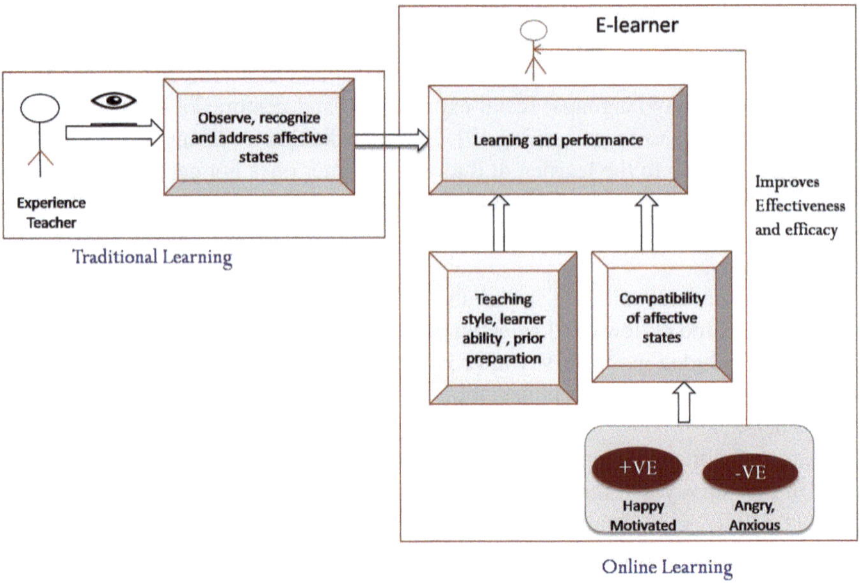

Fig. 3.2 Research motivation

inference from learning viewpoint. A reliable theory of discovering that incorporates effectively cognitive and affective factors are robustly required [2].

Gifted educators alter the learning way and their encouraging style as per understudies input signals (which incorporate intellectual, enthusiastic, and inspirational perspectives), online learning frameworks mostly do not consider these criticism signals resulting too strict and undermined, as they act in similar way for all understudies. Most of the online learning structures focus on data acquiring or intellectual handling.

While building a framework, affective states are viewed with respect to how the content is organized and introduced. To make learning effective, there is a need to build up a learner's model that coordinates the intellectual cycles and inspirational states [9]. Changing an e-learning framework from non-affect sensitive to a framework that incorporates user's affect requires the representation of a circle called as the affective circle. The affective circle envelops recognition of passionate states, proper activities choice for conclusion making, and the fusion of suitable affect by the framework [8]. Students, who get trapped in bad states like depression do not process and hold information proficiently. Through this, it tends to be construed that a user's affect has a noteworthy role in enhancing the viability of e-learning [10]. With this understanding of affect in e-learning, this survey is initiated.

3.3 Research Strategy and Research Questions

The survey related to "affective computing" was carried out in an organized way by following the steps mentioned in Fig. 3.3. Initially, research questions are designed to provide a suitable path to the review. We take it forward with retrieval of allied research papers and then done scrutiny and selected the most relevant research papers. Finally, the findings of the respective authors are studied and cited along with parameter they have measured in their research.

To aim the organized review, the following research questions are identified:

Q1. Which modalities/instruments are used in affective computing study?
Q2. Which emotions and parameters are measured by the researchers?
Q3. Which datasets are used by the researchers for prediction of affect?
Q4. What are the challenges facing this research area?

3.4 Literature Review

Emotions and online learning are connected with one another. The learning execution mirrors, the affective status of online learners in the existing setting of the web-based learning in advanced education. There has been a lot of enthusiasm among computer researchers and designers searching for approaches to progress person and workstation interaction by synchronizing emotion and cognition with task imperative [23]. Affective computing makes the chance of creating computational frameworks for acknowledgment of human affective states and minimize the gap among the profoundly expressive individual and the emotionally challenged computer [23]. Subsequently, affective computing research is an interdisciplinary area that plans to investigate human emotional involvement with technology by combining engineering and computer knowledge with areas like brain science, intellectual science, humanism, education, and morals [16].

These technologies sense the emotions of an individual by using a mix of various information sources – facial expressions, voice recognition, gestures, sensors [13], eye movement, behavioral log analysis, and questionnaires.

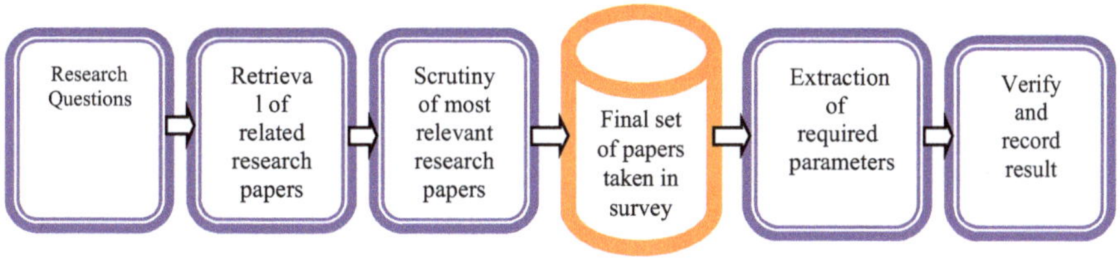

Fig. 3.3 Steps of strategy used for literature survey

Affective computing is a developing field uniting scientists and specialists from different fields, extending from artificial intelligence (AI) and natural language processing (NLP) to cognitive and sociologies [20].

The latest couple of years have seen a flood in the utilization of multidisciplinary procedures to consider the difficult role of emotions in a large number of learning settings. Sentiment analyses has picked up fascination in the computer-based education research network, may be due to the interdisciplinary idea of many research teams that include proficiency in brain science computations and NLP.

T. S. Ashwin and his team have developed system having two modules. In the first module, simply they are identifying emotions, but in the second module, automated inquiry-based instruction teaching strategy is utilized. After analysis, researchers have found that the model using intervention is performing very well and minimizing the rate of inattentive affective states [30]. Here, positive and negative emotions were recognized by utilizing cognitive appraisal approach to affective user modeling (which inferences emotions according to situations experienced by the user as well as the observable behavior of the user) and a physiological approach. The upper part of the model predicts emotions using user profiles, goals, and events and the lower part of the model predicts using sensors like heart rate, skin conductance, blood volume pressure, and EEG brainwaves. These two models converge using the pattern recognition method-Bayesian networks which is employed to model the relations between the emotional states and their causal variables [24].

Sandanayake and team have built up a system to know the students' online learning performance, while estimating the emotional state. Here creators have determined affect from survey and learning performance from standards of conduct in the log record. Few positive and negative emotions were estimated. Critical connection found between the emotions and log record information estimated [17].

Kort has built up an affective model of interaction among emotions and learning.

The flat line is the emotion axis and at right angles is the learning axis. As indicated by Kort in Fig. 3.4, understudies preferably start with quadrant 1 and may enter into further quadrants while learning where their emotion is bad and have unlearning state. If understudies get proper support at this point, they may move toward hopefulness state [18].

Every student has diverse learning style; yet sadly, present e-learning frameworks assess all the students in an equivalent manner. Students' learning relies on their learning style and affective states. Here, authors have recognized students' feeling and learning styles (active or reflex, sensitive or intuitive, etc.) from log document. The Felder Silverman learning style model is referred by authors [19]. As indicated by Felder and Silverman, distinguishing learning styles [11, 12, 19] are portrayed in Table 3.1.

Multimodal combination gives preferable outcomes over the unimodal investigation. There are various sorts of combination strategies utilized for multimodal fusion that are early fusion, late fusion, and hybrid fusion. Finding summed up model and utilization of deep learning for multimodal fusion is significant future work in this area [20].

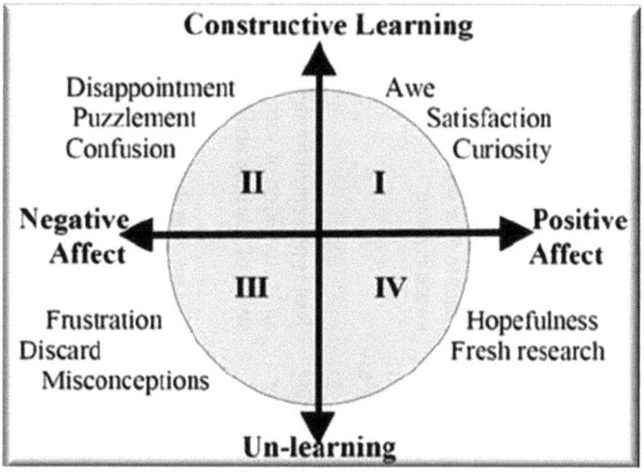

Fig. 3.4 Kort's learning spiral model

Table 3.1 Felder–Silverman learning styles

Learning style	Preference
Sensing-intuitive	How you prefer to perceive or take in information
Visual-verbal	How you prefer information to be presented
Active-reflective	How you prefer to process information
Sequential-global	How you prefer to organize and progress toward understanding information

Affective PC not only gives improved execution in helping individuals, but also improves PC's capacities to decide. In view of this proof, to construct PCs that make intelligent decisions may require construction of computers that have emotions [16].

In online learning environment, feelings majorly affect basic psychological cycles; neurological evidence shows they are not a luxury. Here, authors have featured a few outcomes from the neurological literature which demonstrate that emotions play a fundamental job in human imagination and knowledge, however furthermore in critical human thinking and conclusion making. PCs that will collaborate as expected and intelligently with people require recognizing and expressing affect [15].

Some experiment took data from biosensor wearable gadgets like smart watches and fit band. Authors have utilized here emotionWear framework. In emotionWear framework, a user watches the contents in the smartphone which incorporates VR headset and it is associated with the cloud storage. The user wears remote detecting gloves which produce various signs which are utilized to recognize emotions of the user. The emotions detected were exact, yet there was delay in investigation [27].

Most of the work aims to discover affect using facial expressions [29] and sensors [25]. The overall purpose is to enhance accuracy and take some corrective action to enhance learning. Researchers have performed experiment on data collected at various universities and labs. Different tools were developed by the researchers for data collection and analysis. Table 3.2 summarizes the parameters measured,

Table 3.2 Summary of the parameters measured, modalities used, methodology, and final outcome of the papers

Serial number	References	Parameters measured	Modality/instrument	Methodology	Benefit/outcome
1	Liping Shen [32] 2009	Interest, boredom, confusion, satisfaction, frustration, engagement, hopefulness, and disappointment.	User profiles and learning events and biosensors are used.	Study was guided by Russel's circumplex model and Kort's model. Authors explored evolution of emotions at the time of learning and observed the use of emotion feedback to improve learning.	Experiment confirms performance of emotion aware module over nonemotion aware is more.
2	Farman Ali Khan [33] 2009	(i) Emotions: confidence, confusion, independence, and effort (ii) Learning styles.	Behavioral log file: consist of 16 factors.	Both emotion and learning styles are identified from log file.	Tool is presented for detection and calculation of various styles of learning and affect.
3	Farman Ali Khan [34] 2010	(i) Emotions: confidence, confusion, independence, and effort (ii) Learning styles	Behavioral log file: consist of 16 parameters.	(i) Both emotion and learning styles are identified from log file. (ii) Adaptive course generator is used. (iii) Adaptive affective tactic generator module is developed which will do tagging and provide help according to LS and AS.	The investigation of learner's styles of learning and affect is done for providing them personalized support.
4	T.C. Sandanayake [17] 2013	Few positive emotions: pride, hope, enjoyment, and relief and some negative emotions: anger, boredom anxiety, hopelessness, and shame.	(i) Behavioral log file: evaluating learner's performance. (ii) Questionnaires: measured learner's emotions.	Tool is developed to recognize e-learner's emotions. Multiple regression is used.	(i) Identified learner's learning level and the emotional state. (ii) Correlation between learner's observed behavior and emotions is analyzed.

5	NELL BUISSINK-SMITH [3] 2014	Affective attributes.	Review of primary and secondary source data collection methods are given.	Overview of affective learning in education is given.	Framework for measurement of affective attributes is proposed.
6	M. Feidakis [35] 2014	Anger, boredom, confusion, fatigue, relief/relaxation, curiosity, despair, excitement, disinterest, embarrassment, inspiration, interest, stress, and nothing.	Tool Emotcontrol is developed. Self-reporting is used.	(i) Authors have developed conceptual and computational model. (ii) Implemented model employs self-reporting of affect, feedback, and efficient emotion visualizations.	Users found Emotcontrol tool useful, expressive, usable, and effective.
7	Chih-Hung Wu [14] 2015	Emotions.	Details of physiological signals, multimodal-based approach, and other modalities are given.	For realization of latest trends, here authors have summarized measurements of affective computing in learning field.	(i) The trends of affective computing in learning are discussed. (ii) Multimodal framework for affect recognition is proposed.(iii) Challenges and problems in this field are proposed.
8	Rachel Carlos [36] 2015	Affective states in collaborative environment.	Review paper.	Literature survey is done to find (i) Frequently used emotional states in collaborative surroundings. (ii) Were affective states empirically evaluated? (iii) Types of surroundings which makes use of affective states to support group learning.	Authors verified that 54.84% studies deal with "emotions," 51.61% studies were empirically evaluated, and LMS and e-learning systems are the surroundings which makes use of affective states to support group learning.

(continued)

Table 3.2 (continued)

Serial number	References	Parameters measured	Modality/instrument	Methodology	Benefit/outcome
9	Manuel Rodrigues [37] 2015	Affective states and learning styles.	Only framework is given where authors may use Explicit assumption: self-reporting Implicit assumption: facial agent, keyboard agent, mouse agent, and log agent.	Authors proposed framework in which after linking module with Moodle platform, student's affective states and learning styles get detected and contents were presented accordingly.	(i) In this paper, some problems of Moodle are identified. (ii) The framework is proposed here which includes an affective module in Moodle.
10	Jason Matthew Harley [38] 2015	Review of emotions in computer-based learning.	Pros and cons of (i) Self-reporting (ii) Log files (iii) Facial expressions (iv) Language and vocalization (v) Psychological signals are given.	Survey of all methods and modalities is given. The effectiveness of multimodal approaches is discussed.	Multiple methods used to predict emotions with their pros and cons are given.
11	Arvid Frydenlund [39] 2015	Affect, dominance, favorability, valence, and familiarity.	Multimodal affect detection is done. Sensors and videos are used.	Here, authors have presented a multimodal system for affect prediction. They have used DEAP dataset to train model. They also have introduced a way for extracting video features and creating more training examples from data.	Proposed model performs better for affect, dominance, favorability, and valence than baseline methods.
12	Jyotish Kumar [40] 2015	Disgust, anger, love, and courage.	GSR physiological signal is used to measure electrodermal responses.	Here, GSR of participants was recorded by placing skin electrodes at the palmer surface while watching videos. By analyzing this data, emotions are detected.	Authors have confirmed the performance by comparing it with verbal reported feedback.

13	Ashwin T S [41] 2015	Basic emotions.	Facial expressions detecting multiuser faces in e-learning.	Here, authors have implemented multiuser face detection-based e-learning system using SVM. The method gives 89–100% accuracy using different datasets LFW, FDDB, and YFD.	The method proposed by authors recognizes emotions for several faces in a solo frame.
14	Nigel Bosch [42] 2015	Boredom, confusion, delight, engagement, and frustration.	Facial expressions.	Here, authors proposed a method to detect students' affect. Data were collected in the computer lab. Affect is detected using facial expressions and body movements.	Here, authors have validated face-based detector for learning-centered affect. They have proved, it was possible to detect affect in the wild even though challenges exist.
15	Ishan Behoora [43] 2015	Engagement/interest, delight, frustration, and boredom	Nonwearable sensors like the Microsoft Kinect which captures physical image and motion.	Authors have developed a system. Steps of it are as follows: (i) Data acquisition and feature generation. (ii) Detecting body language from these features and mapping emotional state of design team members (iii) Machine learning modal is trained.	Emotional states are detected with accuracies above 98%.
16	Abhay Gupta [21] 2016	Engagement, frustration, confusion, and boredom.	Video analysis.	Authors present DAiSEE dataset which includes videos of academic emotions.	DAiSEE free-to-use, crowdsourced dataset for modeling four affective states: engagement, frustration, confusion, and boredom in e-learning environments.

(continued)

Table 3.2 (continued)

Serial number	References	Parameters measured	Modality/instrument	Methodology	Benefit/outcome
17	Ramanathan Subramanian [22], 2016	Affect and personality traits.	Physiological signals: GSR, frontal EEG, ECG, and facial features.	Authors present dataset which includes 58 participants and 36 videos. Data is collected using commercial wearable sensors and a webcam.	Authors present novel multimodal affective database ASCERTAIN.
18	Qianyu Hu [44] 2016	Positive, negative, joy, anger, pleasure, irritation, focused, shame, interested, guilt, calm, anxiety, peaceful, stress, comfortable, frustrated, attentive, distracted, and bored.	Questionnaires.	(i) Vary the difficulty of the task. (ii) Perceived affective states and performance time. (iii) Correlation test and performance analysis is done.	(i) No strong correlation between performance time and perceived affective states showed in a simpler task as opposed to some correlations in a more complex task. (ii) Relationship is provided by authors to improve the laboratory performance and safety.
19	Nik Thompson [45] 2016	Enjoyment, activation, and valence.	(i) Questionnaire involved during measurement of perceived learning and for enjoyment. (ii) Sensors are applied to understand activation and valence.	Here, authors have developed affective tutoring system and compared it with a non-affective version. Affective version shows measurable improvements in perceived learning.	Affective tutoring system is more effective than non-affective tutoring system.
20	Bahreini, Kiavash [46] 2016	Neutral, fear, sad, happy, disgust, surprise, and anger.	Vocal and facial expression recognition.	Real-time vocal and facial expressions are captured through videos and analysis is done. Feedback is provided.	Here, developed software allows unobtrusive observation of learners' behaviors and converts it into emotional states.

21	Ray, Arindam [47] 2016	Happiness, sadness, surprise, anger, disgust, and fear.	Sensors and facial expression.	Here, authors proposed affective computing module in e-Learning system. They explore evolution of emotions. Based on the detected emotion, the system delivers lesson using fuzzy logic and artificial neural network.	Achieve more accuracy at decision-level fusion of facial expression and sensor data.
22	Akputu K. Oryina [48] 2016	Learning emotions, hopefulness, engagement, happiness, frustration, surprise, boredom, and confusion.	Facial expressions.	Here, author's present new architecture facial emotion recognition technology (FERT) having three modules: face detection, feature extraction, and emotion recognition.	This paper presents a new architecture FERT. Authors have verified multiple kernel learning (MKL) framework which outperforms traditional classifiers by conducting experiments on contextual emotion datasets.
23	K A Laksitowening [49] 2017	Identified: learning type from three factors, learning styles, motivation, and knowledge ability.	Behavioral log file.	Framework of triple factor approach is provided and illustration of dynamic personalization is given.	Here, authors have used dynamic personalization in which contents are given to learner based on result of triple factor approach.
24	Christian E. Lopez [50] 2017	Performance of learner.	Nonwearable infrared sensors like camera are used.	Here, SVM is used to classify one's performance based on facial key point data which is captured using a nonwearable sensor. Cross-validation approach is used to evaluate the accuracy of model.	Authors have designed system for providing feedback to individual based on task and characteristics of user.

(continued)

Table 3.2 (continued)

Serial number	References	Parameters measured	Modality/instrument	Methodology	Benefit/outcome
25	Shelena Soosay [51] 2017	Psychomotor skills and affective states.	Scores given by evaluators.	Here, authors have prepared an assessment rubric for evaluating different projects based on the psychomotor and affective learning domains aligned with Bloom's taxonomy.	Rubrics designed here is helpful for the students while developing a software project.
26	Tara J. Brigham [13] 2017	Affective states: emotions.	Sensors were used.	Affective computing technology collects data using different sensors and analyzes it to get emotional state. Then system give response based on detected state. This technology is very helpful into future.	While making use of affective computing technology, ethics, privacy and cost-effectiveness are major concerns.
27	Soujanya Poria a [20] 2018	Affect.	Multimodal affect: audio, visual, and text information.	Here, authors had given systematic review of multimodal affect frameworks and different fusion techniques.	This survey has confirmed that multimodal classifiers outperforms than unimodal. Text modality is more important and deep learning approaches are used most widely for extracting features.
28	Jan K. Argasinski [52] 2018	Creation of serious affective game.	Design patterns.	Authors proposed a novel framework for the designing and assessing affective games based on design patterns.	Authors have proposed a novel approach for designing affective serious games.

29	Chih-Hung Wu [53] 2018	Affective states: angry, fear, happy, neutral, contempt, disgust, sad, and surprise.	Multimodal: facial expressions, heart rate monitoring, blood oxygen level, skin conductance response, and electroencephalogram (EEG) signals.	Here, authors have developed the data integration system for integrating data of multiple modalities. Real-time facial expressions are classified into six basic emotions.	Authors have developed integration mechanism for avoiding sampling problem. It averages out sensor's data and rearrange it into same time. System is useful for collecting and integrating multimodal data in affective computing field.
30	Rwitajit Majumdar [54] 2018	Engagement.	Log file in TPS architecture.	Tool is developed to collect data using observational protocol and to analyze results. Analysis and tracking of collected data with respect to time and state of change is done.	Tool is helpful to instructor and researchers.
31	Francesca D'Erricoa [55] 2018	11 cognitive emotions like interest, attention, concentration, surprise, disappoint, frustration, and so on.	Facial expression.	Here, authors have collected data in the form of videos of 10 psychology students during online learning task, and using software detected the emotions.	Authors tested the reliability of software developed for classification of cognitive emotions based on facial expressions.
32	Siddharth [56] 2019	Emotion, valence, arousal, and liking.	Sensors and video data.	Here, authors have applied novel deep learning-based methods on publicly available datasets. Authors have done experiment on each separate dataset as well as by doing fusion of features. They proved that their algorithms outperform for classification on DEAP and MAHNOB-HCI datasets.	Authors proposed here a novel technique for identifying brain sections corresponding to various affective states.

(continued)

Table 3.2 (continued)

Serial number	References	Parameters measured	Modality/instrument	Methodology	Benefit/outcome
33	Yinghui Zhou [57] 2020	Emotional state and learning state.	Expressions, speech, eye movement, physiological signals, text, questionnaires, and multimodal.	Here, authors proposed a framework for online learning supported by Artificial Intelligence, big data, and brain–computer interface. Personalized education was also provided by the system.	The framework proposed helped to obtain more real data. It makes efficient analysis and recognizes emotions more accurately.
34	Troussas C [58] 2020	Affective states.	Sentiment analysis.	Here, authors focus on the concepts of affective computing adapted to social networking-based learning and learners' affective states.	Authors concluded sentiment analysis refers to the use of expert methods to systematically identify, extract, quantify, and study affective states and subjective information.
35	Silvia Ceccacci [59] 2021	Emotion.	Facial coding techniques.	Here, authors have developed a system which takes as input the video capture by the webcam of the device used to attend the course. (i) It performs continuous students' authentication based on face recognition. (ii) Monitor the students' level of attention through head-orientation tracking and gaze detection analysis. (iii) Estimate student's emotion during the course attendance.	Authors describe the overall system design and report the results of a preliminary survey, which involved a total of 14 subjects. Intension of authors is to check user acceptance to continue using such a system.

36	Resham Arya [60] 2021	Affect.	Sensors and facial expressions.	A survey of domains contributing to affective computing is given in detail.	Here, authors have explained different domains in affective computing, presented few existing affective databases, applications are discussed and some challenges along with suggested solutions were given for future work.
37	Saurabh Kumar [61] 2021	Affect.	Text and image.	Here, authors present affective computing model using both text and image data. Experiment was conducted on standard datasets using deep learning models.	The result confirms that the proposed method outperforms the earlier methods.

modalities used, methodology, and final outcome of the papers included in the survey.

Farman Ali Khan [33, 34] has developed a model for prediction of both emotion and learning styles from behavioral log file. They have used adaptive course generator and given content to the learner based on emotion. Adaptive affective tactic generator module is developed which will do tagging and provide help according to LS and AS. Here, authors have predicted emotions like confidence, confusion, independence, and effort along with learning styles of learner. Based on the predicted emotion and learning styles of learner, feedback is given.

Nik Thompson [45] has developed a system to predict enjoyment, activation, and valence. Questionnaire approach is used to predict enjoyment and measurement of perceived learning. Sensors are applied to understand activation and valence. For this purpose, authors have developed affective tutoring system and compared it with a non-affective version. Affective version shows measurable improvements in perceived learning.

Jyotish Kumar [40] has developed an approach for predicting disgust, anger, love, and courage. They have used galvanic skin response (GSR) physiological signal to measure electrodermal responses. Here, GSR of participants was recorded by placing skin electrodes at the palmer surface while watching videos. By analyzing this data, emotions are detected and improvement is observed.

From the literature survey, it is discovered that various modalities like facial expression recognition, questionnaires, interaction log behavior, eye movement, gestures, voice recognition and sensors (galvanic skin response (GSR), electrocardiogram (ECG) [26] and electroencephalography (EEG)) are used by researchers to recognize the emotions.

From the literature survey done, it was revealed that different researchers have developed the tool, emotions were predicted, and the criticism is given to the learner. Due to criticism, the improvements are observed in the learner's performance.

3.5 Survey Outcome and Discussion

As shown in Fig. 3.5, emotions have an enormous role in e-learning. Work previously done in the field of affective computing by various authors is as follows:

- Effect of task complexity on affective state and task execution is identified by giving the mechanical job of cutting piece in rectangular and spherical structure to students.
- According to the learning styles and emotions of learner, course contents will be given to them to upgrade their performance.
- Learning styles of learners were recognized.
- Fundamental emotions were recognized by individual researchers.
- By analyzing the performance based on historical data, feedback is given to learners. Improvement in performance is observed using this method.

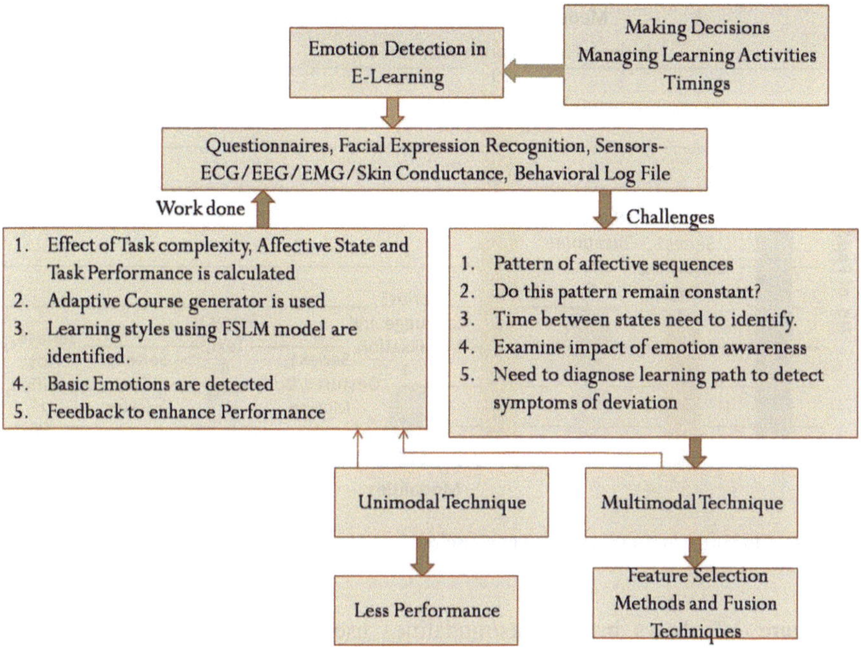

Fig. 3.5 Literature analysis

From the literature survey, it is discovered that various modalities like facial expression recognition, questionnaires, interaction behavior log, eye tracking, gestures/body motions, language and vocalization, text and sensors (galvanic skin response (GSR), electrocardiogram (ECG), and electroencephalography (EEG)) are used by researchers to recognize the emotions.

There are two different techniques, unimodal and multimodal technique, used by researchers to detect emotions in e-learning. There are different feature selection methods and fusion techniques used in multimodal technique. Different fusion techniques are early fusion (fusion at feature level), late fusion (fusion at output level), and hybrid fusion (mixture of early and late fusion). Performance of multimodal techniques is better than unimodal technique.

From this survey, it has been observed that most of the researchers have used facial expressions to recognize the emotion which is followed by sensors modality as shown in Fig. 3.6.

From the literature survey done, it was revealed that researchers have predicted different emotions, developed the tools, and the criticism is given to the learner. Due to criticism, the improvements are observed in the learner's performance. Most of the researchers have worked on academic emotions like interest, engagement, confusion, frustration, boredom, and hopefulness. Affect is generalized term used by the researchers, found in most of the papers.

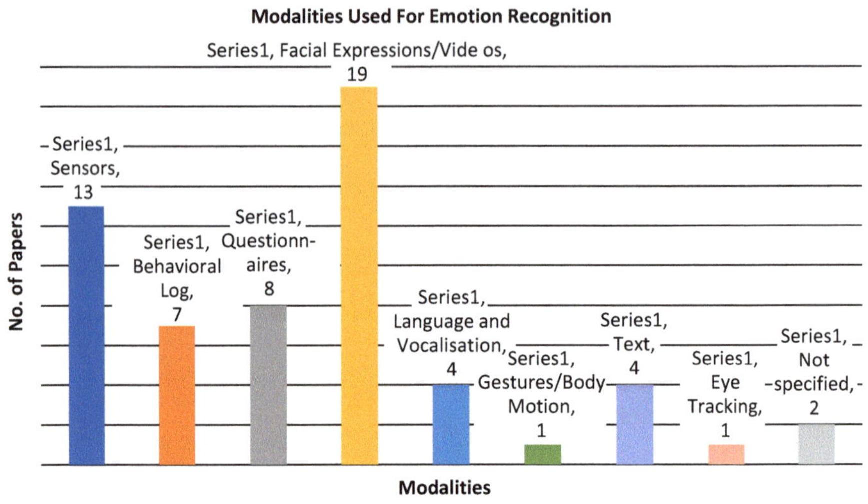

Fig. 3.6 Modalities used for emotion recognition

Figure 3.7 shows instruments/modalities used and parameters identified by researchers in the field of affective computing. After identifying parameters, researchers have taken various actions like use of adaptive course generator, dynamic personalization, giving feedback based on state.

There are various challenges that need to be addressed in this field:

1. Instead of only finding affective state, there is a need to identify "What are the patterns of affective sequences of e-learner during e-learning?"
2. "Do these patterns remain constant throughout the course?"
3. Time span between two different emotions need to identify.
4. Need to examine result of emotion awareness.
5. Need to diagnose learning path to detect symptoms of deviation.
6. Data for few modalities is a major concern.

As given in Table 3.3, there are few affective datasets which are available for affect recognition. But for certain modalities like interaction log, data is a major concern.

3.6 Conclusion

At present, emotion recognition can be carried out by using various modalities. Among them, facial expression and sensors are more widely used for emotion recognition, but emotion recognition based on gestures/body motions and eye tracking are less applied. This study has looked into the latest trends in affective

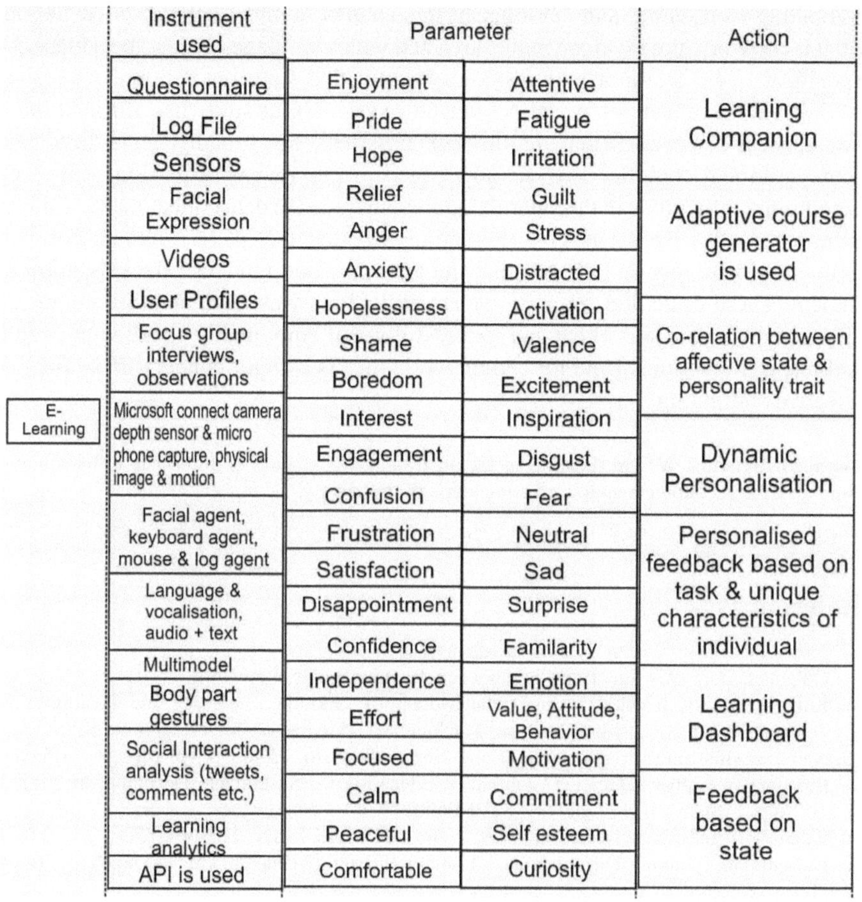

Fig. 3.7 Instruments used, parameters measured, and action taken by researchers in affective computing area

Table 3.3 Affective databases

Sr. No.	Dataset	Modalities	Affect	Purpose
1	DEAP	Physiological	Yes	Implicit affective tagging from physiological signals
2	ASCERTAIN [28]	Face, physiological	Yes	Affect and personality recognition
3	AMIGOS	Audio, visual, physiological, depth	Yes	Personality, mood, and affect recognition
4	DAiSEE	Video	Yes	Dataset for affective states in e-environments
5	DECAF	Face, physiological	Yes	Affect recognition
6	MANHOB-HCI	Face, audio, physiological, eye gaze	Yes	Emotion recognition

computing in teaching and learning. In this chapter, authors have discussed about what is the work already done in this area and what challenges need to be addressed. Specifically, we featured major studies in affective state recognition, which we believe vital segments of any affect detector system. Our study has affirmed other researcher's discoveries that multimodal classifiers can broadly get utilized and beats unimodal classifiers. As acknowledged in this review, a portion of the key exceptional difficulties in this energizing field include (1) data collection, (2) finding patterns of affective sequences, (3) finding whether this pattern remain constant throughout learning, and (4) whether the learning style influence the exhibition of student. These difficulties propose we are still a long way from creating a constant affective state detector which can successfully and emotionally speak with people and feel our emotions. Data for certain modalities is a major concern in the field of affective computing.

Acknowledgments We are thankful to the Vishwakarma Institute of Information Technology, Pune for their encouragement toward completion of this survey.

References

1. Goleman, D. (1995): Emotional intelligence. Bantam Books, New York.
2. Kort, B., Reilly, R. (2001): Analytical Models of Emotions, Learning and Relationships: Towards an Affect-sensitive Cognitive Machine. MIT Media Lab Tech Report No 548.
3. Nell Buissink-Smith, Samuel Mann and Kerry Shephard, How Do We Measure Affective Learning in Higher Education? Journal of Education for Sustainable Development 2011 5: 101 https://doi.org/10.1177/097340821000500113.
4. Picard, R. (2000) "Affective Computing". The MIT Press, ISBN: 0262661152.
5. L. Shulman, "Making Differences: A Table of Learning" Change, vol. 34, no. 6, p. 36–44, 2002.
6. R. J. Marzano, "The Need for a Revision of Bloom's Taxonomy", In The New Taxonomy of Educational Objectives, p. 1–20, 2006.
7. L.W. Anderson, and D.R. Krathwohl, et al (Eds.), "A Taxonomy for Learning, Teaching, and Assessing: A Revision of Bloom's Taxonomy of Educational Objectives". Allyn & Bacon Boston, MA, 2001.
8. D'Mello, S. K., Craig, S.D., Witherspoon, A., McDaniel, B., and Graesser, A., (2008). Automatic Detection of Learner's Affect from Conversational Cues. Journal of User Modeling and User-Adapted Interaction, 18(1–2), 45–80.
9. Cocea, M., Weibelzahl, S. (2007). Eliciting motivation knowledge from log files towards Motivation diagnosis for Adaptive Systems. User Modeling 2007 LNCS Springer Berlin / Heidelberg.
10. Weimin, X., Wenhong, X.(2007). E-Learning Assistant System Based on Virtual Human Interaction Technology, ICCS 2007, LNCS, Springer Berlin / Heidelberg.
11. Graf, S., Kinshuk. (2006): An Approach for Detecting Learning Styles in Learning Management Systems, in Sixth IEEE International Conference on Advanced Learning Technologies, Kerkrade, Netherlands, pp. 161–163.
12. Felder, R.M. and Silverman, L.K., (1988): Learning and teaching styles in engineering Education, Engineering Education, Vol. 78, No. 7, pp. 674–681.
13. Merging Technology and Emotions: Introduction to Affective Computing. Brigham TJ. Med Ref Serv Q. 2017 Oct-Dec;36(4):399–407. https://doi.org/10.1080/02763869.2017.1369289.

14. Wu, Chih-Hung & Huang, Yueh-Min & Hwang, Jan-Pan. (2015). Review of affective computing in education/learning: Trends and challenges. British Journal of Educational Technology. https://doi.org/10.1111/bjet.12324.
15. Jason Matthew Harley, Chapter 5 - Measuring Emotions: A Survey of Cutting Edge Methodologies Used in Computer-Based Learning Environment Research, Editor(s): Sharon Y. Tettegah, Martin Gartmeier, In Emotions and Technology, Emotions, Technology, Design, and Learning, Academic Press, 2016, Pages 89–114, ISBN 9780128018569.
16. R. W. Picard, Affective Computing, M.I.T Media Laboratory Perceptual Computing Section Technical Report No. 321, 1995.
17. T. C. Sandanayake and A. P. Madurapperuma, "Affective e-learning model for recognising learner emotions in online learning environment," 2013 International Conference on Advances in ICT for Emerging Regions (ICTer), Colombo, 2013, pp. 266–271.
18. Kort, B., Reilly, R., and Picard, R. An affective model of the interplay between emotions and learning, In IEEE International Conference on Advanced Learning Technologies, no. 43–46 (2001).
19. Khan, F.A., Weippl, E.R. & Tjoa, A.M. (2009). Integrated Approach for the Detection of Learning Styles and Affective States. In G. Siemens & C. Fulford (Eds.), Proceedings of ED-MEDIA 2009–World Conference on Educational Multimedia, Hypermedia & Telecommunications (pp. 753–761). Honolulu, HI, USA: Association for the Advancement of Computing in Education (AACE).
20. Soujanya Poria, Erik Cambria, Rajiv Bajpai, Amir Hussain, A review of affective computing: From unimodal analysis to multimodal fusion, Information Fusion, Elsevier, Volume 37, September 2017, Pages 98–125.
21. DAISEE: Dataset for Affective States in E-Learning Environments Gupta, A and Jaiswal, R and Adhikari, S and Balasubramanian, Vineeth N (2016) DAISEE: Dataset for Affective States in E-Learning Environments. arXiv. pp. 1–22.
22. R. Subramanian, J. Wache, M. K. Abadi, R. L. Vieriu, S. Winkler and N. Sebe, "ASCERTAIN: Emotion and Personality Recognition Using Commercial Sensors," in IEEE Transactions on Affective Computing, vol. 9, no. 2, pp. 147–160, 1 April-June 2018.
23. Calvo, Rafael & K. D'Mello, Sidney. (2010). Affect Detection: An Interdisciplinary Review of Models, Methods and their Applications, T. Affective Computing. 1. 18–37. https://doi.org/10.1109/T-AFFC.2010.1.
24. Shen, L., Wang, M., & Shen, R. (2009). Affective e-Learning: Using "Emotional" Data to Improve Learning in Pervasive Learning Environment. Educational Technology & Society, 12 (2), 176–189.
25. Ray, A., & Chakrabarti, A. (2016). Design and Implementation of Technology Enabled Affective Learning Using Fusion of Bio-physical and Facial Expression. Educational Technology & Society, 19(4), 112–125.
26. Heart Rate Variability Signal Features for Emotion Recognition by using Principal Component Analysis and Support Vectors Machine, Han-Wen Guo and Yu-Shun Huang, Chien-Hung Lin, Jen-Chien Chien, and Koichi (2016).
27. Coverage of Emotion Recognition for Common Wearable Biosensors, Terence K.L. Hui and R. Simon Sherratt (2018).
28. Emotion and Personality Recognition Using Commercial Sensors, Ramanathan Subramanian; Julia Wache; Mojtaba Khomami Abadi; Radu L. Vieriu (2016).
29. Comparison between Euclidean and Manhattan distance measure for facial expressions classification, Latifa Greche; Maha Jazouli; Najia Es-Sbai; Aicha Majda; Arsalane Zarghili (2017).
30. Ashwin, T.S. and Guddeti, R.M.R., 2020. Impact of inquiry interventions on students in e-learning and classroom environments using affective computing framework. User Modeling and User-Adapted Interaction, pp. 1–43.
31. Wampfler, R., Klingler, S., Solenthaler, B., Schinazi, V.R. and Gross, M.,"Affective State Prediction Based on Semi-Supervised Learning from Smartphone Touch Data," In Proceedings of the CHI Conference on Human Factors in Computing Systems, pp. 1–13, 2020.

32. Shen, Liping & Wang, Minjuan & Shen, Ruimin. (2009). Affective e-Learning: Using "Emotional" Data to Improve Learning in Pervasive Learning Environment. Educational Technology & Society. 12. 176–189.
33. Khan, Farman & Graf, Sabine & Weippl, Edgar & Tjoa, A Min. (2009). Integrated Approach for the Detection of Learning Styles & Affective States.
34. Khan, Farman & Graf, Sabine & Weippl, Edgar & Tjoa, A Min. (2010). Identifying and Incorporating Affective States and Learning Styles in Web-based Learning Management Systems. Interaction Design and Architecture(s).
35. Feidakis, Michalis & Daradoumis, Thanasis & Caballé, Santi & Conesa, Jordi. (2014). Embedding emotion awareness into e-learning environments. International Journal of Emerging Technologies in Learning (iJET). 9. 39. https://doi.org/10.3991/ijet.v9i7.3727.
36. Carlos, Rachel & Reis, Rachel & Lyra, Kamila & Bittencourt, Ig & Rodriguez, Carla & Jaques, Patricia & Isotani, Seiji. (2015). Affective States in CSCL Environments A Systematic Mapping of the Literature. https://doi.org/10.1109/ICALT.2015.95.
37. Rodrigues, Manuel & Fdez-Riverola, Florentino & Novais, Paulo. (2011). Moodle and affective computing: Knowing who's on the other side. Proceedings of the European Conference on Games-based Learning. 2. 678–685.
38. Harley, Jason. (2015). Measuring Emotions: A Survey of Cutting-Edge Methodologies Used in Computer-Based Learning Environment Research.
39. Frydenlund, Arvid & Rudzicz, Frank. (2015). Emotional Affect Estimation Using Video and EEG Data in Deep Neural Networks. Lect. Notes Comput. Sci.. 9091. 273–280. https://doi.org/10.1007/978-3-319-18356-5_24.
40. J. Kumar and J. A. Kumar, "Machine learning approach to classify emotions using GSR," Advanced Research in Electrical and Electronic Engineering, vol. 2, no. 12, pp. 72–76, 2015.
41. T. S. Ashwin, J. Jose, G. Raghu and G. R. M. Reddy, "An E-Learning System with Multifacial Emotion Recognition Using Supervised Machine Learning," 2015 IEEE Seventh International Conference on Technology for Education (T4E), 2015, pp. 23–26, https://doi.org/10.1109/T4E.2015.21.
42. Bosch, Nigel & D'Mello, Sidney & Baker, Ryan & Ocumpaugh, Jaclyn & Shute, Valerie & Ventura, Matthew & Wang, Lubin & Zhao, Weinan. (2015). Automatic Detection of Learning-Centered Affective States in the Wild. International Conference on Intelligent User Interfaces, Proceedings IUI. 2015. 379–388. https://doi.org/10.1145/2678025.2701397.
43. Behoora, Ishan & Tucker, Conrad. (2015). Machine learning classification of design team members' body language patterns for real time emotional state detection. Design Studies. 39. https://doi.org/10.1016/j.destud.2015.04.003.
44. Hu, Qianyu & Bezawada, Shruthi & Gray, Allison & Tucker, Conrad & Brick, Timothy. (2016). Exploring the Link Between Task Complexity and Students' Affective States During Engineering Laboratory Activities. V003T04A019. https://doi.org/10.1115/DETC2016-59757.
45. Thompson, N. & McGill, Tanya. (2016). Genetics with Jean: the design, development and evaluation of an affective tutoring system. Educational Technology Research and Development. https://doi.org/10.1007/s11423-016-9470-5.
46. Bahreini, Kiavash & Nadolski, Rob & Westera, Wim. (2015). Towards Real-time Speech Emotion Recognition for Affective E-learning. Education and Information Technologies. 1–20. https://doi.org/10.1007/s10639-015-9388-2.
47. Ray, Arindam & Chakrabarti, Amlan. (2016). Design and Implementation of Technology Enabled Affective Learning Using Fusion of Bio-physical and Facial Expression. Educational Technology & Society. 19.
48. A. K. Oryina and A. O. Adedolapo, "Emotion Recognition for User Centred E-Learning," 2016 IEEE 40th Annual Computer Software and Applications Conference (COMPSAC), 2016, pp. 509–514, https://doi.org/10.1109/COMPSAC.2016.106.
49. Laksitowening, Kusuma & Santoso, Harry & Hasibuan, Zainal. (2017). E-Learning Personalization Using Triple-Factor Approach in Standard-Based Education. Journal of Physics: Conference Series. 801. 012027. https://doi.org/10.1088/1742-6596/801/1/012027.

50. López, Christian E. and Conrad S. Tucker. "From Mining Affective States to Mining Facial Keypoint Data: The Quest Towards Personalized Feedback." (2017).
51. Nathan, Shelena & Berahim, Mazniha & Ramle, Rosni. (2017). Rubric for Measuring Psychomotor and Affective Learning Domain. Pertanika Journal of Social Science and Humanities. 25. 101–108.
52. Jan K. Argasiński, Paweł Węgrzyn, Affective patterns in serious games, Future Generation Computer Systems, Volume 92, 2019, Pages 526–538, ISSN 0167-739X, https://doi.org/10.1016/j.future.2018.06.013.
53. Wu, Chih-Hung & Kuo, Bor-Chen. (2018). An Exploratory Study of Multimodal Perception for Affective Computing System Design. https://doi.org/10.1007/978-981-10-7398-4_20.
54. Thesis: Visual Analytics of Cohorts in Educational Datasets https://www.it.iitb.ac.in/~sri/students/rwitajit-thesis.pdf
55. D'Errico, Francesca & Paciello, Marinella & de Carolis, Berardina & Palestra, Giuseppe & Vattani, Alessandro. (2018). Cognitive Emotions in E-Learning Processes and their Potential Relationship with Students' Academic Adjustment. International Journal of Emotional Education. 10. 89–111.
56. S. Siddharth, T. Jung and T. J. Sejnowski, "Utilizing Deep Learning Towards Multi-modal Bio-sensing and Vision-based Affective Computing," in IEEE Transactions on Affective Computing, https://doi.org/10.1109/TAFFC.2019.2916015.
57. Yinghui Zhou and Xiaomei Tao. 2020. A Framework of Online Learning and Experiment System Based on Affective Computing. In Proceedings of the 2020 3rd International Conference on E-Business, Information Management and Computer Science. Association for Computing Machinery, New York, NY, USA, 619–624. https://doi.org/10.1145/3453187.3453405
58. Troussas C., Virvou M. (2020) Affective Computing and Motivation in Educational Contexts: Data Pre-processing and Ensemble Learning. In: Advances in Social Networking-based Learning. Intelligent Systems Reference Library, vol 181. Springer, Cham. https://doi.org/10.1007/978-3-030-39130-0_5
59. Ceccacci, Silvia et al. "Facial coding as a mean to enable continuous monitoring of student's behavior in e-Learning." teleXbe (2021).
60. Resham Arya, Jaiteg Singh, Ashok Kumar, A survey of multidisciplinary domains contributing to affective computing, Computer Science Review, Volume 40, 2021, 100399, ISSN 1574-0137, https://doi.org/10.1016/j.cosrev.2021.100399.
61. Kumar, S. (2021), "Deep learning based affective computing", Journal of Enterprise Information Management, Vol. 34 No. 5, pp. 1551–1575. https://doi.org/10.1108/JEIM-12-2020-0536

Chapter 4
Edge Computing: A Paradigm Shift for Delay-Sensitive AI Application

Shalini Nigam and Mandar S. Karyakarte

4.1 Introduction

The invention of the Internet of things (IoT) is making life easier by automating many solutions. The IoT provides smart solutions for many real-world as well as enterprise-related problems. With IoT, we have smart buildings, smart cities, smart vehicles, and many more. Internet of everything (IoE) is an extension of IoT. The traditional IoT devices senses data and transmits this data to the cloud and then the cloud does some data processing and sends back the response to the end device.

Development of IoT devices has achieved remarkable growth in the field of real world and industry. With this rapid development of IoT devices, the continuous interaction for data storage, computation, and communication with the cloud has been increased. Storing, managing, processing and computation of data, response time, and associated cost are the major challenges. Response time is vital factor especially in the delay sensitive and highly interactive applications such as VR gaming and autonomous cars.

Since most of the IoT devices are battery operated, available computation power is also a very important consideration. If the algorithm written for computation takes larger power, then the device will drain off quickly. Another very important challenge is related to privacy and security. Since IoT-enabled devices transmit the data to the cloud, security and privacy of data may be affected in the transit.

The solution for the above-mentioned challenges is edge computing. In edge computing, data generation, data processing, and computation are done at the edge of the network of IoT devices. If required, the data is transferred to the cloud for storage and possible post-event analysis. By processing data near the end devices,

S. Nigam (✉) · M. S. Karyakarte
Vishwakarma Institute of Information Technology, Pune, India
e-mail: shalini.nigam@viit.ac.in; mandar.karyakarte@viit.ac.in

A. Neustein et al. (eds.), *AI, IoT, Big Data and Cloud Computing for Industry 4.0*, Signals and Communication Technology, https://doi.org/10.1007/978-3-031-29713-7_4

significant reduction in latency can be seen. Since in edge computing the massive amount of data is generated and processed at the edge of the network, there is no need to transfer the data to the cloud so security and privacy challenges are handled up to a certain level. Edge computing is capable of providing faster and secure response be it biomedical health-care systems, autonomous cars, or any highly interactive applications. We can make the edge intelligent by implementing machine learning and deep learning methods on edge. By applying deep learning algorithm on edge devices, the latency can be reduced.

Edge computing poses a lot of benefits over cloud computing. In edge computing, edge servers are installed near the network edge so that generated data should be readily available for computation and processing. Edge makes less or no interaction with clouds, thereby reducing the load on the cloud and increases the security and privacy. Edge intelligence ensures that all computation will happen on device only. Edge provides better data management capabilities. "Gartner" [1] says in 2025 more than half a percentage of enterprise-managed data will be generated and processed outside the cloud, that is, at the edge of the end device. There are a lot of deep learning methods that have been developed or being developed for edge implementation. The edge intelligence is making efficient usage of convolution neural network (CNN) and recurrent neural network (RNN) for providing fast computation with less computation power.

Edge intelligence can be implemented with machine learning as well as deep learning methods. Various learning models have been proposed for efficient computing and accuracy. The supervised, unsupervised, and reinforcement learning-based models can be deployed on edge with proper balance of real-time data availability and data computation.

In this chapter, we have proposed a comprehensive study about edge computing. Section 4.2.1 discusses the definition and evolution of edge computing. Section 4.2.2 tells about the difference between edge computing and cloud computing. Section 4.2.3 displays the layered architecture of edge computing. Section 4.2.5 lists out the characteristics of edge computing. Section 4.2.6 discusses the disadvantages of edge computing. Section 4.2.7 discusses the overview of edge AI. Sect. 4.2.8 is about deep learning with edge computing. Sections 4.2.9 and 4.2.10 are about edge intelligence-enabled applications of IoT and challenges in edge-enabled IoT systems, respectively. Section 4.2.11 is about research opportunities in edge AI and Sect. 4.2.12 is conclusion.

4.2 Edge Computing Overview

Edge computing is the approach in which all tasks which an IoT device performs on cloud can be handled on device or at the network edge. The device that works between end node and cloud and which is having computation capabilities more like a cloud is said to be an edge device. The edge device handles many of the requests from the end device without even taking help from the cloud data center.

The network edge is the network of IoT devices such as mobile devices or the IoT devices installed in a smart home which communicate with each other to perform certain tasks.

Edge computing is the next important step after cloud computing. The edge computing which makes the network of IoT devices scalable with better performance and low latency. Edge AI is also a very important era of edge computing which provides effective use of artificial intelligence methods especially for power or battery-limiting applications. Usage of deep learning on edge provides better computation and less delay for delay-sensitive applications.

4.2.1 The Origin of Edge Computing

The idea of edge computing came in 1990 from the development of content delivery network (CDN) [2]. The idea of CDN was back then to introduce nodes near the location of the end devices for fast delivery of cached contents such as images and video.

In 1997, Nobel et al. [3] proposed an idea of how resource-constrained devices can offload certain tasks to the nearby powerful servers. The goal was to mitigate the load of the device to get better performance and fast response. Their later work proposed an idea to improve battery life of the mobile devices. Today, Google, Apple, and Amazon work in a similar fashion for speech recognition applications in mobile phones.

In 2001, Satyanarayanan et al. [4] came up with a little more effective approach that is related to pervasive computing. They proposed a better idea for a peer-to-peer overlay network. These peer-to-peer networks were self-organized overlay networks. The benefit of this approach was found in fault-tolerance capabilities and load-balancing capabilities of the devices in the network. This kind of organization also ensures close proximity of the underlying Internet connection, so the distance between peer-to-peer devices is less. This not only decreases latency but also ensures proper balancing of load.

In 2006, cloud computing was first introduced to the public when Amazon came up with its first "elastic compute cloud." This extended many opportunities in the field of computation, storage, and visualization. But cloud computing was not a better solution for highly interactive applications such as autonomous cars and real-time or delay-sensitive IoT devices.

In 2009, cloudlet was introduced. The purpose of cloudlet was to reduce latency. The latency for the response from cloud to end device is very high as compared to cloudlets. A cloudlet is a small cloud data center. This data center is situated at the network edge.

A cloudlet is a small data center which is installed near the edge of the network to process the request of the nearby mobile devices.

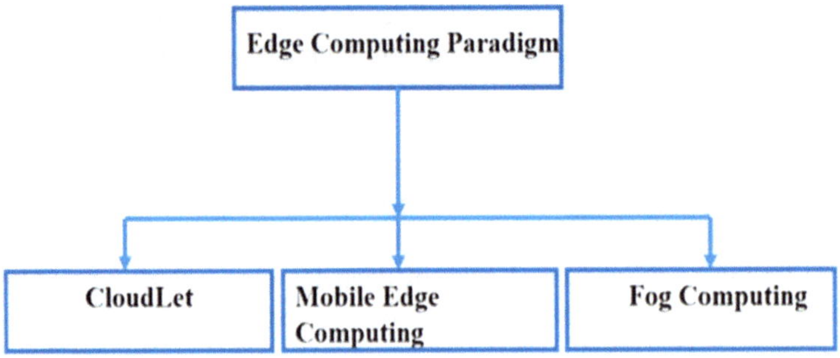

Fig. 4.1 EC paradigms

In 2012, Cisco proposed the idea of fog computing. The idea of fog computing was to handle a huge number of IoT devices with more interaction and less delay. To make the devices more responsive and the network more scalable.

4.2.1.1 Three Paradigms of Edge Computing: Cloudlets, Fog Computing, and Mobile Edge Computing

The evolution of edge computing is followed by cloudlet, mobile edge computing, and fog computing. Following are the paradigms of edge computing with their services and applications (Fig. 4.1).

4.2.1.1.1 Cloudlets

Cloudlets are a small data center installed near the edge of the network. The cloudlets work as the middle tier between cloud and end mobile devices. This is a three-tier architecture. The first layer is of mobile devices, second is of cloudlets, and the third layer is of cloud.

The main purpose of cloudlets is to provide services to resource-intensive and resource-interactive mobile devices with better performance and lower latency. The end device takes computing-related facilities from the nearby cloudlets.

Cloudlets can provide various services like resource management, big data analytics, service management, cloudlet placement, and computing in collaboration with other devices of some other domain.

Application of Cloudlets

At the early stages, cloudlets were applied to resource-interactive and time-sensitive applications such as object detection, virtual reality, and augmented reality. Later on, cloudlets are applied to other areas as well. Cloudlets got popularity in many wearable technologies for low energy consumption and better performance with low latency. For example, a cloudlet-based system Google Glass was proposed.

Cloudlets also find many applications in the field of internet of things. GigaSight is such an application in which heterogeneous videos are stored and analyzed at the cloudlet only, and the end results and other relevant information such as location, time stamp, and date are sent to the cloud. This way, cloudlet reduces data transfer to the cloud and improves bandwidth utilization.

4.2.1.1.2 Mobile Edge Computing

Mobile edge computing (MEC) is a technology which provides more interactive service to the mobile users by making computing done by nearby edge servers. Here delegating computation to the nearby server, MEC provides better resource management. The "European Telecommunication Standards Institute (ETSI)" proposed a standard for MEC in 2014 [5] and highlighted that MEC provides a value-based ecosystem that distributes computation-related tasks from the mobile devices to nearby edge servers.

Key technology of MEC is computational offloading and mobility management. The computational offloading refers to the process of offloading the data on the nearby server for computation. And mobility management means as the device moves the edge servers dynamically get updated for computation. In mobile edge computing, mobility management is a very crucial aspect because when the mobile moves from one location to another it has to communicate to the nearby edge server and then edge servers may communicate to each other in a multi-hop manner to achieve response of the given request. Computational offloading is another crucial challenge of MEC.

The idea behind offloading computation is executing applications which are demanding resources, and also they want a real-time response. Ali et al. [6] proposed a machine learning-based method to provide a better offloading process. They also compared supervised, unsupervised, and reinforcement learning to cope up with offloading-related challenges in an efficient way.

At the edge the resources are very limited; therefore, managing resources is an essential parameter in edge computing paradigm. Therefore, MEC(s) need to be programmed to manage the resources. The edge compared to cloud are resource constrained, heterogeneous, and dynamic in nature. They are resource constrained because they have small processor and less battery life. They are heterogeneous because different edge nodes may have processors of different architecture. They are dynamic because different edge nodes may have different workloads and different applications may compete for the limited resources. Cheol-ho Hong et al. [43] proposed a survey on all possible techniques required for managing the resources at the edge and classified them based on architectures, algorithms, and infrastructures. They have also listed different data aggregation techniques, sharing techniques, offloading techniques, and tenancy to manage the dynamically generated data on the edge.

Yuyi Mao et al. [44] introduced different resource management models for single user, multiuser, and heterogeneous MEC. They have proposed deterministic task model for binary offloading, partial offloading, and stochastic task model in single-user MEC. For multiuser MEC, they have proposed joint radio- and computational resource allocation, MEC server scheduling, multiuser cooperative edge computing, and server selection. They have also discussed techniques to manage resources for MEC with heterogeneous servers such as server cooperation and computation migration.

With enormous data generation by various applications from various mobile devices requiring MEC to be smart enough to provide services at the edge with less delay and more accurate response. MEC can provide more services with less delay by implementing technologies such as software-defined network (SDN), network function virtualization (NFV), and service function chaining (SFC). All these techniques if implemented with MEC makes service management more flexible in terms of computing, storage, access, and fast deployment of new services [45]. Along with this, security and privacy are also vital concern in MEC which can be achieved by employing many communication protocols on to the edge servers and MEC(s).

Applications of Mobile Edge Computing

MEC has applications in many areas such as automation, business, and health care. MEC are widely known for different applications in video streaming with respect to smart city. In video streaming, the monitoring device can collect video locally and MEC servers can be used to analyze and extract meaningful information from that video. This extracted data can be transferred to application servers for further processing thereby reducing traffic. MEC also plays a vital role in AR-based mobile applications. In such applications, low latency and fast data processing are required. Local MEC servers can process the data and can provide better real-time experience.

4.2.1.1.3 Fog Computing

The fog computing standard is mainly charged by the OpenFog consortium. The main objective of this consortium is to motivate standard bodies to create standards so that IoT systems at the edge can communicate securely among each other and with clouds. "OpenFog consortium" announces the release of OpenFog "reference architecture" in 2017. This is a universal technical framework designed to handle challenges related to data management and data processing.

Fog computing is very near to the concept of edge computing. Edge computing and fog computing are used interchangeably. Fog computing states that the end device should ask the nearby edge server to process the request instead of the cloud.

Idea of fog computing was first invented by Cisco. In fog computing, edge servers are responsible for collecting data from many heterogeneous and homogeneous devices, and then compute and process that data at the edge of the network of edge devices. Many servers are installed at the network edge so that real-time response can be provided to the end users. Fog computing provides many services such as resource management, real-time response, big data analytics, security, and privacy.

Application of Fog Computing

Applications of [7] fog computing can be found in many areas such as augmented reality and real-time applications. Implementation of fog computing can also be seen in content delivery and caching for improving web performance. Fog computing can provide elastic resources for better data processing in terms of mobile big data analytics without the high latency drawback of cloud computing.

4.2.2 Criteria-Wise Difference Between Edge Computing and Cloud Computing

Edge computing is a next advanced technology after cloud computing. Edge computing has solved many of the challenges related to cloud with IoT. Cloud computing was doing outstanding job when there was limited usage of IoT devices, but nowadays the rapid use of IoT devices initiating new challenges coping up with that has become essential to get the smooth access to these IoT devices. The difference between advanced cloud technology, that is, edge computing and traditional cloud computing technology on the basis of some important criteria which actually affects the usage of IoT devices is presented. This difference will help understand the flexibility of adapting the emergence of edge computing for IoT devices (Table 4.1).

4.2.3 Layered Architecture of Edge Computing

Architecture of edge computing is a three-layer architecture. The first layer is the device layer, the middle layer is said to be the edge layer, and the third layer is the cloud layer (Fig. 4.2).

4.2.3.1 Device Layer

All the devices that belong to the Internet of things resides at device layer. The device layer includes devices such as mobile phones, smart home appliances, smart transportation equipment, smart traffic control devices, smart meters for smart power grids, smart building sensors, actuators, and laptops. All such devices generate data as they are in close vicinity of the phenomena. This data is offloaded onto the edge for processing and computation. The devices generate the heterogeneous data. This may include images, videos, and numerical data. Whatever data is there, that data is uploaded on the cloud. Because of data heterogeneity, offloading is a key technology of edge computing. Many research works have been done on data offloading on edge. The data offloading promotes another key challenge that is

Table 4.1 Difference between CC and EC

Serial number	Criteria	Cloud computing	Edge computing
1	Latency	High latency	Low latency
2	Data processing	Slow, the cloud has to wait for data and then it can process. If there is delay in data transmission, the processing of data will be delayed.	Fast, as edge servers are near to the devices which are generating data. Sometimes, data generation is on the edge device itself.
3	Data generation	Devices generate data and transmit it to the cloud.	In many cases, data is generated at the edge only.
4	Computation power	Algorithms take more processing power as the cloud needs data from the devices.	Edge intelligence reduces the computation power.
5	Security and privacy	Less security since data may come from the farthest node and may get attacked by the attacker in transit.	Better security than cloud, as distance between device and processing servers is always very less.
6	Energy efficient	Less energy efficient as it performs computation on clouds away from the device.	More energy efficient as it performs computation near the device.
7	Architecture	With cloud, it is a two-layer architecture.	With edge, it is a three-layer architecture (device–edge–-cloud).
8	Targeted users	Targeted users of cloud are large.	Targeted users of cloud are edge users.
9	Scope	Servers are within the Internet, the scope is wide (global).	Servers are located near the network edge, the scope is local.
10	Multi-hop	This could be multi-hop because of the distance between the end device and cloud.	Since edge servers are always near the device, so device-to-server transmission is always single hop but there may be multi-hop communication among edge servers.
11	Mobility support	Limited support	Support
12	Reliable	High	Low
13	Service access	Centralized	At the edge
14	Proximity to the end user	Low	High

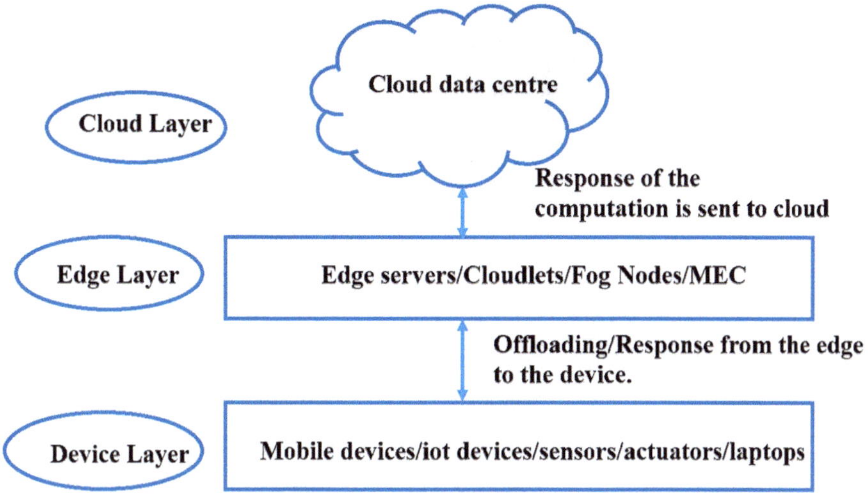

Fig. 4.2 Architecture of edge computing 1

known as close proximity of the edge servers with mobile devices, in case the device which is generating data is a moving device. This challenge is also be mitigated by the edge computing.

4.2.3.2 The Edge Layer

At the edge layer, different servers or cloudlets or MEC are installed. The edge server is responsible for doing all computations data processing and management. The edge layer may sometimes transfer data to the cloud or it performs computation on its own without interacting to the cloud when real-time data is required. It totally depends on the algorithm you apply on the edge. The mobility management, data processing, data management, offloading mechanism, resource management, security, privacy, and all such essential services are implemented at the edge. The edge may take data on permanent or temporary basis. To serve an on-demand service, the server may take the required data temporarily [8]. After serving the request, the edge may discard the data. As shown in the figure, the overall architecture of edge with IoT involves three layers. At the first layer, all data are generated by sensors. These sensors send the data to the second layer where all IoT gateways are installed in terms of edge computing. These IoT gateways could be PC, laptop, cloudlets, or any MEC servers. These gateways are responsible for all communication between end user and clouds. Sometimes when real-time update is required, then IoT gateways do all the data preprocessing and send the required real-time result to the end device. At the third layer, cloud server resides to response all the queries generated by the IoT gateways. Wei Yu et al. [39] have discussed two models for the implementation of edge computing: hierarchical model and software-

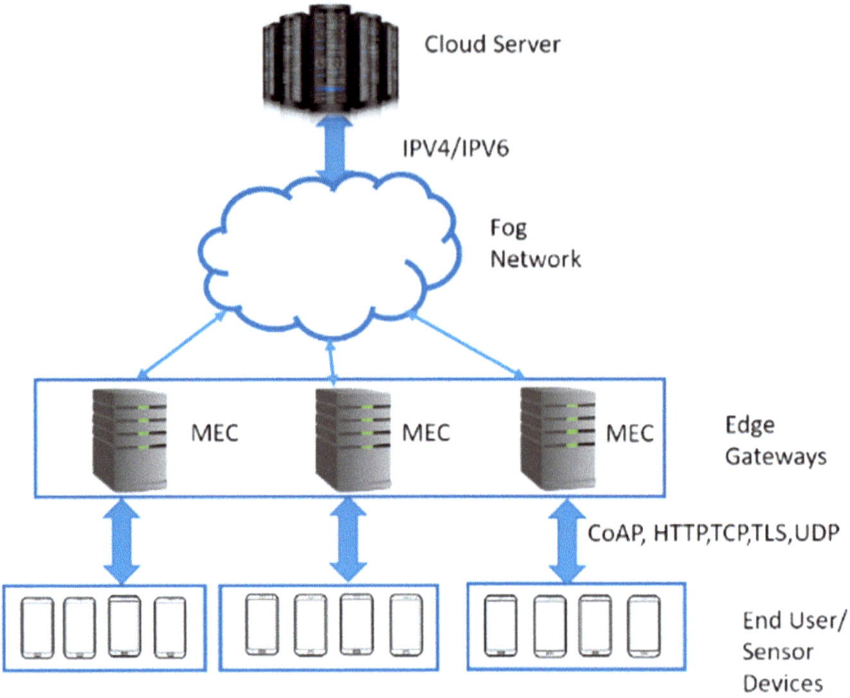

Fig. 4.3 The edge computing model 1

defined model. Jararweh et al. [40] proposed a hierarchical model of edge computing which integrates multiple MEC with cloudlet infrastructure. Here, MEC servers are capable enough to send the response to their corresponding group of sensor device. Tong et al. [41] proposed a hierarchical edge cloud model which can be applied in a situation of peak loads. Software defined model is introduced with which further complexities of edge computing is minimized. Jararweh et al. [40] proposed a software-defined model to integrate the MEC servers and software-defined system capabilities. This integration further reduces the management and administration cost. Manzalini and Crespi [42] proposed an edge operating system which allows available open-source software to achieve powerful network and service platform. Apart from this, security is an important concern when transection of data takes place in the IoT. Therefore, many communication protocols such as CoAP, MQTT, and AMQP are used to implement edge computing (Fig. 4.3).

4.2.3.3 The Cloud Layer

The cloud layer is the last layer of this architecture. The edge layer helps cloud layer to handle many time-sensitive requests to deliver response to the end user on

Table 4.2 SW/HW requirements of edge computing

Serial number	Software/hardware requirement of edge computing	
1	Operating system	Linux, Azure RTOS, Wind River VxWorks/Linux edge computing OS, SWIM distributed edge computing OS, Windows IoT core, Windows Enterprise server for IoT
2	Technology	Machine learning/deep learning
3	Programming language	Python
4	Edge gateways	MEC, LAN, WAN, PC, laptop, cloudlets, mobile phone
5	Communication protocols	TCP, UDP, TSL, CoAP

time. Cloud has delegated most of its work to the edge. But edge does all these tasks at a local level and cloud does the tasks at global level. Traditional clouds were becoming incapable of handling such a huge data in terms of both processing and storage management. Edge servers helped the cloud by managing all these near the device.

4.2.4 Software and Hardware Requirements to Implement Edge Computing (Table 4.2)

4.2.5 Characteristics of Edge Computing

4.2.5.1 Edge Computing Has Many Characteristics [9] as Cloud Computing

This is a booming advanced technology which possesses excellent features than cloud computing for better and efficient performance of IoT devices. Following are some important and unique characteristics of edge computing.

4.2.5.2 Close Proximity to the End Device

In edge computing, the cloud services are always close to the device, which reduces due to which the communication becomes secure. Fast and accurate big data analytics can be performed. Because of the close proximity, the response can be available in real time.

4.2.5.3 Support for Mobility Management

Edge supports better mobility by implementing location ID separation protocol (LISP). This protocol separates host ID and location ID and provides better mobility support. Compared to other mobility protocols such as MIPv4 and MIPv6, LISP provides a better and optimal shortest path to the moving end points and it supports both ipv4 and ipv6. Using LISP ensures mobility in edge computing.

4.2.5.4 Location Awareness

Edge computing devices have optimized to ensure best location awareness implementation. When the device is a moving device and it wants to communicate to its closest edge server then it can do so with less time. Algorithms can be implemented at the edge to search the nearest edge server. Many technologies such as GPS, wireless access points, and cell phone infrastructures can be used to get location of the edge servers. This location awareness characteristics can be used by many edge applications such as autonomous vehicles or vehicle edge computing.

4.2.5.5 Low Latency

Edge computing provides response at a very low latency. Since the edge servers here are near to the device, it takes less time in data transfer and data computation. Sometimes, data is generated on the edge devices itself such as smart watches. So edge computing is highly recommended for delay-sensitive applications. With a combination of low latency, mobility and location awareness provides better possibilities for highly interactive applications.

4.2.5.6 Low Computation Power

By implementing deep learning and machine learning methods on edge devices, we can improve computation of the edge devices. Since IoT devices are mostly battery operated, energy consumption is a very important factor to consider while designing the algorithm. The area of artificial intelligence helps in developing algorithms which consume less power and produce accurate results.

4.2.6 Disadvantages of Edge Computing

We have seen many advantages of edge computing over cloud computing such as low latency in delivering response, efficient data analysis, enables proper security and privacy mechanisms, consumes less power in data transmission, accuracy and fast computation by implementing ML, DL on edge servers, best resource management, and data management techniques.

Despite many advantages, edge computing has some disadvantages as well:

1. Infrastructure cost: establishing edge servers incurs considerable cost.
2. Proper training is needed to use edge servers at enterprise level not much fit for rural areas.
3. Security is still a challenge because each edge device should implement the security policy synchronously. The enterprise which applies a zero trust policies on network security has to keep track of all the server's security policies.
4. Edge computing works on a local level. It still needs a cloud for distributing data globally through the Internet.

4.2.7 Overview of Edge AI

Artificial intelligence is a highly prominent field nowadays. Edge computing can be made more effective by applying deep learning algorithms on edge computing. The edge intelligence means edge with machine learning or deep learning algorithm. Deep learning involves a deep neural network (DNN) that is a hierarchical structure and makes efficient usage of raw data. There are two different types of DNN: the convolution neural network (CNN) and Recurrent Neural network (RNN).

CNNs are widely used computer vision tasks. They are very efficient to process image-based sensory data without explicit programming. With these effective benefits of CNNs, they come at the cost of more memory and computation therefore most of the CNNs come with graphics processing units (GPUs). RNNs can efficiently process sequential information. RNNs make use of Artificial Neural Networks (ANN) for memory cells. Edge computing also makes use of many other deep learning technologies for better computation power and efficiency. For example, Grover et al. [10] has proposed an edge computing-based vehicular network. They also demonstrated how unsupervised deep learning techniques can be used to find any suspicious behavior of the vehicle.

4.2.8 Why Deep Learning with Edge Computing

Deep learning is being applied in many areas such as natural language processing (NLP), computer vision, and artificial intelligence (AI). Deep learning provides efficient computation, powerful information extraction and processing capabilities but requires massive computation resources. The emergence of deep learning has greatly extended the edge computing application. The combination of deep learning and edge computing provided improved performance, greater efficiency, and fast computation response. Wang et al. [11] has discussed many popular deep learning methods for edge computing such as restricted Boltzmann machine (RBM), autoencoder (AE), deep neural network (DNN), convolution neural network (CNN), recurrent neural network (RNN), and deep reinforcement learning (DRL). This data processing and extracting important info out of it is a crucial challenge deep learning algorithms can help to mitigate this issue.

4.2.9 Edge Intelligence-Enabled Applications of IoT

With various prominent features of edge computing, we can find enormous benefits of this advanced cloud technology in the field of Internet of Things. We can find applications of edge computing in extensive domains such as health care, smart city, smart building, real-time applications such as AR and VR gaming, and self-driving cars. We are mentioning some of the use cases of edge computing for IoT devices (Fig. 4.4).

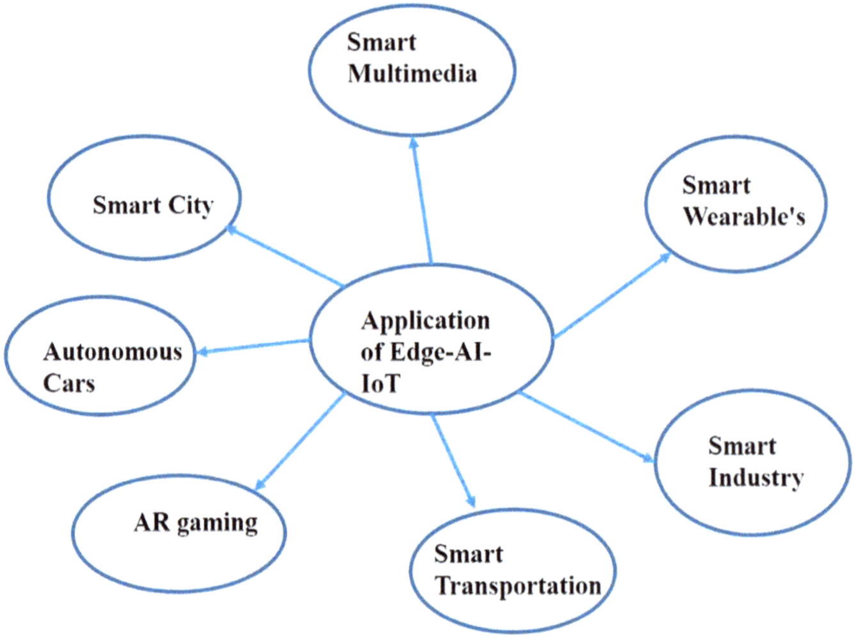

Fig. 4.4 Application of Edge AI-IOT 1

4.2.9.1 Smart Wearables

Zia et al. [12] proposed a wearable sensor-based system for activity prediction using Recurrent Neural Network on an edge device. Here, input is provided through the wearable biomedical sensors such as ECG, magnetometer, gyroscope, and accelerometer. The RNN is trained based on features and later on it is used to predict the activity. The graphics processing unit installed on the edge device provides fast and efficient computation. In this proposed work, the user wears the sensors in different body parts such as chest, right wrist, and left ankle. The edge device (laptop with GPU installed in it) obtains the multimodal data from the sensors through a wireless medium and then performs activity prediction using RNN. GPU installed on the edge device provides fast computation.

4.2.9.2 Smart City

Johan et al. [13] proposed an idea for monitoring real-time transportation using an edge computing device which uses computer vision and deep neural network methods to monitor the real-time multimodal data of transportation of the Australian city of Liverpool (NSW) without compromising privacy of the common people. The edge application of smart city includes smart home, smart building, and smart transportation. The edge computing provides better home automation with maintaining proper privacy.

4.2.9.3 Smart Home

Dhakal and Ramakrishnan [14] had proposed an automatic home or business monitoring system which is based on NFV (Network Function Virtualization) edge server. This monitoring system provides real-time learning of the data streaming from the neighboring business or home. Leveraging the recent advancement in deep learning and machine learning, wireless signals can be efficiently utilized for better human–device interaction. SignFi [15] made efficient usage of the CSI signals of Wi-Fi and therefore it was able to understand 276 sign languages and body gestures including movements of hand, arm, head, and fingers. SignFi made extensive use of CNN (Convolution Neural Network) to achieve this understanding of sign language. This SignFi system can be used by deaf people at home for using electronic appliances installed in a smart home. Wang et al. [16] identified the impact pattern of moving humans on the Wi-Fi signal and leveraged the combination of CNN and LSTM to recognize gestures and activity in the home. Mohammadi et al. [17] found out more possibilities of deep learning modals and methodologies and proposed a supervised deep reinforcement learning model for indoor localization.

4.2.9.4 Smart Building

We can see the effective usage of edge intelligence in smart building as well. There are so many issues with the building such as improved security, less power consumption, and better sensing capability. In this context, edge computing provides a better solution by providing efficient computation and more accuracy. Zheng et al. [18] proposed a "chiller-sequencing" problem which helps in significant reduction of electricity consumption of the building. Yuce and Rezgui et al. [19] proposed a neural network-based model which performs regression analysis on energy consumption relevant data of the building.

4.2.9.5 Smart Grid

Smart grid is a network of electricity based on digital technology. This system supplies electricity to the consumer through two-way digital communication. This system ensures proper monitoring, analysis, and control within the supply chain and provides better energy distribution, reduces energy consumption, and cost. IoT has provided many edge-based solutions for smart grid. Yasir Mahmood et al. [20] has proposed an edge computing-enabled IoT smart Grid. This framework had three layers: the layer at which devices work, the layer of edge servers, and the final layer was of cloud. At the device layer, the IoT devices like sensors, actuators, and controllers are installed and they monitor SG devices such as smart meter and smart appliances. The data generated at smart meters are obtained by the IoT devices. These IoT devices offload the data at the edge node for further computation and processing. Various edge nodes are installed near the SG systems so that real-time data should be available if any request comes and the result can be sent back to the cloud for further processing. Deep learning implementation over edge computing provides efficient control over smart grid. He et al. [21] used deep belief network (DBN) and RBM to find the information of an attacker who tries to inject wrong data in real time in SG systems.

4.2.9.6 Smart Vehicle

Combining edge computing with vehicular network becomes vehicular edge network. Increasing number of IoT devices increases the load on the cloud. Because of this there is delay in the response from the cloud to handle this challenge. It is very important to find something that can provide real-time response. Because the vehicular network is a delay-sensitive network where information should be available readily. Lei Liu et al. [22] provided a comprehensive survey on vehicular edge computing.

4.2.9.7 Smart Multimedia

Nowadays, videos are a very vital source of getting information. In current scenarios, video has become a very widely used Internet application. These days we can find videos for everything, be it for surveillance purposes or for augmented reality. For monitoring traffic as well, videos are taken and analyzed. With this huge development of videos, video analysis has become an important factor to consider. The enormous generation of videos by IoT devices promotes challenges like storage, delayed response, and processing time on cloud data centers so these problems are mitigated by edge intelligence. By combining artificial intelligence with edge computing, we can have faster response and also good quality video analysis can be performed. We can find smart multimedia majorly in three areas: video analytics, adaptive streaming, and caching.

4.2.9.8 Video Analytics

Video analytics is a major part of edge intelligence. Introducing edge intelligence with video analysis reduces latency in processing the video and also reduces the storage cost up to a certain level. The emergence of edge computing near the video-generating device consumes less time in processing the video and provides real-time response. Various work has been proposed in the area of video analytics with edge intelligence. Liu et al. [23] proposed a CNN-based food recognition system. This system ensures low latency and provides real time and accurate information about the food.

4.2.9.9 Adaptive Video Streaming

Adaptive video streaming is becoming a crucial issue nowadays. Streaming of videos affects the quality parameter of the video. Most of the time the quality of video depends on the client devices. Majorly, the bandwidth of the client-side device affects the quality of the video. Real-time and outstanding video quality is a must nowadays. To achieve this edge intelligence has provided many solutions. By offloading video frames on the nearest edge servers, the real-time streaming of video frames will be there so that one can get better quality and real-time response at average bandwidth cost. Wang et al. [24] has proposed an edge intelligent framework which makes use of Deep Reinforcement Learning to intelligently make decisions about which edge server should be assigned to which user for real-time video streaming.

4.2.9.10 Smart Transportation

Emerging edge intelligence on vehicles making it more secure in terms of autonomous cars, traffic monitoring, and traffic signal control. Nowadays, it is not only the Internet of things but also the Internet of vehicles. Leveraging edge intelligence in smart transportation making transportation more secure and easy and hassle free. Smart transportation explored many possibilities of emerging edge intelligence in the area of autonomous driving, traffic analysis and prediction, and traffic signal control.

4.2.9.11 Autonomous Driving

Autonomous driving is a highly real-time-based application. In such applications, delay cannot be tolerated. The delay is highly sensitive in autonomous cars. To mitigate the delay, processing and computation on the edge should be very fast. Implementing deep learning with edge provides fast response. Extensive work has been carried out in this field. Much work has been proposed for proper object detection with accuracy. Chen et al. [26] has proposed a monocular 3D object detection method using CNN approach based on the fact that the object should be on the plane surface or on the plane ground. Bojarski et al. [27] proposed a dynamic learning modal independent from the road features. Directly provide the sensed data to the steering command.

4.2.9.12 Traffic Analysis

Urban traffic management is a very important concern of a well-planned city. Traditional approach of taking the image through a camera is not a state-of-the art solution for large traffic conditions. Traditional approaches make use of "time series analysis" [25] or "probabilistic graph analysis" [28] which may not capture hidden spatiotemporal relationship therein. Deep learning methods provide effective analysis of real-time videos of the traffic and can take decisions accordingly and let the user know about the traffic condition prior. Shaohua Wan et al. [29] proposed an approach which can eliminate various spurious videos from the edge. The basic idea is to find and remove the traffic videos which are having spurious and redundant frames. Based on magnitude of motion detection through spatiotemporal interest point (STIP) and the multimodal linear feature combination and with extensive numerical analysis and verification they eliminated redundant frames, found key frames, segmented the frame of a long video, and identified better object detection.

4.2.9.13 Traffic Signal Control

Traffic signal controlling is also a very important concern for avoiding wasting of time due to traffic congestion and avoiding accidents. Deep learning with edge computing did an outstanding job for efficient analysis and prediction of regular traffic. The edge intelligence can also be applied for finding many efficient techniques for controlling the vehicle at traffic signals. Earlier traffic signal control methods relied on "fuzzy logic" and "genetics." To achieve city-wide traffic signal control the "RL" and "multiagent RL" can be an effective solution. In this case, multiple agents will control the traffic signals in a collaboration.

4.2.10 Challenges in Edge-Enabled IoT Systems

Although deep learning with edge computing provides enormous benefits, there are still some challenges we face in achieving durable real-time solutions. These challenges are mentioned below:

4.2.10.1 Modal Training

The overall performance of deep learning-enabled edge devices highly depends on the fact that how learning models will perform. It is popularly known fact that training models require extensive computation in almost all scenarios and consume massive resources such as CPU and GPU. It is very challenging for the edge servers to collect all the data at one server for modal training. As we know that edge servers often installed for moving devices in such cases data is transferred to the nearest edge server that way it is not always possible to collect data at one server for modal training. Besides this, if edge servers share data among themselves, it may cause quite a good communication cost. So "distributed learning" [30] and "federated learning" [31] are prominent solutions to address this issue. These two emerging technologies are reducing communication cost and improving resource utilization.

4.2.10.2 Modal Deployment

With various advancement in deep learning methods, the learning models are increasing in size and volume, when these big learning models are deployed on the centralized server which is nothing but cloud and these modals receive the input from the distributed end devices which may cause huge delay in receiving the data if the bandwidth for data transmission is not sufficient. The edge server has provided an alternative solution to this by partitioning the deep neural network modal between edge and cloud. In this case, the edge and cloud work simultaneously to execute learning modal. For example, Kang et al. [32] proposed an open scheduler which automatically partitions or shares the DNN between mobile device and data center. In this way, both work together to produce the end result. Huang et al. [33] then found the partitioning problem in the scenario of edge computing where he proposed a partitioning methodology among cloud, edge, and devices. The major aim of this approach was to reduce delay in data transmission from cloud to device and vice versa.

4.2.10.3 Delay-Sensitive Applications

Although edge computing is doing outstanding job in all IoT-related applications, there are still some applications where real-time response is highly essential such as VR gaming and autonomous cars. The edge servers with 5G bandwidth can provide outstanding applications. The 5G bandwidth will mitigate the issue related to data transfer rate. Such a fast Internet can ensure fast response on a real-time basis. 5G technology with edge computing has enhanced the scope of application where the major computation related to analysis and rendering of videos can be performed locally at the edge and the end result can be delivered at real time. But its practical implementation is still a little more complicated; a proper feasibility study is required.

4.2.10.4 Hardware and Software Support

The inclusion of deep learning in edge computing has achieved a remarkable growth in the field of Internet of things. Implementing deep learning on edge requires proper optimization of system-level requirements, since most of the hardware and software platforms and tools and technologies are available for cloud computing. There is a need for optimized hardware and software solutions for edge computing as edge computing requires lightweight architecture, energy efficiency, and edge-based computation frameworks. Much research has been proposed for hardware architecture of edge computing. For example, Du et al. [34] has performed a comprehensive study of Cortex-M microcontrollers and proposed a streaming "hardware accelerator" for better execution of CNN on the edge. Along with this, FPGA-based edge computing platform [35] was also developed to support deep learning-based offloading computation from mobile device to edge. For the software perspective, many tools and technologies have been developed to support edge-based deep learning and computing, for example, "Amazon's GreenGrass" and "Microsoft's Azure IoT edge." From a programming perspective, many frameworks are being developed especially for edge-based scenarios such as "MXNet" [36], "Tensorflow Lite," [ ] and "coreML." Though with these existing systems there is a need to integrate them to get more practical flexibility and better performance.

4.2.10.5 Integration and Heterogeneity

Because of varying types of IoT services, different kinds of IoT devices generate different data in different formats and the data may come from different platforms and from different applications. The edge has to handle such a huge heterogeneity. In the case of cloud technology, if we want to access any services and resources, the cloud assigns those services or resources to the end user. The user might not know from where he is getting services. This is a centralized data cenetr which distributes services. In addition to this there is one common programming language to develop applications for cloud. The application can be developed and deployed onto the cloud and the services can be allocated to different users. On the contrary to this edge nodes are heterogeneous platforms. Despite the benefits of distributed topology, there are some challenging issues in writing applications for edge nodes. Besides this there is no common programming language to develop applications for edge nodes. Developers may face serious issues in writing and deploying applications for edge nodes because of heterogeneous platforms of different edge nodes.

4.2.10.6 Naming

As we know that many of the storage servers opt for different kinds of server operating systems. Then here is a big challenge for file naming, resource management, and resource allocation. Besides this, the massive number of IoT devices generates huge amount of data generating. Uploading of data happens simultaneously, which leads to another challenge which is related to naming of data resources. Many traditional naming strategies are in use such as DNS (Domain Name Server) and URI (Uniform Resource Identifier) but these work well with cloud and other networks but not much fruitful in case of edge computing with IoT. Several new naming schemes are also coming for edge computing such as NDN (Name Data Networking) [37] and MobilityFirst [38]. These naming schemes are especially designed for edge computing. They have their own pros and cons. NDN requires proxy server for integration of different communication protocol and MobilityFirst provides better mobility but requires globally unique identifiers (GUID).

4.2.11 Research Opportunities in Edge AI and or Edge Computing

Integration of edge with cloud and IoT devices has many benefits. As we can see nowadays, many solutions to real-world problems are being automated. We can find the use of sensors everywhere be it health-care system, transportation systems, smart building, or smart city. As the number of IoT devices increases, the data generated by sensors also increases. This huge generation of data demands high quality of storage and accessing facilities.

A good storage scheme can mitigate many issues such as reducing data access time, providing better response time, reducing latency in data delivery, and good data analysis. One can find many opportunities in data management and data storage mechanisms at the edge. Edge cannot store the data for long. Lots of edge data is ephemeral the data will come it will be stored, analyzed, processed, computed, and destroyed.

In recent scenarios, everything is becoming digital. And everything wants to work in collaboration digitally. With this global digitization, the huge enterprise data will be generated and then there will be a demand for better and intelligent data management capabilities.

This huge data generation leads to many new opportunities in the area of maintaining heterogeneous data, offloading of heterogeneous data onto the edge and Delay incurs in processing the data at the edge. Along with this edge devices are not much secure. So security of edge data is paramount while using edge services at the enterprise level. One can find many research opportunities in securing the edge device.

Apart from this, one can find many research opportunities for providing quality of service. There is no doubt that edge computing is providing outstanding benefits for today's IT world, but with the rapid growth of sensory data there may be significant challenges with quality-of-service delivery.

Deep learning and machine learning algorithms are providing efficient and accurate results but at the cost of powerful computation resources. One can also find opportunities in this area that the algorithm should not be computation effective but with proper resource management.

Edge computing is known for its low latency service delivery. This feature of low latency comes because the cloud is now near to the devices. But as we know that edge nodes are also in connection with moving devices. In this case, if the moving device wants to request some service it will request this from the nearest edge server. In such a scenario there may be a case that the given server cannot process the request and it has to forward it to the next edge server in a multi-hop fashion which can cause little delay. Therefore, a proper and systematic approach should be developed to handle such scenarios.

4.2.12 Conclusion

Edge computing is a prominent technology of recent trends. The edge computing provided many essential characteristics to make IT operations easy and handy. Edge-enabled IoT is providing many solutions for difficult/complex IoT problems. The edge is also efficient in handling billions of data generated every second by the millions of IoT devices. Compared to cloud computing, edge computing pushing cloud-related services onto the edge of the network thereby reducing latency and providing better bandwidth utilization. Edge computing provides an energy-efficient solution for IoT devices. Because most of the IoT devices are battery operated so power saving is an important concern in case of IoT-enabled network. By implementing a deep learning algorithm on edge, we can get low computation power on edge nodes. In this chapter, we have provided comprehensive study about edge computing. We have first discussed what are the challenges associated with IoT and cloud, then we have shown the evolution of edge computing such as cloudlet, mobile edge computing, and fog computing. We have also shown major differences between cloud computing and edge computing. Then we have discussed the architecture of edge computing with the various characteristics of edge computing. Then we discussed disadvantages of edge computing with respect to cost and other terms. After this we have explored many application areas of edge computing with IoT. Challenges and opportunities associated with edge computing are also discussed. Various benefits of edge like proximity to the end user, ad hoc infrastructure, auto scalability, multi-tenancy, improved cost-effectiveness, and low latency lead to many research opportunities in the domain of health care, real-time applications, smart city, home automation, car automation, and many more. We hope that this study will initiate more discussion and inspiration on the inclusion of deep learning, cloud computing, and edge computing, and more edge intelligence-enabled applications will be developed in future.

References

1. Thomas Bittman, Bob Gill, Tim Zimmerman, Ted Friedman, Neil MacDonald, Karen Brown, "Predicts 2022: The Distributed Enterprise Drives Computing to the Edge", ID G00757917.
2. J. Dilley et al., "Globally Distributed Content Delivery," IEEE Internet Computing, vol.6, no. 5, 2002, pp. 50–58.
3. B.D. Noble et al., "Agile Application-Aware Adaptation for Mobility," Proc.16-th ACM Symp. Operating Systems Principles (SOSP 97), 1997, pp. 276–287.
4. M. Satyanarayanan, "Pervasive Computing: Vision and Challenges," IEEE Personal Comm., vol 8, no.4 -, 2001, pp.10–17.
5. ETSI, Mobile-edge Computing Introductory Technical White Paper, White Paper, Mobile-edge Computing Industry Initiative, 2014, https://portal.etsi.org/portals/0/tbpages/mec/docs/mobile-edge_computing_-_introductory_technical_white_paper_v1.
6. Shakarami, A., Ghobaei-Arani, M. and Shahidinejad, A., 2020. A survey on the computation offloading approaches in mobile edge computing: A machine learning-based perspective. Computer Networks, p.107496.
7. Yi, S., Li, C. and Li, Q., 2015, June. A survey of fog computing: concepts, applications and issues. In Proceedings of the 2015 workshop on mobile big data (pp. 37–42).
8. Yu, W., Liang, F., He, X., Hatcher, W.G., Lu, C., Lin, J. and Yang, X., 2017. A survey on the edge computing for the Internet of Things. IEEE access, 6, pp.6900–6919.
9. Khan, W.Z., Ahmed, E., Hakak, S., Yaqoob, I. and Ahmed, A., 2019. Edge computing: A survey. Future Generation Computer Systems, 97, pp.219–235.
10. Grover, H., Alladi, T., Chamola, V., Singh, D. and Choo, K.K.R., 2021. Edge Computing and Deep Learning Enabled Secure Multitier Network for Internet of Vehicles. IEEE Internet of Things Journal, 8(19), pp.14787–14796.

11. Wang, F., Zhang, M., Wang, X., Ma, X. and Liu, J., 2020. Deep learning for edge computing applications: A state-of-the-art survey. IEEE Access, 8, pp.58322–58336.
12. Uddin, M.Z., 2019. A wearable sensor-based activity prediction system to facilitate edge computing in smart healthcare system. Journal of Parallel and Distributed Computing, 123, pp.46–53.
13. Barthélemy, J., Verstaevel, N., Forehead, H., & Perez, P. (2019). Edge-computing video analytics for real-time traffic monitoring in a smart city. Sensors, 19(9), 2048.
14. A. Dhakal and K. K. Ramakrishnan, "Machine learning at the network edge for automated home intrusion monitoring," in Proc. IEEE ICNP, Oct. 2017, pp. 1-6.
15. Y. Ma, G. Zhou, S.Wang, H. Zhao, and W. Jung, "SignFi: Sign language recognition using WiFi," Proc. ACM Interact., Mobile, Wearable Ubiquitous Technology., vol. 2, no. 1, pp. 1–21, Mar. 2018.
16. F.Wang,W. Gong, and J. Liu, "On spatial diversity inWiFi-based human activity recognition: A deep learning-based approach," IEEE Internet Things J., vol. 6, no. 2, pp. 2035–2047, Apr. 2019.
17. M. Mohammadi, A. Al-Fuqaha, M. Guizani, and J.-S. Oh, "Semisupervised deep reinforcement learning in support of IoT and smart city services," IEEE Internet Things J., vol. 5, no. 2, pp. 624635, Apr. 2018.
18. Z. Zheng, Q. Chen, C. Fan, N. Guan, A. Vishwanath, D. Wang, and F. Liu, "Data driven chiller sequencing for reducing HVAC electricity consumption in commercial buildings," in Proc. ACM e-Energy, 2018,pp. 236-248.
19. Z. Zheng, Q. Chen, C. Fan, N. Guan, A. Vishwanath, D. Wang, and F. Liu, "An edge based data-driven chiller sequencing framework for HVAC electricity consumption reduction in commercial buildings," IEEE Trans. Sustain. Comput., early access, Jul. 30, 2019, doi: https://doi.org/10.1109/TSUSC.2019.2932045.
20. M. Yasir Mehmood,1 Ammar Oad,2 Muhammad Abrar,1 Hafiz Mudassir Munir,3 Syed Faraz Hasan,4 H. Abdul Muqeet,5 and Noorbakhsh Amiri Golilarz, "Edge Computing for IoT-Enabled Smart Grid".
21. Y. He, G. J. Mendis, and J. Wei, "Real-time detection of false data injection attacks in smart grid: A deep learning-based intelligent mechanism," IEEE Trans. Smart Grid, vol. 8, no. 5, pp. 25052516, Sep. 2017.
22. Lei Liu1 · Chen Chen1 · Qingqi Pei1 · Sabita Maharjan2 · Yan Zhang, "Vehicular Edge Computing and Networking: A Survey", 26, pages 1145–1168 (2021).
23. C. Liu,Y. Cao,Y. Luo, G. Chen,V. Vokkarane, M.Yunsheng, S. Chen, and P. Hou, "A new deep learning-based food recognition system for dietary assessment on an edge computing service infrastructure," IEEE Trans. Services Comput., vol. 11, no. 2, pp. 249–261, Mar. 2018.
24. F. Wang, C. Zhang, F. Wang, J. Liu, Y. Zhu, H. Pang, and L. Sun, "Intelligent edge-assisted crowdcast with deep reinforcement learning for personalized QoE," in Proc. IEEE INFOCOM, Apr. 2019, pp. 910–918.
25. B. Ghosh, B. Basu, and M. O'Mahony, "Bayesian time-series model for short-term traffic flow forecasting," J. Transp. Eng., vol. 133, no. 3, pp. 180–189, Mar. 2007.
26. X. Chen, K. Kundu, Z. Zhang, H. Ma, S. Fidler, and R. Urtasun, "Monocular 3D object detection for autonomous driving," in Proc. IEEE CVPR, Jun. 2016, pp. 2147-2156.
27. M. Bojarski, D. Del Testa, D. Dworakowski, B. Firner, B. Flepp, P. Goyal, L. D. Jackel, M. Monfort, U. Muller, J. Zhang, X. Zhang, J. Zhao, and K. Zieba, "End to end learning for self-driving cars," 2016, arXiv:1604.07316. [Online]. Available: http://arxiv.org/abs/1604.07316.
28. I. Lana, J. Del Ser, M. Velez, and E. I. Vlahogianni, "Road traffic forecasting: Recent advances and new challenges," *IEEE Intell. Transp. Syst. Mag.*, vol. 10, no. 2, pp. 93–109, Summer 2018.
29. Shaohua Wan, Songtao Ding, Chen Chen,Edge computing enabled video segmentation for real-time traffic monitoring in internet of vehicles, Pattern Recognition, Volume 121,2022,108146, ISSN 0031-3203, https://doi.org/10.1016/j.patcog.2021.108146. (https://www.sciencedirect.com/science/article/pii/S0031320321003332).

30. G. Kamath, P. Agnihotri, M. Valero, K. Sarker, and W.-Z. Song, "Pushing analytics to the edge," in Proc. IEEE GLOBECOM, Dec. 2016, pp. 1–6.
31. J. Kone£ný, H. Brendan McMahan, F. X. Yu, P. Richtárik, A. Theertha Suresh, and D. Bacon, "Federated learning: Strategies for improving communication efficiency,"' 2016, arXiv:1610.05492. [Online]. Available: http://arxiv.org/abs/1610.05492.
32. Y. Kang, J. Hauswald, C. Gao, A. Rovinski, T. Mudge, J. Mars, and L. Tang, "Neurosurgeon: Collaborative intelligence between the cloud and mobile edge," in Proc. ACM SIGARCH, 2017, vol. 45, no. 1, pp. 615–629.
33. Y. Huang, F.Wang, F.Wang, and J. Liu, "DeePar: A hybrid device-edgecloud execution framework for mobile deep learning applications," in Proc. IEEE INFOCOM WKSHPS, Apr. 2019, pp. 892–897.
34. L. Du, Y. Du, Y. Li, J. Su, Y.-C. Kuan, C.-C. Liu, and M.-C.-F. Chang, "A reconfigurable streaming deep convolutional neural network accelerator for Internet of Things," IEEE Trans. Circuits Syst. I, Reg. Papers, vol. 65, no. 1, pp. 198–208, Jan. 2018.
35. S. Jiang, D. He, C. Yang, C. Xu, G. Luo, Y. Chen, Y. Liu, and J. Jiang, "Accelerating mobile applications at the network edge with software-programmable FPGAs," in Proc. IEEE INFOCOM, Apr. 2018, pp. 55–62.
36. T. Chen, M. Li, Y. Li, M. Lin, N. Wang, M. Wang, T. Xiao, B. Xu, C. Zhang, and Z. Zhang, "MXNet: A flexible and efficient machine learning library for heterogeneous distributed systems," 2015, arXiv:1512.01274. [Online]. Available: http://arxiv.org/abs/1512.01274.
37. L. Zhang, D. Estrin, J. Burke, V. Jacobson, J. D. Thornton, D. K. Smetters, B. Zhang, G. Tsudik, D. Massey, C. Papadopoulos et al., "Named Data Networking (NDN) project," Relat'orio T'ecnico NDN-0001, Xerox Palo Alto Research Center-PARC, 2010.
38. D. Raychaudhuri, K. Nagaraja, and A. Venkataramani, "Mobility first: A robust and trustworthy mobility-centric architecture for the future internet," ACM SIGMOBILE Mobile Computing and Communications Review, vol. 16, no. 3, pp. 2–13, 2012.
39. Yu, W., Liang, F., He, X., Hatcher, W. G., Lu, C., Lin, J., & Yang, X. (2017). A survey on the edge computing for the Internet of Things. IEEE access, 6, 6900–6919.
40. Y. Jararweh, A. Doulat, O. AlQudah, E. Ahmed, M. Al-Ayyoub, and E. Benkhelifa, "The future of mobile cloud computing: Integrating cloudlets and mobile edge computing," in Proc. 23rd Int. Conf. Telecommun. (ICT), May 2016, pp. 1–5.
41. L. Tong, Y. Li, and W. Gao, "A hierarchical edge cloud architecture for mobile computing," in Proc. 35th Annu. IEEE Int. Conf. Comput. Commun. (INFOCOM), Apr. 2016, pp. 1–9.
42. Manzalini, A., & Crespi, N. (2016). An edge operating system enabling anything-as-a-service. IEEE Communications Magazine, 54(3), 62–67.
43. Hong, C. H., & Varghese, B. (2019). Resource management in fog/edge computing: a survey on architectures, infrastructure, and algorithms. ACM Computing Surveys (CSUR), 52(5), 1–37.
44. Mao, Yuyi & You, Changsheng & Zhang, Jun & Huang, Kaibin & Letaief, Khaled. (2017). Mobile Edge Computing: Survey and Research Outlook.
45. Haibeh, L. A., Yagoub, M. C., & Jarray, A. (2022). A Survey on Mobile Edge Computing Infrastructure: Design, Resource Management, and Optimization Approaches. IEEE Access, 10, 27591–27610.

Part II
Emerging Trends in Artificial Intelligence

Chapter 5
CBT-Driven Chatbot with Seq-to-Seq Model for Indian Languages

Subhash Tatale, Nivedita Bhirud, Priyanka Jain, Anish Pahade, Dhananjay Bagul, and N. K. Jain

5.1 Introduction

According to the World Health Organization (WHO) survey, more than 300 million people are suffering from depression, and about two-fifths of the population suffers from common mental ailments due to the lockdown and the ongoing COVID-19 epidemic [24]. One of the best ways to relieve depression is interacting, but due to fear of being judged most sufferers avoid interacting [1]. Therefore, chatbots came into the picture, and hence the potential of chatbots to provide assistance and answers to queries and questions by users is an interesting proposition and has great potential. Chatbots are one of the latest inventions of digital design after the rise of the web and mobile apps.

Chatbots help to reduce pressure in health systems and are being implemented all over the world, and they can be used as a reliable source of information by professionals in the field to avoid problems with self-medication or the general collective panic [22]. Chatbots are also known as computer-programmed conversational agents. There is also a type of artificial intelligence (AI) interaction between users and machines via natural language processing (NLP). The growing popularity of chatbots is also due to their efficiency in terms of cost savings by replacing human assistants, increasing user satisfaction by shortening response times, and being available 24 hours a day.

Chatbots are intelligent systems that can converse with humans and provide answers using natural language. Chatbots that can reason and learn can scale much

S. Tatale (✉) · N. Bhirud · A. Pahade · D. Bagul
Computer Engineering, Vishwakarma Institute of Information Technology, Pune, Maharashtra, India

P. Jain · N. K. Jain
Artificial Intelligence Group, C-DAC, Delhi, India

A. Neustein et al. (eds.), *AI, IoT, Big Data and Cloud Computing for Industry 4.0*, Signals and Communication Technology, https://doi.org/10.1007/978-3-031-29713-7_5

faster than human personnel [16]. The chatbot accomplishes this by comprehending and responding to the user's inquiries.

The encoder–decoder model using recurrent neural network (RNN) that uses multiple iterations for processing an intermediate state is the most common word in all NLP text-generation models. Overall, the use of NLP and ML in domain-specific chatbots improves their ability to understand what humans want to say and, as a result, provide more accurate responses [14]. Computers can perform a variety of tasks using NLP, including knowledge extraction, sentiment analysis, speech recognition, and fake news detection [18].

As India has a population of more than a billion, this CBT-driven chatbot will help to manage the rush of those seeking medical care. Thereby reducing the high number of walk-ins to hospitals by those who suspect they have been affected by the virus [24]. To provide additional support in mental health treatment, the chatbot intends to help COVID-19 patients and people's mental health from the comfort of their own homes 24 $\times$ 7 [23]. The proposed system also includes a preliminary COVID-19 test in which the user can check for COVID-19 symptoms and their intensity by answering specific questions. The proposed system employs a user-friendly interface based on mobile technology to reach the maximum possible audience and in the future may implement it in web applications and expand it in further different languages.

A chatbot is executed using pattern comparison, in which the sentence order is recognized and a saved response pattern is adapted to select sentence variables. Chatbots are primarily developed as a conversational interface engine written in Python, allowing them to respond based on collections of all known conversations [2]. The Hindi language is used by this chatbot system using the sequence-to-sequence model [4]. This is based on an RNN-based encoder–decoder framework which is one of the most commonly used models for chatbot development. The encoder and decoder components make up the sequence-to-sequence model. The encoder accepts source text as input and processes it to generate a thought vector, which is an intermediate representation of the input text. The thought vector is then fed into the decoder input unit. The decoder now processes the thought vector and generates outputs [15]. One key challenge with sequence to sequence (as in other neural network models) is that there are so many settings and hyperparameters that need to be tuned to get a good performing working model [3].

This proposed work is organized in different sections. Following the introduction to the topic, Section 5.2 deals with the literature survey done on the topic where the authors referred to some important research papers and web resources. Section 5.3 discusses works related to the proposed system. Section 5.4 describes the overall architecture of the proposed system and provides details about the components and subsections of the CBT-driven chatbot. Section 5.5 displays the implementation of the chatbot and provides details about the internal working of the chatbot. Section 5.6 concludes the entire project with the result and discusses future possibilities for improving the effectiveness of the system.

5.2 Literature Survey

Chatbots, which connect patients with health-care providers, health-care information, and treatment, play a role in finding more efficient ways to provide timely medical care, access, and quality treatment to patients. The creation and execution of chatbots is still a developing field that is heavily influenced by AI and ML. The authors attempted to survey the chatbot-related topics in this section.

A parametric comparative review of a survey done on 18 research papers is mentioned in Table 5.1 below:

In this research paper, the reviews have been covered from several papers that have focused on different research work carried out for developing chatbot. After the extensive survey done on the topic and related data, it is found that there is no significant CBT-driven chatbot, or related project done with the Hindi (Indian) language. This CBT-driven chatbot offers a vast variety of options such as a Frequently Answered Questions (FAQ) on COVID-19 and mental illness, a Patient Health Questionnaire (PHQ-9) test to check the early symptoms of illness, and assistance by a Virtual Avatar compiled altogether in one application. Therefore, the authors extended the work for the Hindi language which was previously performed by the authors in the English language.

5.3 Proposed Work

There is an increase in the number of cases of mental stress, severe illness, and other related disorders due to the COVID-19 lockdown situation throughout the globe and to address this situation, the authors proposed a CBT-driven chatbot system for immediate assistance. The proposed chatbot has a well-designed architecture with a structured knowledge engine to provide an intuitive interface. The chatbot also offers mental health counselling and self-assessment exams for mental illnesses, in addition to coronavirus information. The chatbot is designed for an easy self-check for the symptoms of COVID-19 and mental illness. All the important data for answering the question is available in the database from which the chatbot will match and display the result according to the question asked. The input is compared with the inputs saved in the database and a matching response is returned. The Hindi language is used to interact with users.

COVID-19 testing and providing mental assistance to people without physical interaction is essential. Hence, the authors proposed a contactless CBT-driven chatbot system. This chatbot design includes a software algorithm that produces responses based on given input to simulate human conversations in voice or text mode. The chatbot offers users to have a conversation with the voice-chat input feature and then get a system-generated text output. Along with the voice-chat option, this chatbot also has a feature of virtual avatar which acts like a human assistant and has its different features.

Table 5.1 Parametric comparative literature review

Serial number	Paper	Year	Observation/final outcome	Technology used	Interface provided
1.	Chatbot for Depressed People [1]	2021	Helps to overcome depression	Seq2Seq, LSTM	Chat interface
2.	Chatbots for Multilingual Conversations [19]	2019	Multilingual chatbot in actual conversation using different languages	Microsoft Visual Studio	Tutor Mike chatbot
3.	DoctorBot – An Informative and Interactive chatbot for COVID-19 [18]	2020	Developed a chatbot based on COVID-19.	AI, NLP	Chat interface
4.	Conversational Assistant based on Sentiment Analysis [21]	2019	The user can provide input through either text or speech. The proposed model also fetches the latest tweets of the user using the Twitter API.	Seq2Seq model	Web interface
5.	Digital Psychiatry – Curbing Depression Using Therapy Chatbot and Depression Analysis [25]	2018	Specifies depression levels and suggests treatment on the basis of it.	ML and NLP	No interface
6.	Review of Chatbot System in Hindi Language [2]	2019	Chatbot is designed for a normal question–answer interaction in Hindi language.	Python, pattern matching	Chat interface
7.	Mental Health Monitoring Using Sentiment Analysis [11]	2020	The authors discuss the use of mining or extraction of data from various sources such as smart phones, wearable gadgets (smart watches), and health trackers (fitness tracker bands) and propose a model using lambda architecture to analyze the extracted data to predict the mental state of the user maintaining user secrecy and data privacy.	Sentiment analysis, NLP	No interface

8.	Seq2Seq AI Chatbot with Attention Mechanism [8]	2020	A training model is trained for three different conversations. After that they are compared with a baseline seq2seq model. A basic conversational chat interface is created.	RNN, Seq2Seq, Python Flask app	Web interface
9.	Review of Chatbots Design Techniques [7]	2018	The authors reviewed several types of chatbots used for multiple purposes such as for businesses, marketing, and education. The reviews have covered several papers that have focused on chatbot design. The review provides explanations on chatbot's design in today's market.	AI-ML, pattern matching	No interface
10.	A Medical Chatbot [9]	2018	Created a system that allows users to submit complaints and questions about their health. The primary goal in developing this system was to ensure customer satisfaction. The actual benefit of the chatbot is to assist people by providing proper guidance on good and healthy living. The system is beneficial for medical institutes or hospitals because it allows users to freely ask medical dosage-related questions via voice.	NLP, support vector machine (SVM)	Mobile application
11.	Conversational AI Chatbot Based on Encoder–Decoder Architectures with Attention Mechanism [6]	2019	The authors used modern techniques to create a conversational AI chatbot. They created an encoder–decoder attention mechanism architecture. An RNN with LSTM cells is used in this encoder–decoder. The attention mechanism enables the decoder to selectively examine the input sequence while decoding.	Encoder–decoder attention mechanism	Chat interface

(continued)

Table 5.1 (continued)

Serial number	Paper	Year	Observation/final outcome	Technology used	Interface provided
12.	Survey on the Design and Development of Indian Language Chatbots [13]	2021	Developed a chatbot with the conviction that chatbots in Indian languages can be used to enhance communication between Indian users. To operate effectively, the chatbot must have access to a lot of logical resources. The authors thoroughly investigated the methodologies and behavior patterns of various Indian chatbots and compared their accuracy.	NLP	Chat interface
13.	Smart Ubiquitous Chatbot for COVID-19 Assistance with Deep Learning Sentiment Analysis Model During and After Quarantine [5]	2020	This chatbot is developed to help patients during and after the COVID-19 quarantine period.	NLP, LSTM	Web interface, mobile application
14.	Medbot: Conversational Artificial Intelligence Powered Chatbot for Delivering Tele-Health after COVID-19 [12]	2020	The application reduces the barriers for access to health-care facilities and provides information and advice to chronic patients. The authors propose a conversational chatbot "Aapka Chikitsak" on Google cloud. The gap between supply and demand of health-care providers can be bridged by providing services through Google cloud as it increases its accessibility.	NLP, Google cloud	Mobile application

15.	Grammar Checkers for Natural Languages: A Review [10]	2017	Conducted a descriptive study of various grammar checks for global languages as well as some Indian languages such as Hindi, Urdu, Punjabi, and Bangla. The authors examined various grammar checking approaches and methodologies, as well as key concepts and grammar checker internals.	NLP	No interface
16.	Chatbot Using a Knowledge in Database: Human-to-Machine Conversation Modeling [3]	2017	Developed a chatbot in such a way that each of the patterns is paired with chatbot knowledge that has already been processed from various sources. Sentence similarity is calculated using bigram, which divides input sentences into two letters of the input sentence. The data is modeled after the conversation, and the results with the chatbot are investigated.	NLP, bigram	Chat interface
17.	Multilingual Healthcare Chatbot Using Machine Learning [4]	2021	The chatbot application converses with the user using concepts of NLP and also supports speech-to-text and text-to-speech conversion so that the user can also communicate using voice. Five different ML algorithms have been analyzed for disease prediction. The chatbot supports three languages namely English, Hindi, and Gujarati.	NLP, ML	Chat interface
18.	Adaptive Machine Learning Chatbot for Code-Mix Language (English and Hindi) [14]	2020	Implementation of NLU and ML algorithms in the chatbot system being implemented made it smart and user-friendly to make it behave as a virtual friend. Also, algorithms used in the chatbot makes it more efficient, as if it looks like not a virtual system but a human that you are talking to.	NLU, NLG	Chat interface

5.3.1 *Flow of the System*

The proposed COVID-19 chatbot is described in this section. Chatbots can be observed everywhere, replacing website queries and FAQs as well as providing virtual assistance. The authors proposed contactless communication through a chatbot system. Using a sequence-to-sequence network with a many-to-one framework will take the different kinds of symptoms. These different kinds of symptoms would be considered as many inputs as possible and based on the many inputs a single output would be generated.

This section explains the flowchart of the proposed system. Figure 5.1 depicts the same.

(a) *User registration:* To ask queries and to access all other features of the chatbot, the user will have to go through the process of registration and login first. Figure 5.1 shows the representation starting from a user registration followed by login and transfer to the interface page where he/she can ask the query to the chatbot and calculate the probability of having COVID-19.

(b) *COVID-19 test:* A COVID-19 testing system is one of the features the authors included in the chatbot application. The test is developed to predict if the user has the symptoms of COVID-19 or not. The COVID-19 test consists of five

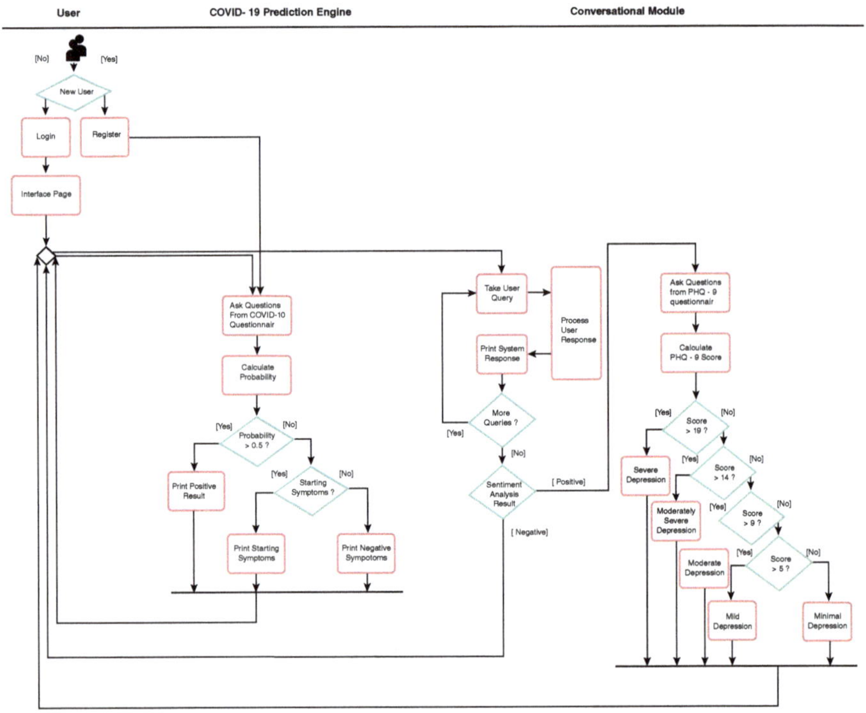

Fig. 5.1 System flowchart

questions, all are yes or no questions. These questions are based on the general symptoms of COVID-19. After answering all of the questions, the user will get to know whether he is holding any possible symptoms of COVID-19 or not. This COVID-19 test produces the best results possible according to the user's answers.

(c) *PHQ-9 test:* The PHQ-9 is a new instrument for making criteria-based diagnoses of depressive and other mental disorders commonly encountered in primary care. The PHQ-9 is a reliable and valid measure of depression severity. It is not a screening tool for depression, but it is used to monitor the severity of depression and response to treatment [20].

Each of the questions has a score ranging from 0 to 3 on the PHQ-9's depression module and there are four options to answer these questions:

1. Nearly every day
2. More than half of the days
3. Several days
4. Not at all

Based on the question answered and the score achieved after having the PHQ-9 test, the final score is calculated and then the user can know the severity of the depression.

- Score greater than 19 – severe depression
- Score less than or equal to 19 – moderately severe depression
- Score less than or equal to 14 – moderate depression
- Score less than or equal to 9 – mild depression
- Score less than or equal to 4 – minimal depression

5.3.2 Architecture

Having a good balancing of cohesion and coupling, the modular architecture of chatbot is organized in three major sections:

(i) COVID-19 prediction engine
(ii) Conversational module
(iii) User interface and virtual avatar specification (Fig. 5.2)

(i) *COVID-19 prediction engine:* After successful login, the authorized user from the interface page will take input as the user query and process the response and print the system response. Chatbot users can self-evaluate his/her symptoms of coronavirus. The proposed test is a simple COVID-19 symptom checking test that allows people to investigate their symptoms on their mobile applications. It is basically a short questionnaire that asks people about various primary and significant corona symptoms and then displays the test results based on their responses. The questions are to be answered: "yes" or "no."

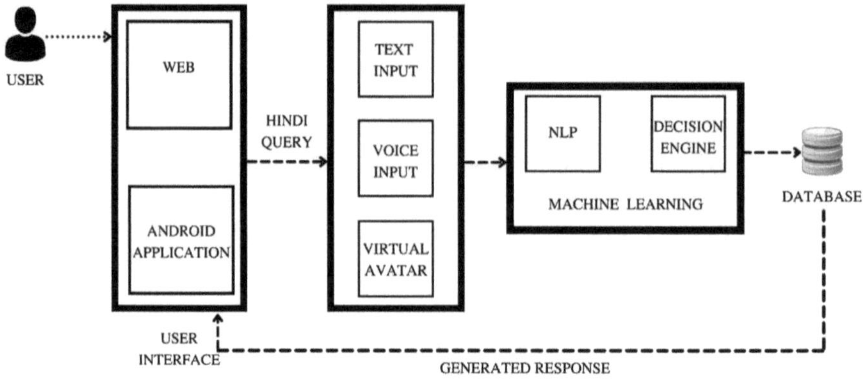

Fig. 5.2 Architecture of chatbot system

(ii) *Conversational module:* A simple messaging interface is used to communicate with the chatbot, and it is designed in similarity with most other messaging apps with an easy-to-understand interface. As a result, the user does not need any special knowledge or skills to use it. The user can have the conversation in the Hindi language. The goal of chatbots is to mimic human dialogue in order to provide the most intuitive user interface for applications or, in some cases, to simply provide entertainment [19].

(iii) *User interface and virtual avatar specification:* The interface aims to provide easy and smooth interaction with the chatbot without expertise for all the users. The application comes with an easy-to-learn and use interface. The interface is also in charge of displaying the user's system-generated results. As a result, the chat interface can be considered the system's face, through which all communication takes place. It acts as a liaison between the system and the user [23].

The virtual avatar is a three-dimensional (3D) human model that enables more human interaction through the audio input. Avatars are visual representations of real or artificial intelligence in the virtual world. They have the appearance of humans, humanlike (often animated) characters, or other living creatures. Using audio input, the user can simply ask a question. The 3D model plays a talking animation after the chatbot creates an answer, and the answer is presented using text-to-speech technology.

Since, it is different for virtual avatars and chatbots, when it comes to digital avatars, in particular, the working of conversational AI is pretty similar to chatbots, apart from a few added things. A physically embodied human provides a more immersive experience than a chatbot because you can see the avatar visually. A virtual avatar talks with the help of expressions and gestures, which gives users a very personalized experience.

The user can utilize the messaging interface or the virtual avatar to communicate with the chatbot. This virtual avatar was developed for the English corpus and now will be updated with the new changes with the Hindi corpus. There are a total of six avatars designed to have a choice for the user as

Male, female, child male, child female, old male, and old female

The avatar design is enriched with some different animations and responses. Some of the essential animations were acquired from the Mixamo website. Blender was used to create these models and setups, and animations are baked in a specific order. Each avatar has three animations: idle, talking, and waving.

The talking animation would be activated when the chatbot generates an output. Callbacks are used to transition between these animations. For interactivity, the virtual avatar interface uses void input and output. The user's voice is captured using the speech-to-text service using the speech-to-text API of Google. The user's voice inquiry is decoded and returned in text format using this API.

5.4 Implementation

In this section, the implementation details and important modules used in the proposed system are given.

(a) *Data deception/preparation:*

Currently, all the data is entered manually by the user while signing in and giving necessary details such as name and email. But in the future, we can have OTP-based login through mobile number or email which will help us to verify the user and prevent from data deception problems.

(b) *Attention mechanism:*

The basic concept behind trying to implement the attention mechanism in the proposed chatbot is to focus on the specific input vectors of the input sequence depending on attention weights. At each decoder step, it determines which input parts are more important. In this configuration, the encoder is not required to condense the complete input into a single vector. The attention function of the transformer is presented in Fig. 5.3 as a method of calculating the significance of a set of information based on specific keys and queries. The basic attention mechanism is nothing more than a dot product of the query and the key. The transformer employs multi-head attention layers, which are made up of multiple scaled dot-product attention layers. The scaled dot-product attention function takes three inputs: Q (query), K (key), and V (value) which is defined in the python code as scaled_dot_product_attention (query, key, value, and mask).

(c) *Sequence-to-sequence model:*

The proposed system uses the sequence-to-sequence transformer to respond to user requests. This transformer takes a character sequence as input and generates a new sequence based on it. For dialogue systems and machine translation, the sequence-to-sequence model was chosen as the most reliable model. This model deals with the problem of sequence generation. It consists of two RNNs: an encoder and a decoder. The encoder takes as input a sequence

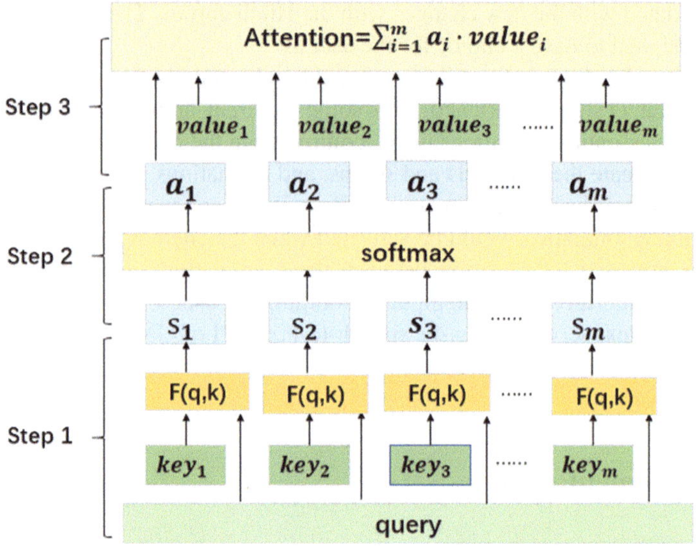

Fig. 5.3 Attention mechanism

(sentence) and processes one symbol (word) at a time. Its goal is to convert a sequence of symbols into a fixed-size feature vector that encodes only the important information while discarding the rest. Data flow can be visualized in the encoder as the flow of local information from one end of the sequence to the other along the time axis. The advantage of using the transformer model is that the information sequence will not be altered. In addition, this model employs the concept of attention heads, which allows it to store almost infinite amounts of information.

(d) *Encoder–decoder framework:*

The standard modeling paradigm for sequence-to-sequence tasks is encoder–decoder. The basic task of an encoder network is to understand the input sequence and produce a lower-dimensional interpretation of it. This interpretation is then passed to a convolutional network, which produces its own output sequence. This framework is made up of two components, an encoder and a decoder, both of which were used in our chatbot (Fig. 5.4).

A. *Encoder layer:* Reads source sequence and produces its representation. Each encoder layer consists of sublayers, two dense layers followed by dropout. For the main encoder layer, we substituted the values in encoder layer as

Encoder = encoder (num_layers = 2, units = 512,d_model = 128,
num_heads = 8,dropout = 0)

We then used tf. keras.utils.plot_model() to visualize our model. Figure 5.5 was generated with the above values in the main encoder layer.

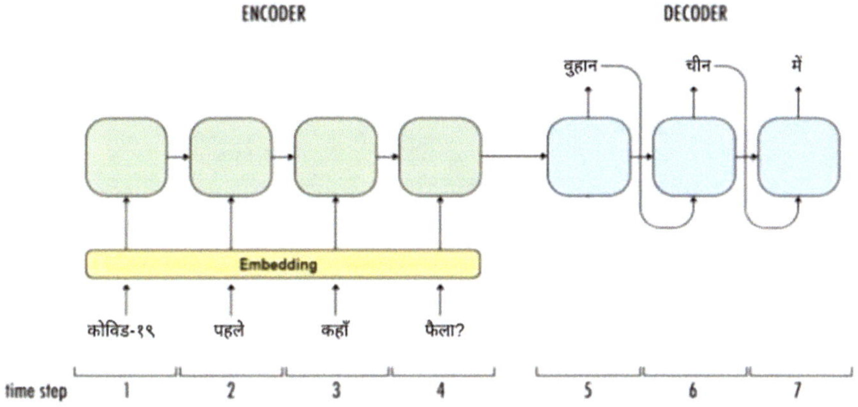

Fig. 5.4 Sequence-to-sequence model

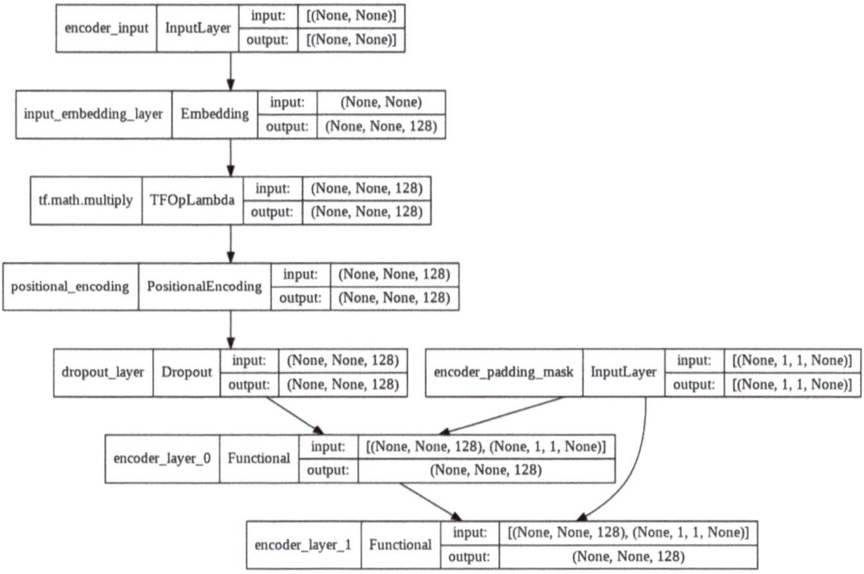

Fig. 5.5 The encoder layer

B. *Decoder layer:* Uses source representation from the encoder to generate the target sequence.

Each decoder layer also consists of sublayers, two dense layers followed by dropout. For the main decoder layer, we substituted the values in decoder layer as

Decoder = decoder(num_layers = 2, units = 512,d_model = 128, num_heads = 8,dropout = 0.1)

We then used tf. keras.utils.plot_model() to visualize our model. Figure 5.6 was generated with the above values in the main decoder layer.

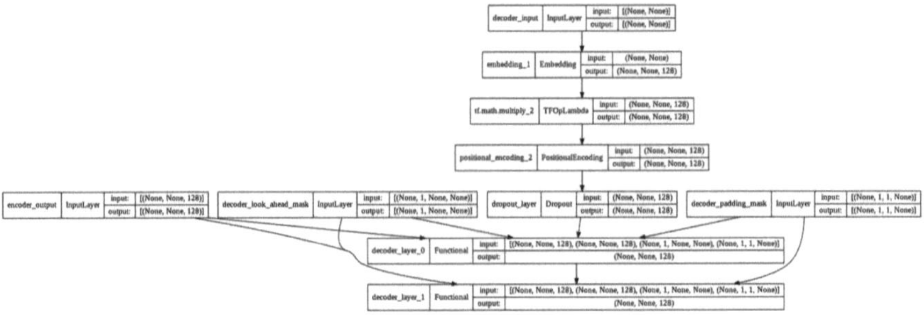

Fig. 5.6 The decoder layer

When a source sentence is passed to an encoder in the backend process, the encoder encodes the entire source sequence into the context vector. This context vector is then passed to the decoder, which generates an output sequence in a target language, such as Hindi. The context vector is responsible for condensing the entire input sequence into a single vector.

(e) *Sentiment analysis:*

The user's overall mental state is determined using all the user's inputs to analyze the sentiment. Sentiment analysis has emerged as a significant research area that aids in the discovery of valuable information which can be applied to a variety of purposes. Sentiment analysis is defined in the code as the recognition of the text to be positive, negative, or neutral, and get the polarity of the sentence. Sentiment analysis using NLP has been used to predict user information such as mental status, depression, anxiety, and stress. According to the output of this module, feedback is given to users, that is, positive, neutral, or negative.

The sentiment analysis model oversees detecting the user's emotion throughout the conversation. TextBlob library, which includes built-in classes for analyzing the sentiment of sentences, examines each user input and categorizes it as positive, neutral, or negative. It produces the following values for any sentence: -1, 0, and 1.

1: when the input is positive
0: when the input is neutral
-1: when the input is negative

At the end of the conversation of the user with the chatbot, the user's total sentiment score is calculated. The user is then asked to appear for a depression assessment based on this sentiment score.

In the proposed system, a sentiment score is generated for each user input. The final score of the user is used to determine whether to take the PHQ-9 depression assessment. If the final score is less than 0, the user is required to give the evaluation. Otherwise, the user has the option of giving or not giving the assessment.

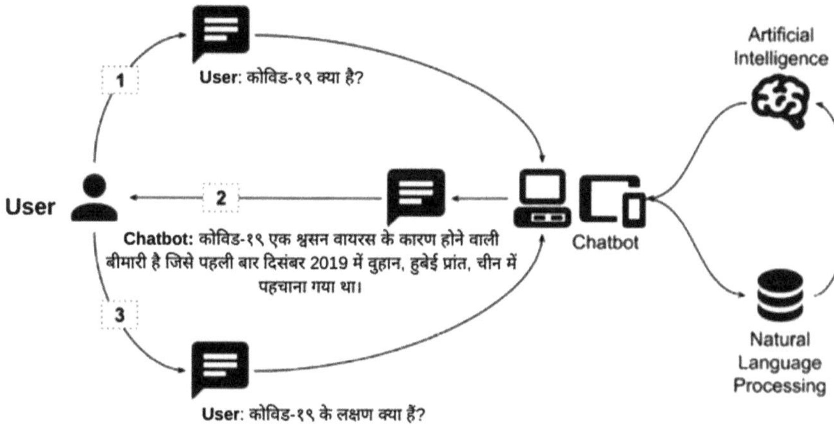

Fig. 5.7 A sample interaction between the user and the CBT-driven chatbot

(f) *Chatbot system:*

A chatbot is a computer programme that can interact and communicate with human users. Chatbots can make mental health interventions more accessible. Dataset for mental health contains questions that users can ask during the conversation. Chatbot uses a sequence-to-sequence transformer for answering user queries. This transformer takes a sequence of characters as input and generates another sequence based on it. Some questions and answers are gathered from an extensive search on the Internet. Figure 5.7 shows a sample interaction between a user and the CBT-driven chatbot.

(g) *Voice-chat interface:*

This CBT-driven chatbot has the feature of asking a query through the voice input and getting the appropriate answer. The user's voice is captured using the speech-to-text service using the speech-to-text API of Google. The user's voice inquiry is decoded and returned in text format using this API. The user's voice is captured using the speech-to-text service using the speech-to-text API of Google. The user's voice query is decoded and returned in text format using this API. This text is then sent to the server. The server responds with a textual answer. The response is converted to voice using the text-to-speech service.

5.5 Results and Discussion

Starting from the initial training, the chatbot progression was trained at several instances to check the overall functionality and accuracy achieved. The data corpus was modified and processed some of the answers accordingly. The length of questions and answers was decreased and again the training accuracy was checked. Table 5.2 contains the results of the training accuracy based on the progression of the dataset.

Table 5.2 Training accuracy of dataset

Serial number	Total number of rows	Modifications done	Accuracy
1.	2500	Very big and lengthy answers	0.37
2.	4500	Increased the corpus and shorten the length of answers	0.43
3.	7000	Added the mental health questions	0.47
4.	10,770	In addition to the Hindi questions and answers, English questions and answers were added.	0.53
5.	14,263	Final dataset with questions and answers related to COVID-19, physical health, and mental health in Hindi and English language.	0.63

Table 5.3 Model statistics

Serial number	Parameters	Output
1.	Vocabulary size	8401
2.	Total questions and answers in the dataset	14,263
3.	Training accuracy reached	0.63
4.	Trained on	Google Colab with GPUs
5.	Time is taken per epoch	12 seconds approximately
6.	Total time is taken for training	40 minutes approximately
7.	Total epochs used for training	153
8.	Loss at the beginning	5.55
9.	Loss at the end	0.02

The model may perform better if increased the corpus size and with the addition of processed data. But still, for better accuracy than achieved, more Hindi data need to be added and also the sentence length should be shortened accordingly (Table 5.3).

The model is trained and tested on Google Colab, and it took approximately 60 minutes to complete the training of about 14,263 rows of data. The accuracy achieved was 63% after the completion of training. By including more relevant knowledge-base data, the effectiveness of general chat queries can be increased.

The training graph per epoch for accuracy and loss is shown in Fig. 5.8.

5.5.1 Training Dataset

Training has been completed on 14,263 rows of questions and answers corpus, where the categories of questions are from different fields of coronavirus, mental health, physical fitness, normal infection problems, and other common viruses. Some of the training data to develop this chatbot was devised from the data available on the WHO website. The complete dataset corpus comprises all the questions and answers in Hindi and English language.

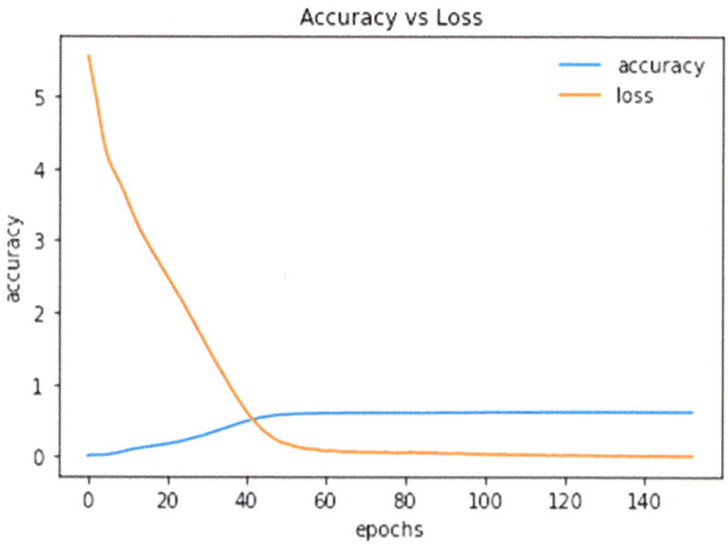

Fig. 5.8 Accuracy versus loss against epochs

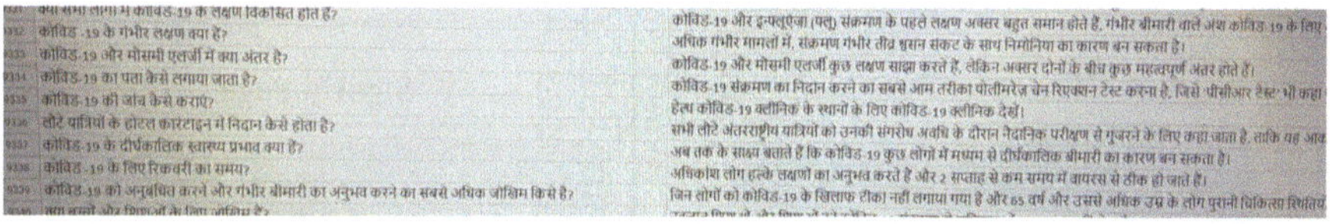

Fig. 5.9 COVID-19 question and answers in the Hindi language

Fig. 5.10 Mental health questions and answers in the Hindi language

Figure 5.9 is a representation of some sample COVID-19 questions and answers. The corpus contains all COVID-19-related questions and their answers. It provides all COVID-19-related information, allowing the user to clear up any concerns or ambiguities about the disease.

Research regarding chatbots in mental health is also emerging. Figure 5.10 represents some sample questions related to mental health and physical fitness.

The CSV format is used to save and access the contents of the dataset. The CSV file containing the dataset is accessed using Python's module, and the data is loaded in a Pandas DataFrame. This dataset can be further expanded in the English language also by converting these Hindi questions and answers into the English language.

5.5.2 Application Overview and the Chatbot Interface Designed

Welcome Page

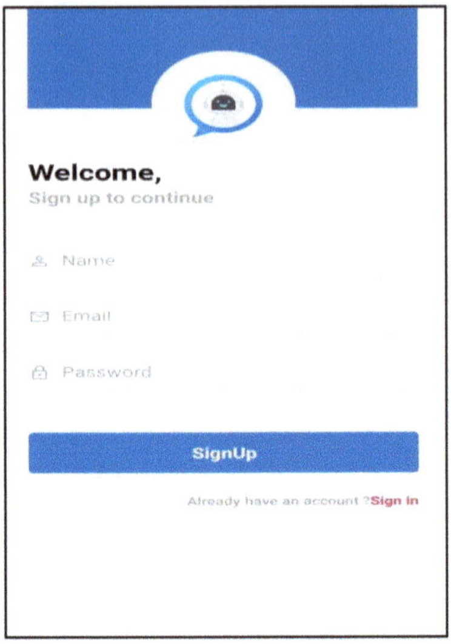

SignUp Page

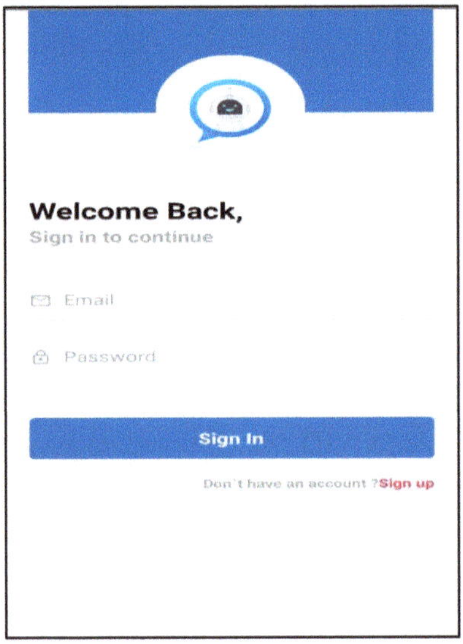

SignIn Page

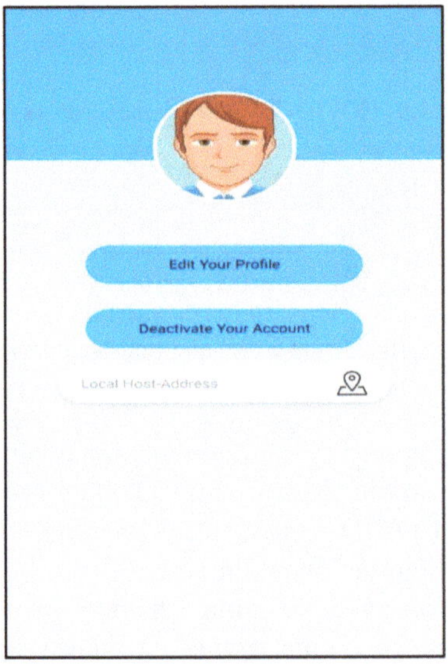

Profile Page

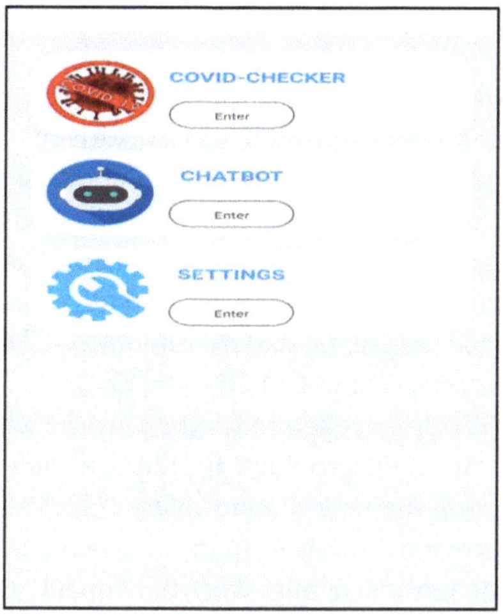

Option Page

Chat Interface page

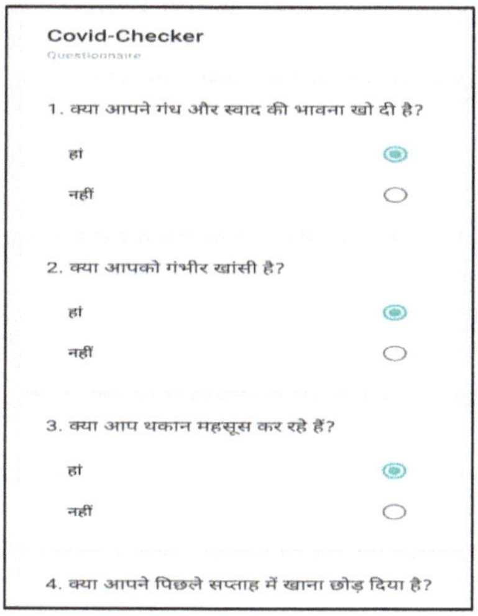

Covid-Checker Screen

Result Screen

5.6 Conclusion

Chatbots can be trusted assistants and provide users with 24/7 support by answering simple, repetitive questions with preprogrammed responses. This CBT-driven chatbot provides patients with a great user experience by simply having a conversation with the bot to get relevant answers to their questions. The main benefit of using chatbots with the help of mobile applications and websites is to reach a broad audience. Besides that, this chatbot platform works positively to provide efficient service in various fields to serve humans in many different and multiple ways. These features make chatbots useful to spread awareness about COVID-19 [17].

The authors recommend the use of this CBT-driven chatbot for the purpose where it would converse with the user based on the input provided by the user and get the output generated by the system based upon the user's input query. The variety of different questions used in the dataset provides the user an appropriate answer and stands out to be one of the important reasons along with the model used. Previously, the authors worked on this project for the English language and now we are expanding the work of the chatbot with the Hindi language. Furthermore, the chatbot can communicate in both English and Hindi, making the user more convenient talking to it and resolving their queries accordingly. Developing a big and appropriate dataset is also an important task for the chatbot systems along with models and implementations. The dataset used for this chatbot was collected from authentic sources, and it manages to cover most of the user queries related to COVID-19 and mental health.

The transformer model's attention mechanism may be the future of NLP, outperforming the success of RNN. Once the COVID-19 pandemic has passed, we plan to reconfigure this CBT-driven chatbot and make it compatible with other epidemics as well as services by utilizing individual APIs or relevant datasets. The authors intend to increase the chatbot language design beyond Hindi and English to include many more regional languages, allowing a wider range of users to use the system. People who do not speak English well can use the chatbot in their chosen language.

References

1. Nagargoje, S., Mamdyal, V. and Tapase, R., 2021. Chatbot for Depressed People. United International Journal for Research & Technology (UIJRT), 2(7), pp.208–211.
2. Bhagwat, V., Nagarkar, M., Pooja, P., Shrikant, J., Kharate, N.G. (2019). Review of Chatbot System in Hindi Language. International Research Journal of Engineering and Technology (IRJET), 6(11), 2395–0072.
3. Setiaji, B., & Wibowo, F. W. (2016, January). Chatbot using a knowledge in database: human-to-machine conversation modeling. In 2016 7th international conference on intelligent systems, modeling, and simulation (ISMS) (pp. 72–77). IEEE.

4. Badlani, S., Aditya, T., Dave, M., & Chaudhari, S. (2021, May). Multilingual Healthcare Chatbot Using Machine Learning. In 2021 2nd International Conference for Emerging Technology (INCET) (pp. 1–6). IEEE.
5. Ouerhani, N., Maalel, A., Ghezala, H. B., & Chouri, S. (2020). Smart Ubiquitous Chatbot for COVID-19 Assistance with Deep learning Sentiment Analysis Model during and after quarantine.
6. Ali, A., & Amin, M. Z. (2019). Conversational AI Chatbot Based on Encoder-Decoder Architectures with Attention Mechanism. Artificial Intelligence Festival 2.0.
7. Ahmad, N. A., Che, M. H., Zainal, A., Abd Rauf, M. F., & Adnan, Z. (2018). Review of chatbots design techniques. International Journal of Computer Applications, 181(8), 7–10.
8. Sojasingarayar, A. (2020). Seq2seq AI chatbot with attention mechanism. arXiv preprint arXiv:2006.02767.
9. Dharwadkar, R., & Deshpande, N. A. (2018). A medical chatbot. International Journal of Computer Trends and Technology (IJCTT), 60(1), 41–45.
10. Bhirud, N. S. (2017). Grammar checkers for natural languages: a review. International Journal on Natural Language Computing (IJNLC) Vol, 6.
11. Shah, A., Shah, R., Desai, P., Desai, C. (2019). Mental Health Monitoring using Sentiment Analysis, International Research Journal of Engineering and Technology (IRJET) ISSN: 2395-0072, Volume: 07 Issue: 07.
12. Bharti, U., Bajaj, D., Batra, H., Lalit, S., Lalit, S., & Gangwani, A. (2020, June). Medbot: Conversational artificial intelligence powered chatbot for delivering tele-health after covid-19. In 2020 5th International Conference on Communication and Electronics Systems (ICCES) (pp. 870–875). IEEE.
13. George, A. S., Muralikrishnan, G., Ninan, L. R., Varrier, P. S., & Dhanya, L. K. (2021, June). Survey on the Design and Development of Indian Language Chatbots. In 2021 International Conference on Communication, Control and Information Sciences (ICCISc) (Vol. 1, pp. 1–6). IEEE.
14. Sancheti, R., Upare, S., Bhirud, N., & Tatale, S. Adaptive Machine Learning Chatbot for Code-Mix Language (English and Hindi).
15. Rus, V. (2018). Intelligent Chatbot using Deep Learning (Doctoral dissertation, University of Memphis).
16. Palasundram, K., Sharef, N. M., Nasharuddin, N., Kasmiran, K., & Azman, A. (2019). Sequence to sequence model performance for education chatbot. International Journal of Emerging Technologies in Learning (iJET), 14(24), 56–68.
17. Patil, A., Patil, K., Shimpi, G., & Kulkarni, M. (2020). COVIBOT: An efficient AI-based Chatbot with Voice Assistance and Multilingualism for COVID-19. Department of Information Technology, 17.
18. Thukrul, J., Srivastava, A., & Thakkar, G. (2020). DoctorBot-An informative and interactive Chatbot for COVID-19. International Research Journal of Engineering and Technology (IRJET), 7(07), 2395–0072.
19. Vanjani, M., Aiken, M., & Park, M. (2019). Chatbots for multilingual conversations. J. Manag. Sci. Bus. Intell, 4, 19–24.
20. Kroenke, K., Spitzer, R. L., & Williams, J. B. (2001). The PHQ-9: validity of a brief depression severity measure. Journal of general internal medicine, 16(9), 606–613.
21. Suraj, D. M., Prasad, V. A., Mitra, S., Rohan, A. R., & Salis, V. E. (2019). Conversational Assistant based on Sentiment Analysis.
22. W. S. Erazo, G. P. Guerrero, C. C. Betancourt and I. S. Salazar, "Chatbot Implementation to Collect Data on Possible COVID-19 Cases and Release the Pressure on the Primary Health Care System," 2020 11th IEEE Annual Information Technology, Electronics and Mobile Communication Conference (IEMCON), 2020, pp. 0302–0307, https://doi.org/10.1109/IEMCON51383.2020.9284846.

23. Bhirud, N., Tataale, S., Randive, S., & Nahar, S. (2019). A Literature Review on Chatbots in Healthcare Domain. International Journal of Scientific & Technology Research, 8(7), 225–231.
24. Coronavirus disease (COVID-19) pandemic, World Health Organization (WHO), emergencies, diseases, novel-coronavirus-2019.
25. B. Sharma, H. Puri and D. Rawat, "Digital Psychiatry - Curbing Depression using Therapy Chatbot and Depression Analysis," 2018 Second International Conference on Inventive Communication and Computational Technologies (ICICCT), 2018, pp. 627–631, https://doi.org/10.1109/ICICCT.2018.8472986.

Chapter 6
A Review of Predictive Maintenance of Bearing Failures in Rotary Machines by Predictive Analytics Using Machine-Learning Techniques

Yasser N. Aldeoes, Prasad Gokhale, and Shilpa Y. Sondkar

6.1 Introduction

Recently, investments in manufacturing machines today have become widespread in various vital fields, especially in production lines, despite being expensive, complex, and very specialized, which means that maintaining the safety of every part of the machine, directly controlling and treating it is to maintain the production chain as a whole and it is good Preserving time caused by machine downtime due to maintenance periods and unsuited failures, which in turn led to idleness and stoppage of the entire production line due to the failure of one of the machine components. To reduce these costs, it is necessary to identify these errors in advance.

Undoubtedly, Industrial 4.0 or "industrial Internet of things" (IIoT) introduced a rampage and fundamental changes in industrial automation, decision-making, and data analysis in a significant and remarkable way, as the "Internet of Things" (IoT) technology played a pivotal turn in the implementation and creation of smart products in various fields [1]. This leads to a smart reality with smart specifications, services, and proper planning for these environments from data collected by several devices located in various areas of industries. In addition, industrial systems capture a huge amount of data that comprises data about events, processes, and alarms that happen along a production line [2]. Furthermore, when these data are processed and evaluated, they can yield useful knowledge and information about industrial processes and system dynamics. Interpretive outcomes for strategic decision-making are feasible by using analytic approaches based on

Y. N. Aldeoes (✉) · P. Gokhale
Department of Computer Engineering, Vishwakarma University, Pune, India
e-mail: yasser.aldeoes-043@vupune.ac.in; prasad.gokhale@vupune.ac.in

S. Y. Sondkar
Department of Instrumentation Engineering, Vishwakarma Institute of Technology, Pune, India
e-mail: shilpa.sondkar@vit.edu

A. Neustein et al. (eds.), *AI, IoT, Big Data and Cloud Computing for Industry 4.0*, Signals and Communication Technology, https://doi.org/10.1007/978-3-031-29713-7_6

data, which provide benefits such as lesser maintenance rates, fewer contrivance faults, fewer repair stops, lower extra parts stock, lengthy spare parts life, upgraded production, improved repair confirmation, operator safety, and overall profit [3]. With the significant increase of many industries in using IoT-supported designs and investigating through how IoT keys can provide novel advantage, IoT has proven to be changing the performance rules for many businesses. The origination of IoT connectivity never ceases to amaze, from self-driving cars to smart homes loaded with voice-activated pieces of equipment [4]. Predictive analytics are used in a variety of ways by companies depending on their needs, ranging from predictive marketing and data mining to the use of machine-learning (ML) algorithms and artificial intelligence (AI) to improve the behavior of business processes and identify trends and new statistical forecasts [5]. The goal of this research is to look at the challenges of using machine learning techniques in predictive maintenance (PdM) and to look at current trends in how failures are forecast. The most essential factors in designing concepts like a predictive maintenance decision support system were failure prediction. It also looked at all frameworks in Industry 4.0 that apply machine learning for predictive maintenance. This post has notably contributed to raising awareness of the difficulties in establishing and applying predictive maintenance [6]. The development and implementation of an online mistake diagnosis system based on ML methods is described and addressed in this chapter. An in-house test setup is used to implement a built fault detection system based on the provided approach, and the consistently discovered results imply that such a system might be commonly used to expect various problems in power drivetrains [7]. Generally, maintenance is divided into three types: run-to-failure (R2F) or reactive maintenance, Predictive Maintenance (PdM) and preventive maintenance (PvM). The following main classifications can be used to categorize techniques for maintenance policies.

- Reactive maintenance or run-to-failure: Maintenance just in the event of a failure, also known as reactive maintenance, is conducted only when something breaks or fails. But this type of maintenance leads to big financial losses due to both equipment repairs to be replaced and downtime of plants that adding a cost [8].
- Preventive maintenance (PvM): This type of maintenance is popular as it is based on preplanned scheduled times (weekly, monthly, or yearly), and is performed on a time period to anticipate process/equipment failure. This approach is effective in avoiding failure. However, through it, unnecessary corrective actions are taken that lead to increased operating costs.
- Predictive maintenance (PdM): It is a strategy that has intelligent predictive tools that include all technologies that rely on continuous machine monitoring and early detection and prevention of potential failures. It includes operational procedures based on historical data (such as ML techniques) and (artificial intelligence) [9].
- Prescriptive maintenance (RxM): The asset maintenance method, known as prescriptive maintenance, employs machine learning to intelligently schedule and plan asset maintenance as well as modify operating conditions for desired outcomes, "what-if" scenarios. On the other side, prescriptive maintenance not

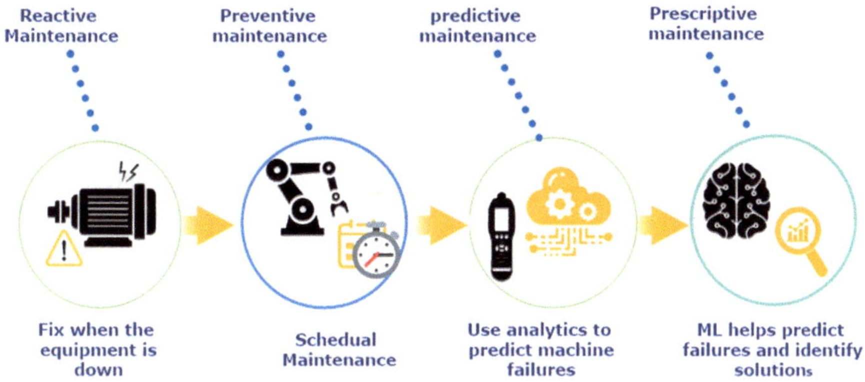

Fig. 6.1 Maintenance types (past and future)

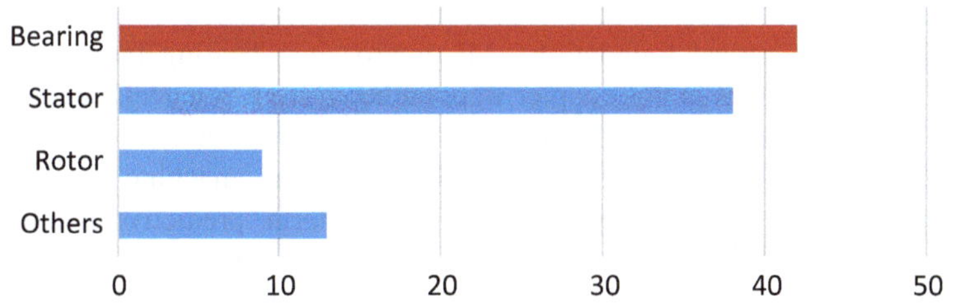

Fig. 6.2 The percentage of rotary machine faults

only for failure signature but also offers guidance on how to postpone or completely avoid equipment breakdown (Fig. 6.1).

Each of the maintenance methods is characterized by a positive aspect from the other; for example, prescriptive maintenance is distinguished from predictive maintenance, as it is more proactive in maintaining the integrity of the equipment and describing the instructions that must be taken to avoid the occurrence of the problem, while on the other hand, predictive maintenance is the best in determining the time of equipment failure.

Improving the dependability, availability, and safety of modern industrial systems and applications is critical for lowering maintenance costs. As a result, monitoring the health of machinery like induction motors is critical. In rotary machines, the bearing is a crucial component [10]. To improve detection, diagnosis, monitoring, and prognosis capabilities, intelligent condition monitoring and defect detection and diagnostic technologies are crucial. Stress, aging, vibration, and lengthy working times impact all parts of general rotary machines (stator, bearing, bar, and rotor). As a result, any failure of any part might result in a major breakdown of the machine, increasing maintenance costs and causing significant losses [11] (Fig. 6.2).

As a result, machine learning (ML) offers powerful prediction methodologies for PdM applications. So, the success of these applications is contingent on the correct deployment of machine learning techniques. As a result, the purpose of this study is to give a review that covers the most widely published PdM methodologies based on ML methods that are collected from Scopus, IEEE databases, and ScienceDirect. The remnant of this chapter will be planned as follows: Section 6.2 defines the survey and analysis for related work. Section 6.3 presents related work technical background. Section 6.4 discusses about the predictive maintenance and machine learning techniques. Section 6.5 presents applications of machine-learning algorithms in PdM and Section 6.6 is about discussion and conclusions.

6.2 Survey and Analysis for Related Work

This chapter includes articles from the database of Scopus, IEEE, ScienceDirect, and Google Scholar. The four databases mentioned above aided in the development of the PdM's theoretical foundations, ML approaches, bearing failures, and ML algorithms. The articles in this chapter were gathered from ScienceDirect and IEEE and divided into four areas. The publications in the first set were gathered from ScienceDirect and IEEE, and the search was based on "machine learning." The articles in the second set were found using the search term "predictive maintenance." The items in the third group were found by a search based on "bearings failure." This group is utilized as supporting work in the contexts of the introduction and the common of research, with data acquired from the databases mentioned above assisting in the development of theoretical foundations for PdM, bearing failure and ML methods. Figure 6.3 depicts the number of documents published annually in the last 5 years between 2017 and 2021.

As can be observed in Fig. 6.4, the PdM-based ML techniques have captivated the interest of researchers recently (last 5 years). Because there were only few papers published before 2017, the focus has been on the number of published articles between 2017 and 2021. As a result, it may be concluded that applying machine learning techniques to the subject of PdM is a novel strategy that is gaining popularity in the scientific community. This could be owing to a rise in the amount of data [12]. (PdM) Predictive Maintenance is an expression that refers to a group of techniques that are applied to monitor and measure the condition of equipment that is in use (e.g., an engine and automobile). So long as analyses is perfect with adequate frequency, PdM technologies permit the user to discover most potential failures well in advance. PdM is a good case of the type of industrial sensor networking applications that have a demonstrable impact in real-world deployments.

6.3 Technical Background

In the wide preponderance of machines, bearings are among the greatest critical components. A new strategy is introduced, atrous convolution (ACDIN) to aimed

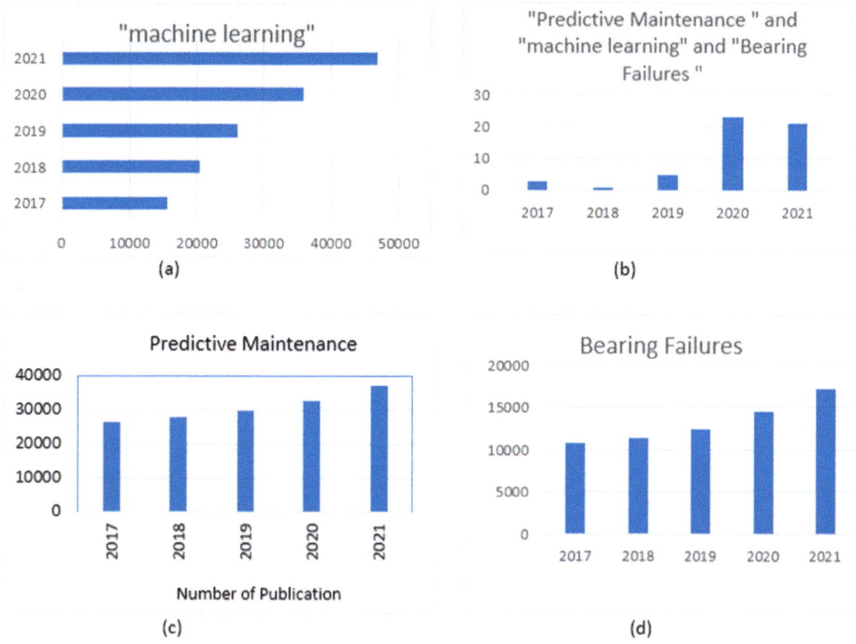

Fig. 6.3 Statistics from ScienceDirect and IEEE and Scopus database (Date: 2017 to 2021), Documents are published with the keywords: (**a**) "Machine Learning". (**b**) "Machine Learning"," Predictive Maintenance", "Bearing failures". (**c**) "Predictive Maintenance ". (**d**) "Bearing failures"

for cope among the problem and improves the accuracy from 75% to 96% on detecting the present bearing errors once skilled with only the data created from false bearing scratches then ACDIN takes signals as new inputs to analyze the mechanism performance among artificial broken bearings and normal spoiled bearings [13]. It is necessary and important to know the technical background to classify the behavior and approaches of machine learning and the processes themselves, for example, the technical background for fault bearings. Therefore, it is necessary to clarify in this portion the dissimilar and common types of defects that is possible to look in bearings [13]. Bearings generally consists of outer and inner ring, cage, and rollers [14].

As indicated in Fig. 6.4, the balls are positioned between the outer and inner rings. The cage is where the balls are kept in their respective locations with relation to one another. A bearing defect can be caused by a fault in any of the four bearing elements; however, the inner and outer rings are estimated that 90% of all faults [15]. For different bearing components rotating inner ring and outer ring and ball shown in Fig. 6.5.

The following is how the fundamental frequencies are determined from the bearing geometry:

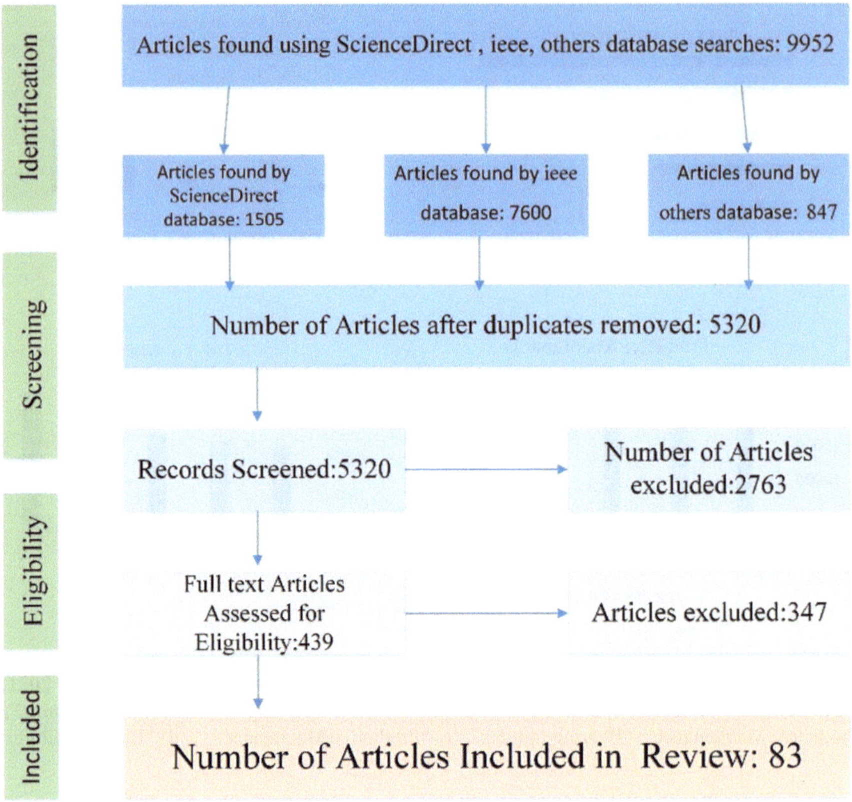

Fig. 6.4 Survey flowchart

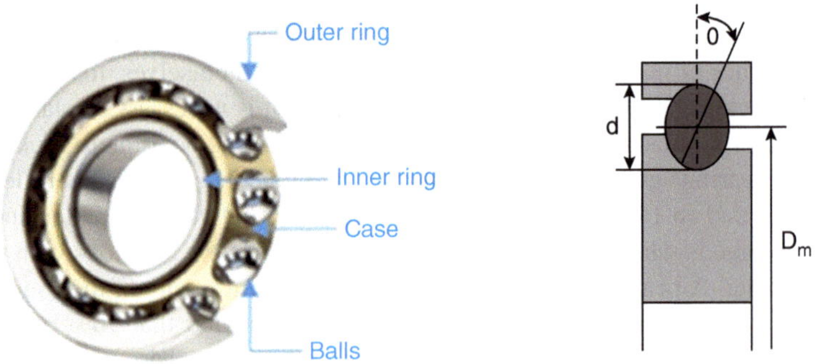

Fig. 6.5 Bearing in side and front view

$$f_{c/o} = f_r/2 \left[1 - \frac{d}{D_m} \operatorname{Cos} \alpha\right] \tag{6.1}$$

$$f_{c/i} = f_r/2 \left[1 + \frac{d}{D_m} \operatorname{Cos} \alpha\right] \tag{6.2}$$

$$f_{b/o} = Z\ f_{c/o} \tag{6.3}$$

$$f_{b/i} = Z\ f_{c/i} \tag{6.4}$$

$$f_b = [fr\ Dm]/2d \left[1 - \left(\frac{d}{D_m} \operatorname{Cos} \alpha\right)^2\right] \tag{6.5}$$

$f_{c/o}$ = frequency of fundamental train (cage) relative to outer ring, f_r = inner ring rotational frequency, d = diameter of roller elements, D_m = pitch circle diameter, α = contact angle, $f_{c/i}$ = frequency of fundamental train relative to inner ring, $f_{b/i}$ = ball pass frequency of inner raceway (BPFI), $f_{b/o}$ = ball pass frequency of outer raceway (BPFO) f_b = frequency of rolling element rotational, and Z = number of rolling elements. From Eqs. (6.1), (6.2), (6.3), (6.4), and (6.5) the bearing fault may be calculated, whereas the frequency created by the balls rolling across the surface of the rings is caused by reiterated impacts every time the rolling parts contacts a flaw.

Whenever there is abnormality on a component, where a frequency component at $f_{c/i}$, $f_{b/o}$, $f_{b/i}$, or f_r appears in case the similar fault happens. Nonetheless, once the errors are in the soon stages, the frequency parts are very tiny which causes complexities for a discovery of the trouble frequencies between a large number of frequencies current and, The actual fault frequencies may differ from those computed. This is because those frequencies are calculated depending on the hypothesis of no skid, which is incorrect [16, 17]. Even so, the most important defy in the implementation of the analyses of these frequencies and how to select them to obtain real results. Nevertheless, the analyses of the historic records to well bearing for contrast purposes can provide suitable outcomes [18]. Detecting bearing problems has been the focus of many articles [19, 20]. There are causes of bearing failure types that can occur in many different forms. The excessive load usually causes premature spalled area in the ball path. Also, overheating causes discoloration of rings, balls, and cages. The temperature rise can also destroy lubricant decreases or degrade the bearing capacity causing early failure. Overwork failure usually pointed to as spalling is the split of the running surfaces and following removal of small discrete particles of body. Pollution is one of the primary reasons of bearing failure, with denting of the bearing raceways and balls resulting in huge vibration and token of damage and clear. Overheating, excessive wear, and consequent bearing failure will occur if the lubricant fails [21–23]. Bearing failure is generally caused by a lack of maintenance and overloading. However, many other things can cause premature bearing failure as well. Some are rising temperatures, wetness, pollution, improper misalignment, lubrication, electric current flow through the bearing, and excessive thrust or overloading [24]. Shi et al. (2021) introduced a unique bearings safe operating technique based on (VMD) variational mode decomposition (VMD) methods and AlexNet neural network for diagnostic and fault detection in a complicated and high-frequency noise environment [25]. When using a low-cost MEMS accelerometer to diagnose bearing faults on complicated rotating equipment, high-frequency noise might be a concern. Ompusunggu et al. (2019) suggested a unique automated diagnostics approach for high-end accelerometer performance that is similar to low-cost accelerometer performance [26]. The interplay in the vibration signals has an impact on the bearing failure characteristics. When a rolling portion passes ended a damaged surface, an unwary force is produced, which causes resonances in the bearing and machine [27]. The location of faults, rotation velocity, and bearing dimensions may all be used to compute the duration of the impulses or characteristic frequency.

6.4 Predictive Maintenance and Machine-Learning Techniques

Over time, many machine learning-based predictive maintenance strategies have proven to be most successful. To do predictive maintenance, the sensors must be connected to the system, and the sensor will track and collect data about the activities of the system. Sensor data for predictive maintenance helps an ML-based application to accurately predict failures with real time. However, the ML algorithms are categorized into three different classes (Fig. 6.6) and are as follows: (1) supervised [28], (2) unsupervised [29], and (3) reinforcement learning [30].

6.4.1 Supervised Learning

6.4.1.1 Classification

A "support vector machine" (SVM) is one of the supervised machine learning technique that can be used when the original method of a real-world system is unidentified, or when obtaining a mathematical relationship is too costly due to the increased influence of a lot of interdependent elements. In a classification job, the samples are often considered to have two modules: negative and positive. An SVM training process creates a model that gives fresh patterns to one of the two groups, building a binary linear classifier that is non-probabilistic. The key distinction between different approaches when utilizing SVMs to classify bearings is the type and number of features employed. Table 6.1 clearly demonstrates this. Medina et al. (2020) illustrate dual algorithms for feature extraction from the Poincaré plot that is constructed with the vibration signals for fault classification measured in gearboxes and bearings by using a multiclass Support Vector Machine. At the

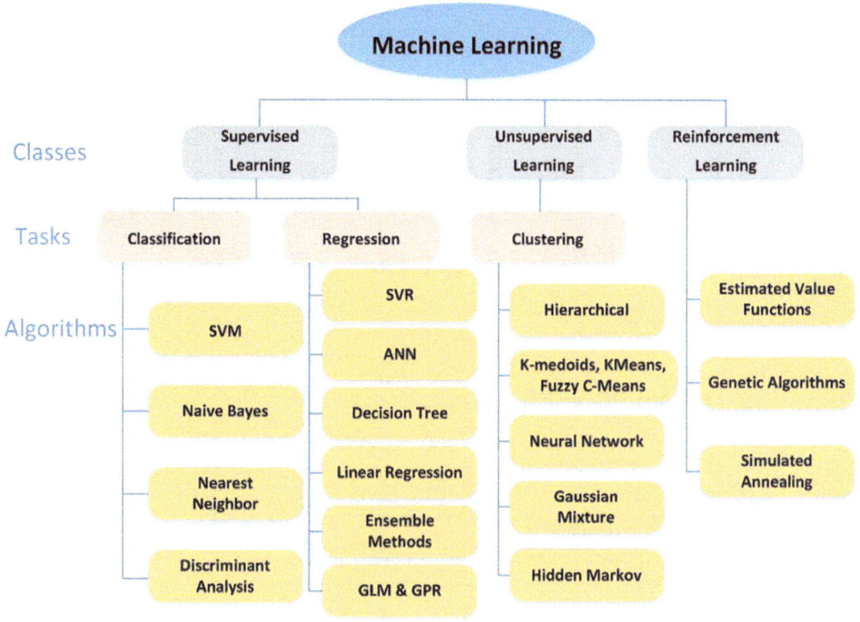

Fig. 6.6 Classification of machine learning techniques

Salesian Polytechnic University in Cuenca, Ecuador, a rolling bearing test was constructed for vibration signal attainment, allowing for the conformation of various bearing health conditions as well as the measurement of vibration signals [31]. For fault categorization of different bearing states, an intelligent IRT-based system was presented [32]. The obtained thermal pictures of various bearing situations were first preprocessed using 2D-DWT, followed by PCA selection of the most appropriate features, which then aids in classification and performance estimate using LDA, SVM, KNN, with SVM outperforming both KNN and LDA [32] (Glowacz et al. 2021). The suggested work offers a fault diagnosis approach based on thermal image analysis. Binarized common areas of image differences (BCAoID) is a technique for extracting features from thermal images. Three electric impact drills (EIDs) were analyzed using thermal images: a healthy EID, an EID with a malfunctioning fan (10 cracked fan blades), and an EID with a damaged gear train. The BCAoID was used to calculate thermal image features [33]. Gangsar et al. (2017) proposed an examination of current and vibration monitoring for effective defect estimate in induction motors (IM) by multiclass support vector machine (MSVM) methods. Mechanical and electrical defects have been identified for maintenance purposes, both collectively and individually. Fault calculations have been made using only the vibration signal, only the current signal, and both the vibration and the current signal at the same time [34].

Kang et al. (2017) proposed SVM classifications for bearing faults and fault sites. Using the relative recompense distance of multiple-domain characteristics and nearby linear embedding, a unique state assessment approach is developed in

Table 6.1 Works related to bearing classification using supervised learning

References	Dataset	Methods of diagnosis	Rate (%)	Faults type	Main points
R. Medina et al. (2020) [31]	Salesian Polytechnic University	SVM	100	Seven different types of roller bearing faults.	Developed a new feature extraction technique for vibration signals based on Poincaré plots, which allows for high-classification accuracy using multiclass SVM.
Mehta et al. (2021) [32]	162 images	KNN, LDA, SVM	88,94,100	Three type of fault: healthy, inner race, and outer race	Identifies a variety of faults. The IRT methodology performs well when compared to other methodologies for fault diagnosis.
Glowacz et al. (2021) [33]	90 thermal images	NN, BNN	10,0 97.91	10 faulty fan, damaged gear-train	The study proposes employing BCAoID of three electric impact drills to analyze thermal pictures for fault identification (EID).
Gangsar et al. (2017) [34]	Indian Institute of Technology, Guwahati	SVM, MSVM	93	Four faults of mechanical parts, five faults of electrical conditions parts, and one healthy	Diagnoses the electrical and mechanical faults. SVM for electrical faults, MSVM better for mechanical faults.
Kang et al. (2017) [35]	CWRU	SVM	99	Healthy, inner, rolling element, outer	In this case, feature reduction using the LLE method yields the top results.
Toma et al. (2020) [36]	University of Paderborn	KNN, DT, RF	97	Different load situations, three types of faults	ML 20 statistical characteristics were studied using a combination of genetic algorithms. Computational complexity is reduced. A high level of precision was attained.
Ali et al. (2019) [37]	Memorial University	KNN, ensemble, SVM	100	Two fault types (Motor1, Motor2), six loadings	Induction motor problems may be accurately detected using the suggested fault diagnosis approach, which can identify single or multiple electrical and mechanical flaws.
Moshrefzadeh (2021) [38]	PoliTo, IMS, FEMTO data sets	S-KNN, SVM	98	Bearing conditions	Compare the performance of two classifiers, S-KNN and SVM. Under continual and changing operational conditions, online monitoring and intelligent diagnostics of rolling bearings is possible.

place of the traditional classification method for effectively assessing various fault locating and levels of performance degradation of a rolling bearing with a united estimate index [35]. In the context of PdM, KNN has been frequently employed as one of the simplest ways for classification. For example, Toma et al. (2020) provides a hybrid motor-current data-driven technique for bearing defect diagnosis that incorporates genetic algorithms (GA), statistical features, and machine learning models. To evaluate the bearing flaws, researchers employed three classification algorithms: random forest, decision tree, and KNN. More than 97% accuracy was attained [36]. Ali et al. (2019) proposed three classification algorithms: K-nearest neighbors (KNN), support vector machine (SVM), and ensemble-based fault diagnosis method for two identical induction motors using trial data. Two signal-processing techniques, matching pursuit (MP) and discrete wavelet transform (DWT), are chosen for feature extraction. Induction motor problems may be accurately detected using the suggested fault diagnosis approach, which can identify single or multiple electrical and mechanical flaws [37]. A combination of SVM and KNN is also used in some solutions. Moshrefzadeh (2021) proposed a new approach for rolling element bearing failure detection and prognosis, as well as online condition monitoring, two approaches for data classification namely subspace k-nearest neighbors (S-KNN), support vector machine (SVM), are used that can distinguish amid different stages of machinery's health, regardless of weight or speed for data classification and diagnosis of the bearings [38]. Naive Bayes is a machine learning classifier that is basic, yet effective and widely used. It calculates the likelihood of each feature happening in each class and revenues the class with the maximum probability [39]. The Bayes rule is as follows:

$$P_{A/B} = \frac{P\left(\frac{B}{A}\right) P(A))}{P(B)} \tag{6.6}$$

Class and characteristics are represented by *A* and *B*, respectively.

P(*A*/*B*) is the possibility of belonging to class *A* with all of the characteristics of class *B*. *P*(*B*) is the probability of all characteristics, and it is used to normalize the data.

6.4.1.2 Regression

ANN is a type of machine learning algorithm that has been proposed in a variety of industrial applications, including soft sensing [40] and predictive control [41]. Bangalore et al. (2018) showed and suggested an artificial neural network (ANN)-based condition monitoring by a new solution employing data from supervisory control and data collecting systems of failures in the gearbox originating from the gearbox bearings to reduce the overall cost of Wavelet Transform maintenance management [42]. (Kolokas et al. 2018) Using the large-scale data processing engine Spark, propose LSTM networks are a form of recurrent ANN to predict the current engine state. Three operational settings and 21 sensors for measurements of the temperature, engine pressure, and fuel, and coolant bleed are included in this case [43]. Among the nonparametric supervised machine learning procedures that may be used for both regression and classification, the decision tree technique is one of the most straightforward techniques for optimal maintenance decision-making [44]. Liu et al. (2020) offered a rolling bearing fault detection technique based on the numerous properties of local mean decomposition (LMD) and random forest. An accurate kind of rolling bearing failure is done when the random forest technique is used. The overall percentage of recognition was 94.4% [45]. Abbasi et al.'s (2018) goal was to create a user-friendly graphical user interface (GUI) program that uses the various linear regression predictive maintenance method to create a user-friendly predictive maintenance data analytic interface. Using the proposed GUI and the multiple linear regression approach, multiple datasets of booster compressor (BC) parameters are used to determine the accuracy of future prediction [46].

6.4.2 Unsupervised Learning

Unsupervised learning (UL) is also famous as information detection makes use of training data that is unlabeled, unclassified, and uncategorized. Unsupervised learning's main purpose is to find hidden and interesting patterns in unlabeled data [47]. Unsupervised learning methods, unlike supervised learning, ca not be immediately applied to a regression or classification problem because the output values are unknown. Kramti et al. (2021) used a novel approach of vibration analysis in an innovative data driven strategy to immediate prognosis of HSSB. The suggested method is based on using spectral kurtosis (SK) to compute classical statistical measures obtained from the frequency domain and time domain. The most appropriate features were then chosen based on metrics (prognosablity, trendability, and monotonicity) to ensure a greater popularization of the prepared Elman neural network. The introduced approach was validated using the center for intelligent maintenance systems' benchmark for training and real-world data from the green power monitoring systems (GPMS) for testing [48]. Yang S et al. (2020) proposed a new rolling bearing defect diagnostics approach depending on the attention mechanism and a two-dimensional convolutional neural network. The results reveal that the upgraded Att-CNN2D network with the attention mechanism has a greater model generalization ability and accuracy than the classic CNN [49]. Jiang et al. (2018) suggested a new smart fault diagnosis strategy for automatically identifying different health problems of Wavelet Transform (WT) gearboxes and detecting efficient failures immediately from vibration signals while classifying the type of faults in a single framework, then providing a learning-based fault diagnosis system for WT gearboxes without diagnostic expertise and extra signal processing [50] (Table 6.2).

Shao et al. (2021) suggested a convolutional neural network (CNN) for rotor-bearing system fault diagnosis under varying working conditions by collecting and characterizing the health condition of the rotor-bearing system using infrared

Table 6.2 Works related to bearing classification using unsupervised learning

References	Dataset	Methods of diagnosis	Rate (%)	Faults type	Main points
Francesca et al. (2019)	Motor data (1400 samples)	DNN	N/A	Four faults, four loads	Detect faults in the motor bearings.
Kramti et al. (2021) [48]	Three bearings of the IMS offline and GPMS for online	ENN, RUL		Three fault types, roller, inner, outer	This study tests their method on two separate bearing datasets using two different models.
Jiang et al. (2018) [50]	Wavelet Transform (20,800 samples)	MSCNN	97	Eight health conditions	Presents a novel MSCNN architecture for intelligent WT fault diagnostics under various operating situations.
Yang et al. (2020) [49]	CWRU (7000 samples of 1HP, 2HP, and 3HP loads)	CNN, Att-CNN2D		Three fault types, roller, inner, outer under three loads of 1, 2, and 3HP	Offer a novel approach for detecting rolling bearing faults using two-dimensional convolution in real working conditions
Pandarakone et al. (2019) [51]	Motor data (three phase, one load)	RF, CNN, DT, KNN, SVM, DL	81–89	Three fault types, healthy, failure with hole, failure with scratch	Detection of bearing faults and continuous monitoring using machine learning algorithms.
Shao et al. (2021) [52]	Thermal images (150 points, 104 target)	CNN	81–95	Eight health states for bearing	Under various operating circumstances, a modified CNN was employed to diagnose thermal images of bearing faults.
Liang et al. (2020) [53]	CWRU 10,000 data points	WT-GAN-CNN	98.38	Four load	The findings demonstrate that the suggested approach works in rotating equipment for single, simultaneous, and early weak failure modes.
Sobie et al. (2018) [54]	CWRU, MFPT and SQ datasets	CNN, NNDTW	99	Healthy and outer race fault	Proposed machine learning algorithms are being used in two new ways to identify faults. CNN and NNDTW applied to ASAs.
Zhiyi et al. (2020) [55]	Hunan University	CNN	98	Eight health states of rotor-bearing system (25 samples)	On datasets of various bearings, several activation functions compared between source and target were minimized.

thermal images, then applied a modified CNN to overcome the training problems in classical CNN to analyze thermal images of the rotor-bearing system collected under varying working conditions [52]. Liang et al. (2020) proposed WT-GAN-CNN, a new intelligent failure detection solution for rotating equipment which includes three parts: Wavelet Transform (WT) that converts 1D vibration data into 2D time-frequency (TF) pictures; generative adversarial nets (GANs) that generate additional picture examples for training; and. CNN that utilize the real training time-frequency pictures and create false training time-frequency images to detect faults in rotating machinery [53]. Korba et al. (2018) applied a novel and automated diagnosis of healthy and faulty bearing defects based on vibratory analysis using SVM, ANN, and neuro-fuzzy network (NFN) classifiers [56]. Pandarakone et al. (2019) used machine learning (ML) algorithms and artificial intelligence (AI) for bearing faults diagnosis of induction motors (IM). For identifying a small bearing defect, multiple diagnosis algorithms from ML and AI are being considered (scratch and hole) [51]. Alberto et al. (2018) provided an in-depth examination of artificial neural s in wind energy systems, identifying the most commonly used methods for various applications and proving that in many circumstances, artificial neural networks can be a viable alternative to traditional approaches [57]. Guo et al. (2017), based on the performance of bearing health indicators, developed a recurrent neural network-based health indicator (RNN-HI) for forecasting bearing remaining useful life (RUL), as proven by two bearing datasets collected from tests and an industrial field [58]. Francesca et al. (2019) employed the stator current from an inductor motor as input for their proposed bearing problem detection and monitoring system. They extracted numerous frequency and time features from the raw current measurements. A DNN was trained using the extracted features. Their proposed approach appears to provide a promising and effective classification, based on the presented data [59]. Sobie et al. (2018), in his paper, discussed how to train algorithms using simulation data, as well as how to use CNN and NNDTW to detect the presence of a bearing race defect using bearing acceleration signals [54]. Gao et al. (2018) suggested a novel fault-type intelligent identification approach depended on a time frequency diagram and Convolution Neural Networks that transform signal recognition into picture recognition for rolling bearing fault diagnosis [60]. Zhiyi et al. (2020) planned an enhanced convolutional neural network (ECNN) constructed since a convolutional autoencoder (CAE) for smart rotor bearing system failure diagnostics utilizing tiny tagged infrared thermal photos. The benefits of the suggested method over current mainstream methodologies are demonstrated by the analysis and comparison of results [55]. Dhalmahapatra et al. (2019) The k-means pattern is a common clustering technique that determines a collection of clusters using an unsupervised strategy [61]. The Euclidean distance was used as a similarity criterion in the development of the k-means clustering algorithm. K-means is used in Eke et al. (2017) to automatically extract groups (clusters) in liquified gases data in a converter's insulating oil. It was the goal to figure out what characteristics of each cluster cause a defect or an alert, so that maintenance actions could be taken [62]. The following are some examples of studies that employ k-means for PdM ([62–64]). Rustam et al. (2017) present the fuzzy kernel k-medoids (FKkMd) algorithm as a strong technique to tackle the anomaly detection-intrusion detection problem by combining fuzzy clustering, medoids, and kernel ideas [65]. The objective function value for the storage optimization problem is overestimated by K-shape, which changes as the number of clusters grows [66].

6.4.3 *Reinforcement Learning*

Reinforcement learning (RL), as one of the disciplines of machine learning, has several applications in many different domains, including intelligent analysis and control and prediction [67]. Goal-directed learning and decision-making are the focus of RL. Reinforcement learning differs from unsupervised learning in terms of goals; in unsupervised learning, the goal is to identify an appropriate action model that maximizes the agent's total cumulative reward.

Design artificial intelligence for robotics and computer games, text-summarizing engines, industrial automation, dialogue agents (speech, text), and other practical applications of reinforcement learning are only a few examples [47]. The main idea and components of the reinforcement learning model are depicted in Fig. 6.7.

In addition, Han et al. (2021) proposed a new method for building an RL environment based on a deep artificial neural network (DANN). It has been proposed, and an effective maintenance decision function has been created [68]. Proposed RL for fault detection and diagnosis motor [69]. In fault classification, reinforcement learning for convolutional neural network (RL-CNN) may automatically modify the learning rate and identify the optimum learning rate values based on its training history [70].

6.5 Challenges

Building accurate algorithms in order to properly and reliably process and prepare data for use as input into models that spend a long time getting the data from each component of the machine is also tricky. You require a resource to record the data

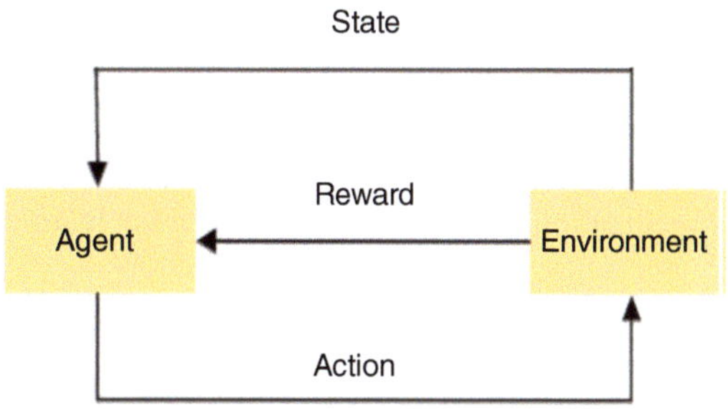

Fig. 6.7 Reinforcement learning

from the equipment and store it in a cloud or particular place. Inadequate access to adequate data leads to inaccurate predictions that may tolerate errors and excessive expectations and fixtures. Difficulty in building machine learning algorithms and data acquisition to apply predictive analytics without signal processing. The major difficulties in putting these strategies into practice in industries are that IoT-based equipment monitoring devices are now only accessible to large businesses and manufacturing facilities. We must look into how these ML-based forecasting methods might be used by small businesses as well in order for them to gain from artificial intelligence.

6.6 Applications of ML Algorithms in PdM

Machine learning is paving the way for smarter and faster data-driven decisions in predictive maintenance (PdM), and its use in industrial plants and infrastructure asset operations is already advancing at a rapid rate. Machine learning is becoming more popular due to the rise of big data and the expansion of the Industrial Internet of Things (IIoT). There are numerous applications for machine learning and PdM in the industrial sector (Table 6.3).

6.7 Discussion and Conclusions

Based on the literature study in last 5 years through this review in many cases for ML techniques are increasingly being utilized to draw the visualizations of PdM, combining PdM and ML yields positive results as well as cost maintenance savings in bearing and generally all rotary machines. However, it is clear that combining PdM methodologies with cutting-edge sensor technology reduces the need for wasteful equipment replacement, saves money, and enhances the safety, availability, and efficiency of the process. The study provides a complete overview of machine learning techniques used in PdM of bearing failures in rotary machines in industrial components that relies on numerous scenarios and forecast what will go wrong and when, and then create alerts based on vast volumes of historical or test data mixed with specific machine-learning algorithms. Several machine learning methods have been studied and presented in recent applications on Predictive maintenance throughout the last 5 years (from 2017 to 2021). Predictive maintenance has been proven to give significant commercial benefits, and Machine learning is a cutting-edge method for doing predictive maintenance. Only 11% of studied organizations have "realized" machine learning-based "predictive maintenance,"

Table 6.3 Works related to applications of ML algorithms in PdM

References	Dataset	Methods of diagnosis	Faults type	Main points
Du et al. (2018) [71]	SVM	Rolling bearing	Vibration signals	Based on the TFIs, a method for fault monitoring in bearings was developed and to analyze nonstationary signals.
Ahmad et al. (2019) [72]	Dynamic regression	Bearing	Vibration measurements data	Using regression model, predicts the state of bearing. As a bearing state indicator, a dimensionless quantity was used. TSP was calculated using ABT. The PRONOSTIA dataset was used to validate and test the model to achieve a fair result in comparison to previous procedures
Jeong et al. (2018) [73]	GA-ANN, SVM	Gear box-rotating machine	Vibration signals and acoustic emission	Investigated a method for detecting prospective rotating machine faults early. An increase in fault can be detected using the feature analysis method.
Kolokas et al. (2018) [43]	ANN, RF, GNB DT	Anode production	Sensor data	PdM method for defect prediction in live time before the equipment failure. 5–10 minutes ahead of time, predicts failures in industrial equipment.
Manfre et al. (2020) [74]	DT, K-nearest NN, neighbors, BNB	Rotating shaft	Vibration, temperature, humidity, noise, pressure	With a precision loss of 5%, the Naive Bayes algorithm reaches its maximum peak. Only the NN algorithm operates differently from the others. The best results were obtained by increasing the training time by 20% with the isolation forest algorithm.
Cheng et al. (2020) [75]	ANN, SVM	Building facilities	Temperature, pressure sensor	The introduction of PdM to increase the feasibility of the FMM process was made easier, thanks to BIM and IoT. For maintenance, four modules are used: (1) prediction of condition; (2) planning of maintenance; (3) monitoring; and (4) evaluation of condition.
Koca et al. (2020) [76]	ANN	Packaging robots	Vibration, temperature, humidity	The MLP framework can withstand unanticipated outages. Can significantly reduce the expenses of unplanned production downtime. Failures are compared both theoretically and practically.

(continued)

Table 6.3 (continued)

References	Dataset	Methods of diagnosis	Faults type	Main points
Gohel et al. (2020) [77]	SVM, LR	Nuclear infrastructure	Temperature, power, current, speed	Maintenance of nuclear infrastructure can be predicted using a machine learning system. The consumption of electricity is kept track of and record the temperature of electrical panels
Janssens et al. (2018) [78]	DNN-CNN, DL	Rotating machinery	Accelerometer, thermocouple, camera measurements	The CNN algorithm was used to detect a variety of rotating equipment situations. Online condition monitoring in Wavelet Transform can be improved with this technology. It can be used to monitor bearings in manufacturing lines. CNN outperforms in machine fault detection and oil-level prediction. When compared to the FE technique, the FL technique performs 6.7% better.
Zhang et al. (2018) [79]	NN	Gas turbine	Gas measurements	For starting deterioration monitoring, an NN and physics-based model were developed. Both models are capable of monitoring the starter's health and displaying indicators of degeneration. When the starters are in good shape, NN offers more accurate outcomes. On starters with increasing degradations, physics-based delivers superior results.
Janssens et al. (2018) [80]	RF	Rotating machine	Vibration	Heat or vibration deficiencies can be corrected with a multisensor system for rotary machines. Increases the accuracy of fault detection by a significant amount.
Hoffmann et al. (2020) [81]	SVM, RNN, k-nearest neighbor	Switchgear	Voltage, temperature, current	The use of cutting-edge sensor technologies in conjunction with machine learning techniques. Predictive maintenance solutions for medium-voltage switchgear are provided.
Cheng et al. (2020) [82]	WD-DTL, CNN, DAN	Bearing	US-speed, US-location	Addressing industrial domain-shift problems by intelligent fault diagnosis-based WD-DTL. WD-DTL helps solve difficulties with unlabeled and inadequately labeled data.

according to a PWC report [83]. According to the literature study, the most common information type employed is real. Only a little research has used simulated data to build machine learning algorithms, for example, Bosch, SECOM, CMAPSS, Case Western Reserve University, and NASA datasets on engine degradation were used in the development of ML algorithm PdM models. Furthermore, some of the research's findings use standard machine learning algorithms with no parameter adjustment. Perhaps this is because PdM is a novel issue for manufacturing specialists, and it is only now being investigated. It is also worth noting that, in order to achieve positive outcomes from a PdM strategy in a plant, the PvM and R2F strategies should already be in place in the process of collecting data for PdM modeling.

Finally, the ultimate objective of this research, in a broader perspective, is to undertake effective predictive maintenance before faults damage the system as a whole. The assessed use cases described in this paper demonstrate that machine learning can successfully forecast failures or irregularities in a variety of applications. We have given several ideas for additional study in this paper:

- Predictive maintenance may be automated with the aid of an intelligent real-time data collecting system.
- When compared to a single ML model, employing several ML models can result in better predictions.
- By combining classification and defect detection methods, PdM may be utilized on equipment or systems that do not have a large data collection.
- All system failures cannot be predicted through one model, and not all machine learning models can be developed using the same technique.

Table 6.4 lists the key terminologies used in this paper.

Table 6.4 Nomenclature

References	Main points
PdM	Predictive maintenance
PvM	Preventive maintenance
R2F	Run-to-failure
SVM	Support vector machine
WT	Wavelet Transform
KNN	k-nearest neighbors
IEEE	Institute of Electronics and Electrical Engineers
ANN	Artificial neural network
CMS	Condition monitoring system
PAT	Predictive analytics toolkit
AI	Artificial intelligence
ML	Machine learning
CBM	Condition-based maintenance
IMS	Intelligent maintenance systems

(continued)

Table 6.4 (continued)

References	Main points
GPMS	Green power monitoring systems
ENN	Elman neural network
RUL	Remaining useful life
DNN	Deep neural network
CWRU	Case Western Reserve University
MFPT	Machine failure prevention technology
IRT	Infrared thermal images
DT	Decision tree
LR	Linear regression
TFI	Time-frequency image
RL	Reinforcement learning
WD-DTL	Wasserstein distance-based deep transfer learning
DAN	Deep adaptation network
NNDTW	Nearest-neighbor dynamic time warping
IMS	Intelligent maintenance systems
RxM	Prescriptive maintenance

References

1. J. Para, J. Del Ser, A. J. Nebro, U. Zurutuza, and F. Herrera, "Analyze, Sense, Preprocess, Predict, Implement, and Deploy (ASPPID): An incremental methodology based on data analytics for cost-efficiently monitoring the industry 4.0" Eng. Appl. Artif. Intell., vol. 82, no. September 2018, pp. 30–43, 2019.
2. R. S. Peres, A. Dionisio Rocha, P. Leitao, and J. Barata, "IDARTS – Towards intelligent data analysis and real-time supervision for industry 4.0," Comput. Ind., vol. 101, no. July, pp. 138–146, 2018.
3. T. P. Carvalho, F. A. A. M. N. Soares, R. Vita, R. da P. Francisco, J. P. Basto, and S. G. S. Alcalá, "A systematic literature review of machine learning methods applied to predictive maintenance," Comput. Ind. Eng., vol. 137, no. September, p. 106024, 2019.
4. R. C. Parpala and R. Iacob, "Application of IoT concept on predictive maintenance of industrial equipment," vol. 02008, pp. 1–8, 2017.
5. P. Ongsulee, V. Chotchaung, E. Bamrungsi, and T. Rodcheewit, "Big Data, Predictive Analytics and Machine Learning," Int. Conf. ICT Knowl. Eng., vol. 2018-Novem, pp. 37–42, 2019.
6. J. Dalzochio et al., "Machine learning and reasoning for predictive maintenance in Industry 4.0: Current status and challenges," Comput. Ind., vol. 123, p. 103298, 2020.
7. J. Sri, L. Senanayaka, H. Van Khang, K. G. Robbersmyr, and S. M. Ieee, "Online Fault Diagnosis System for Electric Powertrains using Advanced Signal Processing and Machine Learning," 2018 XIII Int. Conf. Electr. Mach., pp. 1932–1938, 2018.
8. A. I. Vlasov, V. V. Echeistov, A. I. Krivoshein, V. A. Shakhnov, S. S. Filin, and V. S. Migalin, "An information system of predictive maintenance analytical support of industrial equipment," J. Appl. Eng. Sci., vol. 16, no. 4, pp. 515–522, 2018.
9. M. Calabrese et al., "SOPHIA: An event-based IoT and machine learning architecture for predictive maintenance in industry 4.0," Inf., vol. 11, no. 4, pp. 1–17, 2020.
10. O. Alshorman et al., "A Review of Artificial Intelligence Methods for Condition Monitoring and Fault Diagnosis of Rolling Element Bearings for Induction Motor," Shock Vib., vol. 2020, no. Cm, 2020.

11. S. B. Jiang, P. K. Wong, R. Guan, Y. Liang, and J. Li, "An Efficient Fault Diagnostic Method for Three-Phase Induction Motors Based on Incremental Broad Learning and Non-Negative Matrix Factorization," IEEE Access, vol. 7, pp. 17780–17790, 2019.
12. Nacchia, Fruggiero, Lambiase, and Bruton, "A systematic mapping of the advancing use of machine learning techniques for predictive maintenance in the manufacturing sector," Appl. Sci., vol. 11, no. 6, pp. 1–34, 2021.
13. Y. Chen, G. Peng, C. Xie, W. Zhang, C. Li, and S. Liu, "ACDIN: Bridging the gap between artificial and real bearing damages for bearing fault diagnosis," Neurocomputing, vol. 294, pp. 61–71, 2018.
14. P. Štastniak, R. Kohár, and L. Smetanka, "Dynamic analysis of force interactions in rolling bearings components," AIP Conf. Proc., vol. 2198, no. December, pp. 1–7, 2019.
15. D. Wu et al., "An automatic bearing fault diagnosis method based on characteristics frequency ratio," Sensors (Switzerland), vol. 20, no. 5, pp. 1–12, 2020.
16. I. Attoui, N. Fergani, N. Boutasseta, B. Oudjani, and A. Deliou, "A new time–frequency method for identification and classification of ball bearing faults," J. Sound Vib., vol. 397, pp. 241–265, 2017.
17. T. Haj Mohamad and C. Nataraj, "Fault identification and severity analysis of rolling element bearings using phase space topology," JVC/Journal Vib. Control, vol. 27, no. 3–4, pp. 295–310, 2021.
18. J. J. Saucedo-Dorantes, M. Delgado-Prieto, J. A. Ortega-Redondo, R. A. Osornio-Rios, and R. D. J. Romero-Troncoso, "Multiple-Fault Detection Methodology Based on Vibration and Current Analysis Applied to Bearings in Induction Motors and Gearboxes on the Kinematic Chain," Shock Vib., vol. 2016, 2016.
19. M. Kuncan, "An Intelligent Approach for Bearing Fault Diagnosis: Combination of 1D-LBP and GRA," IEEE Access, vol. 8, pp. 137517–137529, 2020.
20. S. Zhang et al., "Model-Based Analysis and Quantification of Bearing Faults in Induction Machines," IEEE Trans. Ind. Appl., vol. 56, no. 3, pp. 2158–2170, 2020.
21. D. Yang, J. Miao, F. Zhang, J. Tao, G. Wang, and Y. Shen, "Bearing Fault Diagnosis Using a Support Vector Machine Optimized by an Improved Ant Lion Optimizer," Shock Vib., vol. 2019, 2019.
22. X. Qin, D. Xu, X. Dong, X. Cui, and S. Zhang, "The Fault Diagnosis of Rolling Bearing Based on Improved Deep Forest," Shock Vib., vol. 2021, 2021.
23. B. Zheng, H. Gao, X. Ma, and X. Zhang, "Multiteam Competitive Optimization Algorithm and Its Application in Bearing Fault Diagnosis," Math. Probl. Eng., vol. 2021, 2021.
24. J. A. Brumbach, M.E. and Clade, "Industrial Maintenance - Michael E. Brumbach, Jeffrey A. Clade - Google Books," Cengage Learning, 2013.[Online].Available: https://books.google.co.in/books?id=fTc9AAAAQBAJ&pg=PA166& lpg=PA166&dq=The+excessive+load+usually+causes+premture+spalled+area+in+the+ball+path&source=bl&ots=BpgZpa379N&sig=ACfU3U3FHzngdGDq_isw8GScvSpjMB3LA&hl=en&sa=X&ved=2ahUKEwjOyabbneXyAhXiwzgGHdJEAk. [Accessed: 04-Sep-2021].
25. X. Shi et al., "An Improved Bearing Fault Diagnosis Scheme Based on Hierarchical Fuzzy Entropy and Alexnet Network," IEEE Access, vol. 9, pp. 61710–61720, 2021.
26. A. P. Ompusunggu, T. Ooijevaar, B. Kilundu Y'Ebondo, and S. Devos, "Automated bearing fault diagnostics with cost-effective vibration sensor," Lect. Notes Mech. Eng., no. August, pp. 463–472, 2019.
27. S. X. and J. L. Aijun Hu*, Ling Xiang, "Frequency Loss and Recovery in Rolling Bearing Fault Detection.pdf." 2019.
28. A. I. Kadhim, "Survey on supervised machine learning techniques for automatic text classification," Artif. Intell. Rev., vol. 52, no. 1, pp. 273–292, 2019.
29. M. Usama et al., "Unsupervised Machine Learning for Networking: Techniques, Applications and Research Challenges," IEEE Access, vol. 7, pp. 65579–65615, 2019.
30. A. K. Mondal, "A Survey of Reinforcement Learning Techniques: Strategies, Recent Development, and Future Directions," no. September, 2020.

31. R. Medina et al., “Gear and bearing fault classification under different load and speed by using Poincaré plot features and SVM,” J. Intell. Manuf.
32. A. Mehta, D. Goyal, A. Choudhary, B. S. Pabla, and S. Belghith, “Machine Learning-Based Fault Diagnosis of Self-Aligning Bearings for Rotating Machinery Using Infrared Thermography,” Math. Probl. Eng., vol. 2021, 2021.
33. A. Glowacz, “Fault diagnosis of electric impact drills using thermal imaging,” Meas. J. Int. Meas. Confed., vol. 171, no. November 2020, p. 108815, 2021.
34. P. Gangsar and R. Tiwari, “Comparative investigation of vibration and current monitoring for prediction of mechanical and electrical faults in induction motor based on multiclass-support vector machine algorithms,” Mech. Syst. Signal Process., vol. 94, pp. 464–481, 2017.
35. S. Kang, D. Ma, Y. Wang, C. Lan, Q. Chen, and V. I. Mikulovich, “Method of assessing the state of a rolling bearing based on the relative compensation distance of multiple-domain features and locally linear embedding,” Mech. Syst. Signal Process., vol. 86, no. 52, pp. 40–57, 2017.
36. R. N. Toma, A. E. Prosvirin, and J. M. Kim, “Bearing fault diagnosis of induction motors using a genetic algorithm and machine learning classifiers,” Sensors (Switzerland), vol. 20, no. 7, 2020.
37. M. Z. Ali, M. N. S. K. Shabbir, X. Liang, Y. Zhang, and T. Hu, “Machine learning-based fault diagnosis for single- and multi-faults in induction motors using measured stator currents and vibration signals,” IEEE Trans. Ind. Appl., vol. 55, no. 3, pp. 2378–2391, 2019.
38. A. Moshrefzadeh, “Condition monitoring and intelligent diagnosis of rolling element bearings under constant/variable load and speed conditions,” Mech. Syst. Signal Process., vol. 149, p. 107153, 2021.
39. S. Chowdhury and M. P. Schoen, “Research Paper Classification using Supervised Machine Learning Techniques,” 2020 Intermt. Eng. Technol. Comput. IETC 2020, 2020.
40. V. Brunner, M. Siegl, D. Geier, and T. Becker, “Challenges in the Development of Soft Sensors for Bioprocesses: A Critical Review,” Front. Bioeng. Biotechnol., vol. 9, no. August, pp. 1–21, 2021.
41. S. Bayhan and H. Abu-Rub, Predictive Control of Power Electronic Converters, 4th ed. Elsevier Inc., 2018.
42. P. Bangalore and L. B. Tjernberg, “An artificial neural network approach for early fault detection of gearbox bearings,” IEEE Trans. Smart Grid, vol. 6, no. 2, pp. 980–987, 2015.
43. N. Kolokas, T. Vafeiadis, D. Ioannidis, and D. Tzovaras, “Forecasting faults of industrial equipment using machine learning classifiers,” 2018 IEEE Int. Conf. Innov. Intell. Syst. Appl. INISTA 2018, pp. 1–6, 2018.
44. S. Kaparthi and D. Bumblauskas, “Designing predictive maintenance systems using decision tree-based machine learning techniques,” Int. J. Qual. Reliab. Manag., vol. 37, no. 4, pp. 659–686, 2020.
45. N. Liu, B. Liu, and C. Xi, “Fault diagnosis method of rolling bearing based on the multiple features of LMD and random forest,” IOP Conf. Ser. Mater. Sci. Eng., vol. 892, no. 1, 2020.
46. T. Abbasi, K. H. Lim, N. S. Rosli, I. Ismail, and R. Ibrahim, “Development of Predictive Maintenance Interface Using Multiple Linear Regression,” Int. Conf. Intell. Adv. Syst. ICIAS 2018, pp. 1–5, 2018.
47. K. El Bouchefry and R. S. de Souza, Learning in Big Data: Introduction to Machine Learning. Elsevier Inc., 2020.
48. S. E. Kramti, J. Ben Ali, L. Saidi, M. Sayadi, M. Bouchouicha, and E. Bechhoefer, “A neural network approach for improved bearing prognostics of wind turbine generators,” EPJ Appl. Phys., vol. 93, no. 2, 2021.
49. S. Yang, X. Sun, and D. Chen, “Bearing fault diagnosis of two-dimensional improved Att-CNN2D neural network based on Attention mechanism,” Proc. 2020 IEEE Int. Conf. Artif. Intell. Inf. Syst. ICAIIS 2020, pp. 81–85, 2020.
50. G. Jiang, H. He, J. Yan, and P. Xie, “Multiscale Convolutional Neural Networks for Fault Diagnosis of Wind Turbine Gearbox,” IEEE Trans. Ind. Electron., vol. PP, no. c, p. 1, 2018.
51. S. E. Pandarakone, Y. Mizuno, and H. Nakamura, “Algorithm and Artificial Intelligence Neural Network,” Energies, vol. 12, p. 2105, 2019.

52. H. Shao, M. Xia, G. Han, Y. Zhang, and J. Wan, "Intelligent Fault Diagnosis of Rotor-Bearing System under Varying Working Conditions with Modified Transfer Convolutional Neural Network and Thermal Images," IEEE Trans. Ind. Informatics, vol. 17, no. 5, pp. 3488–3496, 2021.
53. P. Liang, C. Deng, J. Wu, and Z. Yang, "Intelligent fault diagnosis of rotating machinery via wavelet transform, generative adversarial nets and convolutional neural network," Meas. J. Int. Meas. Confed., vol. 159, p. 107768, 2020.
54. C. Sobie, C. Freitas, and M. Nicolai, "Simulation-driven machine learning: Bearing fault classification," Mech. Syst. Signal Process., vol. 99, pp. 403–419, 2018.
55. H. Zhiyi, S. Haidong, Z. Xiang, Y. Yu, and C. Junsheng, "An intelligent fault diagnosis method for rotor-bearing system using small labeled infrared thermal images and enhanced CNN transferred from CAE," Adv. Eng. Informatics, vol. 46, no. April, p. 101150, 2020.
56. K. A. Korba and F. Arbaoui, "SVM Multi-Classification of Induction Machine ' s bearings defects using Vibratory Analysis based on Empirical Mode Decomposition," vol. 13, no. 9, pp. 6579–6586, 2018.
57. A. P. Marugán, F. P. G. Márquez, J. M. P. Perez, and D. Ruiz-Hernández, "A survey of artificial neural network in wind energy systems," Appl. Energy, vol. 228, no. April, pp. 1822–1836, 2018.
58. L. Guo, N. Li, F. Jia, Y. Lei, and J. Lin, "A recurrent neural network based health indicator for remaining useful life prediction of bearings," Neurocomputing, vol. 240, pp. 98–109, 2017.
59. F. Cipollini, L. Oneto, A. Coraddu, and S. Savio, "Unsupervised Deep Learning for Induction Motor Bearings Monitoring," Data-Enabled Discov. Appl., vol. 3, no. 1, 2019.
60. D. Gao et al., "A Fault Diagnosis Method of Rolling Bearing Based on Complex Morlet CWT and CNN," 2018 Progn. Syst. Heal. Manag. Conf., pp. 1101–1105, 2018.
61. K. Dhalmahapatra, R. Shingade, H. Mahajan, A. Verma, and J. Maiti, "Decision support system for safety improvement: An approach using multiple correspondence analysis, t-SNE algorithm and K-means clustering," Comput. Ind. Eng., vol. 128, no. June 2018, pp. 277–289, 2019.
62. Eke, S., Aka-Ngnui, T., Clerc, G., & Fofana, I. "Characterization of the operating periods of a power transformer by clustering the dissolved gas data." IEEE 11th International Symposium on Diagnostics for Electrical Machines, Power Electronics and Drives (SDEMPED) (2017).
63. N. Amruthnath and T. Gupta, "A research study on unsupervised machine learning algorithms for early fault detection in predictive maintenance," 2018 5th Int. Conf. Ind. Eng. Appl. ICIEA 2018, no. August 1993, pp. 355–361, 2018.
64. V. Mathew, T. Toby, V. Singh, B. M. Rao, and M. G. Kumar, "Prediction of Remaining Useful Lifetime (RUL) of turbofan engine using machine learning," IEEE Int. Conf. Circuits Syst. ICCS 2017, vol. 2018-Janua, no. Iccs, pp. 306–311, 2018.
65. Z. Rustam and A. S. Talita, "Fuzzy Kernel k-Medoids algorithm for anomaly detection problems," AIP Conf. Proc., vol. 1862, no. July 2017, 2017.
66. H. Teichgraeber and A. R. Brandt, Systematic Comparison of Aggregation Methods for Input Data Time Series Aggregation of Energy Systems Optimization Problems, vol. 44. Elsevier Masson SAS, 2018.
67. W. Dai, Z. Mo, C. Luo, J. Jiang, and Q. Miao, "Bearing Fault Diagnosis Based on Reinforcement Learning and Kurtosis," 2019 Progn. Syst. Heal. Manag. Conf. PHM-Qingdao 2019, no. 1, pp. 1–5, 2019.
68. C. Han, T. Ma, and S. Chen, "Asphalt pavement maintenance plans intelligent decision model based on reinforcement learning algorithm," Constr. Build. Mater., vol. 299, no. February, p. 124278, 2021.
69. W. Zhang and J. Zhu, "A reinforcement learning system for fault detection and diagnosis in mechatronic systems," C. - Comput. Model. Eng. Sci., vol. 124, no. 3, pp. 1119–1130, 2020.
70. L. Wen, X. Li, and L. Gao, "A New Reinforcement Learning Based Learning Rate Scheduler for Convolutional Neural Network in Fault Classification," IEEE Trans. Ind. Electron., vol. 68, no. 12, pp. 12890–12900, 2021.

71. Y. Du, Y. Chen, G. Meng, J. Ding, and Y. Xiao, "Fault severity monitoring of rolling bearings based on texture feature extraction of sparse time-frequency images," Appl. Sci., vol. 8, no. 9, pp. 1–23, 2018.
72. W. Ahmad, S. A. Khan, M. M. M. Islam, and J. M. Kim, "A reliable technique for remaining useful life estimation of rolling element bearings using dynamic regression models," Reliab. Eng. Syst. Saf., vol. 184, pp. 67–76, 2019.
73. Ha, JM., Kim, HJ., Shin, YS. and Choi, BK (2018). "Degradation Trend Estimation and Prognostics for Low Speed Gear Lifetime."*International Journal for Precision Engineering and Manufacturing,* Vol. 19: 1099–1105.
74. M. Manfre, "Creation of a Machine Learning model for the Predictive Maintenance of an engine equipped with a rotating shaft," no. March, 2020.
75. J. C. P. Cheng, W. Chen, K. Chen, and Q. Wang, "Data-driven predictive maintenance planning framework for MEP components based on BIM and IoT using machine learning algorithms," Autom. Constr., vol. 112, no. August 2018, p. 103087, 2020.
76. O. Koca, O. T. Kaymakci, and M. Mercimek, "Advanced Predictive Maintenance with Machine Learning Failure Estimation in Industrial Packaging Robots," 2020 15th Int. Conf. Dev. Appl. Syst. DAS 2020 - Proc., pp. 1–6, 2020.
77. H. A. Gohel, H. Upadhyay, L. Lagos, K. Cooper, and A. Sanzetenea, "Predictive maintenance architecture development for nuclear infrastructure using machine learning," Nucl. Eng. Technol., vol. 52, no. 7, pp. 1436–1442, 2020.
78. O. Janssens, R. Van De Walle, M. Loccufier, and S. Van Hoecke, "Deep Learning for Infrared Thermal Image Based Machine Health Monitoring," IEEE/ASME Trans. Mechatronics, vol. 23, no. 1, pp. 151–159, 2018.
79. Y. Zhang, J. Liu, H. Hanachi, X. Yu, and Y. Yang, "Physics-based Model and Neural Network Model for Monitoring Starter Degradation of APU," 2018 IEEE Int. Conf. Progn. Heal. Manag. ICPHM 2018, pp. 1–7, 2018.
80. O. Janssens, M. Loccufier, and S. Van Hoecke, "Thermal Imaging and Vibration-Based Multisensor Fault Detection for Rotating Machinery," IEEE Trans. Ind. Informatics, vol. 15, no. 1, pp. 434–444, 2019.
81. M. W. Hoffmann et al., "Integration of novel sensors and machine learning for predictive maintenance in medium voltage switchgear to enable the energy and mobility revolutions," Sensors (Switzerland), vol. 20, no. 7, pp. 1–24, 2020.
82. C. Cheng, B. Zhou, G. Ma, D. Wu, and Y. Yuan, "Wasserstein distance based deep adversarial transfer learning for intelligent fault diagnosis with unlabeled or insufficient labeled data," Neurocomputing, vol. 409, pp. 35–45, 2020.
83. Seebo, "Why Predictive Maintenance is Driving Industry 4.0," i4.0 Initiat., pp. 1–13, 2019.

Chapter 7
Crop and Fertilizer Recommendation System Using Machine Learning

Radha Govindwar, Shruti Jawale, Tanmayee Kalpande, Sejal Zade, Pravin Futane, and Idongesit Williams

7.1 Introduction

The agricultural sector is a major pillar of the Indian economy and contributes significantly to the economy each year. However, the sector that has such a significant contribution to the economy faces many problems each year resulting in low yield of crops and farmers' debts. One of the reasons for low production in agriculture is imprecise and incorrect farming practices. A general crop is grown through an area regardless of the soil quality. This practice avoids the need for special care and attention to the soil and crop both. However, it is not possible to know the exact reasons for a bad yield if no analysis is made. This makes it harder to know the reasons for a bad yield. This abets debt and financial losses to the cultivators.

This chapter proposes a crop and fertilizer recommendation system to overcome the aforementioned drawbacks. The proposed system aims to analyze soil data and recommend the most suitable crop and fertilizers for the concerned soil. Machine learning has enabled us to analyze large amounts of data and draw observations from it. The goal is to help the Indian cultivators to make an informed decision and maximize their crop yields. The user has to enter the soil details for their farm, namely, nitrogen, phosphorus, potassium, temperature, humidity, and moisture. Then the pretrained machine learning model will recommend crops to the farmer. For fertilizer recommendation, all the aforementioned parameters, along with the

R. Govindwar · S. Jawale · T. Kalpande · S. Zade · P. Futane (✉)
Vishwakarma Institute of Information Technology, Pune, Maharashtra, India
e-mail: radha.21910261@viit.ac.in; shruti.21910863@viit.ac.in; tanmayee.21911189@viit.ac.in; sejal.21910689@viit.ac.in; pravin.futane@viit.ac.in

I. Williams
CMI, Alborg University, Copenhagen, Denmark
e-mail: idong@es.aau.dk

A. Neustein et al. (eds.), *AI, IoT, Big Data and Cloud Computing for Industry 4.0*, Signals and Communication Technology, https://doi.org/10.1007/978-3-031-29713-7_7

crop to be planted, have to be entered. The pretrained model will recommend fertilizer for the same. The system will be hosted as a website for ease of access to farmers.

Studying different systems and papers for the same purpose, we came to an understanding and tried different approaches to build the proposed machine learning model. Random Forest, decision trees, Support Vector Machine (SVM), Gaussian Naive Bayes, and XGBoost were used and the best performing one was finalized to be the recommendation model.

7.2 Literature Survey

India is a country dependent on agriculture. So, the better the crop yield, the better the production. Finding what crop is suitable for a particular land is usually passed down from generation to generation. But, when finding a technical solution, one comes across many solutions to one problem. Many articles with different methods to recommend crop and fertilizers came into consideration. (Table 7.1) [1].

A journal by the Turkish Journal of Computer and Mathematics Education on crop recommendation predicted with the help of machine learning (ML) and the Internet of things (IoT) (for soil parameters such as soil moisture and temperature using sensors). Random Forest and Naive Bayes Classifier algorithms were used in this experiment. The model actually shows an accuracy of 96.89% [2]. Another paper on crop and fertilizer recommendation from the IRJET publication on a Support Vector Machine (SVM) algorithm model. The dataset has been acquired from the Kaggle website including the required attributes. This project based on the SVM model gave an accuracy of over 90.01%. This system aims to help newcomers in this industry as well by taking into consideration climatic conditions [3]. This paper presents an approach that uses the Internet of Things (IoT) and machine learning (ML) to develop a recommendation system for farmers. The system consists of four stages that start with soil analysis and end with recommendations. The prediction is carried out using the classification and clustering algorithm. They acquired two datasets: the crop requirement dataset and the fertilizer datasets. The performance analysis for nutrient analysis has an accuracy of 84.46% and a precision of 91.04%. The performance analysis for the recommendation system has an accuracy of 93.33%, a precision of 90.60%, recall of 90.00%, and F-score of 90.14% [4]. This paper was presented in a book published by Progress in Advanced Computing. The paper presents an ontology-based system for recommending and assessing crop suitability. It uses ontology to predict the crop's potential profitability and crop suitability based on the soil type and region.

Crop prediction is referred to as random forest, plant food recommendation is finished through k-means formula. The crop data is inherited through the department of agriculture, government of geographic area. The performance analysis shows that the accuracy of the developed system in all fairness is high [5]. A paper published by IJESC proposes an efficient algorithm like Naive Bayes algorithm for crop yield

Table 7.1 A detailed survey of the studied research papers

Reference number	Methodology	Dataset details	Performance measures	Result
[1]	The real-time data from the field area is collected using the IoT system and fed to the trained model and predictions are made. Models used are Random Forest and Naive Bayes.	Self-collected data.	Accuracy of about *96.89%* iis obtained.	The system predicts and suggests the crops to be sown with an accuracy of about 96.89%.
[2]	The data is being first cleaned and preprocessed and then the Support Vector Machine (SVM) algorithm is used.	Kaggle dataset fertilizer_ Recommendation 100 entries. Kaggle dataset for crop recommendation 2000 entries.	Gives the accuracy over *90.01%*.	The Support Vector Machine algorithm helps to predict the crop with 90.1% accuracy.
[3]	The proposed methodology analyzes soil using Arduino and electrodes to gather data. This data is stored in the cloud and recommendation is done using machine learning techniques.	Columns (N, P, K soil type, region, variety, season, NPK ratio, and total ppm value). Entries>1000.	The system has an accuracy of *93.33%*.	On giving crop to be aggregated, state and district, variety of the crop; season and soil type, recommendations are made with 93.3% accuracy.
[4]	An android software wherein prediction is finished by using of Random forest, fertilizer recommendation is accomplished via the k-method algorithm.	Acquired through the department of agriculture, Government of Maharashtra.	Accuracy of *92.14%* for the entire system.	Recommendations of crops and fertilizers were provided with 92.14% accuracy using Random forests and KNN.
[5]	The proposed work implements Naive Bayes algorithm for crop yield prediction and KNN algorithm is used for fertilizer recommendation.	Acquired from the Kaggle site for crop and fertilizer prediction.	Naive Bayes and KNN perform well. Accuracy percent not mentioned.	The recommendation system is made with Naive Bayes and KNN for crop and fertilizer recommendation.

(continued)

Table 7.1 (continued)

Reference number	Methodology	Dataset details	Performance measures	Result
[6]	The proposed system uses the KNN algorithm for crop recommendation through city-wise analysis.	Obtained from the agriculture department of Tamil Nadu in Kancheepuram district.	The accuracy for KNN was *89%* and SVM was *80%*.	KNN is turned into observed to offer the best performance and precision in comparison to decision tree, Random forest, etc., for crop datasets.
[7]	The system architecture proposes dataset collection, feature selection, classification techniques, recommendation system, and crop yield prediction.	Crop yield datasets are obtained from fao.org, soil data for Warangal test area.	Not mentioned.	This paper summarizes a powerful recommendation system for fertilizers and vegetation primarily based on NPK values and the area.
[8]	The soil information is collected through sensors, transmitted from the Arduino through Zigbee and WSN to MATLAB and processing is done with the help of ANN and crop recommendations are done using SVM.	Self-collected dataset.	Not mentioned.	SVM is used to predict crops and recommendations are displayed on web pages.

prediction and K-nearest neighbor (KNN) algorithm for fertilizer prediction. The dataset has been acquired from the Kaggle site for crop and fertilizer prediction. This project uses the data collected by soil analysis and KNN classification algorithm to predict the optimal fertilizer ratio for a given crop. It then helps the end user to make informed decisions about the fertilizer needed for that crop.

Crop recommendation system via town clever evaluation uses the crop and town dataset, the second one module recommends crop for a particular soil nutrient values, and the 1/3 module offers information about the soil vitamins which may be poor for a selected crop. The device uses two datasets: crop dataset and soil dataset. The crop dataset consists of the corresponding pH, N, P, and K values of plant life and the soil dataset consists of the corresponding pH, N, P, and K for soils for various cities in Chennai and different districts of Tamil Nadu. The information is acquired from the agriculture branch of Tamil Nadu in Kancheepuram district. The accuracy received as a result of using KNN algorithm became 89% and by means of using SVM algorithm turned into 80%. The accuracy is calculated primarily based on the effects acquired from check and teaching facts [7]. Another such paper turned into JETIR Org. For crop prediction the use of records mining techniques. This structure is taken into consideration to expect the quality harvest suitable for the agronomists' region. It additionally indicates farming techniques for vegetation inclusive of various farming, spacing, irrigation, sow processing, and so forth along with facet fertilizer and pesticide proposals. This is executed on the premise of historic soil standards of the vicinity and estimating crop and climate fees. Further, cost prediction is carried out based totally on linear regression to resource in ranking the vegetation endorsed. The system architecture of the trouble statements makes use of the following steps to decide the result which can be FactSet series, feature selection, type techniques, advice system, and crop yield prediction. Ensembling is a data mining model used to mix the strength of greater models to gather better predictability and efficiency than any of its models can attain on its own. The dataset consists of specific soil that have taken from warangal location for soil testing in the laboratory. In addition, similar online big crop statistics assets were extensively utilized. The application could be very pleasant and requires less memory [8]. Lastly, from the *European Journal of Molecular & Clinical Medicine*, the proposed approach takes the soil and pH samples because the input helps to expect the vegetation that can be endorsed suitable for the soil and fertilizer that may be used as the solution in the form of the website. So, the soil data is gathered through sensors and the records transmitted from the Arduino with Zigbee and wireless sensor network (WSN) to MATLAB and reading the soil statistics and processing is completed with the help of artificial neural network (ANN) and crop recommendation is carried out with the usage of SVM (Support Vector Machine).

For the dataset, standards like pH, nitrogen, potassium, phosphate, depth, temperature, and precipitation are received. Criteria like pH, nitrogen, potassium, phosphate, depth, temperature, and precipitation are accounted in the research (Table 7.1).

7.3 Proposed System

The proposed system aims to help the farmers cultivate suitable crops and fertilizers for their land according to their geographical location and soil quality. The data used for the recommendation system is a publicly available dataset that comprises unique attributes, such as NPK value, humidity, rainfall, temperature, and pH of the soil.

The architecture of the system can be decomposed into several steps (Fig. 7.1).

1. *Import libraries*

   Various libraries have been used to visualize, analyze, and train the data for the machine learning models to work:

   - Numpy
   - Pandas
   - Matplotlib
   - Seaborn
   - Scikit-learn

2. *Data analysis*

   After importing the libraries and gathering the dataset for the system, the data needs to be analyzed in order to explore and find a pattern in the data. This enables us to find the necessary information and make decisions accordingly.

   Several functions such as df.head(), df.tail(), and df.size() can be used to analyze the data.
3. *Visualize data*

   Visualizing data is an important aspect of machine learning as it allows the data to be represented as charts and graphs, allowing us to better understand and process the data. Data visualization often helps to discover unknown facts and trends within the dataset. Thus, by observing the relationships and trends, one can acquire meaningful information.
4. *Separate features and target label*

   We have to target an attribute in the chosen dataset to be used in the models. The rest are termed as features. The dataset then has to undergo a testing and training process, that is, the dataset has to be divided into variables so as to pass them to the machine learning models for further functioning. This is achieved by train_test_split from scikit-learn to divide features data and target data.
5. *Apply different models*

   After successfully separating features and targeting a label, the data is ready to be fed to the machine learning models for the training aspect. The proposed system trains decision tree, Gaussian Naive Bayes, Support Vector Machine, logistic regression, Random Forest, and XGBoost. Sklearn helps to import these models in order to work upon them. Sklearn being an efficient python library also implements cross-validation and allows us to easily serialize and deserialize a trained model.
6. *Compare accuracy and finalize a model*

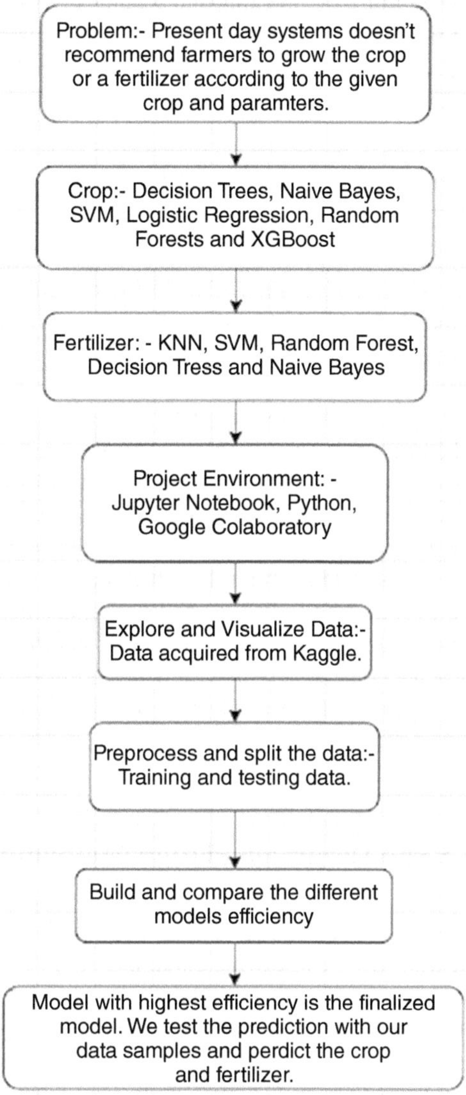

Fig. 7.1 Framework of the system

Finally, the accuracies of all the trained models are compared. The model with the higher accuracy is then finalized and used for the recommendation of crops and fertilizers based on the input.

7. *Make a prediction*
   To confirm the working and accuracy of the finalized model, we can try to predict the output by giving the model some inputs to determine the recommended crop or fertilizer. A correct output confirms the accuracy of the model.

7.4 Implementation and Results

In crop recommendation, users have to give data from temperature, rainfall, N, P, and K values as input and the system will predict the crop they should grow.

In fertilizer recommendation, the users have to input soil data and the type of crop they are growing. The system then will predict what the soil lacks or has in excess and will suggest improvements.

We have used different models and have compared their accuracies. The model with the highest accuracy is chosen for the deployment.

The practical implication of this project will, however, be fulfilled with not a dataset from a Kaggle site but from the real-world dataset authorized by the government of India. Even the sensors will be needed to get the real-time values from the farms and several other attributes including weather and temperature.

In a country like India, where farming has been going on for generations, there are several types of fertilizers including natural as well as artificial products, thus the most accurate and efficient in real-world system will need the resources of all of them. Getting the highest accuracy for an efficient model is still a far-fetched task yet to be accomplished.

7.5 Crop Recommendation Methodology

Data analysis: We acquired a dataset from Kaggle, which contains 2000+ entries. The dataset had the following columns: N, P, K, temperature, rainfall, humidity, and pH level of the soil and a crop label.

The crop label was made the target variable, while all the remaining columns were determined as features.

Data visualization: we have figured out a heatmap for the features. It represents common and unique values while differentiating them with light to dark colours (Fig. 7.3).

Then we separated features and target variables to obtain training and testing datasets and applied different models to determine their accuracies. The most accurate model among them will be selected.

Machine learning models, decision trees, Naive Bayes, SVM, logistic regression, Random forests, and XGBoost were trained using the dataset. The accuracies for each model were 90%, 99%, 90.01%, 95.22%, 99.09%, and 99.31%, respectively.

7.6 Fertilizer Recommendation Methodology

Dataset acquisition: The dataset for fertilizer recommendation was obtained from Kaggle. The dataset contains the following columns: temperature, humidity, moisture, NPK values, and soil type. There are 100 entries in the fertilizer recommendation dataset.

Dataset preprocessing: The dataset was checked for null values or outliers. Categorical variables were encoded using OneHotEncoder and Label Encoder. No null values were found. A large number of potassium entries were 0, so they were omitted before feeding the data to models for reliability concerns.

Proposed algorithm: Observations were made by visualizing data. Different models (KNN, SVM, Random Forest, decision trees, and Naive Bayes) are trained using the dataset, and the most accurate one is finalized. Currently, random Forest classification gives the highest accuracy of 90%. The other models used for this system are KNN, SVM, Naive Bayes classifier, and the decision tree give accuracy of 75%, 55%, 50%, and 85%, respectively (Fig. 7.2).

We created various plots like histogram, bar plot and line plots, and box plots to get the relationship between various parameters, for example, to find the density distribution of potassium across various crops accordingly and checking outliers, if any found using box plot. We similarly tried for nitrogen and phosphorus as well (NPK values). Similarly, we also tried to find out the count of various soil types, for example, sandy, loamy, black, red, and clay and find the relation with output variables (e.g., fertilizer name).

We also generated a heatmap of the crop dataset as shown in Fig. 7.3, where we try to find out the maximum association between parameters, here P and K values shows it with a value of 0.74.

7.7 Conclusion

This system is very promising for the farmers and newcomers in the long run if made into a user-friendly website or an application. It will assist farmers to choose the first-class crops appropriate for his or her land and appropriate fertilizers to provide the most yield in the agriculture enterprise, which in turn can assist the economy of the country. Use of GIS data can be explored ahead as further research. Use of fertilizers specific [7] to the coordinate system will aid in determining sustainable solutions, thereby targeting the zones with defining the quantities and nature of fertilizers that can be used. Overall effect in the yield production can be observed and seems promising in gaining insights from agricultural productivity. Thus, this will result in more earnings for those inside the agricultural area, and this reason will attract more people to contribute to this industry. This will also be helpful for the agriculture department of our country if given a chance to work on more real-world datasets.

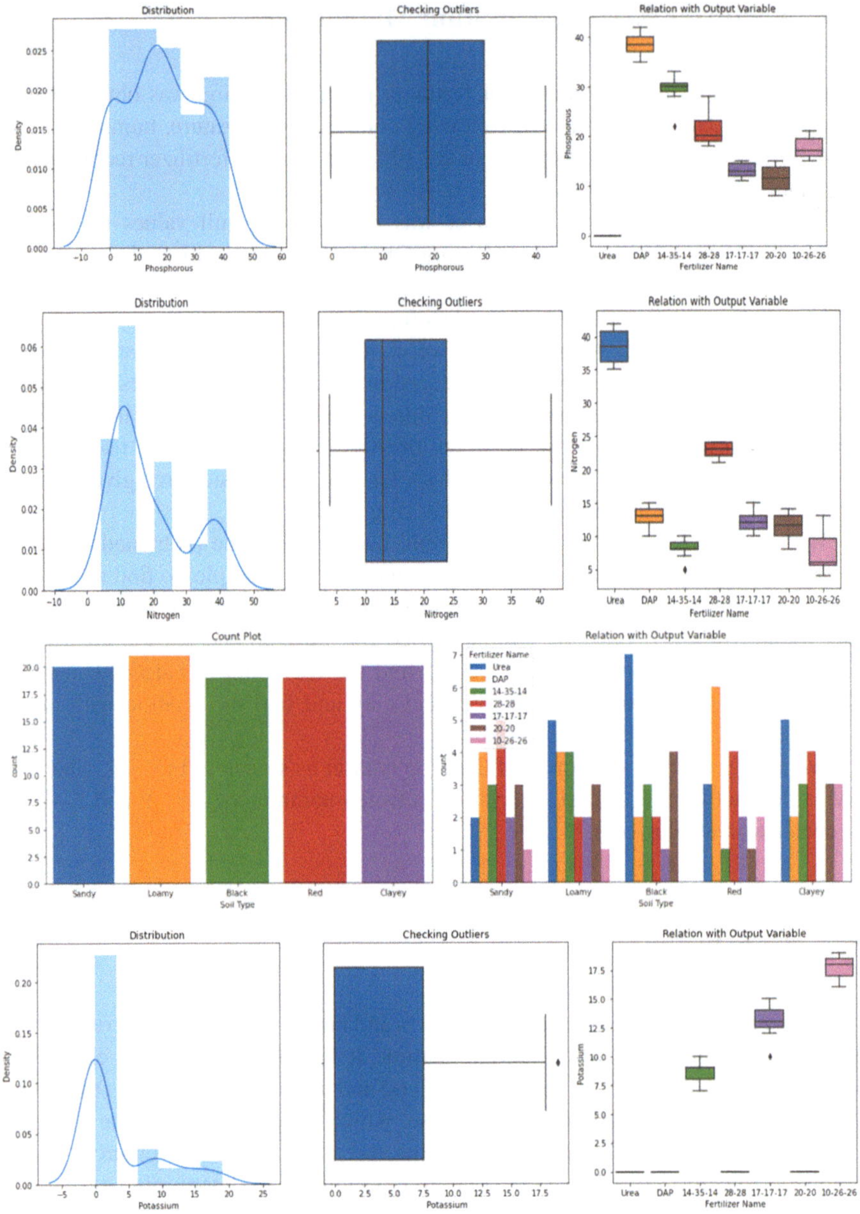

Fig. 7.2 Data visualization of fertilizer dataset

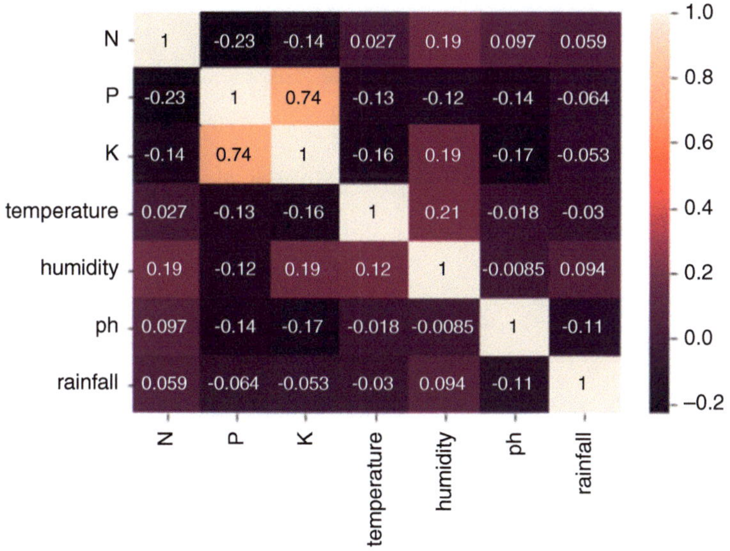

Fig. 7.3 Heatmap of crop dataset

References

1. Anguraj K, Thiyaneswaran B, Megashree G, Preetha Shri J.G, Navya S, Jayanthi J, "Crop Recommendation on Analyzing Soil Using Machine Learning", Turkish Journal of Computer and Mathematics Education, 10 November 2020, Vol.12 No.6 (2021), (pp 1784–1791)
2. Palaniraj, Balamurugan, Durga Prasad, Pradeep "Crop and Fertilizer Recommendation System using Machine Learning" International Research Journal of Engineering and Technology (IRJET).
3. S. UshaKiruthika, S. Kanaga Suba Raja, S.R. Ronak, S. Rengarajan, P. Ravindran, "Design and Implementation of Fertilizer Recommendation System for Farmers", March – April 2020 ISSN: 0193-4120 (pp 8840–8849)
4. Archana Chougule, Vijay Kumar Jha, Debajyoti Mukhopadhyay, "Crop Suitability and Fertilizers Recommendation using Data Mining Techniques", In book: Progress in Advanced Computing and Intelligent Engineering (pp.205–213).
5. Varshini Naresh, Vatsala B., C. Vidya Raj- UG Student, "Crop Yield Prediction and Fertilizer Recommendation", International Journal of Engineering Science and Computing IJESC.
6. A.K.Mariappan, 2Ms C. Madhumitha, 3Ms P. Nishitha, 4Ms S. Nivedhitha, "Crop Recommendation through Soil analysis Using classification in Machine Learning", International Journal of Advanced Science and Technology Vol. 29, No. 03, (2020), (pp. 12738–12747.)
7. Aakunuri Manjula Associate Professor, CSE, G. Narsimha, "Crop Recommendation and Yield Prediction for Agriculture using Data Mining Techniques", 2019 JETIR March 2019, Volume 6, Issue 3 (pp 359–362)
8. Preethi G, Rathi Priya V, Sanjula S M, Lalitha S D, Vijaya Bindhu B," Agro based crop and fertilizer recommendation system using machine learning", European Journal of Molecular & Clinical Medicine Volume 7, Issue 4, 2020 pp(2515–8260).

Chapter 8
Comparative Analysis of Machine Learning Algorithms for Intrusion Detection System

P. Agarwal, D. Sheth, K. Vaghmare, and N. Sakhare

8.1 Introduction

Applications of machine learning (ML) can now be found in most fields of research. As Internet-based activities expand, the need to monitor the network traffic from these activities increases. Organizations use security information and event management (SIEM) to automate this process of monitoring traffic and to bolster their defence against possible threats. It can also generate a response in case of an attack. Such systems are often referred to as security orchestration and adaptive response (SOAR).

An Intrusion Detection System (IDS) is a system that monitors network traffic for anomalous behavior. It sends an alert to the SOC team in case it encounters suspicious activity on the network. It can either be a part of the SIEM or something which consumes the data collected by a SIEM. It uses multiple different tools to classify data as normal or anomalous, and in case of anomalous, it can further identify the exact attack type. The training data used in this case can be derived from historical data collected and analyzed by the organization.

IDS can detect attacks through various techniques such as signature-based detection and anomaly detection. In case of signature-based detection, the IDS has a predefined directory of known threats, and can flag the data as anomalous if it matches a record from the directory. This is coupled with anomaly detection, where any traffic which has inconsistent behavior is reported to the SOC team. Unlike signature-based detection, there is no fixed method to flag data as anomalous, but

P. Agarwal (✉) · D. Sheth · N. Sakhare
Computer Engineering, VIIT, Pune, India
e-mail: nitin.sakhare@viit.ac.in

K. Vaghmare
Data Competency, Persistent Systems Limited, Pune, India
e-mail: kaustubh_vaghmare@persistent.com

A. Neustein et al. (eds.), *AI, IoT, Big Data and Cloud Computing for Industry 4.0*, Signals and Communication Technology, https://doi.org/10.1007/978-3-031-29713-7_8

rule-based classifiers and machine learning models trained on normal data can be used to detect outliers.

Anomaly detection faces numerous challenges, as normal traffic flow can be similar in nature, but different anomalies are different in different ways. This uncertainty for classification often leads to false alarms, and occasional intrusions. Thus, IDS must be fine-tuned to organizations' specific needs and must be fed with new data periodically. Anomalous behavior can be masked using techniques such as packet fragmentation, which fragments the packets to avoid crossing the threshold. Proxying is used in case the original IP address has been blocked. Even a small change in pattern can be used to breach the defence of the IDS.

Considering these issues, IDS is generally trained on normal data. As normal data is mostly similar in nature and generally follows a pattern, a model that has been trained using normal data can flag any activity that does not fit the pattern as anomalous. Based on our analysis, we have understood the fact that such a model has a higher accuracy and precision compared to traditional binary classification models. IDS can only detect anomalous behavior and report it, but it cannot take direct action to prevent it, so it is coupled with the security orchestration, automation, and response (SOAR) systems.

In this chapter, we have used various anomaly detection techniques such as OneClass Support Vector Machine, Isolation Forest, autoencoders (single and stacked), and binary classification techniques using traditional algorithms such as Random Forest classifier, logistic regression, AdaBoost classifier, linear support vector classifier, and stochastic gradient decent classifier. A comparative analysis consisting of the aforementioned machine learning techniques is performed to identify the most suitable technique for anomaly detection.

8.2 Related Work

Gharib et al. used stacked autoencoder which consisted of a normal autoencoder, followed by a sparse autoencoder [1]. Ravipati et al. performed feature extraction and used multiple "weak" classifiers and combined them to form a strong classifier [2]. Dhanabal et al. performed a detailed analysis of NSL-KDD dataset and used J48, SVM, and Naïve Bayes as classifiers after using feature extraction [3]. Hasan et al. used a combination of support vector machines with different kernels for detecting network-based anomalies using the KDD Cup 99 dataset [4]. Homoliak et al. used artificial neural network with back propagation to detect anomalies with 84.20% accuracy [5]. Imran et al. used a combination of neural networks and SVM with different kernels to detect intrusions after reducing dimensions using LDA [6]. Revathi et al. used J48, CART, Random Forest classifier, and Naïve Bayes on NSL-KDD [7]. Tavallaee et al. suggested various methods for preprocessing operations on the NSL-KDD dataset, which reduces wrong detection of records [8]. The goal of this survey, according to Raghavendra Chalapathy and Sanjay Chawla, was to investigate and uncover numerous deep learning models for anomaly detection, as well as analyze their viability for a given dataset. These assumptions can be used as a guide when applying a deep learning model to a certain domain or collection of data to assess the technique's efficacy in that domain. They also state that extending and updating this study in the future may be necessary when more complex methods for deep learning-based anomaly detection become available [9]. This learning problem is interpreted as a binary classification problem by Ingo et al., who compare the resulting classification risk to the conventional performance metric for the density-level problem. The empirical classification risk appears to be a useful empirical performance metric for the anomaly detection problem. This allows them to propose an anomaly detection support vector machine (SVM) for which universal consistency can be easily established [10]. Command sequence learning, according to Terran Lane and Carla E. Brodley, can be a useful technique in the domain of anomaly detection for user recognition in computer security [11]. Sakhare et al. used J48 algorithm to detect anomalous behavior of criminals [12].

8.3 Methodology (Fig. 8.1)

Preprocessing data include the following:

- One Hot Encoding is a method to convert categorical columns into numerical format so that our algorithms can understand the relationship between the column and target variable. It involves creating a binary column for each possible value in the categorical column and assigning a 1 to the corresponding column for each observation and 0 for all other columns
- Standardization
- Labeling 39 different attack types as anomaly

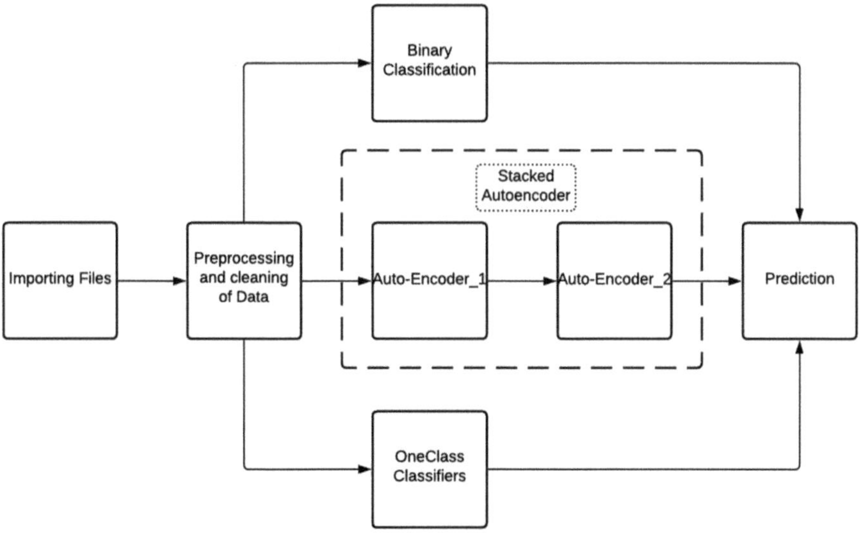

Fig. 8.1 Architecture of proposed system

- Creating a new train dataframe that includes only the normal samples from the train dataset for one-class classifiers and autoencoders

8.3.1 Dataset

NSL-KDD was used as dataset to train our models. It is a preprocessed version of the KDD-CUP 99 dataset. We used the train and test csv files, which had 1,48,517 records in total. First, we changed all attack types into two classes, "anomaly" and "normal." For the training dataset, we used only the records which were labeled as normal for training our autoencoders, OneClass SVM and Isolation Forest. For binary classification, all records were used. One hot encoding was performed on three columns with string values. As the magnitude of difference in values was too high, standard scaling was used on the dataset.

We used a variety of techniques to conduct the analysis, which can be divided into three categories: binary classification using popular algorithms, neural network-based autoencoders, and conventional unsupervised algorithms that train on a single class.

8.3.2 Binary Classifiers

Predicting one of two classes is binary classification. The objective of a binary classification challenge is to develop a machine learning model that can make a prediction when the thing to predict has only two potential values. In our case, it can be either a normal traffic or an anomaly.

We have trained all the binary models on a dataset which had two labels, that is, we transformed the four attack classes into one label as anomaly.

We have used the following binary classifiers:

- Random Forest classifier
- AdaBoost
- Logistic regression
- Linear SVC

8.3.2.1 Random Forest Classifier

Random Forest classifier creates a set of trees from randomly selected subsets of the data. It used bagging and feature randomness when building a tree to try to create an uncorrelated forest of trees whose prediction by committee is more accurate than that of any individual tree.

We got an accuracy of 76% using the Random Forest classifier (Fig. 8.2).

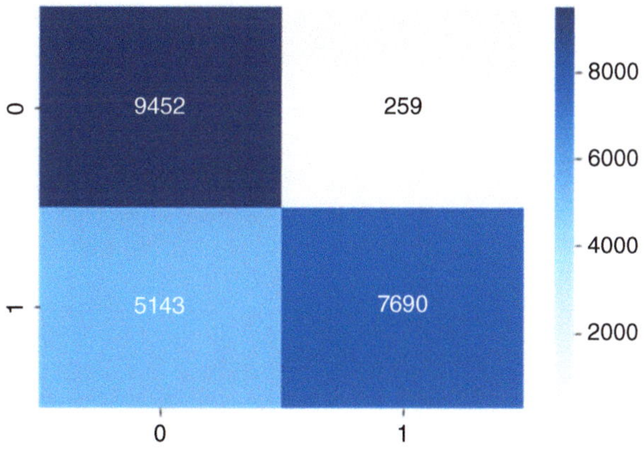

Fig. 8.2 Heatmap of RFC

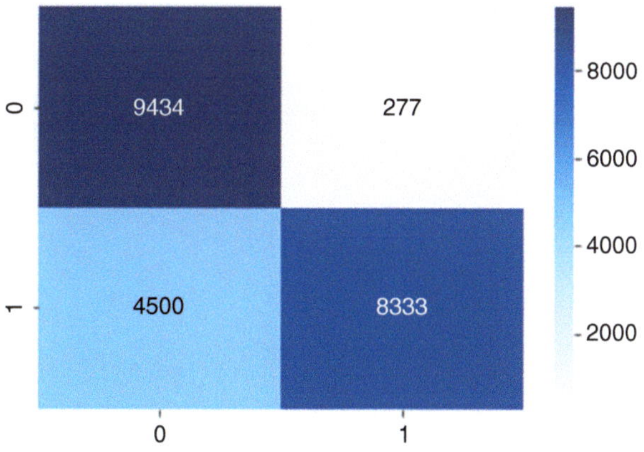

Fig. 8.3 Heatmap of AdaBoost

8.3.2.2 AdaBoost Classifier

Adaptive boosting, or AdaBoost, is a binary classification technique that can only be used for binary classification. The weights are reallocated to each instance, with higher weights applied to improperly identified instances. This is termed as adaptive boosting.

In the binary classification category, AdaBoost did the best.

We achieved a 79.1% accuracy rate (Fig. 8.3).

8.3.2.3 Logistic Regression Classifier

Logistic regression is a linear classifier which uses a linear function

$$f(x) = b_0 + x_1 b_1 + x_2 b_2 + \cdots + x_n b_n \tag{8.1}$$

where $b_1, b_2 \ldots b_n$ are the coefficients of the line of the perfect fit.

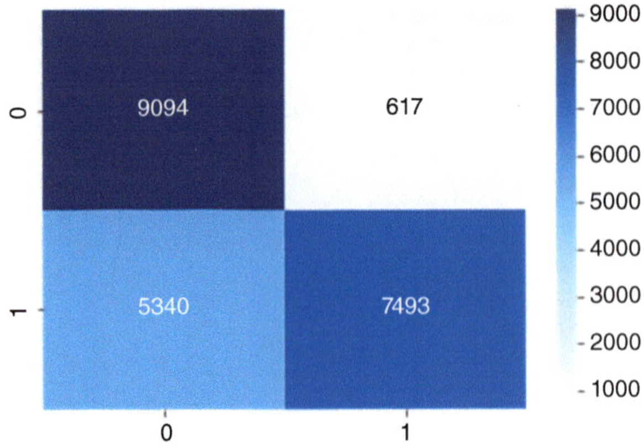

Fig. 8.4 Heatmap of logistic regression

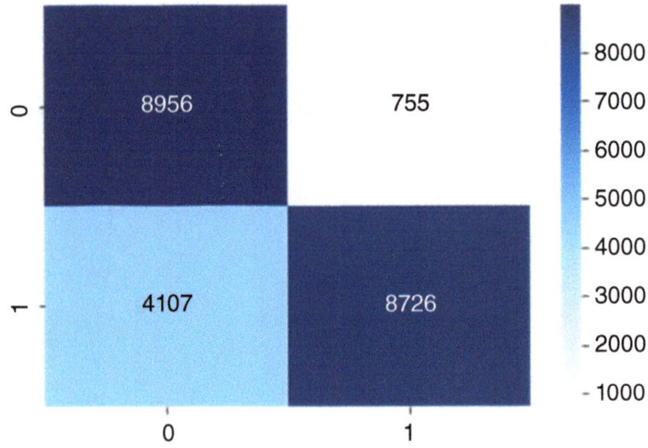

Fig. 8.5 Heatmap of linear SVM

Once the coefficients of the best line are found, one can get the predicted output for any value of x_i.

Logistic regression gives an accuracy of 73.6% (Fig. 8.4).

8.3.2.4 Linear Support Vector Machine

When training the model, linear SVM constructs a line between the two classes. In this case, linear SVM performs fairly and has an accuracy of 78.4% (Fig. 8.5).

8.3.3 One-Class Classifiers

One-class classification is a subfield of machine learning which deals with severely skewed datasets. Binary classification is ineffective when the data is imbalanced beyond a limit. The major application of one-class classifiers is in the field of anomaly detection or outlier detection. Here the model trains on one certain class and classifies anything which does not fit the pattern of the class it was trained on as an outlier or an anomaly. In case of an intrusion detection system where anomalies would be rare, one-class classifiers provide an effective solution. The training data had only one label class, it only included "normal" class of data.

The classifiers we used are as follows:

- OneClass SVM
- Isolated Forest

8.3.3.1 OneClass SVM

The OneClass SVM is a variant of Support Vector Machine. It is an unsupervised outlier identification model that works well for detecting anomalies. OneClass SVM draws a line around a cluster where the majority of the normal data points are located, and all points outside the line are deemed anomalies.

After modifying the parameters to match our data, our accuracy increased to 75.6% (Fig. 8.6).

8.3.3.2 Isolation Forest

The primary concept of Isolation Forest varies from other common outlier detection algorithms in that it discovers abnormalities directly rather than profiling typical data points. Isolation Forest, like any other tree ensemble method, is built on decision trees. In these trees, partitions are created by selecting a feature at random and then choosing a random split value between the feature's minimum and maximum value.

Our accuracy after fine-tuning the model to suit our needs is 80.8% (Fig. 8.7).

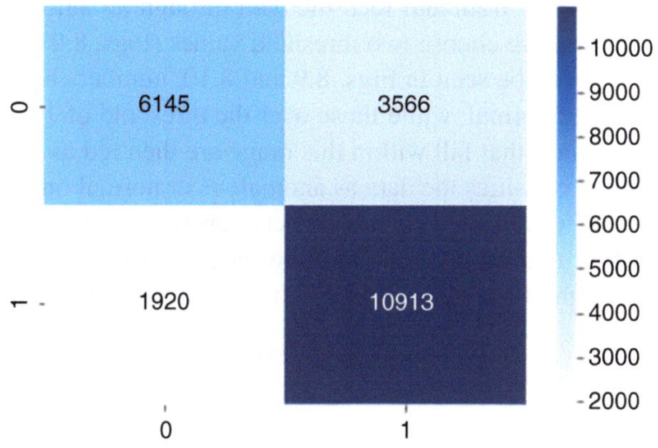

Fig. 8.6 Heatmap of OneClass SVM

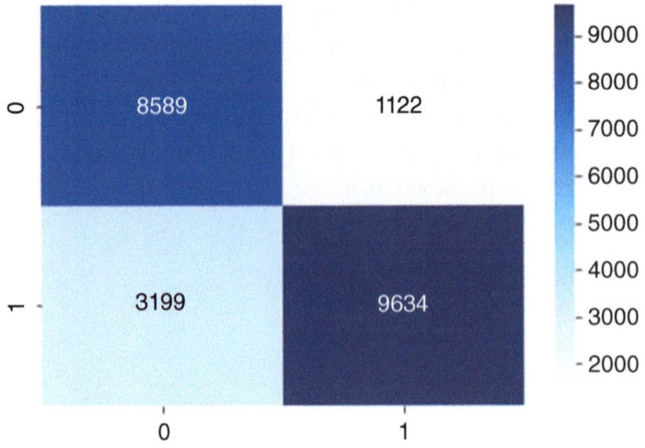

Fig. 8.7 Heatmap of isolation forest

8.3.4 Autoencoders

Autoencoder is an unsupervised artificial neural network that compresses data into lower dimensions (bottleneck layer or code) and then decodes the data to recreate the original input.

In anomaly detection, autoencoders are commonly employed. We used the reconstruction error as the anomaly scores. We consider a data point to be an anomaly if the reconstruction error exceeds a particular threshold.

We employed two different types of autoencoders:

- Simple autoencoder
- Stacked autoencoder (Fig. 8.8)

At the threshold value of 0.56, the maximum accuracy was achieved.

Using the simple autoencoder approach, we were able to achieve an accuracy of 81.9%.

The approach for stacked autoencoder differs from that of the simple autoencoders. We train and feed the data through an autoencoder in this method, except this time we choose two threshold values (Figs. 8.9 and 8.10).

As can be seen in Figs. 8.9 and 8.10, numbers below the threshold of 0.026 are deemed normal, while those over the threshold of 1.75 are regarded anomalies. All the values that fall within this range are then fed as data to the second autoencoder, which classifies the data as anomalous or normal once more (Fig. 8.11).

We established threshold between 0.56 and 1.8 since most normal data falls within that range, and anything outside of it is considered anomalous. We attain the highest accuracy of 87.1% by employing stacked autoencoders.

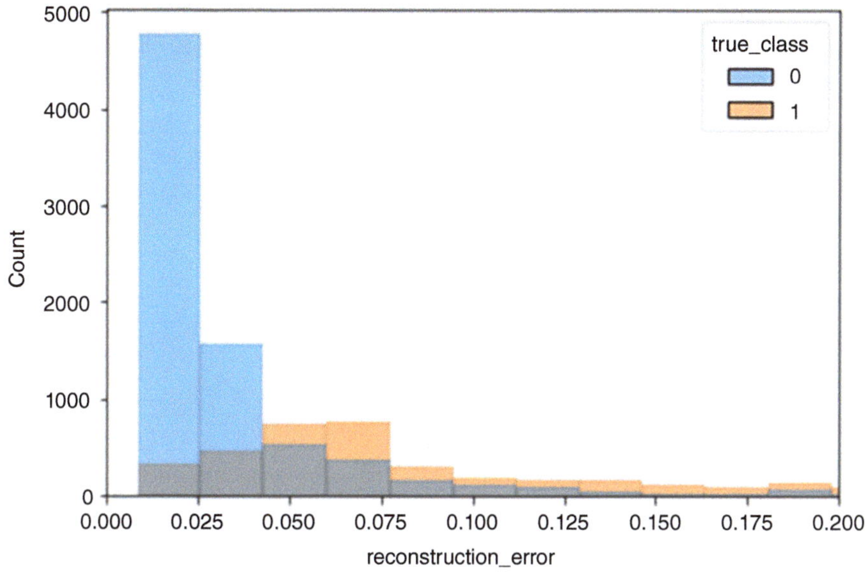

Fig. 8.8 Simple autoencoder—true labels versus the reconstruction error

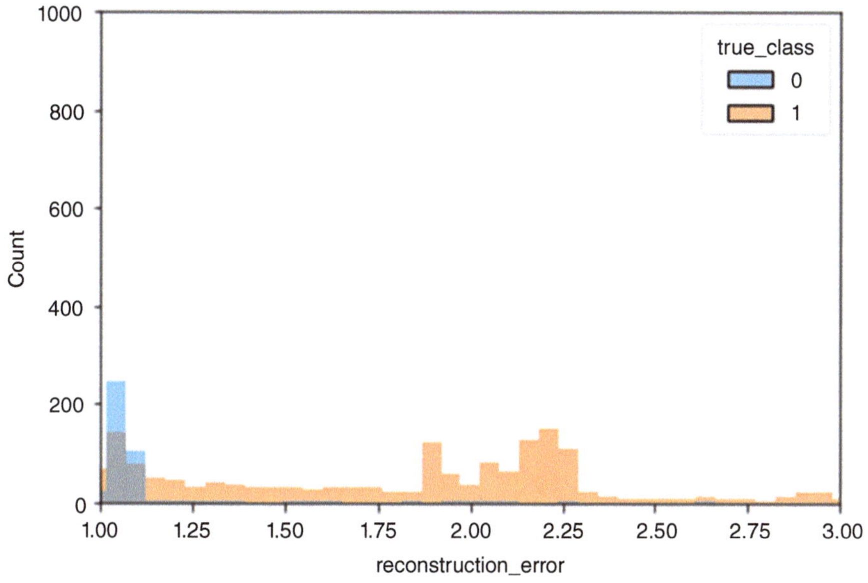

Fig. 8.9 Graph to calculate upper limit

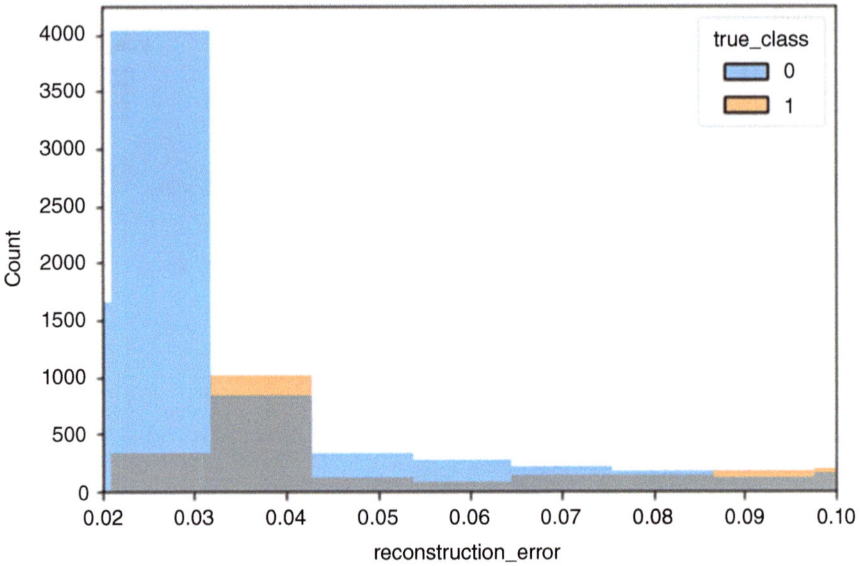

Fig. 8.10 Graph to calculate lower limit

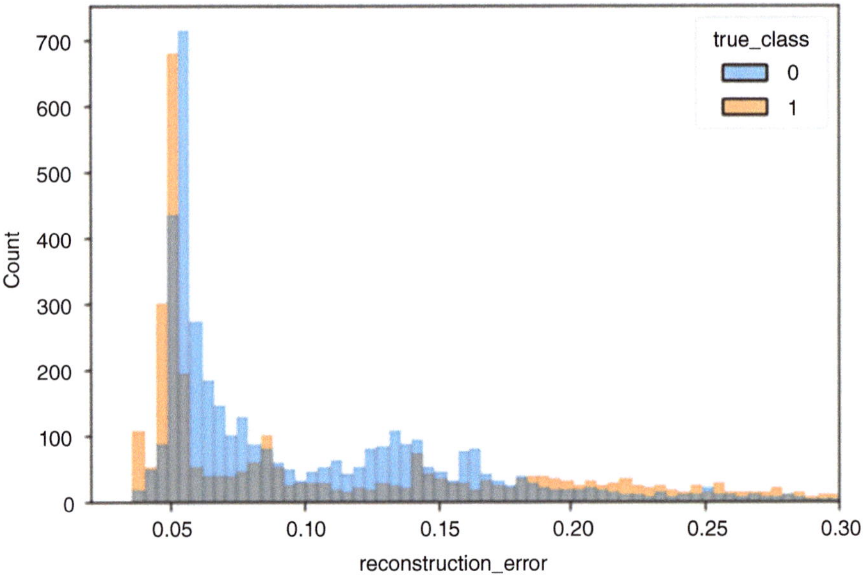

Fig. 8.11 Threshold for stacked autoencoders

Table 8.1 Model evaluation scores

Model	Accuracy	F1-score	Precision	Recall
Random Forest	76.0	74.0	96.7	59.9
AdaBoost	79.1	79.0	96.7	66.8
Logistic regression	73.6	71.5	92.4	58.4
Linear SVM	78.4	78.2	92.0	68.0
OneClass SVM	75.6	75.0	76.0	75.0
Isolation Forest	80.8	81.0	81.0	82.0
Simple autoencoder	81.9	82.8	89.8	76.9
Stacked autoencoder	87.1	88.5	86.8	90.3

8.4 Results and Discussion

In terms of accuracy, stacked autoencoder had the best performance, achieving an accuracy of 87.1% on test data (Table 8.1). AdaBoost had the highest accuracy from traditional binary classification techniques. As we can see from the model evaluation scores, autoencoders trained on one-class data performed better compared to binary classification models. Anomalies have to be detected and thus require an unconventional machine learning approach. The stacked autoencoder was a custom system that used thresholds calculated from graphs of reconstruction error. One of the biggest issues faced by modern IDS are zero-day attacks, which is a broad term to categorize any attack that exploits a previously unknown vulnerability. Although binary classification models do not perform too poorly, they are still prone to zero-day attacks, which are better dealt with through autoencoders using OneClass approach.

8.5 Conclusion

The outcome of the analysis showcases that machine learning models trained using only the normal data from training dataset perform better in terms of anomaly detection, having a considerable difference in terms of accuracy compared to traditional binary classification methods. A system of autoencoder that includes two autoencoders, with one being a sparse autoencoder achieved the highest testing accuracy of 87%. The NSL-KDD, due to the variety of attacks included, is one of the best publicly available dataset to train models for an Intrusion Detection System. Adversarial machine learning techniques can be employed to strengthen our IDS, which can help find weaknesses in the training data. Further classification of anomalies into attack types can be attempted using classification techniques.

References

1. Gharib, M., Mohammadi, B., Dastgerdi, S. H., & Sabokrou, M. (2019). Autoids: auto-encoder based method for intrusion detection system. *arXiv preprint arXiv:1911.03306*.
2. Ravipati, R. D., & Abualkibash, M. (2019). A survey on different machine learning algorithms and weak classifiers based on KDD and NSL-KDD datasets. International Journal of Artificial Intelligence and Applications (IJAIA), 10(3).
3. Dhanabal, L., & Shantharajah, S. P. (2015). A study on NSL-KDD dataset for intrusion detection system based on classification algorithms. International journal of advanced research in computer and communication engineering, 4(6), 446–452.
4. Hasan, M. A. M., Nasser, M., Pal, B., & Ahmad, S. (2013). Intrusion detection using combination of various kernels based support vector machine. International Journal of Scientific & Engineering Research, 4(9), 1454–1463.
5. Homoliak, I., Breitenbacher, D., & Hanacek, P. (2017). Convergence Optimization of Backpropagation Artificial Neural Network Used for Dichotomous Classification of Intrusion Detection Dataset. J. Comput., 12(2), 143–155.
6. Imran, H. M., Abdullah, A. B., & Palaniappan, S. (2013). Towards the Low False Alarms and High Detection Rate in Intrusions Detection System. International Journal of Machine Learning and Computing, 3(4), 332.
7. Revathi, S., & Malathi, A. (2013). A detailed analysis on NSL-KDD dataset using various machine learning techniques for intrusion detection. International Journal of Engineering Research & Technology (IJERT), 2(12), 1848–1853.
8. Tavallaee, M., Bagheri, E., Lu, W., & Ghorbani, A. A. (2009, July). A detailed analysis of the KDD CUP 99 data set. In 2009 IEEE symposium on computational intelligence for security and defense applications (pp. 1–6). Ieee.
9. Chalapathy, R., & Chawla, S. (2019). Deep learning for anomaly detection: A survey. arXiv preprint arXiv:1901.03407.
10. Steinwart, I., Hush, D., & Scovel, C. (2005). A Classification Framework for Anomaly Detection. Journal of Machine Learning Research, 6(2).
11. Lane, T., & Brodley, C. E. (1997, October). An application of machine learning to anomaly detection. In Proceedings of the 20th national information systems security conference (Vol. 377, pp. 366–380). Baltimore, USA.
12. Sakhare, N. N., & Joshi, S. A. (2015). Classification of Criminal Data Using J48-Decision Tree Algorithm. Int. J. Data Warehous. Min, 4(3), 167–171.

Chapter 9
Facial Recognition System Using Transfer Learning with the Help of VGG16

Rajnishkumar Mishra, Saee Wadekar, Suraj Warbhe, Sayali Dalal, Riddhi Mirajkar, and Saurabh Sathe

9.1 Introduction

Facial detection and recognition is a very vital topic studied in the field of Computer Vision (CV) and Artificial Intelligence. Nowadays, applications of Facial Recognition are widespread everywhere due to its feasibility to use, efficiency, and emergence in the past few years. The fascinating challenges & advancements in Facial Recognition Systems have been attracting many engineers, data analysts, and data scientists working in Pattern Recognition, Computer Vision, and biometrics fields. Various facial recognition algorithms are being used in various programs such as compression video and indexing that come under the domain of biometrics. One of the common uses of biometrics is unlocking phones using facial recognition. It is very much popular due to its ease and convenience to use when compared to other unlocking methods like password, pin, pattern, fingerprint, iris scanner, etc. Facial recognition plays are very vital role in biometrics nowadays. Facial recognition techniques are also used to classify the contents of multimedia and to assist in a faster search of items of interest to the end-user. A well-trained facial recognition is widely used in areas such as forensic science and surveillance. It can also be used by law enforcement authorities for enhanced security, and by banking systems. Facial recognition systems have become important after the rise of terrorism and crimes in the last few years. In any case, facial recognition should be used along with other safety and security measures. Face recognition systems are being used at

R. Mishra (✉) · S. Wadekar · S. Warbhe · S. Dalal · R. Mirajkar
Department of Information Technology and Engineering, Vishwakarma Institute of Information Technology, Pune, Maharashtra, India
e-mail: rajnishkumar.21910644@viit.ac.in

S. Sathe
San Jose State University, San Jose, CA, USA
e-mail: saurabh.sathe@sjsu.edu

A. Neustein et al. (eds.), *AI, IoT, Big Data and Cloud Computing for Industry 4.0*, Signals and Communication Technology, https://doi.org/10.1007/978-3-031-29713-7_9

many significant places for security like airports, museums, etc. And the demand is expected to grow exponentially in the coming years.

Despite the rapid growth in the field of Facial Recognition as an important validation method, the algorithms used to detect and identify Human faces have neither been optimized nor improved yet. It is 25+ years since facial recognition systems are being developed but any system has not been able to give desired results in real-life and real-time conditions yet. NIST's (National Institute of Standards and Technology) FRTV (Face Recognition Vendor Test) has shown that the facial recognition methods available don't work well under various real-life conditions.

Modern facial recognition programs aimed at complex areas are becoming very crucial in recent decades. Facial recognition systems are integrated and automated with emerging technologies that have gained very much attention. Color images increase the data complexity as the pixels are mapped into a larger size. This greatly increases the need for computing resources and time while the accuracy of the model decreases. So, the higher the image resolution is, the more difficult it is to deal with. In the past few years, it has been suggested that in-depth reading works best when you have BigData. On the contrary, it has also been recognized that conventional ML methods may work better on comparatively small databases.

The proposed Face Recognition System analyzes color images so that faces can be detected and identified accurately. Convolutional Neural Network (CNN) along with VGG-16 has been used in the proposed study to get improved accuracy along with reducing the amount of time and resources spent. The biggest challenge for a facial recognition system is the recognition and rendering of a feature (i.e. data cleaning). The algorithm that has been suggested here uses in-depth reading strategies to extract features. The proposed algorithm uses in-depth reading to detect faces accurately. The proposed algorithm will be able to detect faces accurately so it can be used to search and identify suspects.

9.2 Literature Review

In this section, various types of facial recognition approaches based on Convolutional Neural networks, Machine Learning, Artificial Neural networks (ANNs), Deep Learning, and different facial algorithms are discussed.

Saypadith and Aramvith [1] proposed a new approach to facial recognition based on the Convolutional and Deep Convolutional Neural Network facial recognition algorithm. This model has been demonstrated on the embedded Graphics Processing Unit system. In 0.23 s, this system can recognize 8 faces simultaneously with an accuracy of above 83.67%.

PCA stands for Principal Component Analysis, which is a statistical technique used to reduce the dimensionality of large datasets while preserving the variation present in the data. In the context of face recognition, PCA can be used to extract the most significant features of a face from an image, such as the distance between the eyes or the shape of the nose.

YCbCr is a color space used in digital image processing and video compression. It separates the brightness (luma) and color (chroma) information of an image, making it easier to process and compress. In the context of face detection, the YCbCr color space can be used to detect skin regions in an image by thresholding the chroma values.

The Skin Colour Model proposed by Chetna Singh et al. [2] combines these two techniques to detect and recognize human faces in images. The YCbCr color space is first used to detect skin regions in the image, and then PCA is applied to extract the most significant features of the detected face regions for recognition. The algorithm has shown effective results for Indian faces, which have little variation in skin texture, complexion, and color, with an overall success rate of 95%.

Shubha and Meenakshi [3] proposed a model based on the Local Binary Pattern algorithm and Image Processing approach that has been implemented in real-time Face recognition. This algorithm is demonstrated on MATLAB software. This model has achieved an accuracy of 89% by which it can be concluded that the Local Binary Pattern Algorithm is better than Principal Component Analysis (PCA), 2D-PCA, and LDA.

Teoh et al. [4] proposed a model for Face Detection and Recognition using Haar feature-based cascade classifiers and the Convolutional Neural Network approach. This model used Haar Cascade for the frontal face and the training is generally done on a server and various stages. This Haar Cascade-based model has shown effective results with an overall success rate of 91.7% in recognizing images and 86.7% in real-time video.

Khan et al. [5] proposed a method to classify almost 1000 different people and identify their faces in an image or a real-time video. The proposed methodology is based on the approach of Transfer Learning. This model has been demonstrated on MATLAB Software using Haar-Cascade Classifiers and trained using the AlexNet Model of Neural Network which shows a success rate of 97.95%.

Deshpande and Ravishankar [6] proposed a method for human face detection and recognition. The developed methodology used the “BioID Face Database” as the standard image database. This method is efficient and is not dependent on distinct facial features like facial emotions, facial gestures, and expressions, face texture, face color and complexion, change in hairstyle, etc., using the of Viola-Jones algorithm, PCA (Principal Component Analysis), and Artificial Neural Network(ANN) and achieved an accuracy of 94%.

Sun et al. [7] proposed a Face Recognition model named “DeepID3”. DeepID3 is a combination of two different types of Deep Neural Network (DNN) Architectures (i.e. inception layers and stacked convolution). This stacked convolution is proposed in GoogleNet Model and specifically planned for face recognition. This DeepID3 model achieved a success rate of 99.53% for “Labeled Faces in the Wild (LFW)” face verification.

Sharma et al. [8] proposed a model based on Deep Belief Network (DBN). This developed model recognizes faces in still images and paintings, real-time videos, the captured images of webcam, blurred images, and side-faced images. In this, Deep Belief Network creates a layer-by-layer architecture that primarily consists of an input layer and an output layer, and a number of different layers are hidden between the input and output layer. This layered architecture should be trained with Restricted Boltzmann Machines (RBMs) to reconstruct input data. By combining all RBMs, they have introduced a collaborating method to get a powerful new model.

Deepesh Raj [9] proposed a model of unsupervised learning (i.e. Principal Component Analysis (PCA) Algorithm) for instantaneous face recognition. This developed methodology majorly consists of three different stages: Acquisition, Feature Extraction, and Face Recognition. The implementation of the proposed system was done on C++ and OpenCV Library, whereas Feature Extraction was performed using PCA Algorithm and three different types of distance classifiers (i.e. Euclidean distance, Mahalanobis distance, and Manhattan distance). This proposed model achieved a success rate of face recognition of about 92.3% whereas the normal Principal Component Analysis (PCA) Algorithm achieved a success rate of 73.1%. This trained model estimated the single queries of face recognition in less than 0.2 s.

Dara and Palanivel [10] introduced a new methodology to perform operations of face detection and face recognition systems using the approach of the VGG16 Neural Network Model. The proposed methodology is implemented in the Tensorflow Library (Keras). It is proposed on the approach of Convolutional Neural Network and Transfer Learning and achieved a success rate of 99.37%.

Swetha et al. [11] proposed a methodology of real-time automated attendance systems for students and employees. The proposed system was developed for organizations such as colleges or small startups to assist them in the basic day-to-day activities. It is proposed on the approach Haar Cascade Classifiers and Local Binary Patterns Algorithm and focuses on capturing images from a live video stream and crediting attendance based on recognition of faces in the image. The main function of the system was to enroll the subject's face against the subject's ID (unique) and Name in the database. It also helped in maintaining attendance records by allotting attendance to the recognized subjects' faces in the database.

Ramesh et al. [12] had proposed the use of surveillance cameras for face recognition using a Haar classifier. The proposed system could successfully recognize multiple faces at once without much computation time. The proposed system consisted of four steps: (i) Training, (ii) Detection and Recognition of faces using the Haar classifier, (iii) Comparison between surveillance camera images and trained real-time images, and (iv) Comparison-based result.

Jaya Prasanna Lakshmi et al. [13] had proposed an automated Smart Security System based on facial recognition using AI/ML algorithms. The proposed system could detect an intruder through surveillance cameras based on facial recognition using AI/ML algorithms. The specified design was implemented with the help of Histogram of Oriented Gradients (HOG) feature extraction and SVM (Support Vector Machine) classifier which takes a video stream as an input and classifies the faces.

Parkhi et al. [14] proposed a facial recognition method from either a single image or from a set of images captured in a video. One of the aims of their project was to show how a very large dataset of 2.6 million images from over 2600 people could be collected using the combination of computer automation and humans. Another aim was to study the different complexities of facial recognition using deep neural networks. The authors evaluated the performance of the method on two commonly used face recognition benchmarks: Labeled Faces in the Wild (LFW) and YouTube Faces (YTF). The results showed that the proposed method achieved state-of-the-art performance on both benchmarks.

9.3 Proposed Methodology

The proposed methodology involves a transferred learning concept on the pre-trained VGG16 model. It involves freezing most of the layers of the VGG16 model, adding and training some extra layers to repurpose the model according to our needs. All these are achieved in five major steps:

(i) Data Collection
(ii) Data Cleaning
(iii) Fine-tuning of VGG16 model
(iv) Model Training
(v) Performance Checking

The first step is a collection of face images of different people. Collection of images has been majorly done through Webcam which is further used as a training and testing dataset. In the second step, collected face images are cropped to specific dimensions and saved to remove unnecessary, irrelevant, or meaningless data (noise) remove noise. In the third step, a very popular image recognition Neural Network model VGG16 is fine-tuned for the specific case of face recognition with the help of a Convolutional Neural Network. The proposed work is implemented using a python module named Keras. Lastly, performance is measured using a test dataset that is different from the training dataset. The proposed system's building block can be seen in Fig. 9.1.

9.3.1 Convolutional Neural Network (CNN)

Convolutional Neural Network also commonly known as CNN, is one of the most powerful deep neural networks. Raw pixel of images acts as data for CNN, the model is trained, then the features are extracted automatically for better classification. It was a revolution as it was very difficult to create a system that would understand visual data. Ever since CNN has played a very important role in the operations of CV. An input layer following multiple hidden layers followed by an output layer is the architecture of convolutional networks. There are various layers in the CNN model which are shown in Fig. 9.2: Input layer, Convolutional Layer, Pooling Layer, ReLU correction layer, Fully-connected layer, and Output layer.

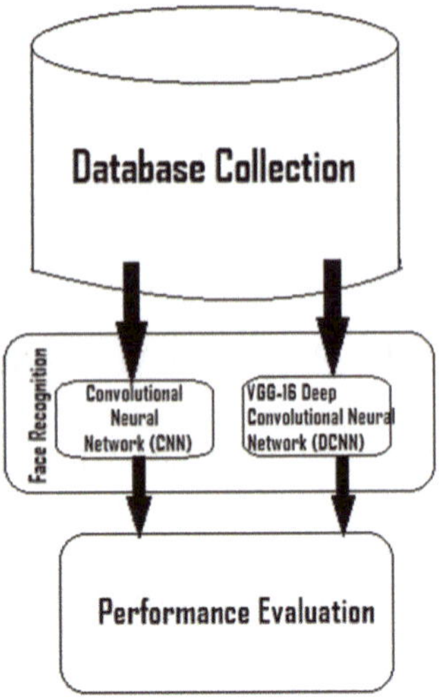

Fig. 9.1 Building block of facial recognition system

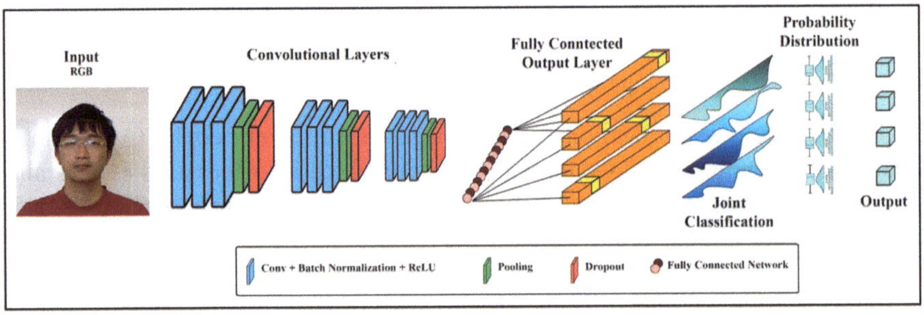

Fig. 9.2 CNN architecture (Source: Google)

9.3.1.1 Convolution Layer

It is the first and most critical layer in CNN. Its purpose is to detect the features present in the images received as input. It receives the image dataset as input and calculates its convolution for each filter. Filters get the same features we want to get in photos. Features are not predetermined but are learned during the training phase.

9.3.1.2 Pooling Layer

This layer lies between 2 Convolution + ReLU layers. It finds various feature maps and pooling operations are applied to each of them. The pooling operation involves preserving features while reducing image size. The main function of this layer is to reduce the number of parameters and reduce calculations. This results in increased efficiency and reduces the chance of over-learning. Different types of pooling functions are shown in Fig. 9.3.

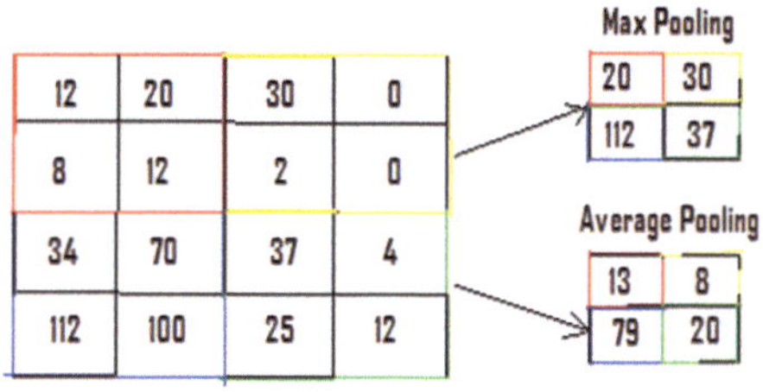

Fig. 9.3 Pooling function

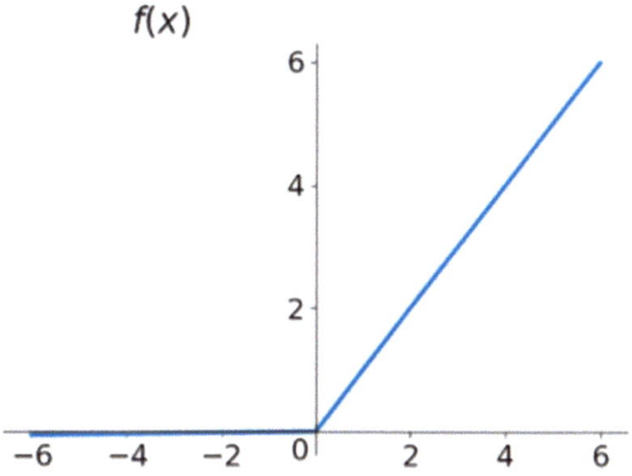

Fig. 9.4 RELU activation function

9.3.1.3 ReLU Correction Layer

ReLU is also known as Rectified Linear Unit and acts as an activation function. It is defined by $ReLU(x) = max(0,x)$. It is a piecewise linear function when the input is positive, and the output is the same as the input, and in all other cases output remains zero. The function is shown in Fig. 9.4.

9.3.1.4 Fully Connected (FC) Layer

It always remains the last layer of any neural network. The job of this layer is to receive an input vector which as a result produces a new output layer that separates the image as a network input: and returns the N-size vector, where N is equal to the number of classes in the image separation problem. Each vector element indicates the probability that the inserted image belongs to a class (Fig. 9.5).

9.3.2 VGG-16 Neural Network Model

VGG16 is one of the modular Convolutional Neural Network models proposed in the paper "Extensive Conversion Networks for Great Image Recognition". It was one of the most popular models submitted in ILSVRC-2014. The model achieved a

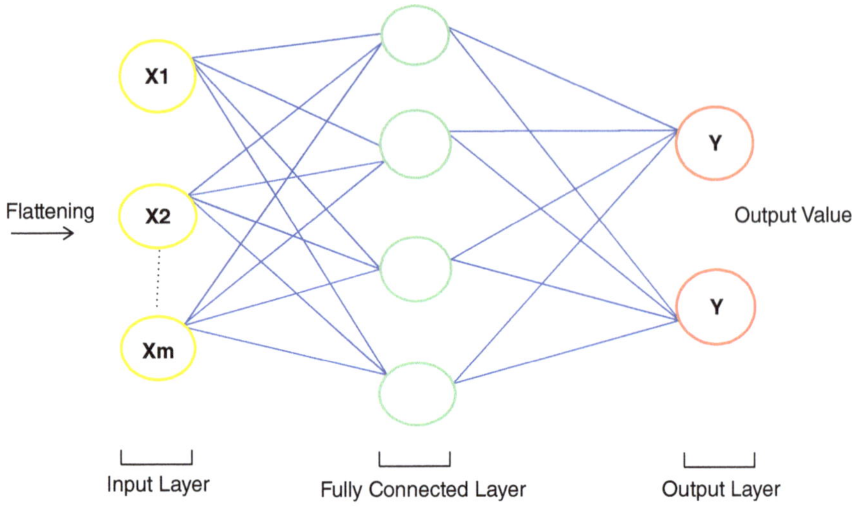

Fig. 9.5 Fully connected layer in CNN

success rate of 92.7% in almost 5 ImageNet, which is 14 million photographic sites of 1000 classes. Accuracy will be increased by changing it with large kernel-sized filters (11 & 5 in the first and second conversion layers, respectively) with multiple kernel-sized filters 3 × 3, respectively. NVIDIA Titan Black GPUs were used to train VGG16 for weeks.

Cov1 layer input is generally a static RGB image of dimension 224 × 224. This image is transferred to a series of convolutional layers, where different filters are applied with the smallest receiving field: 3 × 3 (which is the smallest imaging left / right, top/bottom, center.). In one setting, it also uses 1 × 1 dimensional convolution filters as input channels (followed by non-linear). A conversion action is limited to only 1 pixel; local padding convolutional layer input is similar to the location adjustment saved after conversion (i.e. padding ranges from 1 pixel to 3 × 3 convolutional layers). Consolidation of space is made up of five major layers of sequencing, followed by other convolutional layers (not all the layers of convolution are followed by multiple combinations). Max-pooling is performed with a window of size 2 × 2-pixel, in 2 steps.

There are three Fully Connected (FC) layers that follow a series of some flexible layers (depths vary with structures): the first two FC layers have around 4096 channels each and the third fully connected layer developed a division of ILSVRC which is of around 1000 channels (one channel for each class). The softmax layer is the last fully connected layer. Fully integrated layouts are the same across all networks (see Fig. 9.6).

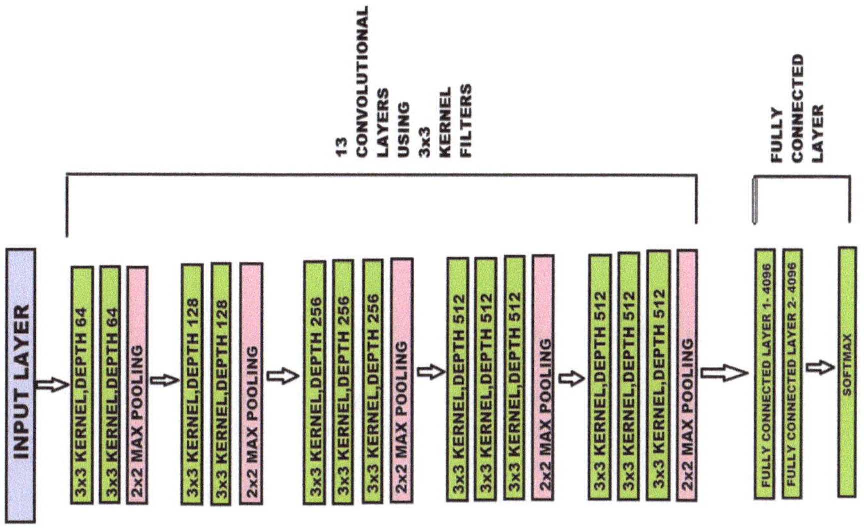

Fig. 9.6 VGG16 architecture (Source: Google)

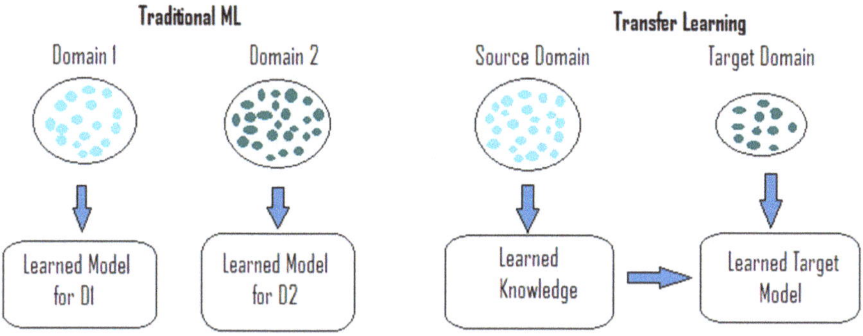

Fig. 9.7 Traditional vs. transfer Learning

9.3.3 Transfer Learning

Transfer learning helps to improve learning performance and reduce computation cost and time to a great extent. The main idea of transfer learning is to borrow labeled facial images to achieve great performance in a particular area of interest. Transfer Learning takes out the data or information from previous source domain functions to identify new target domain tasks from the pre-trained facial images. In transfer learning, the knowledge of an already trained model is used to solve a different but relatable problem (See Fig. 9.7).

Transfer Learning is used in this proposed methodology because training any model takes a lot of time and computing resources and in this fast-paced world, it's not possible to spend very much time on a particular thing. As we are already aware, we have to follow hit-and-trial methodology for finding the right hyperparameters. For that, we have to train a model again and again for different combinations of hyperparameters and check which one provides the best result. So, to avoid that, we use a pre-trained model to attach some new layers, and fine-tune some existing parameters. Using this we don't have to train the whole model, again and again, we just have to train the new and modified layers. This helps in saving a lot of time and computing resources.

9.3.3.1 Dataset Collection and Cleaning

Data can be collected using two ways: manually or by using a script. The manual method for data collection is very time taking and resource-intensive as one has to manually collect data, remove noise, and format all the image data of the same size or dimensions. So instead, a script is used for data collection. It uses the *Haar Cascade model* to detect the face and the area in which the face is present is captured and saved. This saves the majority of the time. The code for this part can be found in Fig. 9.8.

9.3.3.2 Loading the VGG16 Model and Fine-Tuning

The VGG16 model is loaded and all the existing layers are frozen so they don't train again. We add a few layers and fine-tune VGG16 to make this model customized for this specific use case. The last layer is the softmax layer to get the output in the desired class. The code for this part can be found in Fig. 9.9.

9.3.4 Loading Dataset and Training Model

As for training a model, we need Big Data so we can get accurate results in all conditions. But here we have very few amounts of data, so we apply various filters to increase the dataset. The code for this part can be found in Fig. 9.10.

```
# Initialize Webcam
import cv2
import numpy as np

# Load HAAR face classifier
face_classifier = cv2.CascadeClassifier('haarcascade_frontalface_default.xml')

# Load functions
def face_extractor(img):
    # Function detects faces and returns the cropped face
    # If no face detected, it returns the input image

    faces = face_classifier.detectMultiScale(img, 1.3, 5)

    if faces is ():
        return None

    # Crop all faces found
    for (x,y,w,h) in faces:
        cropped_face = img[y:y+h, x:x+w]

    return cropped_face
cap = cv2.VideoCapture(0)
count = 0
```

```
# Collect 100 samples of your face from webcam input
while True:

    ret, frame = cap.read()
    if face_extractor(frame) is not None:
        count += 1
        face = cv2.resize(face_extractor(frame), (224, 224))

        # Save file in specified directory with unique name
        file_name_path = 'C://Users//Rajnish - The Great//Desktop//rajnish' + str(count) + '.jpg'
        cv2.imwrite(file_name_path, face)

        # Put count on images and display live count
        cv2.putText(face, str(count), (50, 50), cv2.FONT_HERSHEY_COMPLEX, 1, (0,255,0), 2)
        cv2.imshow('Face Cropper', face)

    else:
        print("Face not found")
        pass

    if cv2.waitKey(1) == 13 or count == 200: #13 is the Enter Key
        break

cap.release()
cv2.destroyAllWindows()
print("Collecting Samples Complete")
```

Fig. 9.8 Data collection and cleaning

Face Recognition with VGG16

Loading the VGG16 Model

```
In [1]: from keras.applications import VGG16

        # VGG16 was designed to work on 224 x 224 pixel input images sizes
        img_rows = 224
        img_cols = 224

        #Loads the VGG16 model
        model = VGG16(weights = 'imagenet',
                         include_top = False,
                         input_shape = (img_rows, img_cols, 3))
```

```
In [3]: from keras.applications import VGG16

        # VGG16 was designed to work on 224 x 224 pixel input images sizes
        img_rows = 224
        img_cols = 224

        # Re-loads the VGG16 model without the top or FC layers
        model = VGG16(weights = 'imagenet',
                         include_top = False,
                         input_shape = (img_rows, img_cols, 3))

        # Here we freeze all the layers of pre trained model
        # Layers are set to trainable as True by default
        for layer in model.layers:
            layer.trainable = False

        # Let's print our layers
        for (i,layer) in enumerate(model.layers):
            print(str(i) + " "+ layer.__class__.__name__, layer.trainable)

        0 InputLayer False
        1 Conv2D False
        2 Conv2D False
        3 MaxPooling2D False
        4 Conv2D False
        5 Conv2D False
        6 MaxPooling2D False
        7 Conv2D False
        8 Conv2D False
        9 Conv2D False
        10 MaxPooling2D False
        11 Conv2D False
        12 Conv2D False
        13 Conv2D False
        14 MaxPooling2D False
```

```
In [4]: def addTopModel(bottom_model, num_classes, D=256):
            """creates the top or head of the model that will be
            placed ontop of the bottom layers"""
            top_model = bottom_model.output
            top_model = Flatten(name = "flatten")(top_model)
            top_model = Dense(D, activation = "relu")(top_model)
            top_model = Dropout(0.3)(top_model)
            top_model = Dense(num_classes, activation = "softmax")(top_model)
            return top_model

In [7]: from keras.models import Sequential
        from keras.layers import Dense, Dropout, Activation, Flatten
        from keras.layers import Conv2D, MaxPooling2D, ZeroPadding2D
        from keras.layers.normalization import BatchNormalization
        from keras.models import Model

        num_classes = 2

        FC_Head = addTopModel(model, num_classes)

        modelnew = Model(inputs=model.input, outputs=FC_Head)

        print(modelnew.summary())
```

Fig. 9.9 VGG16 and fine-tuning

```
In [11]: from keras.preprocessing.image import ImageDataGenerator

         train_data_dir = 'D://face//train//'
         validation_data_dir = 'D://face//validation//'

         train_datagen = ImageDataGenerator(
               rescale=1./255,
               rotation_range=20,
               width_shift_range=0.2,
               height_shift_range=0.2,
               horizontal_flip=True,
               fill_mode='nearest')

         validation_datagen = ImageDataGenerator(rescale=1./255)

         # Change the batchsize according to your system RAM
         train_batchsize = 64
         val_batchsize = 64

         train_generator = train_datagen.flow_from_directory(
                 train_data_dir,
                 target_size=(img_rows, img_cols),
                 batch_size=train_batchsize,
                 class_mode='categorical')

         validation_generator = validation_datagen.flow_from_directory(
                 validation_data_dir,
                 target_size=(img_rows, img_cols),
                 batch_size=val_batchsize,
                 class_mode='categorical',
                 shuffle=False)

         Found 544 images belonging to 2 classes.
         Found 200 images belonging to 2 classes.
```

```
In [9]: from keras.optimizers import RMSprop
        from keras.callbacks import ModelCheckpoint, EarlyStopping

        checkpoint = ModelCheckpoint("facemodel.h5",
                                     monitor="val_loss",
                                     mode="min",
                                     save_best_only = True,
                                     verbose=1)

        earlystop = EarlyStopping(monitor = 'val_loss',
                                  min_delta = 0,
                                  patience = 3,
                                  verbose = 1,
                                  restore_best_weights = True)

        # we put our call backs into a callback list
        callbacks = [earlystop, checkpoint]

        # Note we use a very small learning rate
        modelnew.compile(loss = 'categorical_crossentropy',
                         optimizer = RMSprop(lr = 0.001),
                         metrics = ['accuracy'])

        nb_train_samples = 544
        nb_validation_samples = 200
        epochs = 1
        batch_size = 64
```

```
history = modelnew.fit_generator(
    train_generator,
    steps_per_epoch = nb_train_samples // batch_size,
    epochs = epochs,
    callbacks = callbacks,
    validation_data = validation_generator,
    validation_steps = nb_validation_samples // batch_size)

modelnew.save("facemodel.h5")

Epoch 1/2
8/8 [==============================] - ETA: 0s - loss: 2.9264 - accuracy: 0.8500
Epoch 00001: val_loss improved from inf to 0.00060, saving model to facemodel.h5
8/8 [==============================] - 160s 20s/step - loss: 2.9264 - accuracy: 0.8500 - val_loss: 5.9737e-04 - val_accura
cy: 1.0000
Epoch 2/2
8/8 [==============================] - ETA: 0s - loss: 0.0166 - accuracy: 0.9917
Epoch 00002: val_loss did not improve from 0.00060
8/8 [==============================] - 206s 26s/step - loss: 0.0166 - accuracy: 0.9917 - val_loss: 0.0047 - val_accuracy:
1.0000
```

Fig. 9.10 Loading dataset and training model

9.3.4.1 Validation and Prediction (Fig. 9.11)

```
In [1]: from keras.models import load_model

        classifier = load_model('facemodel.h5')
```

Testing our classifer on some test images

```
In [*]: import os
        import cv2
        import numpy as np
        from os import listdir
        from os.path import isfile, join

        face_data_dict = {"[0]": "Rajnish",
                          "[1]": "Suraj"}

        face_data_dict_n = {"rajnish": "Rajnish",
                            "suraj": "Suraj"}

        def draw_test(name, pred, im):
            monkey = face_data_dict[str(pred)]
            BLACK = [0,0,0]
            expanded_image = cv2.copyMakeBorder(im, 80, 0, 0, 100 ,cv2.BORDER_CONSTANT,value=BLACK)
            cv2.putText(expanded_image, monkey, (20, 60) , cv2.FONT_HERSHEY_SIMPLEX,1, (0,0,255), 2)
            cv2.imshow(name, expanded_image)
```

```
def getRandomImage(path):
    """function loads a random images from a random folder in our test path """
    folders = list(filter(lambda x: os.path.isdir(os.path.join(path, x)), os.listdir(path)))
    random_directory = np.random.randint(0,len(folders))
    path_class = folders[random_directory]
    print("Class - " + face_data_dict_n[str(path_class)])
    file_path = path + path_class
    file_names = [f for f in listdir(file_path) if isfile(join(file_path, f))]
    random_file_index = np.random.randint(0,len(file_names))
    image_name = file_names[random_file_index]
    return cv2.imread(file_path+"/"+image_name)

for i in range(0,10):
    input_im = getRandomImage("D://face//validation//")
    input_original = input_im.copy()
    input_original = cv2.resize(input_original, None, fx=0.5, fy=0.5, interpolation = cv2.INTER_LINEAR)

    input_im = cv2.resize(input_im, (224, 224), interpolation = cv2.INTER_LINEAR)
    input_im = input_im / 255.
    input_im = input_im.reshape(1,224,224,3)

    # Get Prediction
    res = np.argmax(classifier.predict(input_im, 1, verbose = 0), axis=1)

    # Show image with predicted class
    draw_test("Prediction", res, input_original)
    cv2.waitKey(0)

cv2.destroyAllWindows()
```

Fig. 9.11 Validation and prediction

9.3.4.2 Output (Fig. 9.12)

Fig. 9.12 Output

9.4 Result and Discussion

This section mainly focuses on how we can get an accuracy score for the trained model. It also discusses the comparison between two different facial recognition algorithms (i.e. AlexNet and VGG16 Neural Network Model).

The accuracy of a model can be determined with the help of a confusion matrix. It's used to describe the accuracy of a classification model. The confusion matrix is not only limited to binary classification but also can be used for multi-classification. As facial recognition is also a type of classification (generally multi), we can use a confusion matrix to determine the accuracy. The values of the confusion matrix need to be calculated for each class so that the *ith* confusion matrix considers the g_ia phase considered to be positive and the remaining categories g_j by $j \neq i$ as negative. Since each confusion matrix combines all the visuals labeled with a different category except g_i as a negative category, this method increases the number of real negatives. This gives us:

1. When the event values are predicted correct, it is True Positive (TP).
2. When the event values are predicted incorrect, it is a False Positive (FP) also known as a Type I error.
3. When the non-event values are predicted correct, it is True Positive (TN).
4. When the non-event values are predicted incorrect, it is a False Positive (FN) also known as a Type II error.

The confusion matrix not only helps in describing the accuracy but is also very useful in measuring important parameters like Precision, Recall, and F-measure.

TP*i*, TN*i*, FP*i*, and FN*i* represent real realism, real injustice, false opposition, and false falsehood, in the confusion matrix wrt ith class. The recall is represented by *R* and the accuracy of *P*.

Minimum includes working with the smallest possible unit (total facial images):

$$P_{\text{micro}} = \frac{\sum_{i=1}^{|G|} \text{TP}_i}{\sum_{i=1}^{|G|} \text{TP}_i + \text{FP}_i} \tag{9.1}$$

$$R_{\text{micro}} = \frac{\sum_{i=1}^{|G|} \text{TP}_i}{\sum_{i=1}^{|G|} \text{TP}_i + \text{FN}_i} \tag{9.2}$$

The micro $F1$-score is formulated using the relation of recall, micro-averaged precision, *P_micro*, and, *R_micro*:

$$F1_{\text{micro}} = 2.\frac{P_{\text{micro}}.R_{\text{micro}}}{P_{\text{micro}} + R_{\text{micro}}} \tag{9.3}$$

Assuming the divider gets a large [$F1$] micro, it works very fine. Here, a small amount may not be resistant to all speculative performance. Due to this, if there is an imbalance in class distribution, it can be misleading.

Macro Average is formulated as over large groups and over individual classroom performance rather than viewing by using:

$$P_{\text{macro}} = \frac{1}{|G|}\sum_{i=1}^{|G|} \text{TP}_i/\text{TP}_i + \text{FP}_i \tag{9.4}$$

$$R_{\text{macro}} = \frac{1}{|G|}\sum_{i=1}^{|G|} \text{TP}_i/\text{TP}_i + \text{FN}_i \tag{9.5}$$

Macro $F1$-score is formulated as:

$$F1_{\text{macro}} = 2.\frac{P_{\text{macro}}.R_{\text{macro}}}{P_{\text{macro}} + R_{\text{macro}}} \tag{9.6}$$

If the value of $F1_{\text{macro}}$ is bigger, then it indicated that the classifier performs for individual classes.

Multi-class accuracy is defined as the average of the correct predictions and formulated as:

$$\text{Accuracy} = \frac{1}{N}\sum_{k=1}^{|G|}\sum_{x:g(x)=k} I\left(g(x) = \hat{g}(x)\right) \tag{9.7}$$

where, the indicator function is represented, which returns 1 when there is any match between the classes, and otherwise it will return 0.

VGG16 has several differences from AlexNet. Instead of using large receptive fields like Alexnet, VGG16 uses small receptive fields which allows VGG16 to have a large number of weight layers. This result helps VGG16 to achieve higher accuracy but also makes the training process more computable and time-consuming. The accuracy of VGG 16 is generally seen to be 7–10% higher than AlexNet.

9.5 Future Work

The proposed Facial Recognition System is based on the concept of Transfer Learning and the VGG16 Neural Network. It consists of five main Stages: (i) Data Collection, (ii) Data purification (Data cleaning), Fine-tuning of the VGG16 model, (iv) Model Training, and (v) Performance Checking. As we already discussed how this fine-tuned model works better for this particular situation although the dataset was very small. The model can be tested with much bigger datasets (i.e. BigData) to get a better understanding of how this model will perform in real-life situations. The transfer learning (i.e. fine-tuning methodology) should not be limited only to VGG16. There are various Neural Networks like ResNet, Inception, etc. which perform better than VGG16. Hyperparameters can also be adjusted accordingly to get desired results under exceptional conditions like position scale, camera angle variations, blurred vision, extremely low light, extremely bright light, etc.

With help of this, big leaps of improvement can be achieved. The Artificial Intelligence industry is growing very rapidly and is extensively used almost everywhere in the world. This will take the AI in mobile phone cameras, self-driving cars, Robotic Process Automation (RPA), Healthcare Automation, Social Media platforms, and security systems to the very next level.

References

1. Saypadith S. and S. Aramvith (2018), "Real-Time Multiple Face Recognition using Deep Learning on Embedded GPU System", Asia-Pacific Signal and Information Processing Association Annual Summit and Conference (APSIPA ASC), Pp.1318–1324, 2018.
2. Chetna Singh, Prashant Baheti, and Sarvesh Singh, "Human Face Recognition and Face Detection using Skin Colour Model", Researchgate, 2014.
3. Shubha P. and Meenakshi M., "Human Face Recognition Using Local Binary Pattern Algorithm Real-Time Validation", International Conference On Computational Vision and Bio-Inspired Computing, Pp.240–246, 2019.
4. KH Teoh, RC Ismail, SZM Naziri, R Hussin, MNM Isa, and MSSM Basir, "Face Recognition and Identification using Deep Learning Approach", 2021.
5. Suleman Khan, Shah, Ehtasham Ahmed, Syed A. A. M. Hammad Javed, and Syed Umaid Ali, "Transfer Learning of a Neural Network Using Deep Learning to Perform Face Recognition", International Conference on 24–25 July 2019, IEEE.
6. Narayan T. Deshpande and Dr. S. Ravishankar (2017), "Face Detection and Recognition using Viola-Jones algorithm and Fusion of PCA and ANN".

7. Sun Y., X. Tang, and X. Wang, "Deep learning face representation from predicting 10000 classes", Proc. IEEE Conference Computer Vision Pattern Recognition, pp. 1891–1898, Jun. 2014.
8. Manik Sharma, J Anuradha, H K Manne and G S C Kashyap (2017), "Facial Detection using Deep Learning", IOP Conference Series Materials Science and Engineering 263(4):042092.
9. Deepesh Raj, "A real-time Face Recognition System using PCA and various distance classifiers", CS676: Computer Vision and Image Processing, Pp.1–11, 2011.
10. Showkat A. Dara, S. Palanivel, "Neural Networks (CNNs) and VGG on Real-Time Face Recognition System", Turkish Journal of Computer and Mathematics Education Vol. 12 No. 9 (2021), 1809–1822.
11. S. SwethaK, E. Surekha, S. Sindhuja, Rohini Kailas, "Attendance Management using face detection and recognition", 2020.
12. P. Apoorva, H. C. Impana, S. Siri, M. Varshitha., B. Ramesh, "Identification by Face Recognition using Open Computer Vision Classifiers", IEEE 3rd International Conference on Computing Methodologies and Communication (ICCMC), Pp. 775–778, 2019.
13. K. Jaya Prasanna Lakshmi, Sridevi Warrier, and T. Kumar, "Automated Face Recognition by Smart Security System Using AI & ML Algorithms", Published in 2021 5th International Conference on Trends in Electronics and Informatics (ICOEI), IEEE.
14. Omkar M. Parkhi, Andrea Vedaldi, Andrew Zisserman, "Deep Face Recognition", In BMVC, vol. 1, p. 6 (2015).

Chapter 10
Digitization in Teaching and Learning: Opportunities and Challenges

Sachin R. Sakhare, Nidhi Santosh Kulkarni, Nidhi Deshpande, and Apurva Pingale

10.1 Introduction

Distance learning, also known as online learning, is traditionally defined as an educational or learning procedure in which students are geographically separated. People are not completely unfamiliar with online learning. According to a study, online learning began in the 1960s at the University of Illinois. Due to the increasing importance of education in India, distance learning was introduced in 1962. There are currently platforms such as Coursera, EDX, and others that provide courses from various universities.

This has greatly benefited engineering students. Even though online education is not a new concept, the COVID-19 pandemic has created numerous challenges. Distance learning was once considered a luxury, but it is now regarded as a necessity.

From March 2020, the colleges and universities were online and there were mixed reactions about this system. In [4] Mohanty and Dash discussed the engineering education scenario in INDIA.

Now as online education is mandatory, not everyone has access to the internet or other requirements. Based on the places where the students and teachers live there are more internet issues. Also, there are different challenges department-wise. Engineering is not a theory-oriented subject, it requires a practical approach. Getting this practical knowledge is difficult from this online learning. Conducting practical for Mechanical department, Civil department, etc. online is not so fruitful. Computer

S. R. Sakhare (✉)
Computer Engineering, Vishwakarma Institute of Information Technology, Pune, Maharashtra, India
e-mail: sachin.sakhare@viit.ac.in

N. S. Kulkarni · N. Deshpande · A. Pingale
Vishwakarma Institute of Information Technology, Pune, Maharashtra, India
e-mail: nidhi.21810100@viit.ac.in; nidhi.21810041@viit.ac.in; apurva.21810611@viit.ac.in

A. Neustein et al. (eds.), *AI, IoT, Big Data and Cloud Computing for Industry 4.0*, Signals and Communication Technology, https://doi.org/10.1007/978-3-031-29713-7_10

Engineering, Information Technology departments are the most benefited from this. But others suffer a little due to a lack of practical orientation. Now there are various reasons for this including the software required, internet connection. Also, many students have their practical sessions by teachers sharing their screen or just by the PowerPoint presentation. Not only these but also there are various other problems faced.

The main contributions of this study are as follows:

- The preparation of a questionnaire for students and teachers for noting their experiences during this COVID-19 pandemic
- Preparation of the data set that we got from the responses collected from the survey initiated for teachers and students
- Graphs and visualization to draw useful inferences and interpretations from the data set

With the impact of digitization in teaching and learning opportunities, the same can be applicable and exploited in industry 4.0. It will be a great asset in exploring the learning prospects in industry 4.0, thereby making it easier for imparting training. E-content/e-learning modules can be accessed making employees support the transformation journey in the manufacturing processes.

10.2 Paper Organization

The next section of the chapter is organized as related work which includes the works related to the topic, the next section is of the proposed methodology where the working methods are discussed. The next section is of dataset and dataset description and the last section is of results and discussions.

10.3 Related Work

Since the COVID-19 outbreak in March 2020, it has caused havoc across the world this has brought far-reaching changes in all aspects of our lives and the education system has also been hit hard. Students, schools, colleges, and universities have been strongly affected. During this pandemic, many researchers have also shared their works on teaching and learning in different ways. Edwige Simon in his research employed a mixed-method approach to investigate the impact of online teaching on higher education educators' professional identity, and the role played by technology in this process [1]. The work by Shadnaz Asgari, Jelena Trajkovic, Mehran Rahmani, Wenlu Zhang, Roger C. Lo, and Antonella Sciortino conducts a thorough (quantitative and qualitative) analysis of challenges and factors affecting the online education of engineering courses by conducting surveys among students and faculty members from various engineering subfields at one of the largest and most diverse 4-year US universities (CSULB) [2]. Paul Gorsky and Ina Blau in their work discuss findings and conclusions concerning teaching effectiveness in traditional classrooms [3]. The work by Michał Bączek, Michalina Zagańczyk-Bączek, Monika Szpringer, Andrzej Jaroszyński, and Beata Wożakowska-Kapłon used the survey method and then drew conclusions based on the responses [5]. M.Sandeep Kumar, Prabhu Jayagopal, Shakeel Ahmed, S. S. Manivannan, Kiruba Thangam Raja, and S. Sree Dharinya give us an overview of the adoption of e-learning during COVID-19 in India [6]. Pinaki Chakraborty, Prabhat Mittal, and Savita Yadav, through their work, discuss about the students' opinion about online education during the pandemic [17]. The work by Showkat Ahmad Dar and Ahmad Naseer discusses the importance of online learning and Strengths, Weaknesses, Opportunities, & Challenges (SWOC) analysis of e-learning modes in the time of crisis and also put some light on the growth of EdTech Start-ups during the time of pandemic and natural disasters and includes suggestions for academic institutions of how to deal with challenges associated with online learning [21]. The work by Elizabeth Armstrong-Mensah, Kim Ramsey-White, Barbara Yankey, and Shannon Self-Brown had the purpose of collecting information on how the transition to distance learning impacted undergraduate and graduate students taking courses in public health at GSU and the goal was to identify student academic challenges and unforeseen benefits of distance learning, and to use that information to inform practices that can be implemented during future crises that impact university education [22]. The paper by Amit Joshi, Muddu Vinay, and Preeti Bhaskar aims to identify the challenges faced by the teachers during online teaching and assessment in different home environment settings in India [20]. The paper by Lokanath Mishra, Tushar Gupta, and Abha Shree employs both quantitative and qualitative approach to study the perceptions of teachers and students on online teaching-learning modes and also highlighted the implementation process of online teaching-learning modes [15]. Many educational institutions that were earlier reluctant to change their traditional practices had no choice but to move to online teaching–learning. There is a strong need to innovate alternative educational systems and implement them [7]. It offers how India adopts the e-learning approach in this situation. The outbreak also describes how to handle challenges related to online learning. Online learning and teaching needs only suitable internet connectivity and a laptop or smartphone. Also, students do not have to move from one place to another. Learning from home also provides a good and relaxing place to concentrate, as students can have the best surroundings for themselves [19]. The importance of e-learning is rising, as the academic year has been completely disturbed due to COVID-19. In [8–12] authors mainly presented the pros and cons of online education in India and other countries specially about Engineering Education. Arkorful and Nelly discussed the pros and cons of online education [13]. In [14] comparison of learning outcomes in the context of finance courses is presented. Article [16] analyses the opinions of students about higher education. In [18] authors present the role of professors and technology in online education during pandemic in the country.

This chapter is organized as follows: the proposed methodology, dataset and data description, and results and discussions.

10.4 Proposed Methodology

We have been witnessing these problems and this is what made our research on this topic, especially for engineering students and teachers, find out some solutions to ease the difficulties in online education.

To analyze the situation, we conducted a survey, for both teachers and students. Our work aims questions:

For Students:

1. Is online education more conducive to cheating?
2. At home/place of residence, how many responsibilities do you have?
3. How difficult or easy is it for you to connect to the internet to access your work?
4. How often do you get so focused on activities in your classes that you lose track of time?
5. How difficult or easy is it to use online learning technology (computer, tablet, video calls, learning applications, etc.)?
6. Are you getting all the help you need with your work right now?
7. If you are participating in distance learning, how often do you hear from your teachers individually?
8. Which type of digital approach motivates you to learn?
9. Are the online practical helpful?
10. Do you have the software required for performing practical?
11. How are practical conducted?
12. How difficult or easy is it for you to try hard on your work right now?
13. In your classes, how eager are you to participate?
14. How much time do you spend on online lectures?
15. What technology do you use to attend online classes?
16. What is your department?

For Teachers**:**

1. Do you have high-speed internet at home?
2. What device do you use for distance learning?
3. Which online teaching platform is used by your college?
4. Are you satisfied with the technology and software your university is using for online teaching?
5. How stressful do you find teaching remotely during the COVID-19 pandemic?
6. How well could you maintain a work-life balance while teaching remotely?
   (Consider 5 being extremely well and 1 being not at all)
7. How stressful were your students while learning remotely during the COVID-19 pandemic?
   (Consider 5 being extremely stressful and 1 being not at all)
8. How often do you have a 1–1 discussion with your students?
9. Are your students learning as much now as they were before switching to remote learning?
10. Which online assessment method is better?

11. How effective is an online assessment process? (Consider 5 being extremely well and 1 being not at all).
12. How much time do you spend each day on average on online teaching?
13. What kind of difficulties are you facing regarding the online practical concerning circuit/non-circuit branches?
14. How do you feel overall about online education?

The questions above almost cover all of the issues that students and teachers face. All of the important aspects of online classes, such as effectiveness, interest, access, student engagement, teacher involvement, and practical, could be gleaned from these questions. We were able to thoroughly investigate all of the issues and come up with a solution. The findings of this study could help to improve the efficacy of online learning.

10.5 Dataset and Data Description

The goal of this study was to identify and investigate the magnitude of the impact of online teaching on both students and teachers. This survey was created using "Google Forms." There were other tools available, such as SurveyMonkey, but we chose Google Forms because it has no limit on the number of questions and is completely free. There were 16 multiple-choice questions in this survey. The questions were chosen so that they covered all of the important aspects of online teaching. The student survey included questions not only about technology, such as the software required or the technology used to attend online lectures (PC, Laptop, mobile phones) but also about student involvement. There are concerns about the teaching methods and the methods that students are interested in. This includes both theoretical and practical sessions. The questions about the online exams were also taken into account. The most pressing issue is that of cheating cases. With the general questions, we were also specific about the students' department and the area in which they live. These factors also have a significant impact on student participation in online learning.

To analyze the data we collected, we used the pivot tables. We used Excel and Power BI to create visualizations. From this, we were able to get the exact number of students and teachers choosing options for every question. Also, we were able to get the percentages of each answer choice that was chosen.

By using the above method, we could get the correlation between the major aspects of students and also between students and teachers.

10.6 Results and Discussion

The survey was conducted using google forms.

A total of 1060 students participated in the survey.

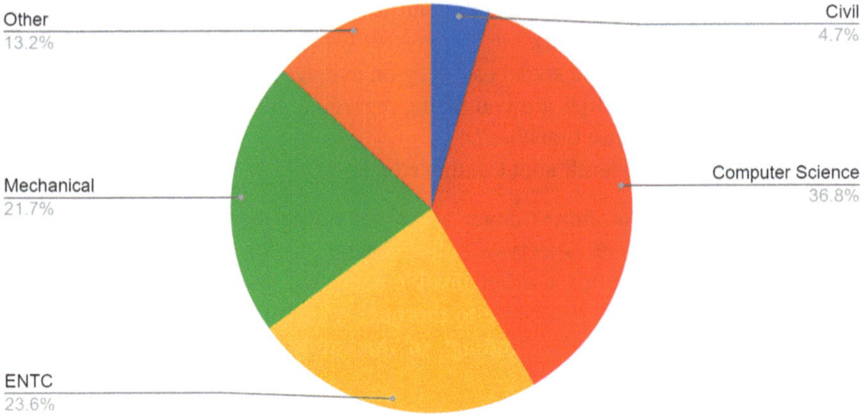

Fig. 10.1 Participant's classification: streamwise

10.6.1 Department

The students were from all the departments. About 36.8% of students were from the computer department, 23.6% of students were from the ENTC department, 21.7% were from the mechanical department, only 4.7% were from the civil department, and 13.2% of students were from other departments like production engineering, industrial engineering, etc.

So numerically we can say that, out of 1060, 390 students were from the computer science department, 250 from ENTC, 230 from mechanical, 140 from other departments, and only 50 students from the Civil department, as shown in Fig. 10.1.

10.6.2 Technology

According to the survey 990 students out of 1060 students use mobile phones and laptops both for online classes, 30 students use phones, 20 students use PC, Laptops, and phones (all three of them), and 20 students use department screen shares. That is 93.4% of students use both laptops and phones for online classes. 2.8% of students use phones (only) to attend classes, shown in Fig. 10.2.

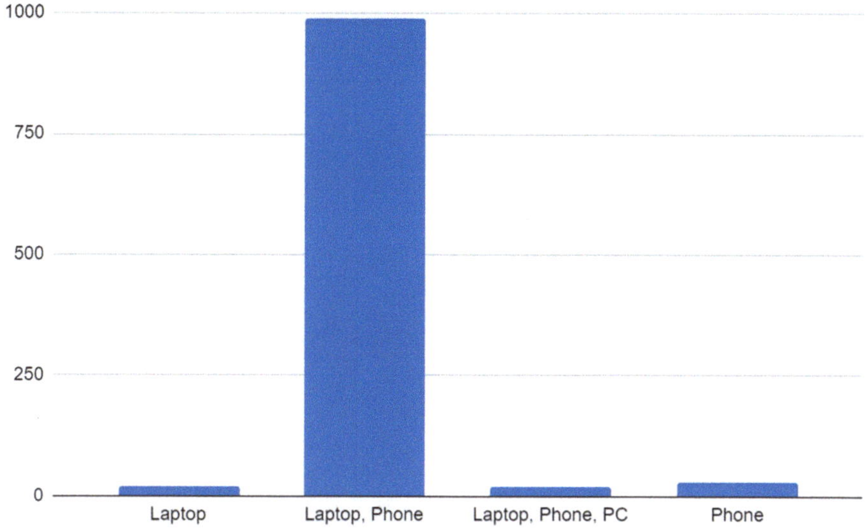

Fig. 10.2 Statistics of equipment used for online learning

10.6.3 Participation

A total of 36.8% of students are often interested in learning, 34% students show interest sometimes, 16% students are always interested, 4.5% are rarely interested, and 8.5% (90/1060) students are never interested in learning, as shown in Fig. 10.3.

10.6.4 Time

From the analysis we can see that 54.7% of students spend 3–5 h, 21.7% spend 1–3 and 5–7 h, and only 1.9% of students spend 7–10 h on online teaching.

10.6.5 Online Practical

For most of the students (i.e., 64.2%), online practical are conducted by screen share, and 26.4% have their practical conducted by the applications, as shown in Fig. 10.4.

Of the students whose practical are conducted using software, 66% have the software required to conduct practical and 23.6% do not have the software required for practical, as shown in Fig. 10.5.

According to the study, 39.6% of students find online practical useful, 1.9% strongly agree that online practical are useful, 28.3% of people are neutral about

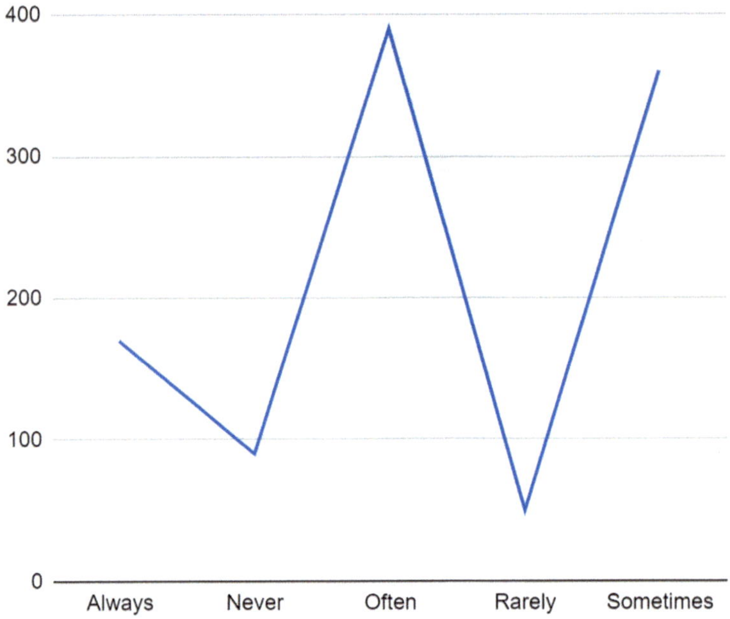

Fig. 10.3 Student's interest of learning online

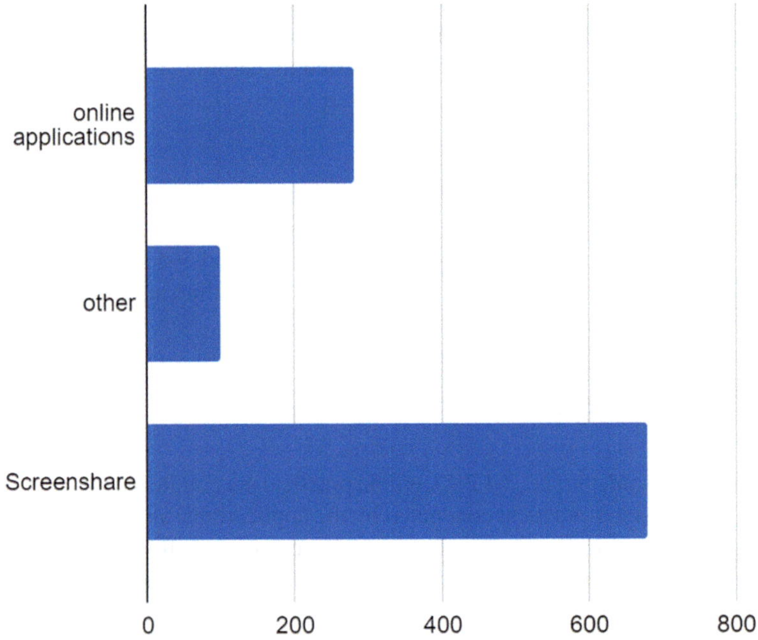

Fig. 10.4 Screen time

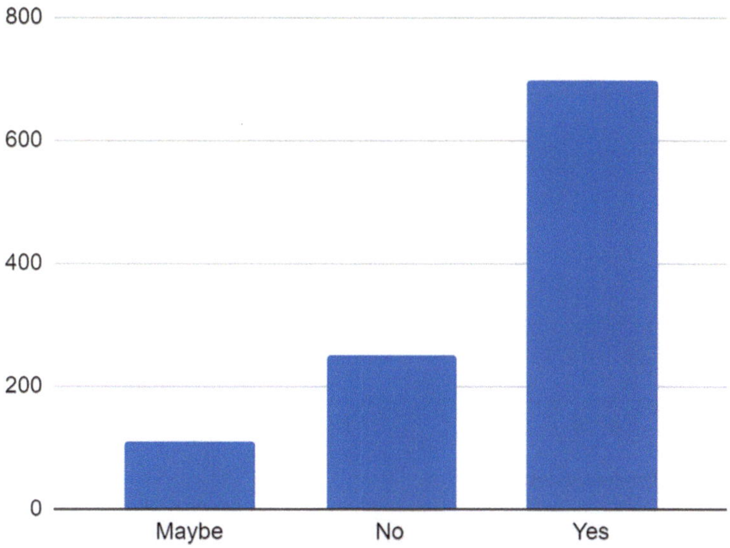

Fig. 10.5 Software availability

this, whereas 17.9% of people disagree with the efficiency of online practical, and 12.3% of students strongly disagree with online practical.

10.6.6 Teacher-Student Interaction

Out of 1060 respondents, only 260 students (24.5%) hear regularly from their teachers, and 9.4% of students do not hear anything from their teachers, 66% of students hear from their teachers occasionally.

Out of 1060 respondents, 420 respondents (39.6%) get the help needed and 33% of respondents do not get the help needed as shown in the Fig. 10.6.

10.6.7 Technology

A total of 46.2% of students found it was moderate to learn new technology, 43.4% of students found it easy to learn new technology, and only 10.4% of students found it difficult to learn new technology.

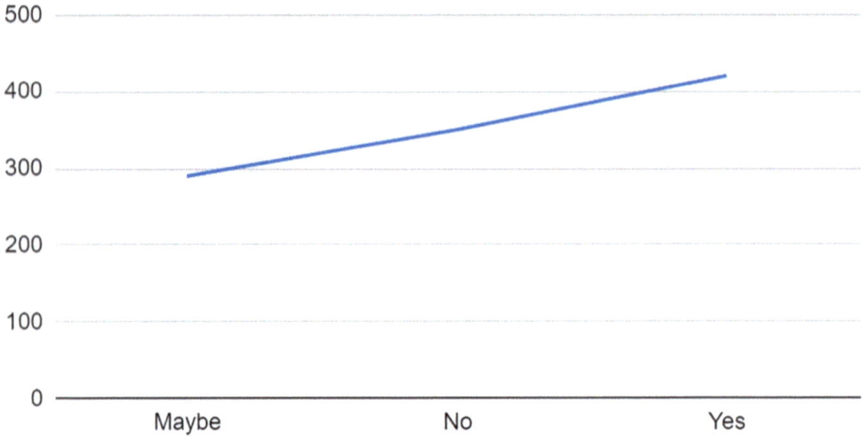

Fig. 10.6 Teacher-student interaction

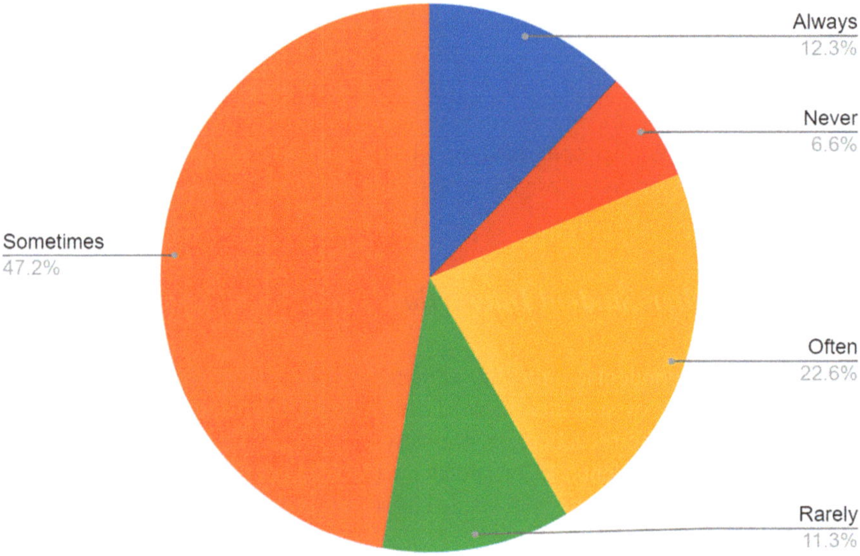

Fig. 10.7 Concentration on learning while online

10.6.8 Focus

Only 12.3% of students can focus very keenly on their work, and 6.6% of students are completely unable to focus on their work. Maximum students (47.2%) are sometimes able to focus on their work keenly, as shown in Fig. 10.7.

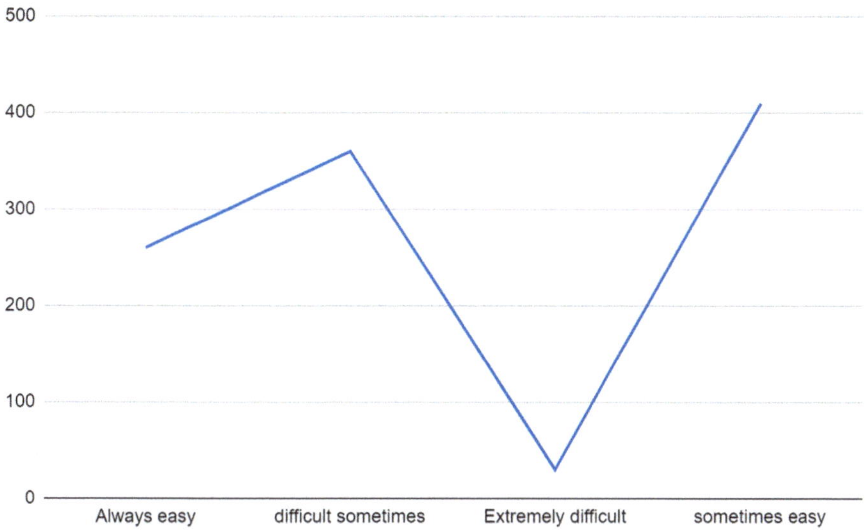

Fig. 10.8 Internet availability

10.6.9 Internet Connectivity

A total of 38.7% of students find it sometimes easy to get internet connectivity, 34% of students find it sometimes difficult to get an internet connection, 24.5% of students find it always easy to connect to the internet, and only 3.8% of students find it extremely difficult to get a proper internet connection, as shown in Fig. 10.8.

10.6.10 Exam

A total of 48.1% of students think that online exams are conducive to cheating, 19.8% do not think this, whereas 32.1% of people have no fixed opinion about this.

10.6.11 Area of Living

Out of 1060 students, 580 students (i.e., 54.7% students) live in rural areas and 45.3% students live in urban areas.

10.7 Results from Teacher's Survey

This survey was also conducted using google forms.

A total of 180 teachers from different universities participated in the survey. The questions and the responses are as follows.

In Fig. 10.9(i), responses of the question regarding equipment used for online learning are shown. In Fig. 10.9(ii), responses regarding high-speed internet and adaptability of teacher for online teaching are shown; most teachers nowadays are equipped with high-speed internet and laptops for conducting online classes and have adapted themselves as the need of the hour. In Fig. 10.9(iii), responses of question regarding online platform used for online education are shown; Most colleges are using Zoom meeting application for the conduction of online classes over Others i. e. Google meet, Microsoft Teams etc.

In Fig. 10.9(iv), responses of the question regarding teacher's satisfaction are shown. The maximum number of teachers was okay with zoom being the mode of communication. But there were around 20% of teachers who were dissatisfied.

Out of all the teachers we surveyed, 60% got accustomed to the stress this pandemic has offered; 30% were finding it difficult to cope up with, as shown in Fig. 10.10(i). Out of 180 teachers, only 70 were able to balance their work-life balance extremely well and others were doing okay too (Fig. 10.10(ii)). It was students, who found it difficult to cope up with the stress as shown in Fig. 10.10(iii). Teaching

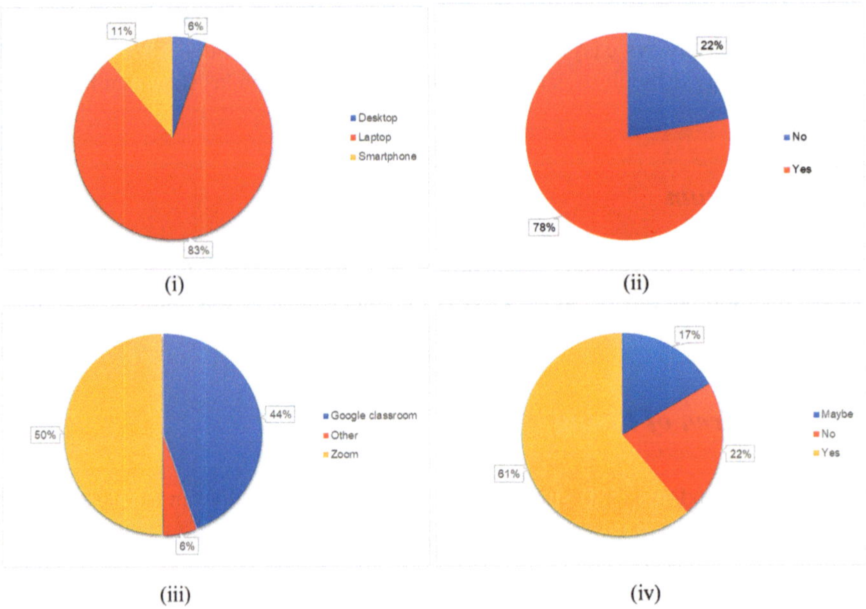

Fig. 10.9 Responses of (i) equipment, (ii) internet, (iii) platform, and (iv) teaching satisfaction

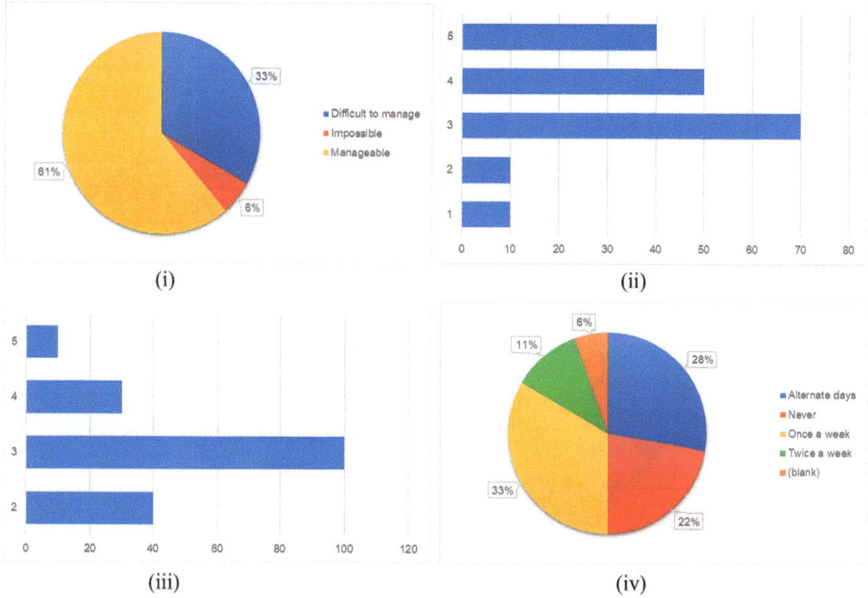

Fig. 10.10 Responses of (i) stress: teachers, (ii) work-life balance, (iii) stress: student, (iv) teacher-student interaction

from home, balancing various responsibilities simultaneously, the teacher-student connection was somewhere lost. But some teachers (here 33%) still managed to interact with students at least once a week, as shown in Fig. 10.10(iv).

Google forms, being more user-friendly, were used immensely to assess students online in comparison to written tests, live presentations, etc., as shown in Fig. 10.11(i). Usually, the college duration is found to be around 7–8 h offline; but it was only around 3–5 h for 50% of the teachers, as shown in Fig. 10.11(ii). While 39% of teachers were unsure of whether students were learning as much as they did in offline mode, the majority of teachers said no, as shown in Fig. 10.11(iii). As can be seen in Fig. 10.11(iv), the majority of teachers feel good about the overall online education.

10.8 Conclusion

According to the responses we have received, the majority of students are from the computer science branch and the least students are from the civil department. Almost all the students use laptops or phones for their online classes. The average time students spend on online classes is about 3–5 h a day. Very few students are rarely eager to participate in online classes and the majority of them are often eager to participate. From the survey, most students feel screen sharing is the best way to

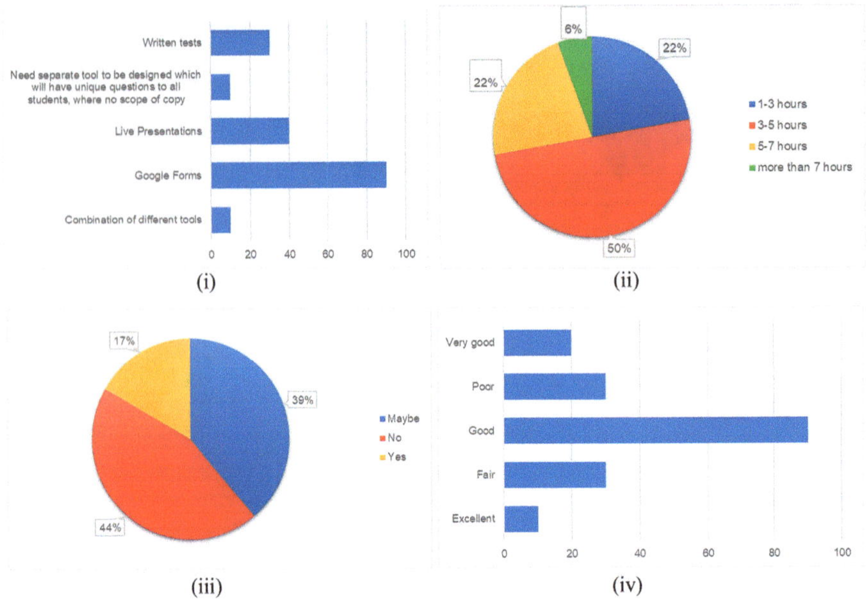

Fig. 10.11 Responses of (i) online evaluation, (ii) online working hours, (iii) online/offline understanding, and (iv) online education: overall feedback

conduct the online practical and most of them also have all the software required for practical, and the rest of the students from civil and mechanical departments find it difficult to have the software. Students find animations and PowerPoint presentations are the best approaches that can motivate them during online classes. Most of the students occasionally hear from their teachers and very few students never hear from their teachers during the course of online learning and students are also getting the proper help with their work. It is moderate for the students to use online learning and most of them are so focused on activities in their classes that they lose track of time. The majority of students find it sometimes easy to connect to the internet and very few students generally from rural areas find it extremely difficult to connect to the internet for their online classes. According to students their family members or neighbors occasionally disturb them and very few students have other responsibilities at their home other than studying. Students feel that online learning is more conducive to cheating.

Looking at these results we can see that many students who face the problem of net issues are from rural areas. We can also say that the students who cannot focus or lack interest in online studies can be from rural areas and this is merely due to the lack of internet and proper technicalities. Also, the students who find that online practical are not useful are from Mechanical, Civil, and other departments. The maximum students from Computer Engineering, Electronics and Telecommunication departments think that online practical are useful. We can say that the overall response and effectiveness of online learning are interlinked between various aspects.

Both students and teachers agreed to the fact that the connection between them was somewhere lost. Most students and teachers agreed to the fact that online assessment is conducive to cheating but many teachers were happy about the online assessment. About 38% of students face difficulty in internet connection sometimes whereas 50% of teachers have a stable internet connection.

The paper highlights and addresses the impact of digitization in the learning and teaching process. Nevertheless, its application with convergence to IoT and Cloud is deeply utilized in industry 4.0. As discussed in the introduction, the main intention of this research is to put forth the findings of online learning in a teaching-centric scenario, but finds a huge space and scope to further accelerate the development in industry 4.0. In on-job training, use of AR/VR will aid and assist skill-effective employees, thereby giving new dimension in the evolving technology.

Acknowledgments We offer our sincere thanks to the R&D wing of VIIT, Pune for allowing us to do a research internship under the guidance of Dr. Sachin R. Sakhare.

References

1. Edwige S. The Impact of Online Teaching on Higher Education Faculty's Professional Identity and the Role of Technology: The Coming of Age of the Virtual Teacher: written by Edwige Simon, 2012.
2. Shadnaz A, Jalena T, Mehran R, Wenlu Z, Roger CL, Antonella S. An observational study of engineering online education during the COVID-19 pandemic. April 15 2021, Journal Pone https://doi.org/10.1371/journal.pone.0250041
3. Paul G, Ina B. Online Teaching Effectiveness: A Tale of Two Instructors, International Review of Research in Open and Distance Learning Volume 10, Number 3. June 2009.
4. Mohanty, A. and Dash, D. (2016) Engineering Education in India: Preparation of Professional Engineering Educators. Journal of Human Resource and Sustainability Studies, 4, 92–101. https://doi.org/10.4236/jhrss.2016.42011
5. Bączek, Michał MD; Zagańczyk-Bączek, Michalina MD; Szpringer, Monika MD, PhD; Jaroszyński, Andrzej MD, PhD; Wożakowska-Kapłon, Beata MD, PhD Students' perception of online learning during the COVID-19 pandemic, Medicine: February 19, 2021 – Volume 100 - Issue 7 – p e24821 doi: https://doi.org/10.1097/MD.0000000000024821
6. Mathivanan, S.K., Jayagopal, P., Ahmed, S. *et al.* Adoption of E-Learning during Lockdown in India. *Int J Syst Assur Eng Manag* (2021). https://doi.org/10.1007/s13198-021-01072-4
7. Sun, A., & Chen, X. (2016). Online education and its effective practice: A research review. Journal of Information Technology Education: Research, 15, 157–190.
8. Jena, Pravat. (2020). ONLINE LEARNING DURING LOCKDOWN PERIOD FOR COVID-19 IN INDIA.
9. Dhawan S. Online Learning: A Panacea in the Time of COVID-19 Crisis. Journal of Educational Technology Systems. 2020;49(1):5–22. doi:https://doi.org/10.1177/0047239520934018
10. Darius, P.S.H., Gundabattini, E. & Solomon, D.G. A Survey on the Effectiveness of Online Teaching–Learning Methods for University and College Students. *J. Inst. Eng. India Ser. B* 102, 1325–1334 (2021). https://doi.org/10.1007/s40031-021-00581-x

11. Naik, G. L., Deshpande, M., Shivananda, D. C., Ajey, C. P., & Manjunath Patel, G. C. (2021). Online Teaching and Learning of Higher Education in India during COVID-19 Emergency Lockdown. Pedagogical Research, 6(1), em0090. https://doi.org/10.29333/pr/9665
12. Bao, Wei, "COVID-19 and online teaching in higher education: A case study of Peking University", Human Behaviour and Emerging Technologies, 07 April 2020.
13. Arkorful V, Nelly A, "The role of e-learning, advantages and disadvantages of its adoption in higher education", International Journal of Education and Research, 12 December 2014.
14. Eddie JA, Christopher WB, "A Comparison of Student Learning Outcomes in Traditional and Online Personal Finance Courses", MERLOT Journal of Online Learning and Teaching, 4, December 2011.
15. Lokanath M, Tushar G, Abha S, "Online teaching-learning in higher education during lockdown period of COVID-19 pandemic", International Journal of Educational Research Open, 3 September 2020.
16. Pinaki C, Prabhat M, Manu SG, Savita Y, "Opinion of students on online education during the COVID-19 pandemic", Human Behaviour and Emerging Technologies, 17 December 2020.
17. Souvik S, "Possibilities and Challenges of Online Education in India during the COVID-19 Pandemic", International Journal of Web-Based Learning and Teaching Technologies.
18. EDWIGE S, "The impact of online teaching on higher education faculty's professor's identity and the role of technology: The coming of age of the virtual teacher", non-Journal, 2012.
19. Amit J, Muddu V, Preeti B, "Impact of coronavirus pandemic on the Indian education sector: perspectives of teachers on online teaching and assessments." Interactive Technology and Smart Education, 4 September 2020.
20. Abinaya S, "Impact of Covid-19 on school education in India", Readers' blog, 25 May 2021.
21. Elizabeth AM, Kim RM, Barbara Y, Shannon SB, "COVID-19 and Distance Learning: Effects on Georgia State University School of Public Health Students", Front. Public Health, 25 September 2020.
22. Anjali A. Dudhe and Sachin R. Sakhare, "Teacher Ranking System to Rank of Teacher as Per Specific Domain", ICTACT Journal on Soft Computing, January 2018, Volume: 08, Issue: 02, ISSN: 2229–6956 (ONLINE).

Part III
AI Based Data Management, Architecture and Frameworks

Chapter 11
AI-Based Autonomous Voice-Enabled Robot with Real-Time Object Detection and Collision Avoidance Using Arduino

Suvarna Pawar, Pravin Futane, Nilesh Uke, Sourav Patil, Riya Shah, Harshi Shah, and Om Jain

11.1 Introduction

Our goal is to create an autonomous robot vehicle that can be easily and effectively operated using voice instructions from a person or user [11]. Speech-Controlled Automation Systems (SCAS) are another name for these systems (SCAS). The autonomous robot we built is a working prototype of the technologies detailed previously. The idea is to create a robot that can be controlled via human voice instructions. The robot will be controlled remotely using a smartphone app called Smart artificial intelligence (AI) for Android. Android applications give a highly good interface for remotely automating the robot. It features a variety of functions that the robot's user may find useful.

In this concept, an Android application with several features is used for the required purpose. The application and the robot can communicate more easily thanks to Bluetooth technology. The specified commands will be delivered across the channel to the Bluetooth module that receives it. VCRVs are voice-controlled robotic vehicles that listen to the user's orders and respond accordingly. The system will require accent training, during which the device will learn to interpret the orders

S. Pawar
School of Computing, MIT Arts, Design and Technology University, Pune, Maharashtra, India
e-mail: suvarna.pawar@mituniversity.edu.in

P. Futane · S. Patil (✉) · R. Shah · H. Shah · O. Jain
Department of Information Technology Engineering, Vishwakarma Institute of Information Technology, Pune, Maharashtra, India
e-mail: pravin.futane@viit.ac.in; sourav.22020060@viit.ac.in; riya.22020082@viit.ac.in; harshi.21911209@viit.ac.in; om.21910089@viit.ac.in

N. Uke
Trinity Academy of Engineering, Pune, Maharashtra, India

A. Neustein et al. (eds.), *AI, IoT, Big Data and Cloud Computing for Industry 4.0*, Signals and Communication Technology, https://doi.org/10.1007/978-3-031-29713-7_11

provided to it and will be incorporated into the Arduino software so that it can do its tasks correctly.

The primary goal of Voice-Controlled Robotic Vehicles is to analyze speech and respond to pre-programmed commands. Backward, forward, right, left, night mode, and halt the robot are the most fundamental commands. The robot will be controlled remotely using an Android application on a smartphone. Our goal is to create a robotic automobile that employs contemporary smartphone technology in a very simple and cost-effective way, employing object recognition with the ESP32 camera module to understand and respond to the distinctions between various sorts of items. We'll also use an ultrasonic sensor to detect things and give the robot's algorithm instructions to avoid colliding with them.

Natural language processing is a new technology that allows us to use our voices to command and control objects. Wireless technology is becoming increasingly significant in today's world. Wired networks are inefficient, confusing, and difficult to comprehend. Wireless technologies have had a huge positive impact on human existence, and the rate of human advancement has greatly quickened. The voice control robot is controlled by vocal instructions delivered directly to the robot by the user. As far as we can tell, this is a wireless robot. On a smartphone that functions as a transmitter, the android app is installed. The orders are issued via the Android application. The Arduino is identified via a Bluetooth connection and the Smart AI application.

A Bluetooth module is connected to the Arduino (HC-05); pairing an Android application (Smart AI) with a Bluetooth module is straightforward. The user provides the Smart AI Android App orders. Voice recognition technology is used in a number of applications to control equipment and contribute to society's modernization. The fundamental idea behind this model is to utilize an Android smartphone to communicate with a robot via Bluetooth network connectivity. This technology is used to assist people with disabilities, in addition to industrial uses such as voice-controlled working robots Every technology has its own set of benefits and drawbacks. On the other hand, Bluetooth-based voice-controlled robot systems have the upper hand. Devices can be connected up to 10 meters apart. Bluetooth also uses the 2.4GHz ISM band, which is available all across the world. Bluetooth services are available at a maximum speed of 2.8 Mbps. Fast advancement in Bluetooth-based speech recognition systems has been possible as a result of these benefits.

Currently, vehicles are operated manually, and the driver is solely accountable. Every movement, including starting and stopping, applying brakes, shifting gears, and accelerating, requires human effort. However, new technologies have arisen in recent years that might be integrated with traditional autos to create new vehicles. The introduction of gesture concepts in the current era is bringing the physical and digital worlds closer together. We prefer technology over people in all perilous situations. Originally operated by hand, these robots may now be controlled through speech and gestures. The interface between a computer and human body language may be characterized as gesture and speech recognition technology. This establishes a communication link between humans and machines. The goal of this study is to improve the security of robots while also simplifying their control systems. A smartphone is required to manage voice directions. The findings of the underlying investigations are used to complete the execution evaluation. The predicted benefits might be used in a number of settings, such as enterprises, medical clinics, and environmental laboratories. The world's most severe problem is the lack of human labor. They may maneuver around in a wheelchair on their own using this gadget by simply speaking orders through Bluetooth. All of the features are integrated into a single module to create a prototype.

So, why is wireless technology used? The offerings and impact of wireless technology for the robot are quicker query responses, less time spent on administrative tasks, more time spent online by users, just-in-time and real-time control, and improved client-host communication, among other advantages. Technology, which offers a collection of fundamental building blocks, and User Applications, which specify a range of tasks that must be carried out effectively on demand, are the two broad factors that control wireless computing. In this project, we have decided to choose wireless technology as our mode of communication between the robot and the smartphone to overcome the bridge of wired technology, to make the robot controlled from far away was our main focus and which also provides us data integrity, speed, protection, and compatibility, if someone found the robot alone they cannot decode it as the other device which has data was wirelessly connected to it which makes it much more safer. Although the idea has been around for a while, it has mostly relied on voice-based communication methods. It is used in places where wired data connectivity would normally be impracticable, rather than as a replacement for wired data communication. The industry has just lately begun to take action to create a standard that is more suited for data transmission.

11.2 Literature Survey

To improve the functionality of AI robot, Licheng Jiao et al. [1]. Specified about the types of detection models and datasets, in a nutshell, metrics and applications. According to their study, domain-specific image object detectors may be divided into two groups: one is a two-stage detector, the most common of which is Faster. One is a one-stage detector, such as YOLO or SSD, while the other is R-CNN. With the advancement of deep learning (DL) and the continual expansion of processing capabilities, great progress has been made in the field of generic object detection. The first CNN-based object detector, R-CNN, made several significant contributions to the area of generic object recognition. They've offered various typical object detection designs to help newcomers get started in this field. To detect objects, deep learning models such as CNN, RNN, and LSTM can be employed. Other computer vision libraries, such as OpenCV, are also free to use. Furthermore, the Internet of Things is built on massive data. The performance of deep learning is reliant on a big amount of data, which functions as rocket fuel.

Humayun Rashid et al. [2] stated that the robot's motions can be controlled by sending vocal instructions to the microcontroller, because the design system incorporates a microcontroller connected to the smartphone through the Bluetooth module for the transmission and receiving of voice commands, these are subsequently converted to text using an android smartphone application. The robot's answers, which are supplied in response to the vocal orders, follow the data. To produce a precise movement in a certain direction, voice instructions are employed. The robot's speaking technology is based on an SD card with a pre-recorded human voice. This audio file is then used by the robot's speaking system. After receiving each of the orders from the device, the robot will pronounce the phrases and operate in accordance with the instructions.

In this paper, by Soniya Zope et al. [3] stated that according to the project's description, voice commands, as well as manual remote operation control, are offered for operating robotic equipment or vehicles. The ATMEGA32 microcontroller and Bluetooth device are used to integrate the control unit interface and detect signals delivered by the android app. The serial data received from the Bluetooth module connected to the ATMEGA32 is sent via the Android app. The study describes how to use a Wi-Fi module and an Android app on a smartphone to drive robot automobiles.

In the given paper, by V. Shivaraju et al. [4] We'll show you how to utilize an Android cellphone and a Wi-Fi module to drive a robot-controlled robot. We'll also show how we can control the appliances without an Android phone by sending regular text messages to the user. The advantage of using a robot-controlled vehicle is that it may be used for a range of tasks. This idea may easily be expanded to include a spy camera that can send movies to the user through Wi-Fi. For the project, solar cells would be employed instead of the typical lithium-ion battery. This robot might potentially be used to move items from one place to another. This project will get better Wi-Fi connectivity, allowing people to do more and communicate across vast distances.

Lokireddy Sai Siddhardha Reddy et al. [5] stated in the paper that the basic goal of a robotic vehicle is to assess voice commands and do tasks without the need for human intervention. The user's voice is used to operate the robot. To run the robot utilizing user voice input instructions, an android-developed program that interacts using Bluetooth modules is necessary. As a result, this book primarily targets physically challenged individuals. As a consequence, users will be able to drive their self-driving robot with more safety and security in the future, as automated braking and slowing capabilities will safeguard them against unexpected hit-and-run attempts. When it comes to hardware, a robotic vehicle's motors will be controlled by a modified Arduino. When the automobile senses any unexpected or abrupt impediment, ultrasonic sensors will cooperate with the Arduino board to help in autonomous braking.

Santoshachandra Rao Karanam et al. [6] mentioned in the paper, the way Image processing is vital in sectors such as medical imaging, medical image analysis and processing, web mining, and so on are all examples of image mining. In the healthcare business, CBMIR (Content-Based Medical Image Retrieval) technologies are critical. The fundamental purpose of medical database graphics is to make searching and retrieving medical data easier. The following article demonstrates how to apply deep learning algorithms for image segmentation in photo processing. This article discusses the many deep learning applications for medical imaging and image mining. It also demonstrates how to use content-based image retrieval to find problems in medical photographs.

M. Naveen Kumar et al. [7] stated in paper that the purpose of image processing is to help a computer understand the information in a picture. OpenCV is a computer library that specializes in image processing. In computer vision applications, it is commonly used as the de facto API. Image processing software can assist us in solving a wide range of real-time issues. This page offers step-by-step instructions as well as real-world examples of OpenCV-based real-time image processing applications.

Mingyuan Xin et al. [8] mentioned in their paper that based on an evaluation of the error backpropagation approach, we suggest a unique depth neural network training criterion for maximum interval minimal classification error. To improve outcomes, the cross entropy and M3 CE are examined and integrated at the same time. Finally, we put our proposed M3 CE-CEc to the test on MNIST and CIFAR-10, two deep learning benchmark datasets. The findings demonstrate that M3 CE can increase cross-entropy and can be used in conjunction with the cross-entropy criterion. M3 CE-CEc performed brilliantly in both datasets.

Prof. P Y Kumbhar et al. [9] stated in the chapter how to use the Haar Classifier on the Raspberry Pi BCM2835 CPU processor to accomplish real-time face detection and tracking of the head postures position from high-quality video, the system uses a mix of SoC and GPU-based architecture. The libraries SimpleCV and OpenCV are used for face detection and tracking the location of head positions. The above-mentioned hardware, together with the SimpleCV and OpenCV framework libraries, were utilized to provide results at 30 frames per second at 1080p resolutions for improved precision and speed in face identification and head tracking.

Wenhuan Wu et al. [10] mentioned that, in an autonomous face recognition system, face detection is crucial. This paper explains how to use a cascade of AdaBoost classifiers to recognize faces and how to set up OpenCV in MCVS. With the aid of OpenCV, we were able to detect faces. A thorough examination of the face detection results is also offered. Experiments show that the method used in this article has a high accuracy rate and excellent real-time performance.

11.3 Proposed System Design and Methodology

The flow to architecture starts with giving input to the robot with the help of android phone, and afterwards user has to connect their device with the Bluetooth module. Later on, the Arduino is connected to the sensors and devices like Camera Module, Servo Motor, Ultrasonic Sensor, and Battery, and then various other components are

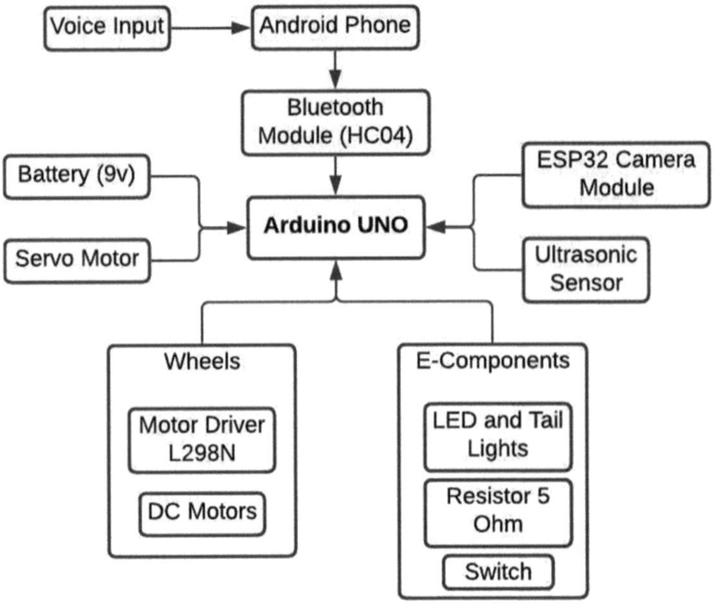

Fig. 11.1 Block Diagram of Robot

connected with the robot to make it more futurized and make all operations flexibly and smoothly like Motor Driver, DC Motors, LEDs, Resistors, and Switch. This was all about the Architecture Diagram. The robot functions in a proper way to give better functional requirements to the user of the robot respectively as per their voice commands (Fig. 11.1).

Speech is the fastest mode of communication, according to research, and it has a wide range of uses, including in-car systems, medical attention, Military personnel, Educate yourself, and Intelligent structures. The current approach focuses solely on controlling the robot, which is capable of comprehending orders and carrying them out. The robot responds to commands sent through an Android device, which is why Arduino is included in the system. Any Android-based smartphone can be used as the controlling device. The transmitter makes use of an Android application that is necessary for data transmission. These commands are read by the receiving end, which then interprets them to operate the robotic vehicle. The android device provides orders to the robot to travel forward, backward, right, and left. Arduino uses the commands to control the motors, which let the robot move in four directions. Serial communication data is transmitted between the android smartphone and the receiver. The Arduino software is intended to control the motor through a motor driver circuit in response to orders provided from an Android device.

11.3.1 Flowchart (Fig. 11.2)

User needs to give voice input, if the input is valid the signal process further and detects objects then the robot moves as per command and avoids object collision. It also controls light and buzzer as per commands of the user and at last stops. Here the user has to follow the following actions:

(i) Firstly, users switch on the robot, then the RED LED of the Bluetooth module will start blinking.
(ii) Users will connect the Bluetooth module with the Android application with HC05.

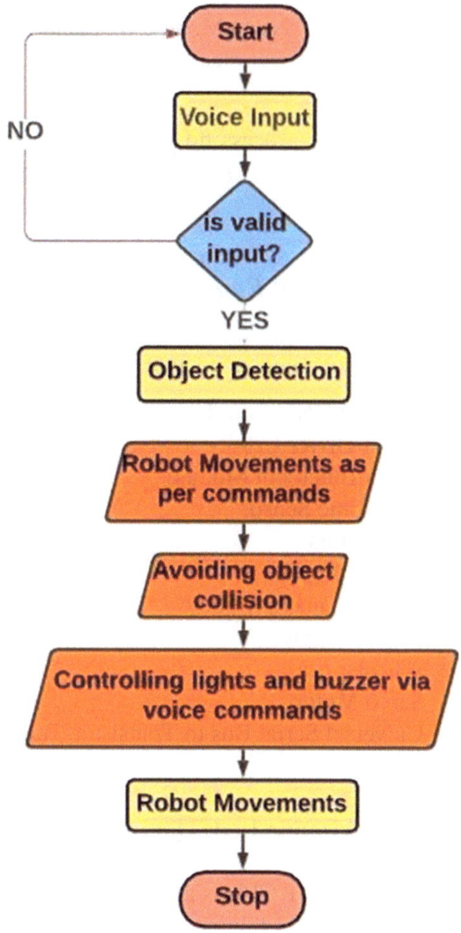

Fig. 11.2 Project Flowchart of Robot

(iii) Users will click on the voice option present in the Android application and will give voice commands; commands are already fed into the Arduino UNO board.
(iv) Once the commands of users match with the command present in the Arduino code the operation will be performed.
(v) When users will use the command "Go Ahead" then the UR sensor will check the object in forward direction. If it detects anything inside 1-meter range then it will not move; instead it will play the buzzer twice and turn on the red parking lights inside the robot. Otherwise, it will move in forward direction up to 8 s as per delay given in the Arduino code.
(vi) Similarly, when users will use a "Left" command, the robot will check for objects in the left direction and will perform accordingly; likewise, for "Right" and "Go Back" Commands.
(vii) Lastly, the "Night Mode" command will turn on the white lights present inside the robot to get clear vision for the camera to detect objects and display the names.

This was the process flow of the autonomous vehicle with all functionality and processes carried out to perform various operations by the user of the robot.

11.3.2 System Requirements

The following components make up the System Design:

(a) Arduino UNO Board
(b) Motor Driver L298N
(c) HC05 Bluetooth Module
(d) Ultrasonic Sensor
(e) BO Motors
(f) Connecting Wires
(g) ESP32 Camera Module
(h) Power Supply
(i) Wheels
(j) Servo Motor
(k) Universal Serial Bus to Transistor-Transistor Logic Module

11.3.2.1 Arduino UNO (Fig. 11.3)

The ATmega328P CPU is used in the Arduino UNO [12], a microcontroller. On this board, you'll find 14 digital I/O pins, six analog inputs, a 16 MHz ceramic resonator, a USB port, a power connection, an ICSP header, and a reset button. Everything you'll need to get started is included with the Arduino Board.

Fig. 11.3 Board for Arduino UNO R3 [13]

Fig. 11.4 Motor Driver L298N [14]

11.3.2.2 L298N Motor Driver (Fig. 11.4)

The L298N is a dual-channel H-Bridge motor driver that can drive two DC motors seamlessly.

11.3.2.3 Bluetooth Module HC05 (Fig. 11.5)

The HC-05 is a Bluetooth module with a 2.4GHz frequency that enables wireless communication between two devices. The HC-05 Bluetooth Module is a straightforward Bluetooth SPP (Serial Port Protocol) module that lets you construct a transparent wireless serial link. It uses serial communication to connect to a controller or a computer, making it simple to set up.

Fig. 11.5 Bluetooth Module (HC05) [15]

Fig. 11.6 Ultrasonic Sensor HC-SR04 [16]

11.3.2.4 Ultrasonic Sensor (Fig. 11.6)

A device that creates ultrasonic sound waves to count the distance to a target item, converts the reflected sound into an electrical signal, and responds is known as an ultrasonic sensor. To calculate the distance between the sensor and the object, the sensor measures the time between the transmitter's sound emission and its contact with the receiver.

11.3.2.5 BO Motors

The DC/BO Motor (Battery Operated) is a tiny DC geared motor with high torque and rpm at low voltages. This motor can spin at roughly 200 rpm in both clockwise and anti-clockwise directions when fueled by a single Li-Ion cell.

11.3.2.6 Connecting Wires

Jumper wires are basic cables having connector pins on both ends that may be used to connect two points without the use of solder. They are available in male-to-

Fig. 11.7 ESP32 Camera Module [17]

male, male-to-female, and female-to-female combinations. Individual jump wires are joined by fitting their "end connectors" into slots on a breadboard, the header connection of a circuit board, or a piece of test equipment. There are solid tips, crocodile clips, banana connections, RF connectors, and more connectors to choose from.

11.3.2.7 ESP32 Camera Module (Fig. 11.7)

The ESP32-CAM module is a full-featured microcontroller that includes video camera streaming and a microSD card reader.

11.3.2.8 Power Supply

Used to supply power to the Arduino so that it may do activities. Batteries are made up of one or more cells, each of which uses chemical processes to create an electron flow in a circuit. The three basic components of all batteries are an anode (the '-' side), a cathode (the '+' side), and some type of electrolyte (a substance that chemically reacts with the anode and cathode).

11.3.2.9 Wheels

Used to effortlessly spin or transfer the robot from one location to another. With a round black grip to keep the robot steady on any surface.

Fig. 11.8 Servo Motor SG90 [18]

11.3.2.10 Servo Motor (Fig. 11.8)

A servomotor is a rotary or linear actuator that has the ability to precisely control angular or linear position, velocity, and acceleration. It consists of a suitable motor and a position feedback sensor.

11.3.2.11 USB to TTL Module (Fig. 11.9)

Prototyping requires USB to TTL module converters because they allow direct interface with the target device. This module connects to your target device by four cords that go into your computer's USB port. For simple serial connections with non-USB devices, the PL2303HX USB to TTL converter module is required.

11.3.3 Project Methodology

Human-robot interaction is one of the most important factors that can help spread the usage of robots in everyday life. Guidance systems are one of the potential uses for guiding visitors or tourists (e.g., at museums, tourist navigation in cities, etc.). We used a simple way to use human speech to operate a robotic vehicle. To begin, all human directions are converted to text using Google's speech-to-text converter, which is then compared to the code presently being fed into the Arduino board. Everything we need is present within the Android app we're using. The textual format command is subsequently sent to the robot's Bluetooth module. The

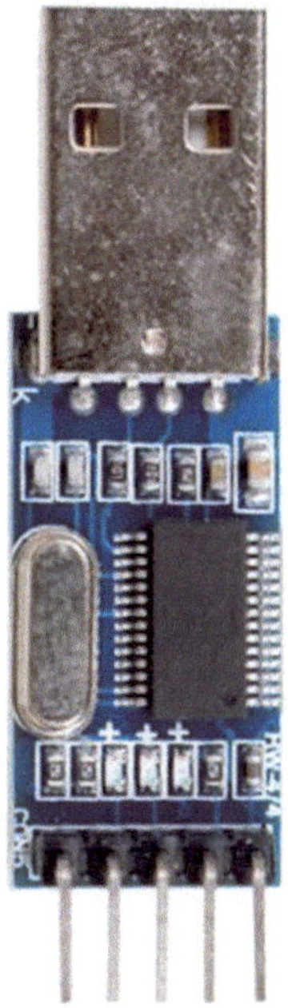

Fig. 11.9 USB to TTL Module [19]

purpose of this research is to provide a streamlined robot hardware architecture with powerful computing platforms so that robot designers may focus on research and testing rather than Bluetooth communication infrastructure. Because students may build their own robots for a low cost and use them as a platform for experimentation in a range of subjects.

11.3.4 Voice-Controlled System

The task of speech recognition is to convert speech to digital data, but the task of voice recognition is to identify the speaker. Voice recognition works by assessing how different people's voices sound. Everyone has a unique speech pattern influenced by physiology (mouth and throat size and shape) as well as habits (voice pitch, speaking style, accent, and so on).

Voice recognition software differs from speech recognition software in some ways. The most typical applications of voice recognition technology are to verify a speaker's identity or to determine the identity of an unknown speaker. Speaker verification and speaker identification are the two types of voice recognition.

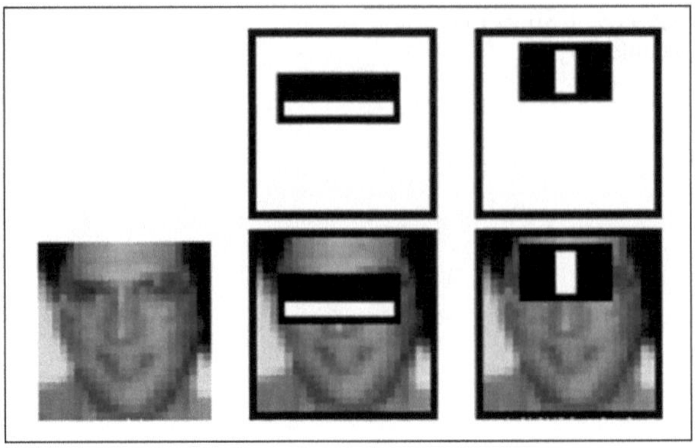

Fig. 11.10 Haar Cascade Working with Face Detection [20]

11.3.5 Algorithm

In their 2001 paper "Rapid Object Detection with a Boosted Cascade of Simple Features," Paul Viola and Michael Jones proposed an effective object identification strategy based on Haar feature-based cascade classifiers. It's a machine-learning method that involves learning a cascade function from a large number of positive and negative images. It's then used to locate items in other photographs.

In this room, we'll be working on facial recognition. The approach requires a large number of positive (face-containing) and negative photographs to train the classifier (images without faces). After that, we'll have to extract attributes from it. The Haar properties illustrated in the image below are used to do this. They resemble our convolutional kernel in appearance. To get a single value for each feature, subtract the total number of pixels in the white and black rectangles from the total number of pixels in the white and black rectangles (Fig. 11.10).

11.4 Robot Implementation

11.4.1 Libraries Used

A reservoir of pre-trained Haar cascades is kept in the OpenCV library. The majority of Haar cascades are employed for one of the following purposes:

- Face detection
- Eye detection
- Mouth detection
- Full/partial body detection
- Object detection
- Adafruit Motor Shield Library
- Servo Motor Library

Fig. 11.11 Android App Logo

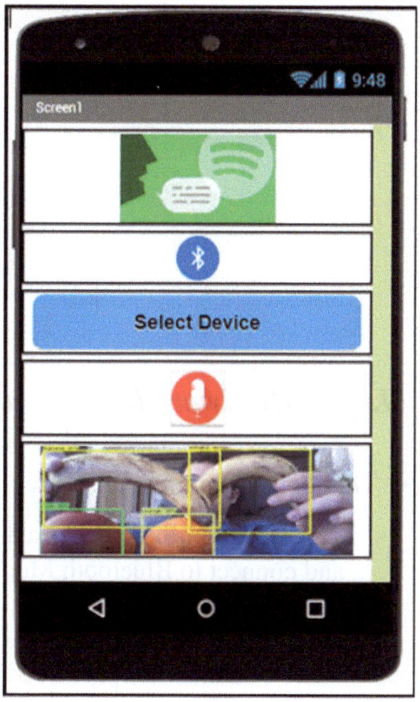

Fig. 11.12 Android Application Design ESP32 Cam detecting the objects as per shape, color, and size

11.4.2 Android Application Design (Fig. 11.11)

Logo of the Android application Smart AI, which is developed in Android Studio to control the AI-based autonomous vehicle through voice commands by App and for surveillance through camera module (Fig. 11.12).

11.4.3 Development Software

Arduino IDE is used to develop the program for Arduino UNO Board to control all things and perform accordingly as per the given input (Fig. 11.13).

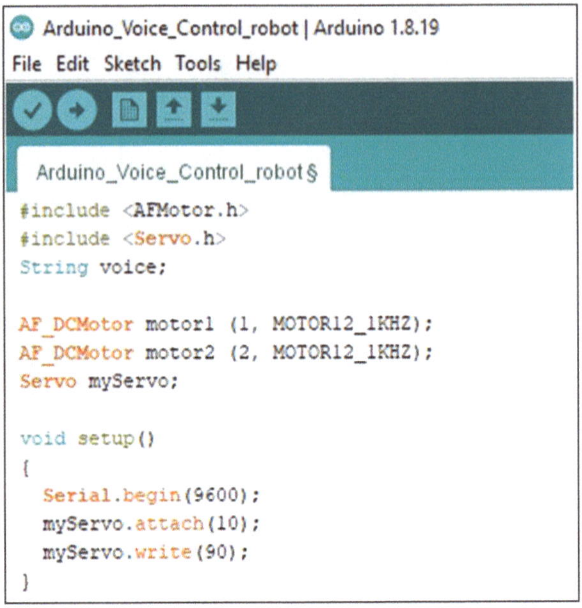

Fig. 11.13 Arduino IDE for Development of Hardware Code

11.4.4 Hardware Implementation (Fig. 11.14)

- The above diagram works as follows:
  1. The Bluetooth button is used to search for available Bluetooth connections and connect to Bluetooth Module HC-04.
  2. The Mic button is used to take voice instructions and perform robot operations based on those inputs by translating the commands to text using Google Voice Assistant's Speech-to-Text API [7], with the robot doing the actions as programmed in the Arduino UNO board.
  3. The background is the video streaming by the robot in front of it detecting and telling the difference between various objects using the ESP32 camera module [8].
  4. The App can also detect other objects like, laptop, keyboard, human, watch, mouse, and many more and display them on the app screen by streaming through the camera module.

11.5 Results and Discussion

When the robot receives a spoken command or instruction from the user, it begins to move.

Fig. 11.14 Implemented Hardware with UR Sensor and Camera Module for Object Detection and Collision Avoidance

1. The user's commands can be used in a variety of ways. "Go Ahead", "Back", "Left", "Right", "Stop", and "Night Mode"
2. The robot follows the user's orders/commands and performs the tasks assigned to it.
3. When an obstruction appears in front of the robot at a distance of less than 1 meter, the ultrasonic sensor detects the impediment and the robot promptly stops by sounding the buzzer twice.
4. The robot will remain in this mode until the user issues the next vocal instruction.
5. The robot operates on Bluetooth Technology with a range of 10 meters which enables it to be controlled from long distances as well.
6. The robot also detects the objects around it and displays the name of the component in the user's Android application Smart AI through video surveillance via ESP32 Camera module.

The robot is given the following vocal commands (Table 11.1):

Case1: When the distance between you and the barrier is more than or equal to one meter, you can utilize voice instructions such as "Forward," "Backward," "Left," "Right," "Stop," and so on.

Case2: You have a few voice command choices when the distance between you and the barrier is less than 1 meter. "Backward," "Left," "Right," and "Stop" are some of the terms used.

Table 11.1 Vocal commands

INPUT (user voice commands)	OUTPUT (execution of commands by robot)
“Go Ahead”	Robot goes in forward direction
“Back”	Robot executes reverse and goes back
“Left”	The robot moves to the left
“Right”	The robot rotates to the right
“Stop”	Terminates all previous commands then turn red light and buzzer on and stops
“Night Mode” or “Night Mode Off”	Starts the front white lights as headlights

11.6 Conclusion and Discussion

This research outlines the prototype and design for an android robot; nevertheless, much more work and research is needed in the future to turn the created robot into a fully working consumer product. The built robot may move in any direction based on a verbal command received from the user using an Android phone and Bluetooth. The robot can be moved forward, backward, left, and right via voice instructions. Thanks to an ultrasonic sensor, the robot can also move completely independently without striking any obstacles.

Using Bluetooth technology, the suggested solution shows how an Android smartphone may be used as a remote controller for robots and other embedded equipment. The suggested technique also demonstrates how image processing may be used to assist any disabled person with little tasks, as well as guide and recognize items.

11.7 Future Scope

The benefits of voice control are that human intervention via programmatic instruction is no longer required, and operations may be completed faster. The robot will be capable of understanding natural language commands. We can do fine-tuning of the robot according to our needs in the future: if the user gives an unrecognized command to the robot it should react by interpreting accordingly. It should auto detect the type of command by checking with the dataset provided and perform the necessary operation told by the user. If the user is giving two commands to the robot, then it should perform both the tasks in sequence as given by the user. Robots will be trained in how to achieve the desired output required by the user. There will be greater implementational challenges to implement, but if it is done in an appropriate way, we will achieve high accuracy in handling the random commands as well. The earlier system was limited to 5 commands so after this sort of fine-tuning it will no longer be limited to specific commands. A collection of control data for tasks like object recognition and collision avoidance is developed when the spoken orders are interpreted. Through robotics, we develop strategies to outsource human functions to machines, and we may assist or even replace humans in doing jobs. Although now limited to high bandwidth areas, we will also use the Lorawan Protocol in our robot to allow it to function in locations with extremely low bandwidth. We may be able to use this technology for military objectives by connecting appropriate sensors because the robot carries a camera module that will show live streaming. We can assign IP addresses to this system in the future so that more than one device can be linked, which will help with live broadcasting of the enemies' suspicious actions and also by training it with different images to understand the difference between the Indian Army and other unregistered people.

Acknowledgments We would like to express our heartfelt gratitude and appreciation to our guide, Suvarna Pawar, for providing us with invaluable guidance, support, and encouragement on a daily basis, which inspired us to work even harder to complete the project. We are also grateful for the efforts that she put in for us to make this project happen.

References

1. Licheng Jiao, Fan Zhang, Fang Liu, Shuyuan Yang, Lingling Li, Zhixi Feng, Rong Qu, "A Survey of Deep Learning-Based Object Detection". ISSN: 2169-3536, Volume 7, 2019.
2. Humayun Rashid, Iftekhar Uddin Ahmed, Sayed Bin Osman, Qader Newaz, Md. Rasheduzzaman, S M Taslim Reza, "Design and Implementation of a Voice Controlled Robot with Human Interaction Ability". ISBN: 978-984-34-2030-5, Paper ID: 65, 26–27 January, 2017.
3. Soniya Zope, Preeti Muluk, Rupali Mohite, Aishwarya Lanke, Megha Bamankar, "Voice Control Robot Using Android Application". ISSN: 2454-1362, Volume-3, Issue-2, 2017.
4. V. Shivaraju, V. Karthik Kumar, "Robot Controlled Car using Wi-Fi Module". ISSN: 2319-8885, Volume-06, Issue: 04, February-2017, Pages: 0759–0762.
5. Lokireddy Sai Siddhardha Reddy, M Sumanth, Maram Venkata Nagasai Teja Bavigadda Purushotham Naidu, Shalini Tiwari, "Voice Based Robotic Vehicle with Obstacle Avoidance". ISSN No. 0976-5697, Volume 11, Special Issue I, May 2020.
6. Santosh Chandra Rao Karanam, Y. Srinivas, M. Vamshi Krishna, "Study on image processing using deep learning techniques". ISSN No. 2214-7853, October 2020.
7. M. Naveen Kumar, A. Vadivel, "OpenCV for Computer Vision Applications". March 20, 2015.
8. Mingyuan Xin, Yong Wang, "Research on image classification model based on deep convolutional neural networks". 11 February 2019.
9. P Y Kumbhar, Mohammad Attaullah, Shubham Dhere, Shivkumar Hipparagi, "Real Time Face Detection and Tracking Using OpenCV". E-ISSN: 2349-7610, VOLUME-4, ISSUE-4, APR-2017.
10. Dr. Wenhuan Wu, Yingjun Zhao, Yongfei Che, "Research and Implementation of Face Detection Based on OpenCV". Vols. 971–973 (2014) pp 1710–1713, 2014-06-25.
11. Shivangi Nagdewani, Ashika Jain, "A Review On Methods For Speech-To-Text And Text-To-Speech Conversion". ISSN: 2395-0072, Volume: 07, Issue: 05, May 2020.
12. Sumeet Sachdeva, Joel Macwana, Chintan Patela, Nishant Doshia, "Voice-Controlled Autonomous Vehicle Using IoT". 21 November 2019.
13. https://www.indiamart.com/proddetail/arduino-uno-ch340-board-19651448730.html
14. https://kuongshun.com/products/l298n-stepper-motor-driver-board-red
15. https://www.electronicwings.com/sensors-modules/bluetooth-module-hc-05-
16. https://www.indiamart.com/proddetail/hc-sr04-ultrasonic-sensor-18101779448.html

17. https://robu.in/product/ai-thinker-esp32-cam-development-board-wifibluetooth-with-ov2640-camera-module/
18. https://www.indiamart.com/proddetail/sg-90-tower-pro-micro-servo-motor-20797318397.html
19. https://www.indiamart.com/proddetail/pl2303-usb-to-rs232-ttl-converter-adapter-module-11086683012.html
20. https://docs.opencv.org/3.4/db/d28/tutorial_cascade_classifier.html

Chapter 12
Real-Time Interactive AR for Cognitive Learning

Priyanka Jain, Nivedita Bhirud, Subhash Tatale, Abhishek Kale, Mayank Bhale, Aakanksha Hajare, and N. K. Jain

12.1 Introduction

Cognitive-communication abilities are human thought processes like orientation, attention, memory, problem-solving, and executive function. It allows functioning successfully and interacting meaningfully with each other. It can be used in human behavior simulation with the advancement in the current technology. By combining simulated elements with a person's real environment; the proposed work will be able to face complex issues related to cognitive rehabilitation. Advancement in new technologies will help them to learn a language with visual cognition in a much simpler way.

Our proposed work on special education based on cognitive behavioral therapy (CBT), provides an effective and trustworthy basis for cognitive learning through visualization. As an extension of our previous work [14], this research has an objective to produce an interactive animated behavior-rich web-based 3D immersive-rich environment, based on the user's input in language form. It proposes a grounding that the visual data has higher bandwidth, and it is easier to comprehend by a person with Specific Learning Difficulties (SLD) or Autism Spectrum Disorder (ASD) [8, 10, 11, 16, 17]. The proposed approach is to understand human language and create a dynamic 3D augmented reality (AR) environment. Learning-disabled children with autism or mental retardation require both quantity and quality education (e.g., dyslexia). Students suffering from this disorder find it difficult to decode new words or break them down to perceive information [27, 28]. The use of

P. Jain · N. K. Jain
Artificial Intelligence Group, C-DAC Delhi, Delhi, India

N. Bhirud · S. Tatale (✉) · A. Kale · M. Bhale · A. Hajare
Department of Computer Engineering, VIIT, Pune, India
e-mail: subhash.tatale@viit.ac.in

A. Neustein et al. (eds.), *AI, IoT, Big Data and Cloud Computing for Industry 4.0*, Signals and Communication Technology, https://doi.org/10.1007/978-3-031-29713-7_12

visual communication has increased because expression as objects in the form of pictures, signs and symbols, gestures, and postures is easier to understand. Also, the visualization approach can serve as a universal language where smooth communication is needed by many different people who speak different dialects and have different cultures in a country like India. The proposed work may enable the end-user to interact with virtual objects/artifacts in his/her real environmental world with the availability of AR. The major contribution in this key technology is the understanding of human language in real-time and performing the set of actions. Significant advancements in Natural Language Processing (NLP) and linguistic analysis have paved the way for computers to understand more complex language phrases and carry out actions as instructed.

Mental imagery e.g., Cognitive visualization is the ability to construct, manipulate, and interpret images in the mind. It has a greater impact to perceive the phenomenon in virtual environments. Using it through augmented reality adds a richer experience where learners can actively pursue their knowledge needs. Therefore, we focus on modeling an easy-to-use and interactive virtual environment specific to the Indian Scenario. It is an immediately deployable solution with large commercial potential and a direct impact on the vision of social welfare. We have worked on this problem statement after understanding the user requirements related to special education. To have a better and deeper understanding, we had consultancy with domain experts and end-user agencies who are associated with mental health rehabilitation issues. With their appreciation for our initiative, we got assurance to get their support as a domain expert in planning and designing the pedagogy. The domain experts and caretakers of patients would support us in assessment and evaluation of the project outcome by providing access to real user of the application. We are now intending to build interactive AR animations to help children to visualize objects in the real world.

The paper is organized into the following sections: Sect. 12.2 presents the literature survey for the language, vision, and AR technologies. With the language erudition and learning paradigm, Sect. 12.3 discusses about the previous work on the subject. Our design goal and an in-stages visual plan are described in detail with clear specimens in Sect. 12.4. The results, evaluation, and research findings are discussed in Sect. 12.5. Section 12.6 provides the conclusion with a discussion on the future scope of research work.

12.2 Related Work

AR and Virtual Reality (VR) technologies have shown tremendous growth with the speed at which the technology is growing. This literature survey shows work especially in the field of AR & VR for educational purposes. We present the critical review with the perception of shortcoming of existing work and contribution by our proposed unique approach that is linguistic analysis for visual cognition. Human behavior and actions by means of associated object management or manipulation can be described with the use of language, as discussed by Guerra-Filho and Aloimonos [6]. According to Gupta [7], linguistic concepts are easy to address the problems related to Computer Vision (CV) and related work using Lexical semantics. A scene contains information that is comparable or like a language which is provided by Sadeghi and Farhadi [24]. It links scenes or artifacts to words, preferably nouns, the activities to action verbs, objects, features, or attributes to adjectives and the prepositions are the relationship between objects. Zhao and Grosky [36] bridged the semantic gap between visual data like pixels or contours to language data like words or sentences. The cognitive theory for studying using multimedia has two different knowledge systems: visual knowledge and verbal knowledge. Funge [2] has mentioned a cognitive modeling application in his automated cinematography using advanced character animation. Singhal and Zyda [26] have represented the avatars of online users in the virtual world in their work. Klcaslan [20] has proposed a concept that uses NLP for children with mind blindness or mental retardation. Schank [25] and Gaddis [3] have proposed a majority of work on duplicating a real environment along with its physical models and interaction between objects. To control the parts of animated humans or animals at the graphical level, joint transformations (for parts like limbs) or surface deformations (for parts like face) is used. Motion synthesis for animated or digital characters is commonly classified between real-time and offline animation generation. The algorithms that were initially considered for offline animations moderately become practical in run-time virtual environments because the speed of processors continues to increase cautiously.

Research in designing a completely independent, mutually interactive, artificial 3D model has also been on the heights. Tu and Terzopoulos [35] implemented a realistic simulation of virtual fishes with their autonomy, simple behaviors, motion generation, and simulated approach. A navigation system was proposed by Noser [22] on vision, memory, and learning for virtual or digital characters. Researchers at Georgia Tech., Brogan and Hodgins [1], have merged assumptions with physically based behaviors for assuming human athletics by designing human running in a 3D environment.

While most research are done in Virtual Reality to make a 3D environment to foster cognitive learning, they often miss out on the point of AR, i.e., inducing only 3D models into the real-world giving the user a more interactive experience. In our research, we improve upon the idea of visual learning by injecting live 3D model animations in the real world captured by our mobile device camera and interacting with it using voice commands. AR is the experience of interaction between the models in the user's real environment and the objects generated by computer graphics. It adds the digital or animated components in the real world with a feeling of real-life experience as a 3D model of a pet animal 'Dog' in the user's room, as shown in Fig. 12.1.

For relevant and correct content, AR uses computer vision, plane detection, and depth tracking, that is, calculating the distance from the object. AR allows the camera and microphone to collect data and process it to get the desired output by the user. AR does not create a virtual environment like VR, but it plays with the existing

Fig. 12.1 3D model of a dog in AR Scene

environment. AR is hugely accessible by users in the growing smartphone market being mobile and personal. AR has its practical applications in almost all domains like automobile, tourism, finance, education, simulation, etc.

12.3 Need and Motivation

Today, the world focuses on cognitive learning (i.e., more constructive use of the brain and learning which is active, effective, and long-lasting). It has been understood that traditional teaching methods are not as effective and sometimes heavily rely on rot learning. However, in recent studies, it is found that technology can help develop cognitive learning among students. The proposed work emphasizes this idea and presents a way to use the latest advancements in AR to facilitate cognitive learning by audio and visual means. Linguistic and Visual Analytics can be understood well in Fig. 12.2.

We have implemented Preksha: A Hindi Text Visualizer [14]. The architecture of Preksha has a good balancing of cohesion and coupling with its modular components. It takes input from the user and produces a dynamic 3D virtual environment, as specified in the input sentence. Figure 12.3 shows the generated out of the Preksha system for input text: "कांफ्रेंस टेबल पर सफेद लैपटॉप, सलि्वर लैपटॉप और सीडी हैं. कमरे में ब्लैक-बोर्ड हैं. एक कोने में पोडियिम के-पीछे महलिा टीचर है. पोडयिम पर

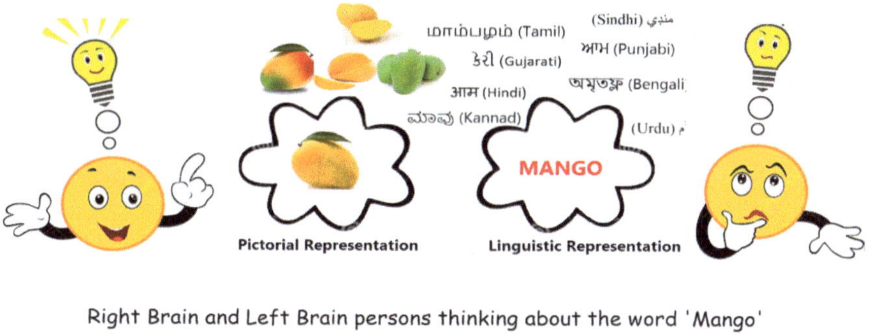

Fig. 12.2 Linguistic and visual analytics

Fig. 12.3 Output generated by Preksha

माइक के-पास पेंसलिस्टैंड और दो पेन-ड्राइव हैं. सफेद लैपटॉप के-आगे किताबों पर चाय की प्याली है. The English text corresponding to the Hindi text is "There are white laptop, silver laptop, and CDs on the conference table. The room has a blackboard. There is a female teacher in the corner behind the podium. Near the mic on the podium, there is a pencil stand and two pen drives. There are cups of tea on the books ahead of the white laptop."

In the current proposed work, the research outcome is achieved using a sophisticated NLP engine in Hindi language processing, a knowledge processor, and a dynamic scene generator. First, language engine pre-processes the text and extracts part-of-speech (PoS) hybrid layered approach. This is done via syntactic analysis, using two parsers: Tree adjoining grammar (TAG) Parser and Context free grammar

(CFG) Parser. The TAG can be used when the elementary element of rewriting is tree-like rather than symbol-based. The processing of a sentence includes morph analyzer, POS tagging, and syntactic parsing. The output of the language engine is in standard XML format, as mentioned in Jain [15].

The next step includes understanding the NLP engine's outcome. It extracts the scene knowledge and prepares a parsed XML file. The parsed XML extracts and prepares another XML for knowledge representation (KR) as an output of the second engine. The KR is collative information of three components. First is Meta information of document as MInfo, second is Environmental information of the scene as EnvInfo, and the third is Entity information of entire 3D models objects EntInfo. In the last step of the dynamic scene generation [9], it selects appropriate 3D objects based on their attributes and suitable background. By conducting a subjective evaluation using an online survey, the Preksha system is evaluated on 10,220 user responses for the parameters of Intelligibility and Fidelity. Here, Preksha is evaluated as Very Intelligible by getting a 3.02 /4 score and Fairly Accurate by obtaining a 2.91 /4 score as discussed in detail in our previous book chapter [18].

12.4 Proposed Work

The architecture of the proposed work is multidisciplinary research covering the domains of a clinical study, computational engines, and software integration with a ready-to-deployable, end-to-end proof-of-concept. The end-user operates the software through mobile or hand-held devices, whereas the computational process takes place at the server. Here, we provide a design of overall work and components for visual and linguistic analytics as a cognitive aid. Cloud platform promises for scalability of Cloud services that are used to keep the client system free from unnecessary clutters and reduce the system computation resources by processing data on servers. Audio and visual tracking supports us to provide interactive animations. The project exploits the capabilities of the popular game engine Unity3D [33] to build AR Scene. Major software development kits used are AR Core, AR Foundation [30], and Vuforia Engine [13]. Backed up by a robust structured resource repository, the basic system consists of 3 main components: (a) language input: text input or Automatic Speech Recognition (ASR) for converting speech into text by STT APIs, (b) language processing engine using NLP on text to generate animated environment by understanding its meaning, (c) AR Scene engine to render rich immersive environment with 3D objects in the mobile scene as shown in Fig. 12.4.

Figure 12.5 shows the overall architecture flow of the proposed research. The input of the system is in language command that is in text or speech form through Android OS-based mobile phone or any hand-held device. The end-user of the system uses the interface 'UIT' of the proposed mobile app to enter text input. For providing speech input, he uses the microphone of the device.

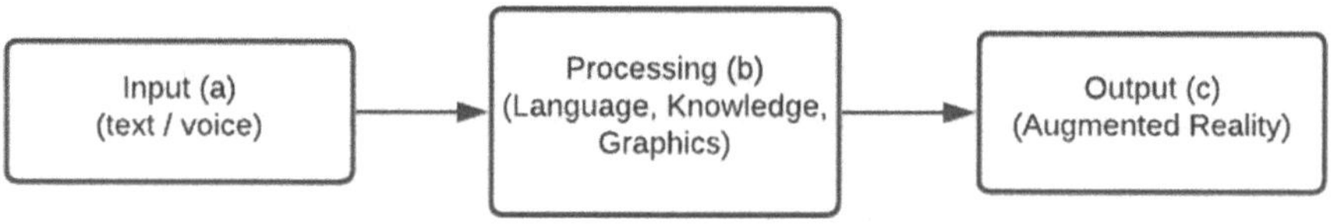

Fig. 12.4 Major components

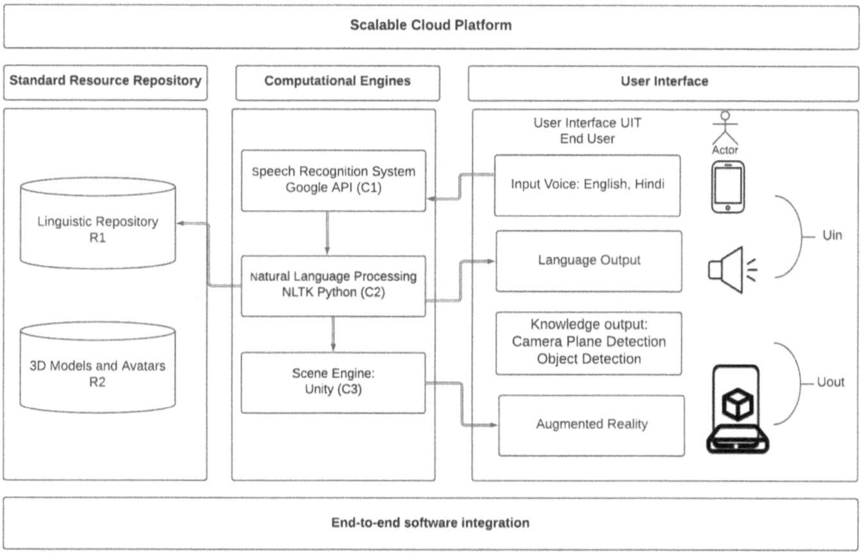

Fig. 12.5 Proposed architecture

It consists of two resource repositories: Linguistic repository R1 and 3D Artifact repository R2. Three major computational engines are used here namely speech recognition engine as C1, NLP engine as C2, and Scene engine as C3. We describe the detailed functioning in subsequent sections after explaining their process flow of work.

Having investigated after receiving the voice input, the engine sends data to different modules. Computational engine speech recognition indicated as C1 (Voice recognition System) receives the voice input from the device camera component (Camera Plane Detection). Camera Plane Detection constantly keeps track of planes, point clouds, and tagged objects. The received audio is fragmented into actual identified words for further processing of language analysis by computational engine C2 (Natural Language Processing). The C2 NLP engine majorly does the shallow parsing using pre-processing and PoS Tagger using R1(linguistic repository). The processed text helps in further operations like the identification of objects and their dependency relation. The engine then checks the availability of models in the database system (R2 - Avatars and Models repository). If the 3D models are available, it is retrieved, and the corresponding animation is performed in the AR Scene. In case of user's not performable input, an appropriate message prompts on the scene with appropriate message. Finally, all modules are combined (C3- Rendering Objects and running Scripts) and rendered onto the mobile device. Figure 12.6 shows the flow diagram of the proposed system. With the parallel processing of visual data, the application will also respond to user interaction with voice commands. It will retrieve the required information from the database located on the cloud. The cloud component is used to reduce the size and computation of the application of mobile devices which have limited resources.

The detailed architecture is described in the following sections. The cloud integration is given in Sect. 12.4.1. Section 12.4.2 discusses linguistic analysis and its processing. The speech is broken down and various parts of it are analyzed including separation of noun, verb, pronoun, and preposition. Once done, nouns are then searched or generated virtually on the screen and the action is performed over it. Section 12.4.3 highlights the visual computing approach in application and sentence spoken into the system and its analysis and conversion of speech to text are explained in Sect. 12.4.4. Section 12.4.5 explains engines and SDKs required for building the application and Sect. 12.4.6 explains the cloud integration required for all modules to run an application at run time.

12.4.1 Scalable Cloud Integration

An efficient software integration enhances the performance of any application for its robustness and scalability. While there are many ways to implement AR applications, Unity 3D software stands out in its efficiency to handle complex data generated in AR Scenes and handling the output with precision. The integrated AR features with the latest and cross-platform technologies such as AR Foundation makes the application portable for multiple devices. The development in.NET [29] environment helps in efficiently handling the hardware letting us focus more on the productivity of the application than the compatibility. Unity 3D is a popular game engine that has expanded its usability in Mixed Reality environments. Therefore, Unity 3D is used in the development of the proposed system. Along with it, we also have used AR Foundation, ARCore, and Unity Barracuda Engine for AR Scene Generation and Object Tracking.

It is not feasible to use all these features in a single standalone app due to the size of the 3D avatars and the NLP engine. If such features are bundled in one app, the system can meet hardware issues. To avoid such run-time issues, the databases have been stored on the cloud and can download this on-demand when required. This will keep the app-based proposed system free from excessive use of unnecessary data and supports the minimal use of resources.

Today, with the advancement in 4G technology, we can download the 3D avatars from the web database and generate them in our scene with minimum lag. The popular cloud engines in the market like Google Cloud and AWS provide database storage for avatars and database storage on the cloud as shown in the R2 module of

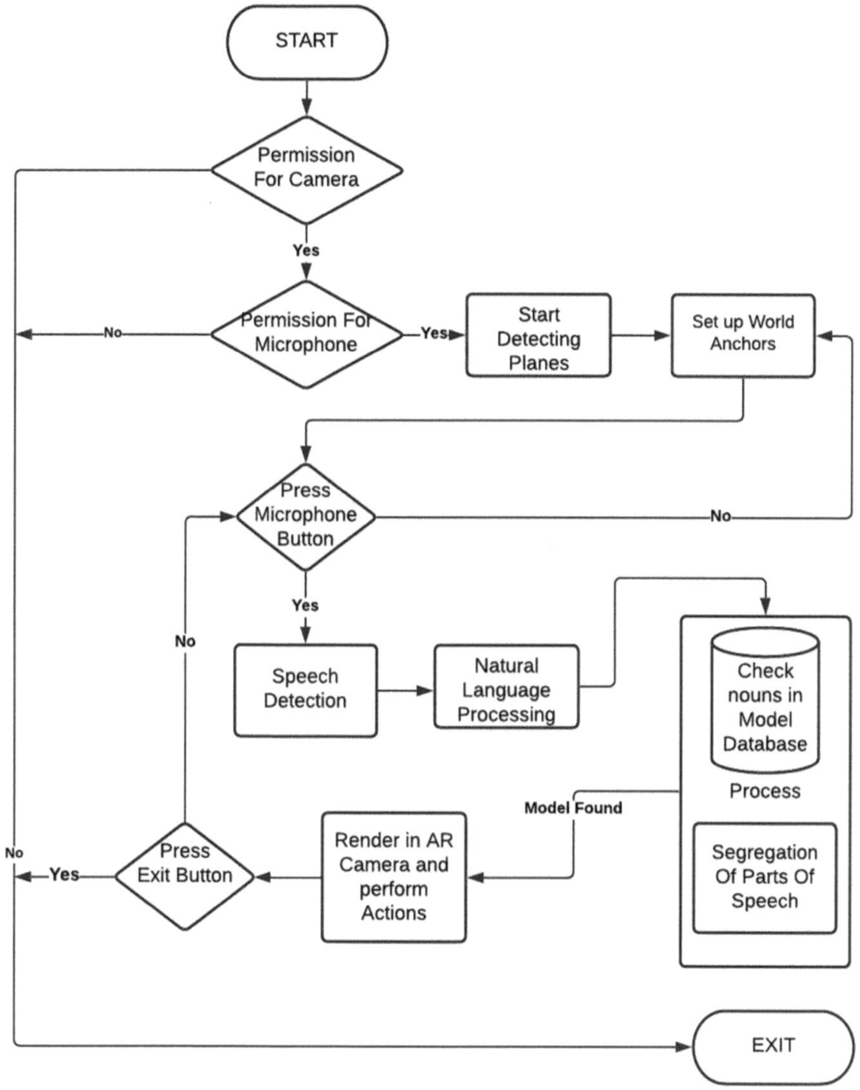

Fig. 12.6 Flow diagram of the proposed system

the architectural diagram. The language processing and the 3D module generation will need cloud integration and for that, the device will need to be connected to the internet all the time while using this app.

12.4.2 Input Interface

User interfaces are the most important aspect of usability when it comes to any user-oriented application. Communication by mobile technology like Smartphones and tablet PCs have revolutionized our life. This also works as an Assistive Technology (AT) market for people with disabilities. However, the computational engines will be residing on servers, and the end product of the system would be an Android-based mobile apps. These can be used by persons, their parents, and educators as a part of learning activities.

The app developed has a very simplistic UI with only a microphone button on top of the Scene. Once tapped, the Google Voice Recognition API [5] opens. On the provided interface, the end-user needs to provide voice input. The API sends the voice input to Google server to process and returns the speech in text format. All these processes are performed on the internet in the back end. Hence, the proposed system needs to be connected to the internet to use this feature. The API can be used on any device capable of connecting to Google servers. For the proposed system, we have used a ready-made plugin of speech recognition for Unity Android [4]. Figure 12.7 shows the pseudo-code for the Android Operating System.

The Android Java Class captures the audio and prepares the data to be sent to Google servers to decipher the words. In Fig. 12.7, we have set the language to US English. We can also set the language for native Indian language support, viz., Hindi. The result is received in another script where it is stored as a string and is shown on the screen in preview builds. Figure 12.8 shows the script for storing the result.

The app then forwards the received text to the language processor as shown in the C2 module of the architectural diagram and explained in part 4.3 in the paper.

```
void TaskOnClick()
    {
        AndroidJavaClass pluginClass = new AndroidJavaClass("com.plugin.speech.pluginlibrary.TestPlugin");
        Debug.Log("Call 1 Started");

        // Pass the name of the game object which has the onActivityResult(string recognizedText) attached to it.
        // The speech recognizer intent will return the string result to onActivityResult method of "Main Camera"
        pluginClass.CallStatic("setReturnObject", "Main Camera");
        Debug.Log("Return Object Set");


        // Setting language is optional. If you don't run this line, it will try to figure out language based on device settings
        pluginClass.CallStatic("setLanguage", "en_US");
        Debug.Log("Language Set");


        // The following line sets the maximum results you want for recognition
        pluginClass.CallStatic("setMaxResults", 3);
        Debug.Log("Max Results Set");

        // The following line sets the question which appears on intent over the microphone icon
        pluginClass.CallStatic("changeQuestion", "Hello, How can I help you???");
        Debug.Log("Question Set");


        Debug.Log("Call 2 Started");

        // Calls the function from the jar file
        pluginClass.CallStatic("promptSpeechInput");

        Debug.Log("Call End");
    }
```

Fig. 12.7 Pseudo-code for Android Application System

```
public class ReceiveResult : MonoBehaviour {

    // Use this for initialization
    void Start () {
        //GameObject.Find("Text").GetComponent<Text>().text = "You need to be connected to Internet";
    }

    void onActivityResult(string recognizedText){
        char[] delimiterChars = {'~'};
        string[] result = recognizedText.Split(delimiterChars);

        //You can get the number of results with result.Length
        //And access a particular result with result[i] where i is an int
        //I have just assigned the best result to UI text
        GameObject.Find("Text").GetComponent<Text>().text = result[0];

    }
}
```

Fig. 12.8 Script for storing the result

12.4.3 Computational Engines

Computational engines are the components accountable to offer virtual surroundings as a User Experience (UX) after processing the models. The structure of the proposed engine includes primary subsequent processing steps: language processing and knowledge processing.

12.4.4 Language Processing Engine

The language considered in the proposed work is Hindi, which is a morphological rich free word order language. It is a relatively challenging and complex task as a single piece of facts may be offered in more than one syntactic construction to ambiguous semantics. In the case of the goal of visualization, this example turns critical. Based on the grasping power of the man or machine, the same piece of facts may be visualized in diverse forms. Therefore, an Automatic text Visualizer (ATV) machine with many-to-many relations of textual content and scene needs a sturdy base for language processing engine. The language engine is familiar with the entered language textual content via means of syntactic evaluation within the first step. The Natural Language Toolkit, or more commonly NLTK [21], is an NLP library that contains collections or packages to make the machine understand human language and extract its meaning to reply with an appropriate answer. Python [23] programming language is used for writing NLP scripts. The input text (query) is processed on the NLP engine(C2) as shown in Fig. 12.5. Language processing is classified into parts (a) Pre-parser processing and (b) Parsing.

Its methods enter textual content in a sentence-by-sentence mode. As mentioned in Jain [15], the processing of a sentence consists of morph analyzer, PoS tagging, and syntactic parsing. The pre-parser processing component starts the language processing to prepare the source text for the normalization process of the parsing. It uses morphological analyzer components like tokenizer, stemmers, and PoS tagger.

```
import nltk
from nltk.tokenize import sent_tokenize, word_tokenize
from nltk.tokenize import PunktSentenceTokenizer

document = 'Monkey is dancing on floor'
sentences = nltk.sent_tokenize(document)
for sent in sentences:
    print(nltk.pos_tag(nltk.word_tokenize(sent)))
```

Fig. 12.9 Pseudocode for tokenization

```
import nltk
from nltk.tokenize import sent_tokenize, word_tokenize
from nltk.tokenize import PunktSentenceTokenizer

document = "कमरे में टेबल पर गमले के पास फोन रखा है."
sentences = nltk.sent_tokenize(document)
for sent in sentences:
    print(nltk.pos_tag(nltk.word_tokenize(sent)))

for sentence in sentences:
    for word,pos in nltk.pos_tag(nltk.word_tokenize(str(sentence))):
        if (pos == 'NNP' or pos == 'VBG' ):
            print(word)
```

Fig. 12.10 Pseudocode for PoS tagging

NLTK library is used for the purpose of the same, the pseudo-code for tokenization is presented in Fig. 12.9.

Input Text: "कमरे में टेबल पर गमले के पास फोन रखा है."

Tokenized output: [कमरे] [में] [टेबल] [पर] [गमले] [के- पास] [फोन] [रखा- है] [.]

The pseudo-code using NLTK library for PoS tagging is presented in Fig. 12.10. PoS-tagged output: कमरे (NNP) में (PSP) टेबल (NN) पर (PSP) गमले (NN) के–पास (PSP) फोन (NN) रखा–है (VBG)'..

Here, the key entities are extracted from the PoS-tagged output and the code snippet provides the output: Monkey (NNP) and Dancing (VBG). The syntactic parsing of an input sentence drives a dependency relation tree with the identity of semantic arguments at the side of their rules and motion verbs. This tree inherits dependency members of the family and linguistic functions primarily based totally on linguistic context. A linguistic knowledge repository (R1) is ready as in step with the rule-primarily based engine system. It is a series of assets required for the computational model. These assets are grammatical lexicons and repository of linguistic rule base. The computational processing of the scene engine is supported through a scene-useful resource repository. A generated entire scene surroundings is the composition of all viable facts obtained from the user's entered textual content. This is implemented in the usage of TAG Parser and CFG Parser as stated in Jain

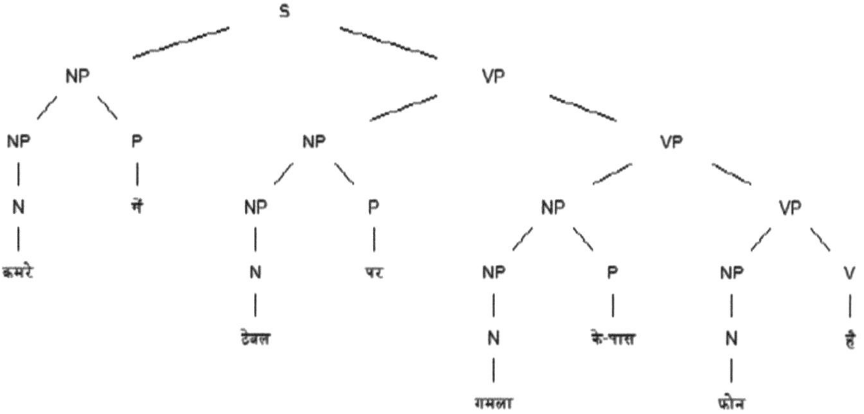

Fig. 12.11 TAG Parsing output on given example sentence

[13] and Jain [19] respectively, as shown in Fig. 12.11. It helps to identify semantic arguments like subject and predicate at the side of their dependency relations.

12.4.5 Knowledge Processing

The current knowledge processing engine is the extension of our earlier work Preksha [14]. Knowledge Extraction (KE) and Knowledge Representation (KR) are the main components of the knowledge engine. The KE engine is used to examine the factors of parsed derivation in a standard XML. Scene knowledge is taken out from the parsed tree, by spotting the semantic and dependency connection of the nodes. By spotting semantic and dependency connection, scene knowledge is taken out from parsed tree. The entered textual content is a set of Text Clausal (TC). Local Word Groups (LWG) which includes TC and one or more semantically associated textual content. An Entity is a bit of 3-tuple records correlated to physically bodily objects. Using the systems from the language-parsed output, the task of the knowledge engine can be done while resolving anaphora and other co-references. The postposition is phrases for prepositions in Hindi. पूवर्सगर्, विभक्ति, or सम्बन्धसूचकअव्यय that are greater complicated Jain [12] to address and to evaluate to pre-position phrases of English or comparable European languages. The premise of placement strategy is offered from these relations. The extracted knowledge which is processed for KR is saved in some other XML document layout that is the output of the knowledge engine.

Scene Knowledge consists of MInfo, EnvInfo, (EntInfo1, EntInfo2,....., EntInfop). The tag <Scene> contains the data about the XML file with the knowledge of an instance of digital surroundings. After MInfo of the XML file, the tag <Scene> incorporates sub-tags viz. <EnvInfo> and < SceneInfo>. The tag <EnvInfo> holds the surroundings statistics of the scene being rendered. It is part of entered textual input which is difficult to visualizable however is vital for enough knowledge transfer. The Tag <EntInfo>has sub-tags like <Background>and < Objects>tags. The <Objects> contains objects which refer to a physical unit and a 3D model in a scene. One example of <EntInfo> is an entire representation of a single item entity of a 3D model within the scene such as its attributes and relations. The Object Attribute Relation (OAR) version is used for <EntInfo> tag as mentioned by Wang [34]. In the OAR model, objects 'O', attributes 'A', and relations 'R' suit the manner of human cognition. It also can be defined as a triplet, that's OAR = (O, A, R). In the entered text, the noun entities are the objects 'O' for rendering as a 3D model within the virtual surroundings. The corresponding noun has its attribute 'A' as it is an adjective. The attribute 'R' is the relation among the objects. The <Objects> contains a subtag <Relation>to symbolize the relation call that is related to objects or parents. The item positioned in a virtual space gets its coordinates by<Position>tag.

We used Unity Game Engine that supports C# [33] language. Using IronPython script, we can implement python scripts in C# programming language. PoS tag is used to extract the noun (object) and verb (action word) from a sentence. It further fetches the respective 3D character and action file from the repository and sends it over to the Unity Game Engine for the further rendering process.

12.4.6 Output Interface

AR experiences can be built by ARCore of Google's platform, which enables a hand-held device such as a mobile device to sense its environment, understand the world, and interact with information using different APIs. ARCore is available for both Android as well as IOS devices. It, therefore, supports cross-platform development. To integrate virtual content with the real world it has three key capabilities as seen through mobile.

- Environmental estimation to estimate the relative size and spatial location of surfaces, be it horizontal or vertical.
- Light estimation to estimate the Environment's current lighting conditions from the phone.
- Motion tracking in understanding and tracking its position relative to the world.

ARCore understands the position of the mobile device in the world and builds its understanding of objects around it. The motion tracking uses feature points in the scene and tracks its movement over time and combines it with a mobile device's internal sensor such as an accelerometer to determine the phone's orientation and position. It can also detect flat surfaces like floor, table-top, and help estimate light and environmental conditions around it. This allows the end-user to place 3D models into our scene and look at it from every angle even if the camera of the phone points towards another direction briefly. The 3D models get anchored to the point we place them.

The most striking feature in today's AR application is the ability of the application to identify objects around it in a spatial region. Unity originally used the TensorFlow [31] Sharp plug-in with their Machine Learning Toolkit, ML-Agents to serve an environment to train intelligent agents. These are not just useful to game developers but also for AI researchers. These can also be used to train and identify objects in the AR Scene.

Initially, TensorFlow was used often to train image models and identify cats from dogs in Python. Now with powerful mobile devices, we can bring object recognition to the AR platform, where users can identify and recognize common objects such as tables and chairs. This will allow for far more seamless interaction between real and virtual objects. Unity also allows cross-platform development using the new Barracuda inference Engine [32]. Barracuda can run Neural Nets both on GPU and CPU. These can be used to load trained models in the AR Scene to identify objects. Vuforia [34] engine also allows scanning a specific 3D object and then using it in AR Scene to recognize that particular object and perform the associated task with it.

A dataset of many 3D images of regular annotated objects such as 'chairs', 'tables', and 'cars' is trained using Neural Networks and the trained model is then loaded into the Barracuda Engine. Each frame of the scene is then chopped into several small frames and is compared with the trained knowledge to check if any object can be identified in the scene. Once detected, a small red box showing what the mobile device thinks the object in the frame can be and its probability is shown. With a simple tap, the plane beneath the red box gets tagged with the name of the object identified.

In the background, the Unity Engine will keep on detecting planes and anchor points in the scene and understand the spatial organization of objects as shown in the architectural diagram. This will help to place the 3D avatars in the scene concerning the objects detected. An example is given in Figs. 12.12 and 12.13. As shown in the above figure the applications think that the object in Fig. 12.12 has a 49% probability of being a bottle. And the object in Fig. 12.13 has a 41% probability of being a TV monitor. The SDKs are still in Beta mode. The high-resolution video captured by phone makes the data-heavy process and occasionally may not recognize objects with 100% accuracy. However, it is just enough to tag the plane just beneath the identified objects with the same name.

Fig. 12.12 Object (Bottle) Detection

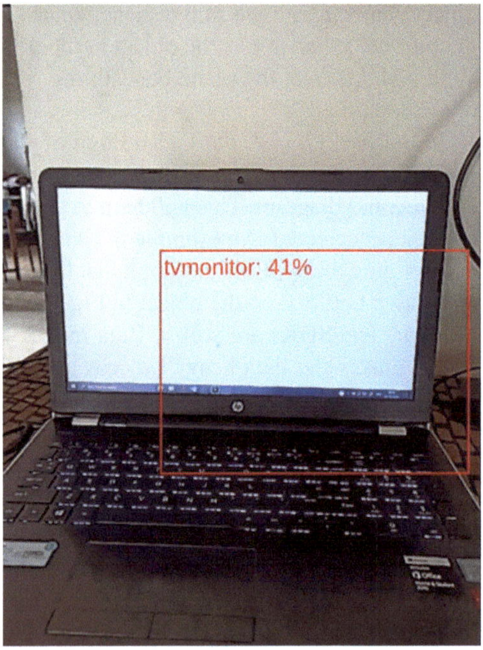

Fig. 12.13 Object (TV Monitor) Detection

12.5 Result and Future Discussion

The outcome of the proposed work is in the form of a mobile app. It takes input from the user as a keyboard-typed text and voice input using mobile's mike. It processes the input text command and responds by generating the intended scene in AR World. Figure 12.14 presents two outputs generated by our system. Given an input in the text area of the application interface – "The boy is standing on the table," the first image in Fig. 12.14a presents the virtual Avatar tagged as Boy (Human) performing a standing pose on the table. It takes care of identification of a real-world object "Table" and placing a 3D model of a boy from an annotated resource repository. The second image in Fig. 12.14b shows that by giving Hindi language input in the voice text form using mobile mike - "टेबलपरबिल्लीहै (A cat is on the table)," the system identifies the spatial relation between these two objects and places the "बिल्ली (Cat)" on the "टेबल (Table)" as per the input command given by the user in a linguistic form.

(a) Command by Keyboard input
(b) Command by Voice input.

This mobile app as the research outcome has been tested on mid-range devices with 6 GB Ram and MediTek G90T processor on Android Nougat with at least 3 GB of Ram and a dual-core processor with support for Google ARCore. The execution performance of the implemented system was found pretty 'OK' with minimum lag in rendering the AR environment. After running many test cases with more complex scenes, it was found that although powerful, today's mobile devices are not as capable to identify objects in the real world in high resolution. The application runs perfectly with object detection in low resolution, however, misses the point of AR World. The application performs very best without object recognition and gives no lag whatsoever. However, with the advancement in technology every day, the day is not far to see handheld devices perform complex tasks like object detection.

We have worked to aid in capability enhancing to learn functional computational skills like money, time, capacity, weight, mass, length, distance along with reading, and writing aids. The system would handle simple language having references of physical objects used in day-to-day life like including animals, fruits, transports, home appliances, etc. The narration should have references for shape, size, color, texture, position, distance, and spatial relations. Extendible to other Indian languages also, currently, this system would be capable of handling natural language in speech and textual form for code-mixing language (Hindi and English).

12.6 Conclusion

By providing the grounding effect of an immersive-rich environment in mental imagery, we have shown that the proposed research work as an AR application

Fig. 12.14 AR model generated in the scene. (**a**) Command by Keyboard input. (**b**) Command by Voice input

is very useful, especially for students who have difficulty understanding concepts or abstract data which are hard to understand or visualize. Persons with learning disabilities like dyslexia have difficulty learning and often learn very slowly. We have studied the importance of automatic text visualization and concluded that generating an enriched visual immersive environment by linguistic analysis may help the user to get the information to comprehend easily. We have worked with its detailed plan by the consultancy of domain experts and provided an end-to-end solution fabricated with its detailed design and architecture for text-to-scene conversion process. All the internal components have been discussed in detail along with their technology and implementation aspects. Results have been presented and elaborated in depth.

Having accessed the research outcome in a form of an application on their handheld devices, the user can now visualize and comprehend their complex machines to simpler items in their smartphone right as they work in the real world. With simple voice commands, the mobile device can now generate lifelike imagery. The project can be further extended with advanced animation and incorporating human gestures of the virtual 3D avatars. Fast network availability can make learning more meaningful, attractive, and challenging than traditional teaching by using the interactive, visual, and envelope features available in the virtual environment. Real-time interaction with virtual reality on smartphones for cognitive health needs more exploration. Also, it is necessary to investigate the effect and use of serious cognitive rehabilitation games that allow players to connect with other players using an Internet connection, especially with mobile devices.

References

1. Brogan, D. C. and Hodgins, J. C.: (1995) 'Group behaviours for systems with significant dynamics'. In Proceedings of IEEE/RSJ International Conference on Intelligent Robots and Systems, pp. 528–53
2. Funge, J., Tu, X., Terzopoulos, D.: (1999) 'Cognitive Modeling: Knowledge Reasoning Planning intelligent characters'. In Proceedings of the 26th annual conference on Computer graphics and interactive techniques SIGGRAPH 1999, Los Angeles, CA., Annual Conference Series, ACM SIGGRAPH. pp. 2938. DOI:https://doi.org/10.1145/311535.311538
3. Gaddis, T.: (1998) 'Virtual reality in the school'. In Virtual reality and Education laboratory, East Carolina University. DOI: https://doi.org/10.1089/109493103322278772
4. Github, Unity Android STT, https://github.com/gsssrao/UnityAndroidSpeechRecognition [Accessed: Oct 14, 2020
5. Google Cloud, Speech-to-text, https://cloud.google.com/speech-to-text [Accessed: Nov. 3, 2020
6. Guerra-Filho, G., and Aloimonos, Y.: (2007) 'A language for human action'. Computer, Volume 40, Issue 5, pp. 4251.
7. Gupta, A.: (2009) 'Beyond nouns and verbs. Ph. D. Thesis, University of Maryland.
8. Jain P, Bhavsar RP, Lele A, Kumar A, Pawar BP, Darbari H (2017a) Knowledge acquisition for automatic text visualization. In: Proceedings of national conference on advances in computing NCAC-2017
9. Jain P, Bhavsar RP, Pawar BP, Darbari H (2018b) VRML for automatic generation of 3D scene. Int J Comput Appl 2(8):1797– 2250. https://doi.org/10.26808/rs.ca.i8v2.01
10. Jain P, Darbari H, Bhavsar VC (2013) Text visualization as an aid to language learning disability. In: Proceedings of national conference on e-learning and e-learning technologies ELELTECH, India
11. Jain P, Pawar P, Koriya G, Lele A, Kumar A, Darbari H (2015) Knowledge acquisition for language description from scene understanding. In: IEEE international conference on computer, communication and control conference IC4-2015. IEEE explores
12. Jain P, Pawar P (2016) From pre-position to post-position. In: Proceedings of international journal of modern computer science (IJMCS), vol 4, pp 66–71. ISSN: 2320-7868
13. Jain, P., Bhavsar, R. P., Kumar, A., Pawar, B. P., Darbari, H.: (2018) 'Tree Adjoining Grammar based Parser for a Hindi text-to-scene conversion system'. In Proceedings of 4th International Conference for Convergence in Technology I2CT-2018, IEEE Xplore, pp. 1–7. DOI:https://doi.org/10.1109/i2ct.2018.8529491

14. Jain, P., Bhavsar, R., Shaikh, K., Kumar, A., Pawar, B. V., Darbari, H., and Bhavsar, V. C.: 'Virtual Reality: An aid as Cognitive Learning Environment. A case study of Hindi Language'. In Virtual Reality 24, pp 771–781 (2020). https://doi.org/10.1007/s10055-020-00426-w
15. Jain, P., Darbari, H., and Bhavsar, V. C.: (2014) 'Vishit: A Visualizer for Hindi Text'. In Proceedings of Fourth International Conference on Communication Systems and Network Technologies CSNT-2014, IEEE Xplore. pp. 886–890.
16. Jain, P., Darbari, H., and Bhavsar, V. C.: (2017) 'Cognitive support by language visualization: A case study with Hindi language'. In Proceedings of 2nd International Conference for Convergence in Technology I2CT-2017, IEEE Xplore. pp. 110–115.
17. Jain, P., Darbari, H., and Bhavsar, V. C.: (2017) 'Spatial Intelligence from Hindi Language Text for Scene Generation'. In Proceedings of 2nd International Conference for Convergence in Technology I2CT-2017, IEEE Xplore. pp. 132–138.
18. Jain, P., Shaikh, K., Bhavsar R. P., Kumar, A., Pawar, B. P., Darbari, H., Bhavsar, V. C.: (2020) 'Evaluation of Automatic Text Visualization Systems: A Case Study'. In: Hassanien A., Bhatnagar R., Darwish A. (eds) Advanced Machine Learning Technologies and Applications. AMLTA 2020. Advances in Intelligent Systems and Computing, vol 1141. Springer, Singapore. https://doi.org/10.1007/978-981-15-3383-9_3
19. Jain, P., Shaikh, K., Bhavsar, R. P., Kumar A, Pawar B. P., Darbari H, Bhavsar V. C.: (2019) 'Cascaded finite-state chunk parsing for Hindi language'. In Proceedings of international conference on communication and information processing (ICCIP-2019) and in Elsevier-SSRN 2019 (https://papers.ssrn.com/sol3/papers.cfm?abstract_id=3422328)
20. Klcaslan, Y., Ucar, O., and Guner, E. S.: (2008) 'An NLP-Based 3D Scene Generation System for Children with Autism or Mental Retardation'. In Proceedings of the 9th international conference on Artificial Intelligence and Soft Computing (ICAISC '08) pp929–938. DOI:https://doi.org/10.1007/978-3-540-69731-2_88
21. NLTK 3.5 documentation, https://www.nltk.org, [Accessed: Nov. 1, 2020]
22. Noser, H., Renault, O., Thalmann, D., and Thalmann, N. M.: (1995) 'Navigation for Digital Actors based on Synthetic Vision, Memory and Learning'. In Computers & Graphics, vol. 19, pp.7–19.
23. Python, https://www.python.org/doc/ [Accessed: Nov. 1, 2020]
24. Sadeghi, M. A., and Farhadi, A.: (2011) 'Recognition using visual phrases. In IEEE Conference on Computer Vision and Pattern Recognition CVPR 2011, Colorado Springs, CO, USA, pp. 17451752. DOI:https://doi.org/10.1109/CVPR.2011.5995711
25. Schank, R. C.; (2000) 'The Virtual University'. In CyberPsychology& Behavior Volume 3, Issue 1, pp. 9–16. https://doi.org/10.1089/109493100316184
26. Singhal, S. and Zyda, M.: (1999) 'Networked Virtual Environments: Design and Implementation'. In Siggraph Series, Addison-Wesley, Reading, MA. DOI https://ci.nii.ac.jp/naid/10011789465/en/
27. Tatale, S., Bhirud, N., Parmar, R., Pawar, S.: (2019) 'A review on Virtual Reality for educating students with learning disabilities'. In Proceedings of the 5th International Conference on Computing, Communication, Control, And Automation (ICCUBEA).pp1–6. DOI: https://doi.org/10.1109/ICCUBEA47591.2019.9128570
28. Tatale, Subhash, Nivedita Bhirud, Rishab Parmar, and Shubham Pawar. "Education using Virtual Reality for students with learning disabilities." In 2019 5th International Conference On Computing, Communication, Control And Automation (ICCUBEA), pp. 1–7. IEEE, 2019. DOI: https://doi.org/10.1109/ICCUBEA47591.2019.9128515 [31]. Unity, Scripting backends in unity,.NET: https://unity.com/how-to/programming-unity [Accessed: Nov. 3, 2020]
29. Unity3D, AR Foundation, https://docs.unity3d.com/Packages/com.unity.xr.arfoundation@4.0/manual/index.html [Accessed: Nov. 3, 2020]
30. Unity3D, Machine Learning, Tensorflow: https://unity3D.com/machine-learning [Accessed: Nov. 2, 2020]
31. Unity3D, Unity Barracuda: https://docs.unity3D.com/Packages/com.unity.barracuda@0.3/manual/index.html [Accessed: Nov. 2, 2020]

32. Unity3D, Unity Documentation, C#, https://docs.unity3D.com/Manual/CreatingAndUsingScripts.html [Accessed: Nov. 1, 2020]
33. Unity3D, Unity Documentation, https://docs.unity3D.com/Manual/index.html, [Accessed: Nov. 1, 2020]
34. Wang, Y.: (2006) 'The OAR model for Knowledge representation', In Proceedings of Canadian Conference on Electrical and Computer Engineering, IEEE Xplore. pp. 1727–1730. DOI: https://doi.org/10.1109/CCECE.2006.277686
35. X. Tu and Terzopoulos, D.:Artificial fishes: Physics, locomotion, perception, behavior'. In Proc. SIGGRAPH '94, Andrew Glassner, Ed. ACM SIGGRAPH, July 1994, Computer Graphics Proceedings, Annual Conference Series, pp. 43–50, ACM Press, ISBN 0-89791-667-0.
36. Zhao, R., and Grosky, R. I.: (2002) 'Bridging the semantic gap in image retrieval'. In Distributed multimedia databases: Techniques and applications, pp. 1436. DOI: https://doi.org/10.4018/978-1-930708-29-7.ch002

Chapter 13
Study and Empirical Analysis of Sentiment Analysis Approaches

Monish Gupta, Sumedh Hambarde, Devika Verma, Vivek Deshpande, and Rakesh Ranjan

13.1 Introduction

"Data is the new oil", in the twenty-first century. But just like oil is an immensely valuable asset but of no value when unrefined, the huge amount of data when processed can yield valuable information which can help in various fields and applications. Considering the huge size of the data being generated every day, it is not possible to process all this data manually. Various techniques have been employed to extract useful information from the data, one of them being sentiment analysis or opinion mining. Sentiment analysis is the process of finding the sentiment, mostly positive, negative, and neutral, given a portion of text. Apart from these categories, there are other categories in which the text can be classified into, e.g., happy, angry, sad, surprised, enthusiastic [16].

With the popularization of Web 2.0, there has been an explosion in the amount of user-generated data on the internet. Sentiment analysis has been used in applications in a variety of fields. People express their opinions on popular social media sites as well as product review sites. Along with this, there are other sources of data like news articles related to all kinds of events like politics, entertainment, sports, finance, etc. Businesses can analyse the product reviews of their products posted online by customers and make suitable changes [11]. Analysis of product reviews posted on online shopping sites can help identify products which a customer is more likely to buy, boosting profits for the company, as well as increasing customer ease

M. Gupta (✉) · S. Hambarde · D. Verma · V. Deshpande
Vishwakarma Institute of Information Technology, Pune, Maharashtra, India
e-mail: monish.17u611@viit.ac.in; sumedh.17u570@viit.ac.in; devika.verma@viit.ac.in; vivek.deshpande@viit.ac.in

R. Ranjan
Charles W Davidson College of Engineering, San Hoje State University, San Jose, CA, USA
e-mail: rakesh.ranjan@sjsu.edu

A. Neustein et al. (eds.), *AI, IoT, Big Data and Cloud Computing for Industry 4.0*, Signals and Communication Technology, https://doi.org/10.1007/978-3-031-29713-7_13

[31]. Sentiment analysis of reviews on travel and tourism websites [19] can help in recommending better hotels and destinations to tourists and boost the tourism industry, while sentiment analysis of news articles [26] can help in creation of customized news feeds sent to users. Movie review analysis has also been done to increase quality of movie recommendations [20, 23]. Sentiment analysis has also been done to improve accuracy of predictions, like election results [27] and movement in stock prices based on financial news [12, 25]. In this work, we review and implement different approaches for sentiment analysis; cover the working, advantages, and limitations of each of these methods and conduct an experiment to find out which of these methods is best suited for sentiment analysis in today's scenario. Paper is organized as follows: Section 13.2 covers the related work summarizing popular sentiment analysis techniques. This is followed by experiment design in Sect. 13.3. We discuss results in Sect. 13.4 followed by conclusion and future work in the last sections.

13.2 Literature Survey

Sentiment analysis process usually includes three common steps as shown in Fig. 13.1. The first step is pre-processing of the dataset, which involves removing stop words, null rows, html tags, url, and converting text to lowercase. Once the data is pre-processed, and features are extracted sentiment classification is performed on it. Finally, the results of the analysis are evaluated by using confusion matrix to calculate precision, recall, accuracy, and F1score.

The various methods available for sentiment classification can generally be divided into lexicon-based methods, machine-learning-based methods and deep-learning-based methods, with each having its own advantages and drawbacks.

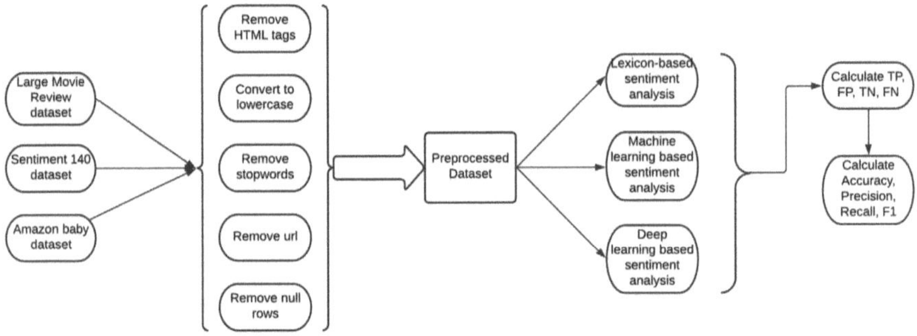

Fig. 13.1 Sentiment Analysis Architecture

13.2.1 Lexicon-Based Methods

Lexicon-based methods consist of creating a lexicon of sentiment words, and the sentiment scores associated with them. Sentiment words are those positive or negative words which are deemed to be important in calculating sentiment. Hu and Liu [9] proposed an approach to predict sentiment based on frequency of occurrences of positive and negative words in a sentence; this method can be broadly described as "lexicon-based method". Popular lexicons are SentiWordnet, Liu and Hu opinion lexicon, and SentiWords [30]. SentiWordnet assigns to each sentiment word a positivity score, a negativity score, and an objectivity score. SentiWords assigns to the words a score between -1 and 1 based on the positivity and the negativity.

The principal advantage of the lexicon-based methods is that as they are unsupervised methods, they have no need for a training phase or a labelled dataset. But the disadvantage is that the contextual information is ignored, e.g., in the sentence "The product is cheap and easy-to-use, but it stopped working within 1 month", there are more positive words (cheap, easy-to-use) than negative words (stopped), so probability of it being classified as positive is higher, though from the context we can see that it is a negative review.

Certain work has been done to overcome this disadvantage of lexicon-based methods by combining it with other deep-learning-based methods. Yang et al. [31] proposes a model for sentiment analysis by combining lexicon-based methods and deep learning. Their method uses lexicon for enhancing sentiment features and using deep learning, feature for sentiment analysis are extracted. Neural networks and SentiWordnet lexicon were combined to perform sentiment analysis [23]. Eighty-one major non-English language lexicons were gathered by [5].

13.2.2 Machine-Learning-Based Methods

Several research highlight the recent shift towards statistical approach for sentiment analysis, relying on sufficiently large datasets for supervised training. A hybrid sentiment classification model of adaboosted decision tree and Support Vector Machine (SVM) was proposed in [21]. The preprocessed data is given as input to the SVM and adaboosted decision tree followed by their outputs further given as input to the decision tree. The Naïve Bayes and SVM machine-learning models were trained on Amazon dataset which contains reviews of various gadgets and electronics products like Camera, smartphones, etc., and [11] reported an accuracy of 98.17% for Naïve Bayes and 93.54% using SVM for sentiment analysis of Camera Reviews. The authors of [8] compared three sentiment analysis lexicon methods named SentiWordnet, TextBlob, and W-WSD along with the machine-learning techniques, Naïve Bayes and SVM. A study and comparison between multinomial Naïve Bayes, SVM, Maximum Entropy, Bernoulli Naïve Bayes, and

Decision Trees was carried out in [20]. Gomathi [7] show a comparative study of Naïve Bayes, SVM, Random Forest, and Maximum Entropy on the google reviews dataset. Borade et al. [17] proposed the use of Naïve Bayes method for sentiment analysis and to tag them as positive or negative whereas [24] highlight SVM outperforms compared to Naïve Bayes and maximum entropy methods.

13.2.3 Deep-Learning-Based Methods

Studies highlight that with smaller dataset although machine-learning methods are preferable, provided availability of sufficiently large dataset, deep-learning algorithms have outperformed other approaches to sentiment analysis. For sentiment analysis task, results obtained in [28] show multilayer perceptron outperforms over other algorithms like SVM and random forests. Jangid et al. [12] proposed the use of Bidirectional Long Short-Term Memory (Bi-LSTM) units for extracting aspects from microblog and headline. A novel sentiment analysis methodology based on the lexicon and combining attention-based Bidirectional Gated Recurrent Unit (BiGRU) and Convolutional Neural Network (CNN) is proposed in [31]. Bi-LSTM model was proposed on the Military life board on PTT website data in [14] along with the support of a self-developed military sentiment dictionary. Souma et al. [25] proposed the hybrid model combining recurrent neural network (RNN) with LSTM to train the News Archive Data from Thompson Reuters. Latest studies [1] show the implementation of various deep-learning algorithms and use of SentiCNN model [10] for solving several challenges of sentiment analysis.

13.2.4 Datasets

Datasets used for sentiment analysis differ based on the field of application. Most of the research involves using publicly available data from online product review websites. The Movie Review Dataset issued by Stanford consists of 50,000 IMDb movie reviews which are labelled positive or negative [23]. Work done by [19] consolidates reviews from popular hotel review website TripAdvisor. Zabha et al. [31] uses product reviews from famous Chinese shopping website dangdang.com, while lot of studies make use of Amazon reviews dataset, BBC news dataset [26], and tweets from Twitter [8, 21, 32] for sentiment analysis. The benefits of using a twitter dataset are more information and sentiments expressed in lesser words, availability of a vast dataset, and possibility of real-time analysis.

13.3 Experiment Methodology

For empirical analysis of different sentiment analysis approaches, we set up an experiment over three different datasets over a variety of methods described in Fig. 13.2.

Considering the wide variety of applications of sentiment analysis, we selected the following datasets to be used in the experiment;

1. Large Movie Review Dataset [15]: This is a dataset provided by the Stanford University and consists of 25 k movie reviews for training testing separately each. It consists of two classes (positive = 1 and negative = 0). Out of the 25,000 reviews each provided for training and testing, 12,500 reviews are positive and the remaining 12,500 reviews are negative.
2. Sentiment 140 Dataset [2]: This dataset consists of 1.6 million tweets extracted from twitter and consists of two classes (0 = negative and 4 = positive)
3. Amazon Baby Dataset: This dataset consists of reviews from Amazon and has 3 columns (name, review, and rating). Here name field contains the subject of the reviews, review field contains the text data, and ratings field consists of score of review between 1 and 5, 5 being highest rated.

13.3.1 Pre-processing

To extract the features from text which are not explicitly available like in structured data, we need to clean data to remove unwanted characters and perform pre-processing on all the corpus.

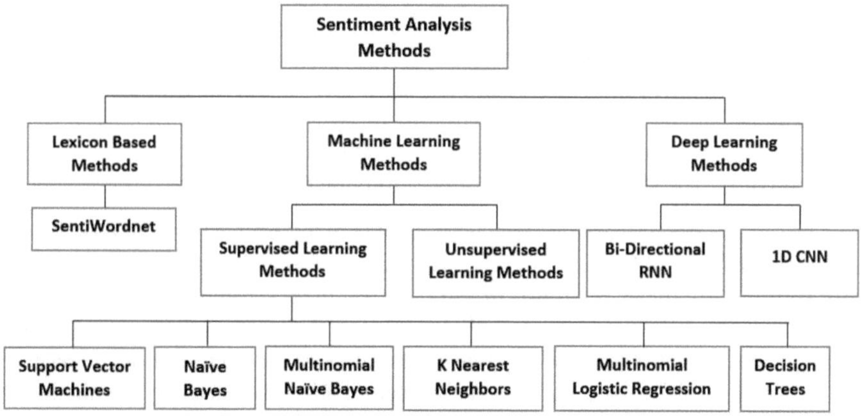

Fig. 13.2 Sentiment Classification Methods Used for Experimentation

13.3.1.1 Pre-processing on Large Movie Reviews Dataset

1. Removing HTML tags: HTML tags contained in reviews field are removed.
2. Lowercase: Converted the reviews to lower case.
3. Stopwords: Removed the stopwords contained in reviews.

13.3.1.2 Pre-processing on Sentiment140 Dataset

1. Trimming the dataset: Trimmed the dataset to contain 12,500 positive and negative reviews each.
2. Removing HTML decoding: Removing HTML decoding like &amp, &quot, etc.
3. Removing @username: Removed the username contained in the tweets.
4. Removing URL links: Removed URL links contained in the tweets.
5. Hashtags: Hashtags might contain important information which is necessary to understand the sentiment so we just removed the hashtag symbol and kept the text under it in our reviews.
6. Lowercase: Converted the tweets to lowercase.
7. Stopwords: Removed the stopwords contained in the tweets.

13.3.1.3 Pre-processing on Amazon Baby Dataset

1. Remove null rows: Remove the rows containing empty fields.
2. Add sentiment field based on ratings: We assigned the review a negative sentiment if it has rating 1 or 2 and assigned the review a positive sentiment if it has rating 4 or 5. We dropped the reviews containing the rating as 3 since we only wanted two sentiment classes, positive and negative.
3. Trimming the dataset: We trimmed the dataset to contain 12,500 positive and negative reviews separately each.
4. Combined the name and review field: Since both the columns are important in sentiment classification, we combined them into one column to pass it to our model.
5. Lowercase: The reviews are converted to lowercase.
6. Stopwords: The stopwords are removed from the reviews.

Most of the pre-processing steps were implemented using source python library re [29] and NLTK [4]. Additionally, we even performed vectorization and train test split on all the 3 datasets. In case of machine learning for vectorization, we used the tf-idf vectorizer and the values of their attributes are as follows. (max_features = 5000, min_df = 5 and max_ df = 0.7). In case of deep learning, we first padded the reviews and then for vectorization we used word embeddings provided by Keras [6]. Then we perform a 70:30 train test split of the dataset.

13.3.2 Methods

Lexicon-based In this method, we used the SentiWordnet lexicon [3], which is the most commonly used lexicon for lexicon-based sentiment analysis. A lexicon positivity, negativity, and objectivity score is assigned for every synset within the Wordnet. If the positivity score is greater than the negativity score, then the predicted sentiment is "positive" and vice versa.

The labelled and predicted sentiments are then compared to calculate accuracy, precision, recall, and f1-score for the method.

Machine Learning We train the sentiment classification model using six different machine-learning algorithms (i.e., SVM, Naïve Bayes, Decision Trees, Multinomial Naïve Bayes, KNN, and Multinomial Logistic Regression (MLR)). All the six models have been implemented using the open-source python sklearn library [18]. The attributes of each of these methods are described below.

1. SVM: Regularization parameter is set to 1, kernel = linear, and gamma = 'auto'.
2. Naïve Bayes: Default attributes provided by sklearn.
3. Multinomial Logistic Regression: Attribute multi class = 'multinomial' and solver = 'newton-cg'.
4. Multinomial Naïve Bayes: Default attributes provided by sklearn.
5. KNN: Attribute n_neighbours = 7.
6. Decision Trees: Default attributes provided by sklearn.

Deep Learning For deep learning, we implement Bidirectional Recurrent Neural Network [22] and 1D Convolutional Neural Network [13]. Both of these models have been implemented by using the Keras library.

13.3.3 Evaluation Metrics

The performances of the different sentiment classification techniques are measured using the following metrics:

True Positive (TP): Positive text classified as positive.
False Positive (FP): Negative text classified as positive.
True Negative (TN): Negative text classified as negative.
False Negative (FN): Positive text classified as Negative.

$$\text{Accuracy} = \frac{\text{TP} + \text{TN}}{\text{TP} + \text{TN} + \text{FP} + \text{FN}} \times 100$$

Precision: The ratio of true positives to the total reviews classified as positive.

$$\text{Precision} = \frac{\text{TP}}{\text{TP} + \text{FP}} \times 100$$

Recall: The ratio of True Positives to total actual positives.

$$\text{Recall} = \frac{\text{TP}}{\text{TP} + \text{FN}} \times 100$$

F1: The weighted average of precision and recall.

$$\text{F1} = \frac{2 * \text{precision} * \text{recall}}{\text{precision} + \text{recall}} \times 100$$

13.4 Results

13.4.1 Results of Lexicon-Based Methods

The results of lexicon-based models are shown in Table 13.1. The average F-score achieved using lexicon-based method is 62.33%.

13.4.2 Results of Machine-Learning-Based Methods

The results of training models using different machine-learning algorithms are shown in Tables 13.2, 13.3, 13.4, 13.5, 13.6, and 13.7. The average F1 score using SVM is 84%, using Naïve Bayes is 74.50%, using Multinomial Naïve Bayes is 82%, using Decision Trees is 71%, using Multinomial Logistic Regression is 84.33% and using K Nearest Neighbours is 71.16%. The result highlights that we achieved the best performance using the Multinomial Logistic Regression algorithm.

Table 13.1 Results of SentiWordnet

	SentiWordnet Method			
Dataset	Accuracy (%)	Precision (%)	Recall (%)	F1-Score (%)
Large Movie Reviews Dataset	64.54	66.5	66.5	66.5
Sentiment 140 Dataset	58.46	59.5	58.5	57
Amazon Baby Dataset	63.31	63.5	62	63.5

Table 13.2 Results of Support Vector Machines

	Support Vector Machines			
Dataset	Accuracy (%)	Precision (%)	Recall (%)	F1-Score (%)
Large Movie Reviews Dataset	87.94	88	88	88
Sentiment 140 Dataset	72.92	73	73	73
Amazon Baby Dataset	90.56	90.5	91	91

Table 13.3 Results of Naïve Bayes

	Naïve Bayes			
Dataset	Accuracy (%)	Precision (%)	Recall (%)	F1-Score (%)
Large Movie Reviews Dataset	80.03	80	80	80
Sentiment 140 Dataset	65.2	65.5	65	65
Amazon Baby Dataset	78.66	79.5	78.5	78.5

Table 13.4 Results of Multinomial Naïve Bayes

	Multinomial Naïve Bayes			
Dataset	Accuracy (%)	Precision (%)	Recall (%)	F1-Score (%)
Large Movie Reviews Dataset	86.04	86	86.5	86
Sentiment 140 Dataset	73.17	73.5	73.5	73
Amazon Baby Dataset	87.18	87	87.5	87

Table 13.5 Results of Decision Trees

	Decision Trees			
Dataset	Accuracy (%)	Precision (%)	Recall (%)	F1-Score
Large Movie Reviews Dataset	71.12	71	70	71
Sentiment 140 Dataset	67.02	67	66.5	67
Amazon Baby Dataset	74.93	75	75.5	75

Table 13.6 Results of Multinomial Logistic Regression

	Multinomial Logistic Regression			
Dataset	Accuracy (%)	Precision (%)	Recall (%)	F1-Score (%)
Large Movie Reviews Dataset	88.69	88.5	88	88.5
Sentiment 140 Dataset	73.36	73.5	72.5	73.5
Amazon Baby Dataset	90.84	90.5	90.5	91

Table 13.7 Results of K Nearest Neighbours

	K-Nearest Neighbours			
Dataset	Accuracy (%)	Precision (%)	Recall (%)	F1-score (%)
Large Movie Reviews Dataset	75.6	76.5	76	75.5
Sentiment 140 Dataset	57.3	57.5	57.5	57
Amazon Baby Dataset	81.22	82	81	81

Table 13.8 Results of Deep-Learning Methods

	Bidirectional RNN (%)	1D- CNN (%)
Large Movie Reviews Dataset	88.42	83.7
Sentiment 140 Dataset	70.21	69.67
Amazon Baby Dataset	90.46	87.61

13.4.3 Results of Deep-Learning-Based Methods

We have used accuracy as an evaluation metric for deep-learning models. The results obtained are shown in Table 13.8. The average accuracy for Bidirectional Recurrent Neural Network is 83.03% and the average accuracy for 1D Convolutional Neural Network is 80.32%. The results of deep-learning methods highlight that Bidirectional Recurrent Neural Network provides slightly better results than 1D Convolutional Networks.

The results obtained from the above experiment show that lexicon-based method does not give the best results mostly due to the inability of the lexicon-based method to integrate the contextual information into the process; treating the sentence as a bag-of-words and ignoring the word order of the sentences; apart from that negation handling is a feature that needs to be dealt with to improve the accuracy of the model. Lexicon-based method can be integrated with deep-learning methods to best handle these problems. Amongst machine-learning algorithms, Support Vector Machines and Multinomial Logistic Regression gave higher results compared to other algorithms we experimented with. For deep-learning approach, both algorithms have shown similar performances with the Bidirectional RNN giving slightly better results than 1D CNN.

The precision value highlights the ability of the trained model not to classify a negative review as positive. The recall value is the ability of the trained model to find all the positive samples. From the obtained results it is observed that the difference between precision and recall values is less which shows that equal amount of dataset has been classified as false positive by the algorithm as it has classified false negatives. This is valid as the dataset had approximately equal sets of positive and negative reviews.

13.5 Conclusion

This paper compares the best models under the three main types of sentiment analysis methods which are lexicon-based, machine-learning, and deep-learning methods. It was found that machine-learning and deep-learning methods provide better results than the lexicon-based methods, provided the availability of relevant dataset. Through the experiment conducted, we identify that SVM and Multinomial logistic regression provided better results than the other machine-learning methods mentioned. For deep-learning approach, both algorithms show similar performances with bidirectional RNN giving slightly better results than 1D CNN.

13.6 Future Work

In this paper, we applied some of the most popular sentiment analysis methods across a variety of datasets including the large movie reviews dataset, the sentiment 140 dataset, and the Amazon baby dataset and compared their performance using f1-score, accuracy, precision, and recall as performance metrics. We noticed that the algorithms provided significantly better results on large movie reviews dataset and the Amazon baby dataset than on the sentiment 140 dataset. For future work, we aim to improve the performance of the models on the sentiment 140 dataset. This dataset contains tweets extracted from twitter and the tweets usually contain hashtags, emoticons, links, and short forms or slangs which are very difficult to detect due to ambiguity present in the language and hence affect the performance of the models.

Sarcasm detection is another challenge of sentiment analysis in which further work needs to be done; as people convey negative emotions using positive words, there is a high probability of the review being misclassified as positive instead of negative.

While there has been a considerable amount of research done in sentiment analysis in non-English languages, the major challenges that face us are the lack of availability of quality lexicons for these languages and the usage of mixed languages. Further work can be done to introduce a standard cross-lingual approach framework and develop quality lexicons for other languages.

Most of the work done so far focuses on classifying the text into positive or negative. But for specific projects, a more detailed categorization of the text may be needed (e.g., angry, disappointed, happy, enthusiastic, etc.). Work needs to be done to make the classifications more detailed and accurate.

References

1. Ain, Q.T., Ali, M., Riaz, A., Noureen, A., Kamran, M., Hayat, B., Rehman, A.: Sentiment analysis using deep learning techniques: A review. International Journal of Advanced Computer Science and Applications 8(6) (2017). https://doi.org/10.14569/IJACSA.2017.080657
2. Alec Go, R.B., Huang, L.: Twitter sentiment classification using distant supervision. CS224N Project Report, Stanford, 1 (2009)
3. Baccianella, S., Esuli, A., Sebastiani, F.: Sentiwordnet 3.0: An enhanced lexical resource for sentiment analysis and opinion mining. In: Calzolari, N., Choukri, K., Maegaard, B., Mariani, J., Odijk, J., Piperidis, S., Rosner, M., Tapias, D. (eds.) LREC. European Language Resources Association (2010), http://nmis.isti.cnr.it/sebastiani/Publications/LREC10.pdf

4. Bird, S., Klein, E., Loper, E.: Natural language processing with Python: analyzing text with the natural language toolkit. O'Reilly Media, Inc. (2009)
5. Chen, Y., Skiena, S.: Building sentiment lexicons for all major languages. In: ACL (2014)
6. Chollet, F., et al.: Keras. https://keras.io (2015)
7. Gomathi, D.S.: Sentiment analysis of google reviews of a college (2019)
8. Hasan, A., Moin, S., Karim, A., Band, S.: Machine learning-based sentimental analysis for twitter accounts. Mathematical and Computational Applications 23, 11 (02 2018). https://doi.org/10.3390/mca23010011
9. Hu, M., Liu, B.: Mining and summarizing customer reviews. In: Proceedings of the Tenth ACM SIGKDD International Conference on Knowledge Discovery and Data Mining. p. 168–177. KDD '04, Association for Computing Machinery, New York, NY, USA (2004). https://doi.org/10.1145/1014052.1014073
10. Huang, M., Xie, H., Rao, Y., Liu, Y., Poon, L.K.M., Wang, F.L.: Lexicon-based sentiment convolutional neural networks for online review 12 Gupta et al. analysis. IEEE Transactions on Affective Computing pp. 1–1 (2020). https://doi.org/10.1109/TAFFC.2020.2997769
11. Jagdale, R., Shirsat, V., Deshmukh, S.: Sentiment Analysis on Product Reviews Using Machine Learning Techniques: Proceeding of CISC 2017, pp. 639–647 (01 2019). https://doi.org/10.1007/978-981-13-0617-4_61
12. Jangid, H., Singhal, S., Shah, R.R., Zimmermann, R.: Aspect-based financial sentiment analysis using deep learning. pp. 1961–1966 (04 2018). https://doi.org/10.1145/3184558.3191827
13. Kiranyaz, S., Avci, O., Abdeljaber, O., Ince, T., Gabbouj, M., Inman, D.J.: 1d convolutional neural networks and applications: A survey. Mechanical Systems and Signal Processing 151, 107398 (2021). https://doi.org/10.1016/j.ymssp.2020.107398; https://www.sciencedirect.com/science/article/pii/S0888327020307846
14. Liang-Chu Chen, C.M.L., Chen, M.Y.: Exploration of social media for sentiment analysis using deep learning. Soft Computing 24, 8187–8197 (2020). https://doi.org/10.1007/s00500-019-04402-8
15. Maas, A.L., Daly, R.E., Pham, P.T., Huang, D., Ng, A.Y., Potts, C.: Learning word vectors for sentiment analysis. In: Proceedings of the 49th Annual Meeting of the Association for Computational Linguistics: Human Language Technologies. pp. 142–150. Association for Computational Linguistics, Portland, Oregon, USA (June 2011), http://www.aclweb.org/anthology/P11-1015
16. Nandwani, P., Verma, R.: A review on sentiment analysis and emotion detection from text. Soc. Netw. Anal. Min. 11 (august 2021). https://doi.org/10.1007/s13278-021-00776-6
17. Omkar Borade, Kaushike Gosavi, A.G., Shinde, A.: "sentiment analysis of college reviews". vol. 5, pp. 319–322 (04 2017), http://www.ijedr.org/papers/IJEDR1702054.pdf
18. Pedregosa, F., Varoquaux, G., Gramfort, A., Michel, V., Thirion, B., Grisel, O., Blondel, M., Prettenhofer, P., Weiss, R., Dubourg, V., Vanderplas, J., Passos, A., Cournapeau, D., Brucher, M., Perrot, M., Duchesnay, E.: Scikit-learn: Machine learning in Python. Journal of Machine Learning Research 12, 2825–2830 (2011)
19. Prameswari, P., Zulkarnain, Surjandari, I., Laoh, E.: Mining online reviews in Indonesia's priority tourist destinations using sentiment analysis and text summarization approach. In: 2017 IEEE 8th International Conference on Awareness Science and Technology (iCAST). pp. 121–126 (2017). https://doi.org/10.1109/ICAwST.2017.8256429
20. Rahman, A., Hossen, M.S.: Sentiment analysis on movie review data using machine learning approach. In: 2019 International Conference on Bangla Speech and Language Processing (ICBSLP). pp. 1–4 (2019). https://doi.org/10.1109/ICBSLP47725.2019.201470
21. Rathi, M., Malik, A., Varshney, D., Sharma, R., Mendiratta, S.: Sentiment analysis of tweets using machine learning approach. In: 2018 Eleventh International Conference on Contemporary Computing (IC3). pp. 1–3 (2018). https://doi.org/10.1109/IC3.2018.8530517
22. Schuster, M., Paliwal, K.: Bidirectional recurrent neural networks. IEEE Transactions on Signal Processing 45(11), 2673–2681 (1997). https://doi.org/10.1109/78.650093
23. Shaukat, Z., Zulfiqar, A.A., Xiao, C., Azeem, M., Mahmood, T.: Sentiment analysis on imdb using lexicon and neural networks 2 (2020). https://doi.org/10.1007/s42452-019-1926-x

24. Shivaprasad, T.K., Shetty, J.: Sentiment analysis of product reviews: A review. In: 2017 International Conference on Inventive Communication and Computational Technologies (ICICCT). pp. 298–301 (2017). https://doi.org/10.1109/ICICCT.2017.7975207
25. Souma, W., Vodenska, I., Aoyama, H.: Enhanced news sentiment analysis using deep learning methods. Journal of Computational Social Science 2 (2019). https://doi.org/10.1007/s42001-019-00035-x
26. Taj, S., Meghji, A., Shaikh, B.: Sentiment analysis of news articles: A lexicon based approach (02 2019). https://doi.org/10.1109/ICOMET.2019.8673428
27. Tjong Kim Sang, E., Bos, J.: Predicting the 2011 Dutch senate election results with Twitter. In: Proceedings of the Workshop on Semantic Analysis in Social Media. pp. 53–60. Association for Computational Linguistics, Avignon, France (Apr 2012), https://www.aclweb.org/anthology/W12-0607
28. Valencia, F., Gómez-Espinosa, A., Valdés-Aguirre, B.: Price movement prediction of cryptocurrencies using sentiment analysis and machine learning. Entropy 21(6), 589 (Jun 2019). https://doi.org/10.3390/e21060589.
29. Van Rossum, G.: The Python Library Reference, release 3.8.2. Python Software Foundation (2020)
30. Warriner, A., Kuperman, V., Brysbaert, M.: Norms of valence, arousal, and dominance for 13,915 English lemmas. Behavior research methods 45 (02 2013). https://doi.org/10.3758/s13428-012-0314-x
31. Yang, L., Li, Y., Wang, J., Sherratt, R.S.: Sentiment analysis for e-commerce product reviews in Chinese based on sentiment lexicon and deep learning. IEEE Access 8, 23522–23530 (2020). https://doi.org/10.1109/ACCESS.2020.2969854
32. Zabha, N., Ayop, Z., Anawar, S., Erman, H., Zainal, Z.: Developing cross-lingual sentiment analysis of Malay twitter data using lexicon-based approach. International Journal of Advanced Computer Science and Applications 10 (01 2019). https://dx.doi.org/10.14569/IJACSA.2019.0100146

Chapter 14
Sign Language Machine Translation Systems: A Review

Suvarna R. Bhagwat, R. P. Bhavsar, and B. V. Pawar

14.1 Introduction

The most important aspect of computer science is problem-solving, an essential skill for life. Since the evolution of computing mechanisms, various computational paradigms have been successively utilized to develop software and hardware for problem-solving. The technological and computational efforts taken in the initial period were more meant for scientific and commercial aids rather than real-life applications. An inclination towards the utilization of technology for the development of tools that can be employed in health, business, and social contexts has surely been observed at a later time. One of the significant areas where technology has been brought into being in the early 1990s is 'Assistive Technology.' It is a field where the design and development of equipment or product systems are done to increase, maintain, or improve the functional capabilities of persons with disabilities.

Persons with disabilities form a countable portion of the population. In a scene where every distinguished individual in society seeks acceptance and provision of equal levels of opportunities, physically deprived people are not any exception. Assistive technology can aid towards special needs of such people and help them to raise their capabilities to the level of ordinary people. Since its establishment, this kind of technological aid can be found in the following abstract categories [1].

- Mobility impairments aid: Wheelchairs, transfer devices, walkers, prosthesis

S. R. Bhagwat (✉)
Vishwakarma Institute of Information Technology, Pune, Maharashtra, India
e-mail: suvarnarbhagwat@viit.ac.in

R. P. Bhavsar · B. V. Pawar
School of Computer Sciences, KBC North Maharashtra University, Jalgaon, Maharashtra, India
e-mail: rpbhavsar@nmu.ac.in; bvpawar@nmu.ac.in

A. Neustein et al. (eds.), *AI, IoT, Big Data and Cloud Computing for Industry 4.0*, Signals and Communication Technology, https://doi.org/10.1007/978-3-031-29713-7_14

- Visual impairments aid: Screen readers, Braille embossers, refreshable Braille display, desktop video magnifier, Screen
- Magnification software, large-print, and tactile keyboards, Navigation assistance, wearable technology
- Hearing impairment aids: Hearing aids, infrared listening devices, amplified telephone equipment, and speech-generating devices.

So as listed, in the near of beginning, assistive equipment was more electronic and mechanical kinds of devices. Available electronic, and mechanical devices are useful in increasing mobility capacity, amplification of visual and audio signals, etc. But in recent years, IT-enabled assistive products to be used in the field of education, sports, computer accessibility, memory aids, and home automation are also being investigated.

One significant difference between hearing-impaired persons and persons with other kinds of disabilities is that the people in the latter category hardly face any kind of communication problem. Apart from not being able to hear and eventually not being able to speak, the major challenge faced by the hearing impaired is the communication barrier between them and the rest of society. As we know, communication requires a common medium of Language, signs, or behavior to exchange information between individuals. Communication can become a trouble if the medium used by communicators differs. In the case of two different spoken languages, bilingual dictionaries and basic knowledge of each other's language can come to the rescue. But the trouble in communication may intensify if communicators are using languages of different modalities, viz. spoken language (Oral-auditory mode) and sign language (Gestural-visual mode).

Many times, Hearing Impaired persons are encouraged to be able to make use of Lip-reading. It can aid only up to a certain extent, as he/she should have a good grasp of spoken Language. Also, keen observation of slow and signer's clean articulation can aid a hearing person to overcome the barrier. Still, this is also inadequate, as Sign Languages are as complex as spoken languages, also they do not confide in spoken Language.

Sign Language Interpreters can be utilized to make a way through this communication barrier. They can provide intermediary services to both ends. But several practical issues can limit the use of Sign Language interpreters now and then. Some of these issues can be:

- Low availability of Sign Language interpreters
- High hiring cost
- Futile for confidential communication
- Impractical for short communication (e.g., shopping, booking, etc.)
- Not viable for the reading of documents, newspapers, and websites daily
- Physical and time-bound constraints of Sign Language interpreters for long conversations

The use of computer-aided technology to lessen the language barrier was pioneered in the 1960s. 'Machine Translation,' which is the computer-aided translation

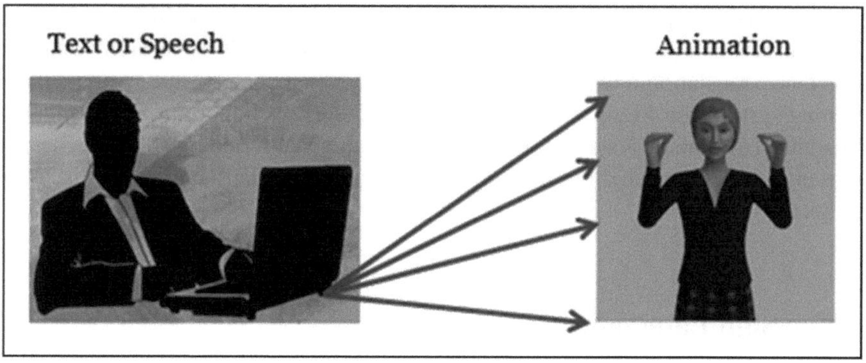

Fig. 14.1 A typical sign language machine translation system

from one Spoken Language to another, has been significantly employed for bridging the communication gap between spoken languages. Various approaches from rule-based, data-driven to machine learning have been successfully developed for a variety of spoken languages by researchers all over the world since then.

After the boost in linguistic research of Sign Languages, Machine Translation is also being utilized for reducing the barrier between Spoken Language and Sign Language. The first vital attempt at the translation of Spoken Language to Sign Language was made in the 1990s using a Rule-based approach. It has been a thriving research area since then. This variation of Machine Translation is known as 'Sign Language Machine Translation.' Such a system accepts spoken language text or speech and translates them to sign language gloss or visual interpretation. Figure 14.1 gives an outlook of a typical Sign Language Machine Translation system.

Existing Sign Language Machine Translation systems translate Spoken Language text/speech to Sign Language GLOSS/articulation. Here, the source and target languages are of different modalities, so correlated challenges can be seen.

We are taking efforts the translation of a simple sentence in the Marathi Language to its equivalent in Indian Sign Language. Marathi is one of the 22 official languages listed in the eighth Schedule of the Indian Constitution. In actuality, there are about 75 major spoken languages in India and including dialects, the count rises to 325 [29].

Sign language is used by hard-hearing communities to express their ideas or views. Like other languages, it is also an independent language with complex grammar. It consists of naturally evolved visual-manual signs influenced by social-cultural factors. Indian Sign Language is the official/standard sign language used by the deaf community in India. Some of the reported dialects of Indian Sign Language are listed as Bangalore-Madras, Bombay, Calcutta, Delhi, and Shillong Indian Sign Language [30]. These dialects are not standardized or crisply defined. Due to these reasons and the scarcity of computational resources, we are currently focusing on official Marathi and official Indian Sign Language as a 'source-target language pair' for Machine Translation.

This survey paper is arranged in the following way. Section 14.2 briefs about Sign language, along with its essential characteristics, which make Sign Language Machine Translation a challenging task. It also gives a concise survey of available writing systems for Sign Language. Section 14.3 surveys important Sign Language Machine Translation systems for Foreign Sign Languages. Section 14.4 specifically presents ventures carried out for Indian Sign Language. Gaps analysis is discussed in Sect. 14.5.

14.2 Sign Language

From the literature survey, it has been observed that the realization of Sign Language Machine Translation systems had been done quite late in the 1990s as compared to that of Machine Translation systems since the 1960s. It might be because of the late venture of sign language linguistic studies. One can surely say that Sign Languages are natural languages, as they have evolved over a period of time just like any spoken language. Sign languages do have their grammatical structure, and Sign Language Machine Translation systems have to consider them.

In Sign Language, information is conveyed through the meaningful articulation of hands, arms, head, and shoulders. Such an articulation, containing a unique composition of hand shape, palm orientation, location of hands, and movements form a manual component of a sign. Facial expressions like eye gaze, raised eyebrows, and puffed cheeks also play an essential role. They form a non-manual element of a sign. One sign may represent a single word, a single phrasal structure, or even a small sentence in Spoken Language. Sign languages exhibit grammatical characteristics at all linguistic levels from phonetics, phonology, morphology, syntax, and semantics to even pragmatics.

Similar to spoken languages, there is no universal sign language, and almost every country has got its official sign language. Through a literature survey, it has been observed that American Sign Language is the most studied and well-documented Sign Language. Linguistic studies on Indian Sign Language were started around 1978 by Vasishta. Ulrik Zeshan does remarkable field research and linguistic documentation in 2004 [2]. Though Indian sign language is being used by HI as well as interpreters, its linguistics study is gaining momentum in the last decade [3]. Presently online Video-based dictionary of words used in daily life is developed by Ramakrishna Mission Vivekananda Educational and Research Institute, Coimbatore, in collaboration with C-DAC, Hyderabad [4].

14.3 Challenges of Sign Language Machine Translation

Although both Spoken Language and Sign Language are natural languages, there are some interesting characteristics of Sign Language, which make them challenging for machine translation. Sign languages make use of manual gestures (hand movements), non-manual features (facial expressions), and the use of space simultaneously. This results in the simultaneous nature of morphology. In this section, such features are aligned against that of Spoken Language [5].

14.3.1 Simultaneity in Articulation

In the acoustic channel, it is hard to hear more than one thing at a time. Also, we only have one vocal track, so spoken speech is essentially linear. So, for Spoken Language, there can be just one serial stream of phonemes. Thus, spoken words are said to be sequential at the level of articulation.

On the other side, in the case of sign languages, the visual system can perceive many things at once also multiple visible articulators: (two hands, head, shoulders, and facial expressions) can articulate at a time. Thus, in the case of Sign Language, multiple things can be articulated & perceived at a time. So, simultaneity is observed at the articulation level [26].

14.3.2 Non-manual Features

As already stated, non-manual features like facial expressions, eye gaze, and head movements play an important role in Sign Language grammar. They are used to express emotions, the manner of the act. They are also used as morphological and syntactical markers to express interrogative sentences, negations, and imperative sentences. It is again interesting to note that non-manual features are used simultaneously with manual features.

14.3.3 Signing Space

Generally, signs are articulated within the area in front of the sign's upper body. Space is also of prominent grammatical importance in Sign Languages. The signer's backside and front side spaces are associated with past and future tenses [27]. Space is used to denote pronouns. Space is also used to build up location points, also known as anaphoric reference points and verb agreements.

14.3.4 Morphological Incorporation

In the literature of sign languages, the process of 'Incorporation,' is a way of combining two signs. In this process, a phonetic parameter of one sign is replaced by that of another sign. It results in a complex sign. Sign languages exhibit various kinds of incorporation, like numeral incorporation, classifier incorporation, and Ad-position incorporation.

14.4 Sign Language Writing/Representation Systems

There are currently 7139 living spoken languages in the world. Out of them, approximately 3900 languages are unwritten (i.e., they don't have their writing systems or they are unwritten) [28].

On the other hand, one may think that sign languages cannot be written or read. But writing systems and sign languages are available. The more interesting thing about them is that they are language-independent. As they describe the articulation done while signing, one can write any sign language using any of the writing systems. During the literature survey, four prominent writing systems are observed viz. GLOSS, STOKOE notation, SignWriting, and Hamburg Notation System (HamNoSys). More on these writing systems are discussed in the next subsection.

As shown in Fig. 14.1, the output of Sign Language Machine Translation systems is sign language in written and/or visual form. Choosing an appropriate writing system for the representation of the target Sign Language is an eventual and important step in the development of Sign Language Machine Translation.

Articulation and perception of sign languages are Visual-spatial. The information that they convey cannot be written and read as that spoken languages. It took time to accept sign language as a formal language, even writing systems for sign language were lately recognized. This section attempts a survey of existing sign language writing.

14.4.1 Annotation Systems

One can view it as a one- or two-word translation in one Language for a word or morpheme in another language. These systems transcribe the information observed from sign articulation. Sign Language linguistic knowledge is necessary for writing or reading annotations. In literature, such systems are named 'GLOSS.' It is generally used by linguists to represent signing sequences along with grammatical, spatial, and non-manual features (e.g., tenses, emotions, types of sentences, manner, etc.).

Spoken language stems are normally used to gloss over what is being articulated and gestured. Such a system can be employed for computational experimentation of sign language, as spoken language scripts can be used to write annotations.

GLOSS does not have universal standards. A transcriber decides the level of detail at which the Sign Language articulations can be described. It just portrays the signing sequence but does not give any idea about what the signing will look like. Because of these shortcomings, this kind of system generally cannot be used for communication purposes but used for transcription [5].

14.4.2 Pictorial Systems

As discussed, annotation systems reveal morphological and syntactical features of sign language rather than phonological aspects. A pictorial scripting system, like SignWriting, SiS5 can be used to represent phonological level characteristics of signs. Such systems use compact icons having simple drawing elements like a circle, rectangles, lines, etc. to represent articulation as well as facial gestures [6].

14.4.3 Symbolic Systems

These are text-based systems for which ASCII fonts are available. So, they can be used for computational purposes. Some noticeable systems observed in this category are STOKOE, SignFont, SLIPA, ASL Ortho, and HamNoSys [7].

The available sign-writing systems are studied based on various features like font type, universality, linearity, machine compatibility, etc. Table 14.1 depicts a comparative study of the studied Sign Language writing systems against various characteristics.

14.5 Overview of Sign Language Machine Translation at the Global Level

Machine Translation systems research for spoken languages can be observed since the 1960s. But similar kinds of efforts for sign language endeavored in the 1990s. More or less 20 systems have been deployed in Sign Language Machine Translation. Some systems for sign language generation through graphical avatars are observed. Several types of research for sign language recognition also have been done. This section delineates earlier attempts in Machine Translation for sign language. Along with the demonstration of how systems in this paradigm have progressed, it also discusses their pros and cons. Wherever feasible these systems are analyzed for

Table 14.1 Comparison of imminent Sign Language writing systems

Writing system	Can it be used to represent all Sign Languages?	Support for the representation of Facial Expressions	Font type	Iconicity	Arrangement	Machine compatibility
Stokoe	No	No	Custom	Non-Iconic	Linear	ASCII
SignWriting	Yes	Yes	Pictorial Clip-art	Iconic	Nonlinear	ASCII, Unicode
HamNoSys	Yes	Yes	Custom	Iconic	Linear	Unicode, Convertible to Animation
SignFont	No	Some	Custom	Non-iconic	Linear	ASCII
ASLphabet	No	No	Iconic, Custom	Iconic	Linear	Online interface only
ASL Ortho	No	No	Roman	Non-iconic	Linear	ASCII, Unicode
SiS5	No	Yes	Pictorial	Iconic	Nonlinear	Handwritten
SLIPA	Yes	Yes	Roman	Non-iconic	Linear	ASCII, Unicode
ASLSJ	No	Yes	Roman	Non-iconic	Linear	ASCII, Unicode
SignScript	No	Yes	Custom	Non-iconic	Linear	ASCII
Gloss	Yes	Yes	Roman	Non-iconic	Linear	ASCII, Unicode

- Source and target languages
- Translation Methodology
- Set of grammatical features of languages
- Domain
- Format of data
- Amount of data
- System architecture
- Details of graphical avatar generation
- Evaluation methods & results
- Handling of simultaneous morphological features of sign languages

In addition, it is also demonstrated how, over time, systems within this paradigm have progressed, as well as the more recent move toward more empirical methodologies. Discussion for foreign sign languages and Indian sign language is done separately in two subsequent sections.

14.5.1 The Zardoz System

It was the very first prominent attempt at using Machine Translation approaches for the translation of spoken Language to Sign Language. It was carried out by Veale et al. (1998) as a multilingual system for the translation of English text into Japanese Sign Language (JSL), American Sign Language, and Irish Sign Language [8]. The system has used the interlingua approach and includes multiple task-based modules. PATR-based unification grammar is used to represent interlingua. It also presents the design of a complete generation of Sign Language & a detailed avatar animation phase. DCL: Doll Control Language was used for animation generation [9].

Although this system was theoretically strong, it was minimally implemented. Also, no evaluation process is noted.

14.5.2 Translation from English to American Sign Language by Machine (TEAM)

Zhao et al. (2000) employed the TEAM system using a syntactic Transfer-based approach. It was initially proposed for English to American Sign Language, but later on, also implemented for South African Sign Language. It was supported by the US Air force and the Naval office of the USA. The system architecture was divided into two phases. In phase 1, Synchronous tree-adjoining grammars (STAGs) are used to create source and target language parse trees. The output of phase 1 is the American Sign Language gloss, which can incorporate American Sign Language morphological and adverbial aspects. The output of Phase-1 is used for graphical Avatar-based animation generation during phase-2. Human modeling & simulation technology is used to generate animation using a 3D graphical avatar [10].

This system is not domain-specific, but only limited morphological aspects have been experimented with. The project was successful in incorporating emotive capabilities in animation. Evaluation not noted.

14.5.3 Visual Sign Language Broadcasting (ViSiCast)

Ian Marshall and Eva S'afar have utilized the Semantic Transfer-based approach for the translation of English text into British Sign Language. Link Parser was used to analyze an input English text, whereas Prolog declarative clause grammar rules were used to convert this linkage output into a Discourse Representation Structure (DRS), which can store proposition level information (e.g., tenses, aspect, etc.). Sign language output was represented using Head-driven Phrase Structured Grammar along with HamNoSys notations. Eventually, 'Signing Gesture Markup Language (SiGML)' was used to generate animation [11, 22].

It was a domain-specific system. Animated weather forecast reports and simple informative messages at the post office are successfully generated. It lacks functionality for non-manual features. Again, no evaluation is noted.

14.5.4 A Multi-path Architecture

None of the systems discussed till now have considered classifier predicates (CPs). It is a phenomenon in which signers use special hand movements to indicate the location and movement of invisible objects (representing entities under discussion) in space around their bodies. A system by Matt Huenerfauth, University of Pennsylvania, USA 2006, proposed an interlingua approach where the interlingua is 3-D visualization of the objects in an English sentence. It also incorporates transfer and direct approaches into a single system. The translation process can follow one of the multiple paths offered by the system. Viz. Interlingua, transfer, or direct translation [12].

Evaluation of the system was done by native signers in terms of grammatical correctness, the naturalness of movement, and understandability. The system was also evaluated against animations created by a signer articulating CPs wearing motion-capture suits.

14.5.5 Research by RWTH Aachen Group

Statistical Machine Translation approach for Sign Language Machine Translation was utilized by Jan Bungerothe at RWTH Aachen university group in 2006. The system was employed for the German-DGS language pair with translation both to and from Sign Languages. Initially, the system was developed for whether report domain and later adopted for the Airline travel system. The Phrase-based SMT system already developed at RWTH was used for this purpose [13].

Output was not animated, so no manual evaluation was needed. Automatic evaluation is carried out on the gloss output, where BLEU scores range between 0.17 and 0.22.

14.5.6 Project Web-Sign

Web-sign project 2004 was carried out at the University of Tunis, Tunisia to develop tools that will be able to make information over the Web accessible for the deaf. Under this project, Mohamed Jemni and Mehrez Boulares attempted an

Example-Based machine learning system associated with genetic algorithms for the development of a Web-based interpreter of Sign Language to convert English to American Sign Language & French Sign Language. This system initially divides paragraphs into sentences. Then genetic algorithms were used for detecting various similar sequences existing in the initial sentence. The system will recognize the sentence if the system already learns it. If the system doesn't recognize this sentence, then the proximity search is launched to get a better sentence. Fuzzy logic was used to determine the emotional interpretation of the sentence [14, 15].

A tool developed by a peer (Oussama El Ghoul) at the University of Tunis was used to create animations. Sign Markup Language (SML) was developed for the same.

14.5.7 Machine Translation Using Examples (MaTrEx)

An approach that combines EBMT, SMT, translation memory (TM), and information retrieval (IR) technologies for improving the quality & scale of Marker-based EBMT was successfully implemented at Dublin City University. Sara Morrissey et al. (2008) employed the same here for Sign Language Machine Translation. In this system, aligned sentences, words, and chunks are input to SMT for which a small corpus of 595 sentences was utilized. A working model of the system was created using the Air Travel Information System (ATIS) corpus [16].

An automatic evaluation was done for annotations/GLOSS. The decrease was observed in 'Word Error Rate' and 'Position Independent Word Error Rate' because of chunking. A manual evaluation was done for animation considering understandability, fidelity, naturalness, etc.

14.5.8 Japanese to Japanese Sign Language (JSL) Glosses Using a Pre-trained Model

This project has used a huge bilingual corpus of 13,000 sentences to pre-train the language model for developing a transformer-based Neural Machine Translation Model. It has experimented with a domain-free vocabulary of size 6000 words. The bilingual corpus contains the morphological le 2features of JSL like nodding, pointing, and classifiers inscribed. The system output is JSL gloss and it is evaluated using the BLEU score. However, the visual generation of sign language is not reported [23].

14.5.9 Sign Language Production Using Generative Adversarial Networks

Spoken-to-sign language translation systems have generally used a graphical avatar or prerecorded videos for the generation of sign language. But this is the first approach that has not used any graphical avatar for continuous video generation. The Neural Machine Translation along with the motion graph is used for the translation of spoken language sentences to sign language gloss. Then the generative model is used to create realistic videos. The system is evaluated using the BLEU score and WER [24].

Table 14.2 summarizes the main features of Sign Language Machine Translation Systems for Foreign Sign Languages, which are already discussed in detail.

14.6 Overview of Sign Language Machine Translation for Indian Languages

Very few attempts are noted in the area of research performed in the paradigm of Sign Language Machine Translation for Indian Sign Language. Mainly, they have employed Rule-based approaches because of the scarcity of Indian Sign Language linguistic resources. This section discusses these attempts.

14.6.1 INGIT: Limited Domain Formulaic Translation from Hindi Strings to Indian Sign Language

This has been the very first attempt at Indian Sign Language. Purushottam Kar et al. 2008, have employed Rule-based machine translation for Hindi strings to Indian Sign Language strings. The system takes input from the reservation clerk and translates it into Indian Sign Language. It is based on a hybrid formulaic approach. Here, Fluid Construction Grammar is used to implement Formulaic grammar [17].

The output of the Schematization module forms a structured sequence of signs for considering the temporal, location, and person queries. The ellipsis resolution module extends Fluid Construction Grammar (FCG). The generated Indian Sign Language string is interpreted using HamNoSys & SiGML.

Table 14.2 Sign Language Machine Translation Systems for Foreign Sign Languages

Project	Source – target language	Sign representation	MT approach	Remark related to the handling of Simultaneous Morphology handling
ZARDOZ System (1994)	English – American, Japanese Irish SL	Doll Control Language, Graphical display	Knowledge-based interlingua	Spatial Features and Non-Manual Features are handled.
TEAM project (2000)	English – American Sign Language	GLOSS/3-D Human modeling & simulation	Syntax level transfer based	STAG Trees are used to embed non-manual features in Gloss
ViSi-CAST (2002)	English – British Sign Language	SiGML 2-D Animation	Semantic level transfer based	Simultaneous morphological features are handled using manual intervention
Multi-path (2004)	English – American Sign Language	3-D virtual reality Avatar (NLI) software	Hybrid: Interlingua, Direct & Transfer	Non-Manual Features are handled
RWTH Statistical Translation (2007)	German – German Sign language (DGS)	SiGML 2-D Animation	Phrase-based Statistical Translation	Not reported
Web-Sign Project (2006)	English – American & French Sign Language	Text Adapted Sign Modeling Language	Statistical Translation	Non-Manual Features are generated using Time Place, subject, and action
MaTrEx (2008)	English – Dutch Sign Language	HamNoSys/SiGML 2-D Animation	Hybrid of SMT & EBMT	Not reported
Japanese to JSL using Pre-trained Model (2020)	Japanese to Japanese Sign Language	Gloss	Neural Machine Translation	Nodding, pointing, classifiers inscribed in Gloss.
Sign Production using GAN (2020)	English to British Sign Language	Gloss, Realistic Videos	NMT, Generative Adversarial Networks for video generation	Non-manual features are generated using motion Graph Technology

14.6.2 Dictionary-Based Translation Tool for Indian Sign Language

Tirthankar Dasgupta (2008), has developed a dictionary tool for the conversion of Hindi/Bengali/English to Indian Sign Language. Dictionary can store Hindi/English words, their meaning (with the help of Word-Net), POS, and corresponding HamNoSys. A sentence is accepted as input by the tool. Part of Speech tagging is done with the help of nlp.stanford.edu/software/tagger.shtml tagger. This tool can be used to associate signs to the words, phrases, or sentences of a spoken language text. The sign associated with each word is composed of its related Part-of-speech and semantic senses. HamNoSys & SiGML are used for interpretation [18].

In the later phase of this venture, a prototype for English to Indian Sign Language translation is employed [19]. The system can handle morphological functionalities like discourse, directionality, and classifier predicates minimally. Only simple sentences in English can be utilized as input.

14.6.3 Indian Sign Language Corpus for the Domain of Disaster Management

In a joint venture of C-DAC (Pune) and Ali Yavar Jung Institute of Hearing Impaired (Mumbai), a bilingual corpus for English and Indian Sign Language has been developed. Corpus contains around 4000 words of disaster domain and around 600 sentences [20]. Initially, all the sentences in English were signed by a single signer. The synthetic avatars are created using a trace around the original video. Challenges concerning contiguity, non-manual signs, and co-articulation were addressed. This system can be used only for giving disaster-related messages to Deaf people, and cannot be utilized to learn Indian Sign Language by people who do not have any sign language background.

14.6.4 Indian Sign Language from Text

A team at Thapar University, Patiala, India has made this admirable attempt under principal investigator Dr. Prateek Bhatia. They have developed an online multilingual multimedia Indian Sign Language dictionary. This dictionary contains 2000 English and 3286 Hindi words along with their Corresponding HamNoSys Codes [21].

Table 14.3 summarizes the main features of Sign Language Machine Translation Systems for Indian Sign Languages, which are already discussed in detail.

Table 14.3 Sign Language Machine Translation Systems for Foreign Sign Languages

Project/author	Source – target language	Sign notation /MT approach	General remark	Remark related to the handling of Simultaneous Morphology handling
INGIT (2008) Purushottam Kar, IIT Khargpur	Hindi – Indian Sign Language	HamNoSys, SiGML 2-D Animation /Rule-based Approach	Domain-specific. Small corpus (230 utterances, 90 words)	Simultaneous Morphological features are not handled graphically.
Dictionary-based prototype system (2008) Tirthankar Dasgupta, IIT, Kanpur, India	English – Indian Sign Language	HamNoSys/SiGML 2-D Animation Rule-based Approach	Sign videos incur memory overhead.	Morphological functionalities are handled minimally.
Disaster management Corpus building (2014) C-DAC, Ali Yavar Jung Institute for Hearing Impaired, Mumbai.	Disaster management messages ISL	Videos Tracing Corpus Size: 600 sentences 4000 words	Domain-specific. It can be used only for giving disaster management-related messages to Hearing Impaired.	Not reported
Indian Sign Language from Text (2016) Dr. Pratik Kumar, Thapar University, Patiala.	Hindi, English Indian Sign Language	HamNoSys/SiGML 2-D Animation Rule-based Approach	It is a User-friendly system for the translation of Hindi and English sentences. Stanford Parser is used for parsing source sentences.	Any handling of simultaneous Morphological features is not reported

14.7 Gap Analysis

Though there are ample NLP applications available, the languages they can process is a small subset of existing languages used worldwide. Many languages are still in need of NLP tools and computationally feasible resources that can be utilized by existing approaches/applications. The circumstances apply to many regional languages in India. The rich heritage of the regional languages of India has provided a massive scope for research in computational linguistics.

It has been noted from the available systems in the literature that, most of the spoken-to-sign language machine translation systems are developed using a rule-based approach. The lack of bilingual parallel corpora and other computational resources has led to very few efforts using statistical and knowledge-based approaches.

Also, to our observation, most of the systems lack the handling of simultaneous morphological features during Machine Translation. The linguistic study of Indian Sign Language has confirmed the fact that it exhibits Simultaneous Morphology, like other sign languages. The literature on Sign Language Translation Systems has considered the fact that it is necessary to handle the intrinsic features like grammatical use of signing space, classifier incorporation, numeral incorporation, ad position incorporation, and simultaneous use of non-manual features, during the generation of sign language. We also didn't find any system for the translation of Marathi to Indian Sign Language.

14.8 Discussion on the Designed Prototype and Proposed Enhancement

We have reported a novel approach to translate a Marathi sentence in text form, to Indian Sign Language Gloss and Animation in [25]. Figure 14.2 shows the system architecture.

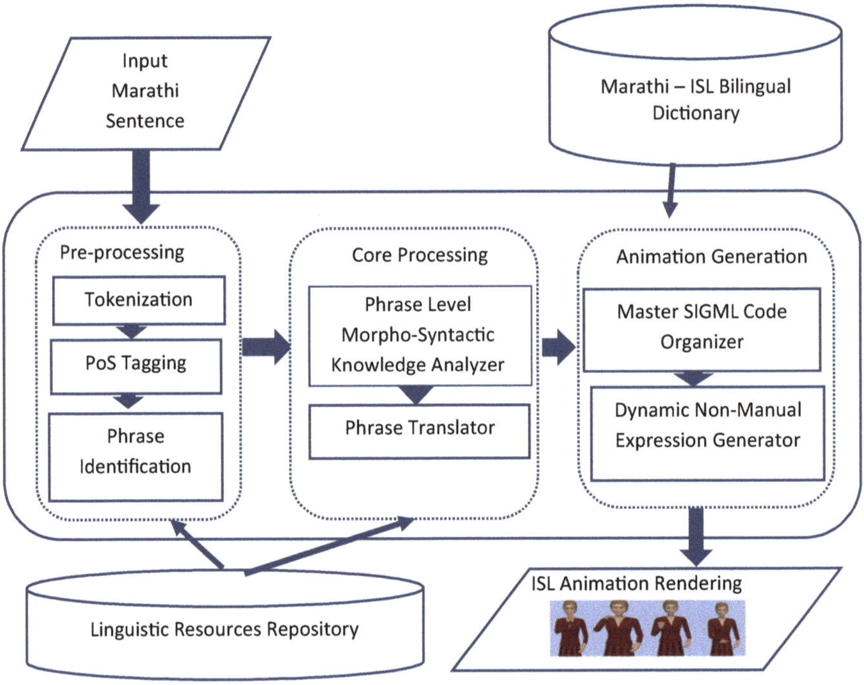

Fig. 14.2 System architecture for Marathi Simple Sentence to Indian Sign Language Machine Translation System

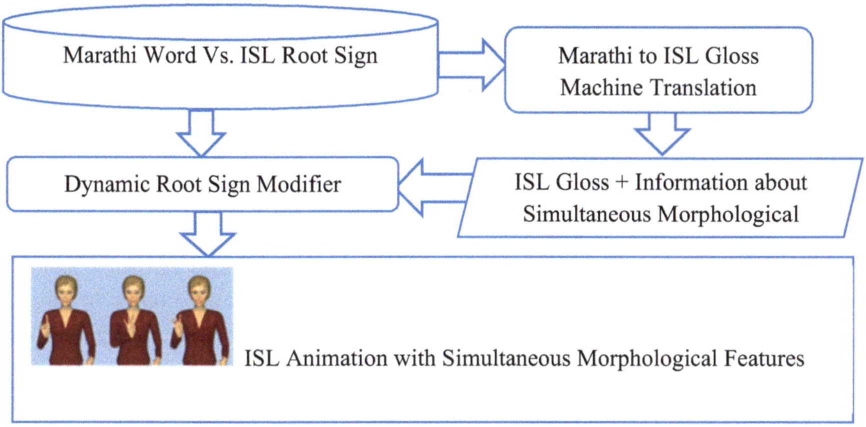

Fig. 14.3 A proposal system to handle simultaneous morphology of Sign Languages during Machine Translation

The detailed morphological and syntactical comparative analysis of Marathi and Indian Sign Language has led us to practice the phrase-level rule-based approach for designing this system. The Marathi input sentence is preprocessed using tokenization, PoS Tagging, and phrases are identified. The sentence is then translated into Indian Sign Language Gloss. The source and target language grammar rules are explicitly utilized in both steps. As sign languages are visual languages, it was much necessary to visualize the output. We have used HamNoSys and SiGML to render the animation of generated gloss.

Now we are proposing enhancement to our prototype by adding a new module for handling simultaneous morphological features at run time. The system architecture of the proposed extension is illustrated in Fig. 14.3.

As shown in Fig. 14.3, a module, 'Dynamic Root sign Modifier,' is needed to incorporate simultaneous morphological features in graphical animation. This module will gather the required information from the Machine Translation unit. It will make appropriate changes in the phonetic parameters of root signs stored in the database.

14.9 Conclusion

This work has described the preliminaries of Machine Translation. It has discussed the basics of Sign Language linguistics in brief. We have studied and noted the challenges faced during spoken-to-sign language machine translation. Most of the challenges faced are due to the simultaneous use of space, and manual and non-manual gestures.

We have surveyed reported Sign Language Machine Translation Systems and presented a report on important systems. It observed that most of the sign languages all over the world are computational-resource scary languages. So mainly rule-based approaches are used for Machine Translation. Recently few systems have reported the implementation of knowledge-based approaches using a comparable size of corpora.

As Marathi and Indian Sign Language lack compatible linguistic resources, we have developed a prototype system using a rule-based approach for translation. During the literature survey, we observed that reported systems have not handled simultaneous morphological features of Indian sign language comprehensively. So, we find a need for a module that can uphold these features during machine translation. A brief idea of the proposed system is also discussed.

14.10 Future Scope

We are looking forward to developing this system as a web-based application so that it can be availed in the process of teaching and learning ISL. It can also be used as an online interpreter from Marathi to ISL.

Construction of Marathi Vs. Indian Sign Language bilingual corpora is the natural future proposal with the main challenge of human expertise and extensive man hours. The availability of bilingual corpora would make it feasible to use statistical and knowledge-based approaches from machine translation.

References

1. Assistive Technology – Wikipedia. Available: https://en.wikipedia.org/wiki/Assistive_ technology, accessed March 2019.
2. Ulrik Zeshan, Sign Language in Indo Pakistan: A description of signed language, John Benjamins Publishing Company, 2000.
3. Samar Sinha, "A Grammar of Indian Sign Language", Ph.D. dissertation, Centre for Linguistics, School of Language, Literature & Culture Studies Jawaharlal Nehru University, New Delhi, India, 2012.
4. Ian Marshall, and Éva Safar, "Sign Language Generation using HPSG", 9th International Conference on Theoretical and Methodological Issues in Machine Translation, Keihanna, Japan, 2002.
5. Sara Morrissey, "Assessing Three Representation Methods for Sign Language Machine Translation and Evaluation", 15th Conference of the European Association for Machine Translation, Leuven, Belgium, pp. 137–144, May 2011.
6. Sign Writing Site, Available: http://www.signwriting.org, accessed November 2018.
7. Thomas Hanke, "HamNoSys, An introductory guide", International Studies on Sign Language & Communication for Deaf, vol. 5.
8. Tony Veale and Cunniengham, "Competitive Hypothesis Resolution in TWIG: A Blackboard-Driven Text-Understanding System", 10th European Conference on Artificial Intelligence, Chichester, UK, 1992.

9. Tony Veale, Alan Conway and Brona Collins, "The Challenges of Cross-Modal Translation: English to Sign Language Translation in the Zardoz System", Machine Translation, vol. 13, no. 1, pp. 81–106, 1998.
10. Liwei Zhao, Karin Kipper, William Schuler, Christian Vogler, Norman Badler, and Martha Palmer, "A Machine Translation System from English to American Sign Language", Lecture Notes in Computer Science, Volume 1934, Envisioning Machine Translation in the Information Future 4th Conference of the Association for Machine Translation in the Americas, pp. 54–67, 2000.
11. Thomas Hanke, ViSiCAST Deliverable D5-1: Interface Definitions Manuscript, Hamburg, Germany, 2002.
12. Matt Huenerfauth, "Generating American sign language classifier predicates for English-to-ASL machine translation", Ph.D. Dissertation, Faculties of the University of Pennsylvania Philadelphia, 2006.
13. Ney, Jan Bungerothe and Hermann, "Automatic Generation of German Sign Language Glosses from German Words", Springer-Verlage Berlin eelberg LNAI, vol. 3881, pp. 49–52, 2005.
14. Mohamed Jemni and Achraf Othman, "Statistical Sign Language Machine Translation: from English written text to American Sign Language Gloss", IJCSI International Journal of Computer Science, vol. 8, no. 5, September 2011.
15. Oussama El Ghoul, Mohamed Jemni, "WebSign: A system to make and interpret signs using 3D Avatars", Sign Language Translation and Avatar Technology, Dundee, UK., 2011.
16. Sara Morrissey, "Data-Driven Machine Translation for Sign Languages," Ph.D. Dissertation, Dublin, Ireland, 2008.
17. Purushottam Kar, Madhusudan Reddy, Amitabha Mukerjee and Achla M. Raina, "Ingit: Limited domain formulaic translation from Hindi strings to Indian Sign Language", ICON, Neidle, Carol, 2000.
18. Tirthankar Dasgupta, "Multilingual Multimedia Indian Sign Language Dictionary Tool", 6th Workshop on Asian Language Resources, Hyderabad, India, 2008.
19. Tirthankar Dasgupta, "Prototype Machine Translation System from Text-To-Indian Sign Language", IJCNLP-08 Workshop on NLP for Less Privileged Languages, Hyderabad, India, 2008.
20. Mahesh Kulkarni and Mathew Martin, "An Indian Sign Language (ISL) Corpus of the Domain Disaster Message Using Avatar", 3rd International Symposium on Sign Language Translation and Avatar Technology, Chicago, USA, 2013.
21. Paras Viz and Prateek Kumar, "Mapping Hindi Text to Indian sign language with Extension Using Wordnet", International Conference on Advances in Information Communication Technology & Computing, Bikaner, India, 2016.
22. Iva Marshall and Safar E., "A Prototype Text to British Sign Language (BSL) Translation System", 41st Annual Meeting of the Association of Computational Linguistics Conference, Sapporo, Japan, 2003.
23. S. Stoll, N. Camgoz, S. Hadfield and R. Bowden, "Text2Sign: Towards Sign Language Production Using Neural Machine Translation and Generative Adversarial Networks," International Journal of Computer Vision, 4 2020.
24. T. Miyazaki, Y. Morita, and M. Sano, "Machine Translation from Spoken Language to Sign Language using Pre-trained Language Model as Encoder," in Proceedings of the LREC2020 9th Workshop on the Representation and Processing of Sign Languages: Sign Language Resources in the Service of the Language Community, Technological Challenges and Application Perspectives, Marseille, 2020.
25. S. R. Bhagwat, R. P. Bhavsar and B. V. Pawar, "Translation from Simple Marathi sentences to Indian Sign Language Using Phrase-Based Approach," 2021 International Conference on Emerging Smart Computing and Informatics (ESCI), 2021, pp. 367–373, https://doi.org/10.1109/ESCI50559.2021.9396900.
26. Aronoff, Mark et al. "The Paradox of Sign Language Morphology." Language 81 (2005): 301–344.
27. Alkoby, Karen. (1999). A Survey of ASL Tenses.

28. Ethnologue, Languages of the world, Available: https://www.ethnologue.com/categories/development, accessed October 2022.
29. Indian Languages, Available: https://www.education.gov.in/hi/sites/upload_files/mhrd/files/upload_document/languagebr.pdf, accessed October 2022.
30. Glottolog, Available: https://glottolog.org/resource/languoid/id/indi1237, accessed October 2022

Chapter 15
Devanagari Handwritten Character Recognition Using Dynamic Routing Algorithm

Savita Lonare, Rachana Patil, and Renu Kachoria

15.1 Introduction

Nowadays, Handwriting character Visibility is one of the active research fields in handwriting recognition. Many handwritten character recognition systems have been researched, experimented with, and adapted for actual use in the past couple of years, seeking high accuracy and trustworthiness. Cursive writing patterns, overlapped letters or numbers, personal style of writing, and variation in strokes from person to person increase the recognition system's complexity. Several supervised methods have been experimented with and published for handwritten digit and character recognition, such as the deep neural network (e.g. convolutional neural networks (CNN)) [1]. Identifying Marathi handwritten text using deep learning techniques is done in [2–5].

S. Impedovo et al. [6] proposed a handwritten digit recognition algorithm using multi-objective optimization. They used a non-dominant sorting genetic algorithm for the classification of digits. In another approach, Anuran Chakraborty et al. proposed an autoencoder and transfer learning-based image recognition [7]. In [8], Field Programmable Gate Arrays were used for accurate feature extraction; then multilayer perceptron neural network classifier was used for Farsi handwritten digit classification. CNN was the most popular choice for an image classification task among all these approaches.

S. Lonare (✉)
Dr. D. Y. Patil Institute of Technology, Pune, India

Research Scholar at GITAM, Visakhapatnam, India

R. Patil · R. Kachoria
Pimpri Chinchwad College of Engineering, Pune, India
e-mail: rachana.patil@pccoepune.org; renu.kachoria@pccoepune.org

A. Neustein et al. (eds.), *AI, IoT, Big Data and Cloud Computing for Industry 4.0*, Signals and Communication Technology, https://doi.org/10.1007/978-3-031-29713-7_15

The majority of the research attained high performance for classification accuracy, more than 98% or 99% [9]. The benchmark dataset MNIST for image classification accuracy is achieved by combining various CNNs in an ensemble learning technique [10]. This work was further extended to achieve 99.77% recognition accuracy [11, 12] outperformed with an accuracy of 98.75%. Niu et al. proposed a hybrid CNN-SVM classifier for recognizing handwritten digits [13]. All these works reported outstanding recognition precision. However, CNN is the primary reason why deep learning is so widespread. CNN worked well for the vast dataset. But CNN has a drawback in its basic construction. Due to this, they are failing in some applications. CNN recognizes features in images and gets trained to recognize the objects with the gained information about the features. The primary hidden layers in CNN detect specific features such as edges, and as CNN gets more profound, the hidden layer can see more complex features. This learning that CNN gains are later used in making a final prediction [14]. The problem with CNN is that CNN cannot learn spatial information about any image. The process of max pooling used in CNN loses most of the vital information as it chooses the most active neuron's weight to be transferred to the next higher layer. At this point, the crucial spatial information gets lost.

Sara Sobour et al. [14] proposed a dynamic routing-based algorithm called capsule to address this issue. A Capsule network is a type of ANN. A Capsule network is designed to reuse a few capsules' outputs to create stable output for higher-level capsules.

15.2 Literature Review

A noteworthy numeral of research published in handwriting recognition [15–19]. SVM has been used for handwritten digit recognition for the first time in [20]. Later, many supervised classification problems, such as face detection [21] and object recognition [22] were solved using SVM. SVM became the best choice because of its accuracy. In [23], a Freeman Chain Code approach has been used for feature extraction, and a handwritten digit recognition system is proposed using SVM.

In recent years, the availability of more powerful machines, substantial training data, and deep learning methods have become more popular. Particularly for image data CNN achieved enormous success. Many handwritten digit recognition methods were proposed using CNN. These methods achieved high accuracy with MNIST datasets [24–27]. CNN has also been proven to identify handwritten Chinese typescripts [28]. LeNet5 CNN architecture-based feature extractor is used on the MNIST database [29]. This approach is demonstrated with excellent recognition accuracy. The impressive performance of the research work proves the efficiency of the CNN feature extraction operation. Further remarkable research on handwriting recognition is given in [30, 31].

Although CNN worked fine, they have some drawbacks. The pooling layer in CNN is responsible for losing valuable information [14]. Recently, the capsule network proposed by Sabour et al. [14] for handwritten digits classification gained the researcher's attention due to its capability to overcome the pooling layer in CNN. In this research, the authors worked on MNIST data. This research proposed a novel architecture called capsule Network (CapsNet).

Further, some capsules may contain noise or redundant data in the dynamic routing [14] process. To ease the disturbance of those capsules, Zhao et al. [32] proposed three approaches. A multi-label classification framework was proposed by N. Zhang et al. [33], where the capsule network with attention mechanism was experimented with. The experiments showed improved results. B. Zhang et al. [34] proposed a CapsNet framework for sentiment analysis in domain adaptation settings with semantic rules, boosting comprehensive sentence representation learning. This model pays distinct consideration to the typical information throughout the training process, efficiently regulates the neural network parameters for various features, and extracts more unseen characteristic details. M. Z. Hasan et al. [35] used Capsule Networks and CNN for Changma digit recognition. The results of using CapsNet are superior to that of CNN. The paper is about using CapsNet to classify the Devanagari handwritten characters and digits over the traditional convolution layer. The experiments show that CapsNet has a better capability of relative spatial features and gives better accuracy than the CNN models.

15.3 Problem with CNN

CNN is essentially a structure where numerous neurons are stacked together. CNN has been extraordinarily outstanding at handling image classification problems. It is very tough to have a neural network map for all the image pixels since it's computationally expensive. So, convolution is a technique that supports you abridge the computation cost to a great level without losing the crucial features of the data. Figure 15.1 shows a typical structure of a CNN model. The CNN model performs convolution that involves matrix multiplication and summation of those results, employing nonlinear activation functions for improved performance. It also has a pooling layer to extract the essential information to pass it to the next layer. The problem with CNN is that the CNNs perform remarkably well when classifying images that are very similar to the data set.

If the pictures have tilt, rotation, or any other diverse positioning, CNNs have poor performance. This issue gets solved by introducing different variations of a similar image while training. Each layer comprehends an image at a much more fundamental level in CNN. Suppose you are trying to classify trucks and cars. The very basic curves and boundaries are recognized by the first layer. The second layer might recognize the straight lines or the smaller shapes, like front look of the car and truck. Higher layers understand more complex shapes like the whole car or truck.

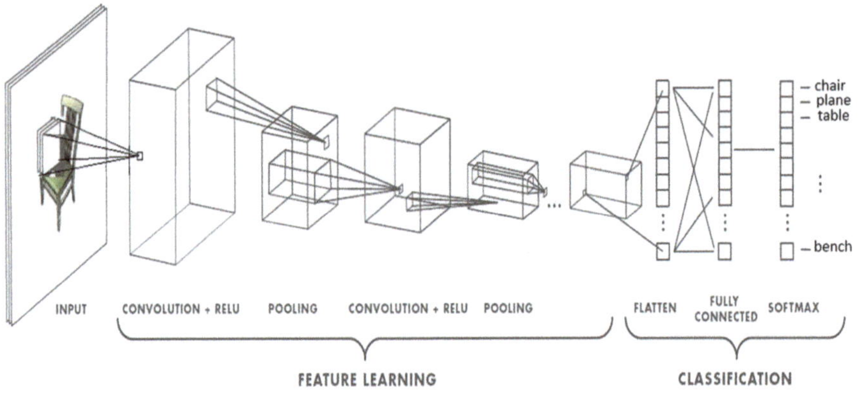

Fig. 15.1 CNN structure

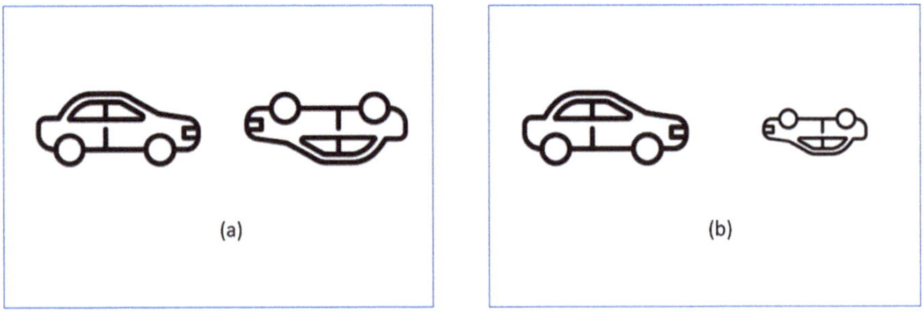

Fig. 15.2 (**a**) Disfiguration transformation. (**b**) Proportional transformation

CNN uses pooling after each layer for computations in realistic time frames. But in reality, it also loses out the positional data. The pooling layer helps in creating positional invariance. Without a pooling layer, CNNs would fit only the data close to the training set. This invariance also triggers false positives for images that have the components of a car but are not in the exact sequence. So the structure can trigger the right to match the left in the images shown in Fig. 15.2a. An observer can see the difference. The pooling layer also adds this kind of invariance. It wasn't the purpose of the pooling layer. The pooling was supposed to add a proportional, positional, and orientational invariance instead of adding all types of positional invariance. It leads to the problem of detecting the car in the right image from Fig. 15.2a as a correct car. In this case, we needed equivariance instead of invariance. Due to invariance, CNN becomes tolerant of minor changes in the viewpoint. Because of this issue, CNN could not realize that the image on the right is similar to but rotated car image.

On the other hand, equivariance makes a CNN realize the alternation or slight change and adjust itself accordingly to avoid losing spatial positioning inside an image. A car will still be smaller, but the CNN will decrease its size to identify the image. This problem leads to the recent progression of CapsNet.

15.4 Capsule Network

In terms of neuron networks, network structure, dissemination, and interlayer distribution methods, there are noteworthy differences between CapsNet and CNN architectures [36]. The operations performed in a CNN are given in Sect. 15.3 (i.e. weighting of inputs, sum of weighted inputs, and nonlinearity); the operations slightly differ in CapsNet. The capsule is an essential structure in CapsNet. Unlike the typical system of neuron networks, the parameters used as inputs and outputs in CapsNet are the vectors. Therefore, the internal parameters correspond to the vectors, which differ from the activation functions.

Generally, CapsNet is designed to get inverse graphics. Every capsule in CapsNet receives an input image, and it identifies the object in that image with initial parameters [14]. Thus, the capsules are defined as functions that can predict the initial parameters of the chosen object in a particular area. The length of the activation vector represents the predicted probability of an object. Similarly, the activation vector's orientation shows the object's parameter instantiation. CapsNet is strong enough to integrate flexibility. CapsNet can translate image measurement parameters like rotation, slope, extension, thickness, etc. In essence, the operations performed within the CapsNet performed as follows:

1. Matrix Multiplication: Input vectors get multiplied with weight matrices to encode the essential spatial relationship of low-level and high-level features within the image.
2. Scalar Weighting of the input: The input vector weights are calculated in this step. Later, these weights are used to select the higher-level capsule where current lower-level weights need to be routed. Here dynamic routing algorithm is used as given in [14].
3. Dynamic Routing Algorithm: It permits the routing of the information among different components in the CapsNet.
4. Nonlinearity using the "squash" function: The "squash" function takes all the information vector and converts it into a vector of length less than or equal to 1 while maintaining the direction of the vector. Squash function is nothing but the reshaping of the vector into 1D.

The structure of CapsNet for handwritten Devanagari character recognition is simple. The model for a CapsNet is shown in Fig. 15.3, which consists of different layers. CapsNet initially has two convolution layers followed by a primary capsule. After the primary capsule (PriCaps), there is a second layer comprised of DigitCaps. The last layer of CapsNet is fully connected layers. The CapsNet model shown in Fig. 15.3 is used for identifying handwritten character images. For this research, Devanagari character images with dimensions $28 \times 28 \times 1$ are given as input to the

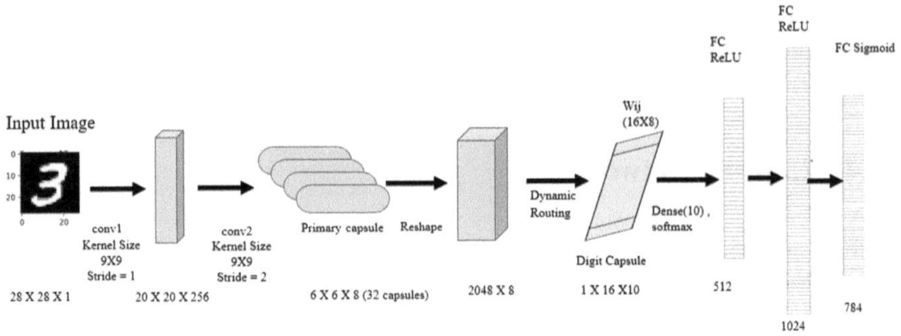

Fig. 15.3 CapsNet Model

CapsNet. The initial convolution layers have the same kernels but differ in stride. The ReLU activation function activates convolution layers that produce different feature maps as layer outputs. These feature maps are then reshaped to form the primary capsule layer. The PriCaps layer constructs the corresponding vectors and works as the capsule input layer. This layer also reshapes the input, then fed to an input vector for the subsequent layers. The DigitCaps layer shown in Fig. 15.3 is the output layer for the capsule. The loss function for the classification, which encodes the image into the character, is present after the DigitCaps layer. The fully connected (FC) layers act as a decoder responsible for reconstructing the image. The FC layers also prevent overfitting. DR algorithm [14] works between the PriCaps and the DigitCaps. The DR algorithm updates the necessary parameters along with the traditional back-propagation algorithm.

15.5 Dynamic Routing Between Capsules

The CapsNet consists of a Convolutional network layer called PriCaps layer and DigitCap layer shown in Fig. 15.3. The PriCap is initial capsule layer which is trailed by multiple capsule layers until the final capsule DigitCap. Conv1 extracts features from the input image and the output of conv1 is then fed into the PriCap layer as shown in Fig. 15.3. Each ith capsule ($1 \leq i \leq N$) in l^{th} layer has output vector u_i $\in\mathbb{R}$ that encodes the spatial features as instantiation parameters. The output vector u_i of the ith capsule is fed to all capsules in the next $(l + 1)^{\text{th}}$ layer. The jth capsule at $(l + 1)^{\text{th}}$ layer will receive vector u_i and find its multiplication with corresponding weighted matrix W_{ij} as shown in Eq. 15.1. The subsequent vector $\hat{\mathrm{u}}_{\mathrm{j|i}}$ is an ith capsule at l^{th} level transformation of the entity signified by j^{th} capsule at level $l + 1$. The prediction vector of priCap, $\hat{\mathrm{u}}_{j|i}$ specifies how much the PriCap i contributes to the class capsule j.

$$\hat{\mathrm{u}}_{j|i} = W_{ij} u_i \tag{15.1}$$

$$s_j = \sum_{i=1}^{N} c_{ij} \hat{\mathbf{u}}_{j|i} \tag{15.2}$$

A product of the prediction vector $\hat{\mathbf{u}}_{j\,|\,i}$ and a coupling coefficient c_{ij} which represents the agreement between these capsules is carried out to find a single PriCap i's prediction to the class capsule s_j. If the agreement is high, then the two capsules are considered to be relevant to each other. Consequently, the coupling coefficient will be increased, otherwise, it will be reduced. A weighted sum (s_j) is calculated using Eq. 15.2. Later these individual primary capsule predictions for the jth class capsule is calculated to obtain the candidates for the squashing function (v_j) given in Eq. 15.3. This squashing function makes sure that the length of capsule output should lie between 0 and 1.

$$v_j = \frac{\|s_j\|^2}{1 + \|s_j\|^2 \ \|s_j\|} \tag{15.3}$$

The coupling coefficient c_i is calculated using Eq. 15.4 to ensure that the prediction i for l layer should be related to that of prediction j in $l + 1$ *layer*.

$$c_{ij} = \frac{\exp(b_{ij})}{\sum_k \exp(b_{ik})} \tag{15.4}$$

The coupling coefficient is updated after every iteration using dot product of v_j and $\hat{\mathbf{u}}_{j\,|\,i}$.

15.6 Introduction of Devanagari Character Set

Devanagari is one of the viral scripts in India because it is the foundation for languages such as Marathi, Hindi, Nepali, and Sanskrit [37]. Unlike other foreign languages such as English, Japanese, and Russian text, there is a horizontal line on the word called "Shirorekha." This horizontal line also denotes the end of the word. There are 58 characters in the Devanagari script. The writing ways of the Devanagari are a combination of characters, numerals, and some modifiers. It follows the phonetic principle, where many characters follow a combination of vowels and consonants. Devanagari script is also known as a phonetic script because writing also considers the sounds of the characters. Along with consonants, it has 14 modifiers that are used with consonants. For example, a modifier '◌ा' used with consonant 'ग' is written like 'ग+आ=गा.'

15.7 Challenges in Recognizing Devanagari Character

The notable feature which makes Devanagari more multifaceted is the resemblance of characters. For example, रा and शा; प and ष, घ and ध etc. Different individuals' writing styles and strokes make it challenging to differentiate characters, mainly when they are handwritten. This research aims to take advantage of CapsNet for detecting the spatiality of every character feature.

15.8 Experiment

The data used in this research is available on Kaggle [38]. The dataset contains 5800 samples with 58 character classes and ten numerals for the Devanagari script. Each character and digit have 100 sample images. The samples are pre-processed to create an image of size 28 × 28 × 1. Figures 15.4 and 15.5. show the sample numeral and character images present in the dataset.

A feature map at each layer of the CapsNet (encoder) is shown in Fig. 15.3. The model accepts handwritten character images with a size of 28 × 28 × 1 and encodes them into a 16-dimension vector of instantiation parameters called Digit Capsule in various classes. The Convolution layers are classic convolution layers with the activation function ReLU to extract local features from the image. There are two convolutional layers. First Conv1 with 9 × 9 convolutional kernel, a stride of 1, has ReLU activation function. And another Conv2 with a 9 × 9 convolutional

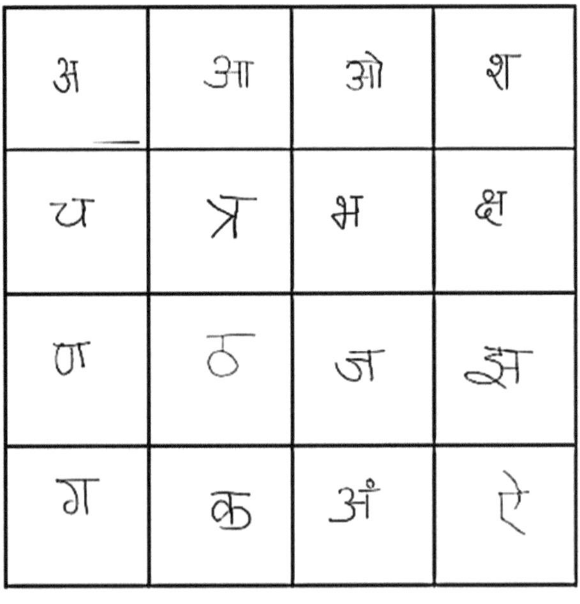

Fig. 15.4 Devanagari character dataset samples

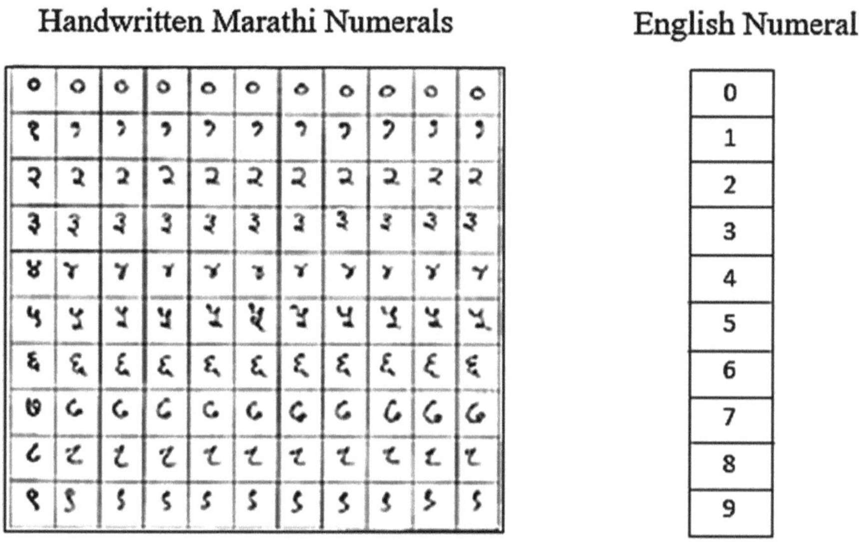

Fig. 15.5 Handwritten Devanagari digits and equivalent English digits

kernel, a stride of 2, and a ReLU activation function. Conv1 results in 20 × 20 × 256 features, and Conv2 results in 6 × 6 × 8 × 32 features of scalars. The output is then reshaped to obtain 2048 × 8 features of vectors. DigitCaps then construct a 16D vector that has all the features required for rebuilding the image. This 16D vector is then decoded into an image using last three fully connected layers as shown in Fig. 15.3. The evaluation metric used for this work is Mean Absolute Percentile Error (MAPE) given by Eq. 15.5.

$$\text{MAPE} = \frac{\sum (\text{actual} - \text{predicted}) / \mid \text{actual} \mid}{n} \times 100 \tag{15.5}$$

15.9 Results and Discussion

The proposed model experimented with the Devanagari characters dataset, containing 5800 Devanagari characters. The training set contains 80% images for every character; the remaining 20% is used for the testing set. The total number of classes for the character is 48, and for the number 10. The model was executed for 30 epochs with a batch size of 20. Table 15.1 shows the CapsNet model's performance over other handwritten character prediction models. The outcomes show slight improvement over the CNN model. Figure 15.6a, b show the accuracy and loss graph over the iterations. The testing accuracy is improving with the decrease in the loss for 30 iterations. For image processing, this work used the Keras ImageDataGenerator class. The CapsNet model achieved 94.56% training

Table 15.1 Testing accuracy

Model	Accuracy (%)
Reddy et al. [39]	97.33
Deore et al. [38]	97.45
Saha et al. [40]	93.00
CapsNet Model	**97.88**

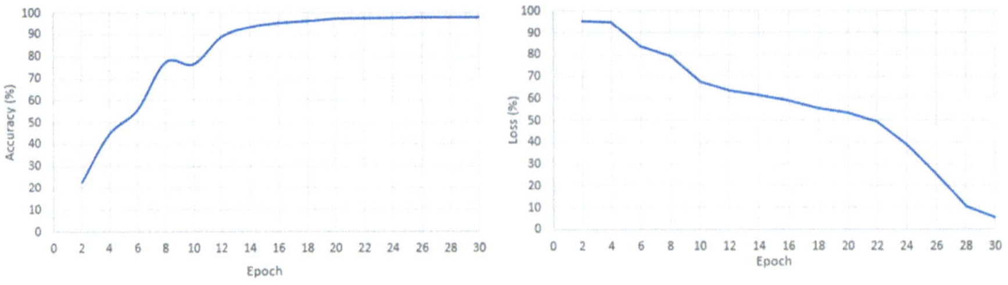

(a) (b)

Fig. 15.6 (**a**) Testing accuracy. (**b**) Testing loss

and 97.88% testing accuracy. The experiment is also conducted for the resemblance of characters. The CapsNet successfully recognized the spatial relationship of features and could differentiate the characters with similar characteristics with some increased processing cost due to complexity of the CapsNet architecture.

15.10 Conclusion and Future Scope

Individual's writing style makes character detection in Indian languages difficult and complex. Many characters in Devanagari differ slightly from other. In the case of CNN, these features get lost while training the model due to the pooling layer. CapsNet can learn these slight differences and understand every spatial phenomenon. This work uses the CapsNet model to classify the isolated handwritten Devanagari characters. This work used a Kaggle dataset for the Devanagari font. CapsNet has given an accuracy of **97.88%.** In the future, exploiting CapsNet for more complex languages is planned.

References

1. S. Ahlawat and A. Choudhary (2020) Hybrid CNN-SVM Classifier for Handwritten Digit Recognition. *Procedia Computer Science*, 167, pp. 2554–2560. https://doi.org/10.1016/j.procs.2020.03.309.
2. V. T. Lanjewar and R. N. Khobragade (2021) Transfer learning using Pre-trained AlexNet for Marathi Handwritten Compound Character Image Classification. *International Conference on Intelligent Technologies, CONIT 2021*. https://doi.org/10.1109/CONIT51480.2021.9498418.

3. Y. Gurav, P. Bhagat, R. Jadhav, and S. Sinha (2020) Devanagari Handwritten Character Recognition using Convolutional Neural Networks. *2nd International Conference on Electrical, Communication and Computer Engineering, ICECCE 2020.* https://doi.org/10.1109/ICECCE49384.2020.9179193.
4. Prashant Misal, Prof. A.J. Patankar and Dr. Nitika Singhi (2021), Handwritten Marathi Character recognition using Deep learning - Inpressco.
5. P. M. Kamble and R. S. Hegadi (2015) Handwritten Marathi Character Recognition Using R-HOG Feature," *Procedia Computer Science*, vol. 45, no. C, pp. 266–274. https://doi.org/10.1016/J.PROCS.2015.03.137.
6. S. Impedovo, G. Pirlo, and F. M. Mangini (2012) Handwritten digit recognition by multi-objective optimization of zoning methods," *Proceedings - International Workshop on Frontiers in Handwriting Recognition, IWFHR*, pp. 675–679. https://doi.org/10.1109/ICFHR.2012.209.
7. A. Chakraborty, R. De, S. Malakar, F. Schwenker, and R. Sarkar (2020), "Handwritten digit string recognition using deep autoencoder based segmentation and ResNet based recognition approach," *Proceedings - International Conference on Pattern Recognition*, pp. 7737–7742, 2020. https://doi.org/10.1109/ICPR48806.2021.9412198.
8. M. Moradi, M. A. Poormina, and F. Razzazi, "FPGA implementation of feature extraction and MLP neural network classifier for Farsi handwritten digit recognition," *EMS 2009 - UKSim 3rd European Modelling Symposium on Computer Modelling and Simulation*, pp. 231–234, 2009. https://doi.org/10.1109/EMS.2009.13.
9. K. Jarrett, K. Kavukcuoglu, M. Ranzato, and Y. LeCun, "What is the best multi-stage architecture for object recognition?," *Proceedings of the IEEE International Conference on Computer Vision*, pp. 2146–2153, 2009. https://doi.org/10.1109/ICCV.2009.5459469.
10. D. C. C. Cireşan, U. Meier, J. Masci, L. M. Gambardella, and Jürgen Schmidhuber, "Flexible, High Performance Convolutional Neural Networks for Image Classification".
11. D. Ciregan, U. Meier, and J. Schmidhuber, "Multi-column Deep Neural Networks for Image Classification," *Proceedings of the IEEE Computer Society Conference on Computer Vision and Pattern Recognition*, pp. 3642–3649, Feb. 2012. https://doi.org/10.1109/CVPR.2012.6248110.
12. G. E. Hinton, S. Osindero, and Y.-W. Teh, "A fast learning algorithm for deep belief nets *".
13. X. X. Niu and C. Y. Suen, "A novel hybrid CNN–SVM classifier for recognizing handwritten digits," *Pattern Recognition*, vol. 45, no. 4, pp. 1318–1325, Apr. 2012. https://doi.org/10.1016/J.PATCOG.2011.09.021.
14. S. Sabour, N. Frosst, and G. E. Hinton, "Dynamic Routing Between Capsules," Oct. 2017, [Online]. Available: http://arxiv.org/abs/1710.09829
15. K. M. Sayre, "Machine recognition of handwritten words: A project report," *Pattern Recognition*, vol. 5, no. 3, pp. 213–228, Sep. 1973, https://doi.org/10.1016/0031-3203(73)90044-7.
16. A. Graves, "Offline Handwriting Recognition with Multidimensional Recurrent Neural Networks."
17. Online and off-line handwriting recognition: a comprehensive survey | IEEE Journals & Magazine | IEEE Xplore. https://ieeexplore.ieee.org/document/824821 (accessed Feb. 17, 2022).
18. C. N. Manisha, E. S. Reddy, and Y. K. Sundara Krishna, "Role of Offline Handwritten Character Recognition System in Various Applications," *International Journal of Computer Applications*, vol. 135, no. 2, pp. 975–8887, 2016.
19. J. A. Sánchez, V. Bosch, V. Romero, K. Depuydt, and J. de Does, "Handwritten text recognition for historical documents in the transcriptorium project," *ACM International Conference Proceeding Series*, pp. 111–117, 2014. https://doi.org/10.1145/2595188.2595193.
20. C. Cortes, V. Vapnik, and L. Saitta, "Support-vector networks," *Machine Learning 1995 20:3*, vol. 20, no. 3, pp. 273–297, Sep. 1995. https://doi.org/10.1007/BF00994018.
21. P. J. Phillips, "Support Vector Machines Applied to Face Recognition".

22. M. Pontil and A. Verri, "Support vector machines for 3D object recognition," *IEEE Transactions on Pattern Analysis and Machine Intelligence*, vol. 20, no. 6, pp. 637–646, 1998. https://doi.org/10.1109/34.683777.
23. A. Boukharouba and A. Bennia, "Novel feature extraction technique for the recognition of handwritten digits," *Applied Computing and Informatics*, vol. 13, no. 1, pp. 19–26, Jan. 2017, https://doi.org/10.1016/J.ACI.2015.05.001.
24. K. Jarrett, K. Kavukcuoglu, M. Ranzato, and Y. LeCun, "What is the best multi-stage architecture for object recognition?," *Proceedings of the IEEE International Conference on Computer Vision*, pp. 2146–2153, 2009, https://doi.org/10.1109/ICCV.2009.5459469.
25. D. C. Cireşan, U. Meier, J. Masci, L. M. Gambardella, and J. Schmidhuber, "Flexible, high performance convolutional neural networks for image classification," *IJCAI International Joint Conference on Artificial Intelligence*, pp. 1237–1242, 2011. https://doi.org/10.5591/978-1-57735-516-8/IJCAI11-210.
26. D. Ciregan, U. Meier, and J. Schmidhuber, "Multi-column deep neural networks for image classification," *Proceedings of the IEEE Computer Society Conference on Computer Vision and Pattern Recognition*, pp. 3642–3649, 2012. https://doi.org/10.1109/CVPR.2012.6248110.
27. G. E. Hinton, S. Osindero, and Y. W. Teh, "A fast learning algorithm for deep belief nets," *Neural Comput*, vol. 18, no. 7, pp. 1527–1554, 2006. https://doi.org/10.1162/NECO.2006.18.7.1527.
28. X. Qu, W. Wang, K. Lu, and J. Zhou, "Data augmentation and directional feature maps extraction for in-air handwritten Chinese character recognition based on convolutional neural network," *Pattern Recognition Letters*, vol. 111, pp. 9–15, Aug. 2018. https://doi.org/10.1016/J.PATREC.2018.04.001.
29. F. Lauer, C. Y. Suen, and G. Bloch, "A trainable feature extractor for handwritten digit recognition," *Pattern Recognition*, no. 6, pp. 1816–1824, 2007. https://doi.org/10.1016/j.patcog.2006.10.011ï.
30. M. D. Zeiler and R. Fergus, "LNCS 8689 - Visualizing and Understanding Convolutional Networks," 2014.
31. A. Krizhevsky, I. Sutskever, and G. E. Hinton, "ImageNet classification with deep convolutional neural networks," *Commun ACM*, vol. 60, no. 6, pp. 84–90, May 2017. https://doi.org/10.1145/3065386.
32. W. Zhao, J. Ye, M. Yang, Z. Lei, S. Zhang, and Z. Zhao, "Investigating Capsule Networks with Dynamic Routing for Text Classification," *Proceedings of the 2018 Conference on Empirical Methods in Natural Language Processing, EMNLP 2018*, pp. 3110–3119, Mar. 2018. https://doi.org/10.18653/v1/d18-1350.
33. N. Zhang, S. Deng, Z. Sun, X. Chen, W. Zhang, and H. Chen, "Attention-Based Capsule Networks with Dynamic Routing for Relation Extraction," *Proceedings of the 2018 Conference on Empirical Methods in Natural Language Processing, EMNLP 2018*, pp. 986–992, Dec. 2018. https://doi.org/10.18653/v1/d18-1120.
34. B. Zhang, X. Xu, M. Yang, X. Chen, and Y. Ye, "Cross-domain sentiment classification by capsule network with semantic rules," *IEEE Access*, vol. 6, pp. 58284–58294, 2018. https://doi.org/10.1109/ACCESS.2018.2874623.
35. M. Z. Hasan, K. M. Z. Hasan, S. Hossain, A. al Mamun, and M. Assaduzzaman (2019) Handwritten Changma numerals recognition using capsule networks, *2019 5th International Conference on Advances in Electrical Engineering, ICAEE 2019*, pp. 386–391. https://doi.org/10.1109/ICAEE48663.2019.8975468.
36. M. Khodadadzadeh, X. Ding, P. Chaurasia, D. Coyle, and S. Member, (2021) A Hybrid Capsule Network for Hyperspectral Image Classification; A Hybrid Capsule Network for Hyperspectral Image Classification, *IEEE Journal of Selected Topics in Applied Earth Observations and Remote Sensing*, vol. 14. https://doi.org/10.1109/JSTARS.2021.3126427.
37. U. Pal and B. B. Chaudhuri (2004) Indian script character recognition: A survey, *Pattern Recognition*, vol. 37, no. 9, pp. 1887–1899, https://doi.org/10.1016/J.PATCOG.2004.02.003.

38. S. P. Deore and A. Pravin (2020) Devanagari Handwritten Character Recognition using fine-tuned Deep Convolutional Neural Network on trivial dataset, *Sadhana - Academy Proceedings in Engineering Sciences*, vol. 45, no. 1. https://doi.org/10.1007/S12046-020-01484-1.
39. R. V. K. Reddy and U. R. Babu (2019) Handwritten Hindi Character Recognition using Deep Learning Techniques, *International Journal of Computer Sciences and Engineering*, vol. 7, no. 2, pp. 1–7. https://doi.org/10.26438/IJCSE/V7I2.17.
40. P. Saha and A. Jaiswal (2020) Handwriting Recognition Using Active Contour, *Advances in Intelligent Systems and Computing*, vol. 1056, pp. 505–514. https://doi.org/10.1007/978-981-15-0199-9_43.

Part IV
Security for Industry 4.0

Chapter 16
Predictive Model of Personalized Recommender System of Users Purchase

Darshana Desai

16.1 Introduction

Personalized recommendation is used as an effective instrument to moderate the cognitive load of users on cybernetic markets in e-commerce websites by suggesting information about product or services. Users' personal information, purchase behaviors, preferences, geo-location search, and website trail are monitored and latest information are stored by e-commerce sites to serve users better. Real-time personalized recommendations are presented by understanding implicit tastes, surfing of websites, behavioral intents like time spent on the page, number of the visit to pages, frequency of item viewed, and search and purchase history with now and current data analytics. By continuous monitoring of the user's activities to create personalized recommendations and suggest highly relevant content, users' detailed information is fetched on a real-time basis upon constant monitoring of users' activities, which have raised privacy concerns and trust issues in e-commerce websites. Identification of users' personality traits, information, activities fuel data breaching which may lead to high privacy concerns, reduction in trust amongst users, and affect their purchase decisions. The personalization process needs more data with monitoring users' activities with the right context to generate highly relevant personalized recommendations. Users experience more privacy concerns with tracking of their activities and interaction. Different privacy policy is introduced to regulate the business process, data acquisition, and its usage for commercial or other purposes including Do-Not-Track initiative and limiting data collection by providing users control for data collection, which have reduced privacy concerns in users [14, 19, 27].

D. Desai (✉)
MCA Department, Indira College of Engineering and Management, Pune, Maharashtra, India

A. Neustein et al. (eds.), *AI, IoT, Big Data and Cloud Computing for Industry 4.0*, Signals and Communication Technology, https://doi.org/10.1007/978-3-031-29713-7_16

The research will provide a roadmap for business to use personalization features while considering privacy concerns and trust issues with personalized contents. This chapter is organized as follows: Section 16.2 defines prior work in the context of personalized recommendation, trust, privacy concerns, satisfaction, and its effect on purchase behavior. Section 16.3 represents the proposed hypotheses and Sect. 16.4 explains the research methodology adopted to address the research question and analysis using Exploratory Factor Analysis (EFA) and Confirmatory Factor Analysis (CFA). The section also shows results of model validity and Structural Equation Modeling (SEM). Subsequently, Sect. 16.5 presents the result analysis, with hypothesis testing results and discussion. Finally, Sect. 16.6 represents findings and future scope of research.

16.2 Related Work

Personalized recommendation produced in e-commerce websites on a real-time basis is a continuous process addressing users' requirements with highly relevant information [6, 7]. The user's needs are identified implicitly by analyzing search history, website clicks, purchase history, demographics, likings, and explicitly producing customization options and choices of filters [3, 9]. Research suggests that recommendation services reduce users' cognitive load producing ease of use, perceived usefulness, and enjoy the shopping experience, subsequently leading to positive behavior of purchase and revisit to the e-commerce websites [10, 17, 18]. Users experience higher satisfaction, more trust with greater interaction with an e-commerce website, and show positive behavioral intention and purchase [4, 7]. On the contrary, users experience higher privacy concerns and lower satisfaction with the recommendation associated with higher financial risks [24]. Research also suggests that recommendations of sensitive products generate higher privacy concerns in the user and are likely to experience lower trust in the e-commerce websites affecting negatively in behavioral intentions like revisit and purchase.

Personalized recommendation systems adopt different approaches and techniques to learn about users' preferences by analyzing users' behavioral data like collaborative filtering, user profiling, and content-based and hybrid model-based recommendations [9, 29]. Research has found significant performance improvement in business and customer experience with the personalized recommendation in e-commerce websites like Amazon.com. Collaborative filtering technique is used to generate product recommendation by analyzing like-minded people's purchase behavior and suggest the users related items which are bought frequently together [16, 21].

Frequent interaction with the e-commerce websites and increased use of recommendation services produce increased satisfaction, higher trust, and greater purchase intentions. However, trust in the recommendation is strongly associated with the users' desire to avail personalization services [4] and higher relevance of personalized recommendation to users' implicit need. And the accuracy of recommendation induces more satisfaction in users, which instigates positive behavioral intentions like revisiting the website and purchase ([7, 10, 24]; Liang et al. 2007).

16.3 Gap Analysis

Earlier research clearly states that highly personalized recommendation and user behavioral tracking have raised more concerns towards privacy and security of the users' information [6]. Recently users' privacy concerns have increased to prominently develop trust before purchasing on e-commerce websites. Personalization strategies have been used extensively over a decade to lower cognitive load on the user and influence for high returns in business, but little is focused on customer data privacy and their concerns on information disclosure. Though users' satisfaction and cognitive behavior with personalized information is researched towards its effect on purchase intentions [7], studies have shown a deficit in addressing the issue of impact of personalized recommendations on users' purchase intention through intrinsic feeling of trust and privacy concerns. In this view to address the research gap, our research explores personalized recommendations used in e-commerce website Amazon.com on users' purchase behavior through privacy concerns and trust in the website.

16.4 Research Hypotheses

Research in personalization has shown a significant relation between information personalization and users' intention to revisit the website [8]. Our research framework proposes interrelation of personalized recommendation, privacy concerns, trust, satisfaction, and behavioral intention of user-like purchase in e-commerce websites.

16.4.1 Personalized Recommendation, Privacy Concerns

Users' personal information is used in order to produce personalized content [15, 23]. Personalized recommendations are offered through website by apprehending users' implicit requirement by tracking users' activities on a website, by observing website usage and website trailing [9]. Users' explicit behavior like product purchase history and preferences help to improve recommendations quality; however, users may feel intrusion of privacy with such behavior. Users may develop higher privacy concerns and consciousness about tracing their activities by the website in the process of personalization. Website users may have concerns about privacy of users with website personalization in case users are unaware about the motive after personalized information which eventually diminishes trust in ecommerce sites. Users' involvement develops privacy concerns whenever recommendation process utilizes their personal information without consent, or prior knowledge generating adverse emotion with the personalized ecommerce website. So, research postulates the following:

H1: User exhibits more privacy concerns with highly personalized recommendation.

16.4.2 Personalized Recommendation and Satisfaction

Users experience greater satisfaction when the produced information has high relevance, which fulfills their implicit needs leading to perceived ease of use and usefulness of the recommendation [7]. Explicit personalization or customizations offered to users produce a feeling of control and participation in personalized recommendation process. This generates higher satisfaction in the user, and subsequently the interaction with the personalized ecommerce website becomes more joyful [5, 10, 11]. So, this research work hypothesizes:

H2: Users' satisfaction is positively associated with personalized recommendations.

16.4.3 Privacy Concern with Trust

Personalized websites lessen users' cognitive exertions taken to information search, which leads to satisfaction with highly relevant personalized information [5, 7, 11] and eventually developing trust towards websites. Users' with higher privacy concerns need more control over personalization which induce trust in the ecommerce websites [2, 6]. Users' desire to maintain privacy of the information build higher trust when fulfilled with implicit requirements. Personalized recommendation generates cognitive benefits to users with compromise in information privacy. Personalization process requires users' demographic, personal and behavioral information including their likings and needs based on search history. This information disclosure produces privacy concerns with the associated privacy risk in regard to information or content personalization [24]. So, research hypothesizes the following:

H3: Users show higher trust when privacy concerns are addressed properly with explicit control of users' information sharing.

16.4.4 Privacy Concerns and Purchase Intention

Users' privacy concern is the intrinsic desire to control information disclosure and its use. Users show higher privacy concerns when their information privacy is breached while interacting with the website or during online transaction [2]. Users' concerns about privacy and trust are preserved when their anonymity is preserved and have choices to control through customization options to tailor users' environment. Prior research works have established that users' control over information sharing is the key factor in the development of concerns regarding privacy and website environment [22, 26]. Users identify lower risk about privacy of their information when provided with higher control over website and customization options along with disclosure to privacy policies and users are likely to purchase more when their privacy concerns are addressed properly with the policies and security of the information ([20, 28]; Desai 2020). So researcher postulates:

H4: Users' purchase intention is affected by privacy concerns.

16.4.5 Satisfaction, Trust, and Purchase Intention

Trust is inherent feeling of the user and preparedness to be susceptible to the actions with prospect and addressed expectations [12, 25]. Users having control over information distribution, display, accessibility of personal information, and security of the personal details are more likely to develop higher satisfaction and confidence in e-commerce web portals. Personalized information having high relevance like recommendations, location-based services, and targeted commercials produce intrinsic feeling of trust and satisfaction to motivate purchase from the website. So, we propose the following:

H5: Users' trust is positively affected by satisfaction.

H6: Users' purchase intention is positively with users' trust.

H7: Users' purchase intention is positively users' satisfaction.

16.5 Methodology of Research

16.5.1 Data Collection and Sampling

Our research provides predictive model of users' purchase intention on Amazon.in e-commerce website who are using this site for more than 2 years. The data was

Table 16.1 Sample demographics

Measure	Item	Frequency
Gender	Male	328
	Female	172
Age (years)	18–25	392
	26–35	68
	36–50	38
	>50	2
Education	Undergraduate	183
	Graduate degree	120
	Post Graduate	195
	Doctorate	2
Occupation	Student	363
	Service	106
	Self-employment	12
	Homemakers/retired	19
Experienced	Directly	428
Personalization	Indirectly	72

collected based on real purchase of the users on the Amazon.in site. The constructs that were recognized from the preceding studies are personalized recommendation, privacy concerns, satisfaction, trust, and purchase intention identified using a questionnaire having a five-point Likert scale (Table 16.1).

16.5.2 Data Analysis and Measurement Model

Pilot study of the research with the 50 responses gave confirmation about the construct reliability and validity of the questions in survey. The respondents were all over India who have purchased online on Amazon.in e-commerce websites and have experienced personalization in the form of the personalized recommendation of product, service, or information directly or indirectly. The data was collected through stratified sampling technique. A total of 441 usable responses were extracted out of 500 received responses through data preprocessing. Data cleaning was done after the removal of noisy, improper, unfinished, and inconsistent data. In order to retrieve final responses researcher considered data after removal of responses below standard deviation 0.30 which was further used for data analysis. Extracted factors with exploratory factor analysis were further confirmed with CFA Confirmatory Factor Analysis. SPSS Amos 21.0 was used for Structural Equation Modeling (SEM) technique to identify the best fit of the suggested model.

KMO and Bartlett's test result value of 0.804, which is above 0.7, shows the sample size adequacy of data collection. Exploratory Factor Analysis techniques discovered 5 factors, namely personalized recommendation, trust, privacy concerns,

KMO and Bartlett's Test

Kaiser-Meyer-Olkin Measure of Sampling Adequacy.		.804
Bartlett's Test of Sphericity	Approx. Chi-Square	3757.126
	df	120
	Sig.	.000

Fig. 16.1 KMO and Bartlett test result

Total Variance Explained

	Initial Eigenvalues			Extraction Sums of Squared Loadings			Rotation Sums of Squared Loadings[a]
Factor	Total	% of Variance	Cumulative %	Total	% of Variance	Cumulative %	Total
1	4.866	30.413	30.413	1.915	11.971	11.971	3.044
2	2.439	15.242	45.655	3.812	23.822	35.793	3.345
3	1.894	11.840	57.495	1.583	9.897	45.689	1.891
4	1.155	7.220	64.715	1.760	11.000	56.689	3.013
5	1.073	6.705	71.420	.740	4.623	61.312	2.808
6	.635	3.970	75.390				
7	.585	3.654	79.044				
8	.559	3.497	82.541				
9	.539	3.366	85.907				
10	.452	2.822	88.729				
11	.401	2.509	91.238				
12	.385	2.406	93.644				
13	.369	2.305	95.948				
14	.324	2.025	97.974				
15	.245	1.532	99.505				
16	.079	.495	100.000				

Extraction Method: Maximum Likelihood.

a. When factors are correlated, sums of squared loadings cannot be added to obtain a total variance.

Fig. 16.2 Total variance

satisfaction, and purchase intentions through the extraction method Maximum Likelihood. Figure 16.2 shows 5 factors with 61.312% of the cumulative load.

The Cronbach's Alpha coefficient standards of construct items were observed between the range of 0.70 and 0.90 displaying higher constructs internal consistency and consistency of questionnaire items or constructs. Construct items' factor-loading values were identified above 0.5 as shown in Fig. 16.3, showing satisfactory factor loading with higher construct validity and reliability.

Figure 16.4 shows the factor correlation matrix and the interrelation of the factors identified after EFA.

Pattern Matrix[a]

	Factor				
	1	2	3	4	5
ECIP2	.793				
ECIP3	.763				
ECIP4	.725				
ECIP5	.652				
ECIP1	.609				
ECPC2		.865			
ECPC1		.699			
ECPC3		.691			
ECPC4		.615			
ECSAT1			.992		
ECSAT2			.928		
ECPUR2				.866	
ECPUR1				.755	
ECPUR3				.516	
ECTrust2					.985
ECTrust1					.690

Extraction Method: Maximum Likelihood.
Rotation Method: Promax with Kaiser Normalization.
a. Rotation converged in 6 iterations.

Fig. 16.3 Pattern matrix

Factor Correlation Matrix

Factor	1	2	3	4	5
1	1.000	.229	.029	.309	.368
2	.229	1.000	.082	.569	.508
3	.029	.082	1.000	.098	.073
4	.309	.569	.098	1.000	.445
5	.368	.508	.073	.445	1.000

Extraction Method: Maximum Likelihood.
Rotation Method: Promax with Kaiser Normalization.

Fig. 16.4 Factor correlation matrix

16.5.3 Confirmatory Factor Analysis and Validity Test

Confirmatory Factor Analysis technique was used to check the proposed model fit with SPSS AMOS 21.0. Best fit of the model is shown when the fit indices of the mode are within the accepted range and suggested values. The values shown in the table below displays acceptable fit as the model fit indexes surpasses the suggested value.

Composite Reliability (CR) displays constructs internal consistency of the item which is above 0.7 representing good reliability and measurement items constancy of every construct item. Each CR scored above 0.8, which is higher than the value recommended by Fornell and Larcker [13], signifying good reliability and constancy for the measurement items of each construct.

Bagozzi and Yi [1] suggested standards for the convergent validity of proposed measurement model as mentioned: (1) construct factor loadings should be higher value than 0.5; (2) Composite Reliability (CR) should be above 0.7; and (3) Average Variance Extracted (AVE) score of each construct should exceed 0.5 [13]. Table 16.2 result shows that factor loading of construct values of Composite Reliability is above 0.7 and AVE score of the constructs between 0.5 and 0.96 shows good convergent validity and strong interitem correlations with the constructs.

Fornell and Larcker [13] suggested that the constructs' AVE score should exceed correlation coefficients of the construct for the discriminant validity of the model. The correlation coefficient value shown in Table 16.2 in the proposed research Model, the diagonal elements represent square roots of the constructs' Average Variance Extracted value. The AVE score for the constructs is higher than the square root of the correlation coefficient values for any two constructs. Research shows decent discriminant validity of the constructs identified in proposed model measurement as all the constructs are distinct from each other. Construct reliability, discriminant validity, and convergent validity of the research shows good fit values of the measurement model.

Table 16.2 CR, AVE, and discriminant validity

	CR	AVE	PI	PR	PC	SAT	Trust
PI	0.788	0.555	**0.745**				
PR	0.836	0.507	0.311	**0.712**			
PC	0.821	0.534	0.612	0.235	**0.731**		
SAT	0.984	0.969	0.096	0.027	0.064	**0.984**	
Trust	0.834	0.715	0.505	0.385	0.540	0.068	**0.846**

CR Composite Reliability, *AVE* Average Variance Extracted, *PR* Personalized Recommendation, *PC* Privacy Concerns, *SAT* Satisfaction, *PI* Purchase Intentions

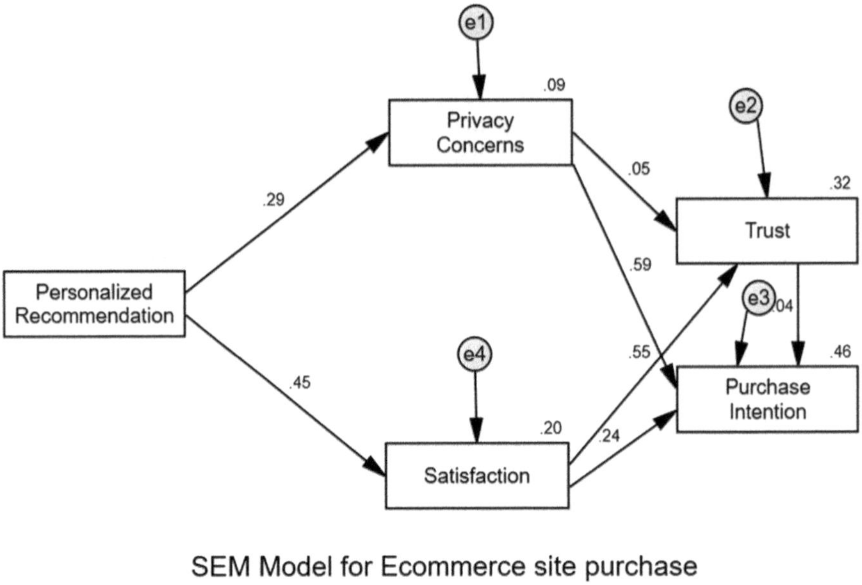

Fig. 16.5 Structural Equation Modeling result

16.5.4 Structural Equation Modeling (SEM)

The result of Structural Equation Modeling shows good model fit satisfying all the criteria of fit indices of the model having values v2/df = 1.772, GFI = 0.961 AGFI = 0.94, NFI = 0.95, CFI = 0.98, RMSEA = 0.039. Values shown for the fit indices of the model in structural equation modeling indicate a good fit for the model.

In Figure 16.5, the variance is explained (R^2) and the standardized path coefficient illustrates that all the proposed hypotheses are proven except users' privacy concerns and trust, which shows that users are not aware of privacy concerns. It also shows that users' purchase is not dependent of trust on the personalized websites. R^2 variance explained that the values of purchase intentions and Trust are 0.32 and 0.46, respectively, which is higher than 0.3, indicating that the proposed research model is good and accurate.

16.6 Result and Discussion

Hypothesis testing result shown in Table 16.3 shows that users demonstrate concerns about privacy with the increase in personalized recommendations and feel satisfied with the highly relevant recommendations which address their needs. Research results show that users' trust is not affected by privacy concerns but it is

Table 16.3 Hypotheses testing result

Hypotheses testing	Estimate	S.E	C.R.	*P*-value
H1: User exhibits more privacy concerns with higher personalized recommendation	0.296	0.043	6.835	***
H2: User experience satisfaction is positively associated with personalized recommendations	0.493	0.044	11.246	***
H3: Users show higher trust when privacy concerns are addressed properly with explicit control of users' information sharing.	0.076	0.054	1.406	0.160
H4: Users' purchase intention is affected by privacy concerns.	0.740	0.050	14.826	***
H5: Users' trust is positively affected by satisfaction	0.246	0.040	6.129	***
H6: Users' purchase intention is positively with users' trust	0.646	0.036	17.809	***
H7: Users' purchase intention is positively users' satisfaction	0.027	0.030	0.890	0.373

*** indicates p value <0.001

highly related with the satisfaction with the personalized recommendations. Users' purchase behavior with ecommerce websites depends on privacy concerns and trust with the site. Users are likely to purchase if users' privacy concerns are addressed properly while offering personalized recommendations. The result also shows the willingness of the user to purchase, which is not affected by Satisfaction.

Our study collected data from Amazon users experience on personalized recommendations generated with collaborative filtering mechanism. As mentioned in hypothesis 1, user exhibits more privacy concerns with higher personalized recommendation as shown in the result. Users also experience satisfaction positively associated with personalized recommendations, which is supported with the result of p-value in hypothesis 2. When users' concerns about privacy are addressed through customization and users' control over website personalization produce trust and satisfaction with the ecommerce website which is in line with the proved hypothesis 3 & 5. Users' purchase intentions with the website are positively associated with the trust and satisfaction which is in support of proposed hypothesis; purchase intentions are affected by users' privacy concerns.

16.7 Conclusions and Future Scope of Research

Users' privacy concerns and satisfaction in ecommerce websites' personalized recommendation play a significant role in developing trust and positive behavioral intention to purchase with the e-commerce website. Personalized recommendations on ecommerce websites lead to higher satisfaction in case users' implicit needs are satisfied with the high relevance with expectation. On the other hand, it may lead to high privacy concerns if recommendations are not addressed properly. Users' purchase intention is affected by privacy concerns and users' satisfaction with the personalized recommendations leads to trust development towards the site, resulting in positive purchase intentions. Result shows that personalized recommendation is positively associated with users' satisfaction and privacy concerns. Users experience lesser trust towards websites with high privacy concerns, but the result shows that users' purchase intentions are not affected by satisfaction but are positively affected by trust and privacy concerns.

The proposed research shows a qualitative study of users' purchase intention on an e-commerce website (viz., Amazon.in). This research adopted survey based online data collection through questionnaires. Researcher further aim to explore users' purchase intent in a lab with controlled environment. This Research can be further enhanced with the personalized recommendation attributes with focus on deep insight of optimization of recommendations with high relevance using realtime analytics of the users' interactions data. Real-time predictive analytics model can be further proposed to address users' dynamic needs in future in considering both types of data using big data analytics techniques and real-time personalization.

References

1. Bagozzi, R.P., Yi, Y. (1988). On the evaluation of structural equation models. JAMS 16, 74–94 https://doi.org/10.1007/BF02723327
2. Barth, S., & de Jong, M. D. T. (2017). The privacy paradox – Investigating discrepancies between expressed privacy concerns and actual online behavior – A systematic literature review. In Telematics and Informatics (Vol. 34, Issue 7, pp. 1038–1058). Elsevier Ltd. https://doi.org/10.1016/j.tele.2017.04.013
3. Chellappa, R.K., Sin, R.G. (2005). Personalization versus Privacy: An Empirical Examination of the Online Consumer's Dilemma. Information Technology Manage 6, 181–202 https://doi.org/10.1007/s10799-005-5879-y
4. Dabholkar, P. a., & Sheng, X. (2012). Consumer participation in using online recommendation agents: effects on satisfaction, trust, and purchase intentions. The Service Industries Journal, 32(9), 1433–1449. https://doi.org/10.1080/02642069.2011.624596
5. Desai, D. (2022a). Hyper-Personalization: An AI-Enabled Personalization for Customer-Centric Marketing. In S. Singh (Eds.), Adoption and Implementation of AI in Customer Relationship Management (pp. 40-53). IGI Global. https://doi.org/10.4018/978-1-7998-7959-6.ch003
6. Desai D. (2021). Role of Privacy Concern and Control to Build Trust in Personalized Social Networking Sites. In: Sheth A., Sinhal A., Shrivastava A., Pandey A.K. (eds) Intelligent Systems. Algorithms for Intelligent Systems. Springer, Singapore. https://doi.org/10.1007/978-981-16-2248-9_10 (Scopus Indexed)
7. Desai D. (2019). "Personalization Aspects Affecting Users' Intention to Revisit Social Networking Site" Published in International Journal of Trend in Scientific Research and Development (ijtsrd), ISSN: 2456-6470, Volume-4 | Issue-1, December, pp. 612–621 URL: www.ijtsrd.com/papers/ijtsrd29631.pdf
8. Desai D. (2018). An Empirical Study of Website Personalization Effect on Users Intention to Revisit E-commerce Website through Cognitive and Hedonic Experience. In: Balas V., Sharma

N., Chakrabarti A. (eds) Data Management, Analytics, and Innovation. Advances in Intelligent Systems and Computing, vol 839. Springer. https://doi.org/10.1007/978-981-13-1274-8_1
9. Desai D. Kumar S. (2015). "Web Personalization: A perspective of design and implementation strategies in Websites" Khoj: Journal of Management Research & Practices, 109–119 ISSN No: 0976-8262.
10. Desai D.(2016). "A study of personalization effect on users' satisfaction with ecommerce Websites" Sankalpa- Journal of Management & Research, Volume 6, Issue 2, 51–62 ISSN No. 2231-1904.
11. Desai, D. (2022b). Hyper-Personalization: An AI-Enabled Personalization for Customer-Centric Marketing. In Adoption and Implementation of AI in Customer Relationship Management (pp. 40–53). IGI Global.
12. Dwyer, C., Hiltz, S. R. and Passerini, K. (2007). 'Trust and privacy concerns within social networking sites: A comparison of Facebook and MySpace', Americas Conference on Information Systems, Proceedings of the Thirteenth Americas Conference on Information Systems, Keystone, 9–12 August, Colorado, USA, p. 339. https://aisel.aisnet.org/amcis2007/339
13. Fornell, C., &Larcker, D. (1981). Evaluating Structural Equation Models with Unobservable Variables and Measurement Error. Journal of Marketing Research, 18(1), 39–50. doi:https://doi.org/10.2307/3151312
14. Ghahtarani, A., Sheikhmohammady, M., &Rostami, M. (2020). The impact of social capital and social interaction on customers' purchase intention, considering knowledge sharing in social commerce context. Journal of Innovation and Knowledge, 5(3), 191–199. https://doi.org/10.1016/j.jik.2019.08.004
15. Gupta, A., &Dhami, A. (2015). Measuring the impact of security, trust, and privacy in information sharing: A study on social networking sites. Journal of Direct, Data and Digital Marketing Practice, 17(1), 43–53. https://doi.org/10.1057/dddmp.2015.32
16. Konstan, J. A., Riedl, J., Konstan, J. A., &Riedl, J. (2012). Recommender systems: from algorithms to user experience. 22, 101–123. https://doi.org/10.1007/s11257-011-9112-x
17. Liang, T.-P., Chen, H.-Y., Du, T., Turban, E., & Li, Y. (2012). Effect of Personalization on the Perceived Usefulness of Online Customer Services: a Dual-Core Theory. Journal of Electronic Commerce Research, 13(4), 275–288. http://www.ecrc.nsysu.edu.tw/liang/paper/2/79 Effect of Personalization on the Perceived (JECR, 2012).pdf
18. Mahrous, A. A. (2011). Antecedents of privacy concerns and their online actual purchase consequences: a cross-country comparison. In Int. J. Electronic Marketing and Retailing Vol. 4, Issue 4
19. Mayer, J.R., Narayanan, A.(2011) Do not track iab/w3c/ietf position paper. Tech. rep., W3C
20. Mohamed, N. and Ahmad, I. H. (2012). Information privacy concerns, antecedents and privacy measure use in social networking sites: Evidence from Malaysia, Computers in Human Behavior, Vol. 28, No. 6, pp. 2366–2375 DOI: https://doi.org/10.1016/j.chb.2012.07.008
21. Pu P, Chen L, and Hu R (2011). A User-Centric Evaluation Framework for Recommender Systems. Proceedings of the Fifth ACM Conference on Recommender Systems (ACM), 157–164.
22. Sheehan, K.B., Hoy, M.G. (1999). "Flaming, Complaining, Abstaining: How Online Users Respond to Privacy Concerns", Journal of Advertising, (28), 3: 37–51.
23. Shin, D.(2010). The effects of trust, security, and privacy in social networking: A security-based approach to understand the pattern of adoption, Interacting with Computers, 22(5), 428–438 https://doi.org/10.1016/j.intcom.2010.05.001
24. Stevenson, Darren, and Pasek(2015). Privacy Concern, Trust, and Desire for Content Personalization. The 43rd Research Conference on Communication, Information and Internet Policy. https://doi.org/10.2139/ssrn.2587541
25. Sutanto, J., Palme, E., Tan, C., &Phang, C. (2013). Addressing the Personalization-Privacy Paradox: An Empirical Assessment from a Field Experiment on Smartphone Users. MIS Quarterly, 37(4), 1141–1164. Retrieved March 5, 2021, from http://www.jstor.org/stable/43825785

26. Taylor, D.G., Davis, D., &Jillapalli, R. (2009). Privacy concern and online personalization: The moderating effects of information control and compensation. Electronic Commerce Research, 9, 203–223. DOI:https://doi.org/10.1007/s10660-009-9036-2
27. Wang, Y., Kobsa, A.(2007). Respecting users' individual privacy constraints in web personalization. In: Conati, C., McCoy, K., Paliouras, G. (eds.) UM07, 11th international conference on user modeling, Berlin– Heidelberg–New York, pp. 157–166. Springer-Verlag, Corfu
28. Yerizal, Y., &Abror, A. (2019). The Influence of E-Wom and Image Destination on Revisit Decision Moderated by Trust: A Literature Review. https://doi.org/10.2991/piceeba2-18.2019.58
29. Zhang S, Yao L, Sun A, and Tay Y. (2019). Deep Learning Based Recommender System: A Survey and New Perspectives. ACM Computing Surveys (CSUR) 52(1): 1–38.

Chapter 17
Rethinking Blockchain and Machine Learning for Resource-Constrained WSN

Nilesh P. Sable and Vijay U. Rathod

17.1 Introduction

A wireless sensor network (WSN) is a collection of sensor nodes that can communicate and interact to discover a physical environment using a wireless connection. Gross mechanical operations are translated into subtle sensory responses by this WSN in order to turn these physical objects into equivalent electrical values. The information gathered by the various sensor nodes is compiled and transmitted to the base station (BS). Memory unit, processor, power unit, sensor group, and transceiver are the five modules that make up the sensor node. Among these modules, the processor exercises command control. WSN's primary challenge is providing high throughput with incomplete network resources such as processing power, capacity of the transmission channel (i.e. bandwidth), and storage interplanetary. The general scenario at WSNs is shown in Fig. 17.1.

WSNs can be used for a variety of applications. A maximum number of sensor nodes are installed in a hostile atmosphere. WSN nodes are energy constricted, finding it challenging to recharge the sensor nodes' energy sources once they are installed. One of the most promising strategies for trying to overcome the energy limitation issue is the routing algorithm used in the network. WSN routing algorithms are in charge of sustaining and determining the best routes in the network.

Investigates with wireless sensor networks (WSNs) have shown that sensors can be linked to an individual, such as objects, animals, or people, in order to keep

N. P. Sable (✉)
Department of Information Technology, Bansilal Ramnath Agarwal Charitable Trust's, Vishwakarma Institute of Information Technology, SPPU, Pune, Maharashtra, India

V. U. Rathod
Department of Artificial Intelligence & Data Science, GH Raisoni College of Engineering & Management Wagholi, Pune, SPPU, Pune, Maharashtra, India

A. Neustein et al. (eds.), *AI, IoT, Big Data and Cloud Computing for Industry 4.0*, Signals and Communication Technology, https://doi.org/10.1007/978-3-031-29713-7_17

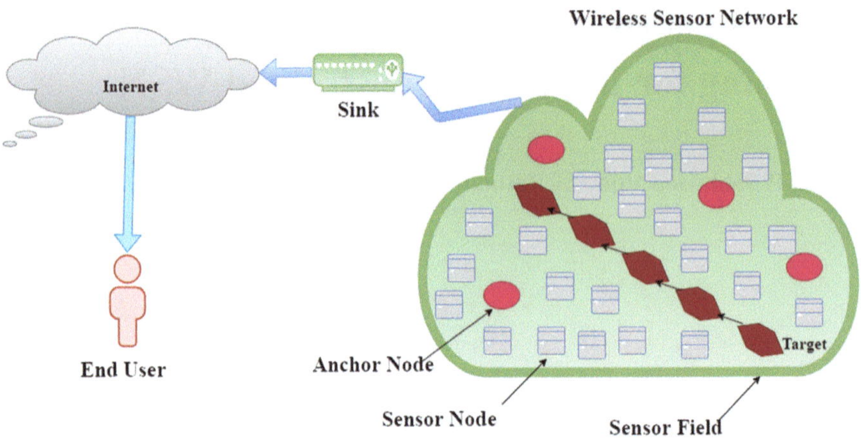

Fig. 17.1 WSN's general situation [1]

a physical mobile under surveillance in its wireless network setting. Regardless of how the sensor nodes are powered, it's critical to use energy-efficient wireless network routing (WNR) strategies to extend the sensor's life as much as possible.

The most significant issue with WSN design is the sensor node's battery energy, which is limited and consumed in a short period of time. In time-varying and dynamic wireless sensor network environments, it is critical to provide efficient wireless network routing schemes that distinguish malicious nodes. Because the radical change in routing evidence amongst two routing nodes is challenging to differentiate appropriately, existing routing strategies struggle to detect such abnormal nodes.

WSNs are a promising technology that uses a self-organizing network and a single and multi-relay to gather and send data to customers, with applications in military defence spending, atmospheric engineering, related industries, farmland intelligent systems, as well as other fields. WSN consists of an enormous amount of micro-combined sensor nodes that collaborate to comprehensive monitoring systems, awareness, and information gathering. One of the key WSN technological innovations is multi-hop routing, which is responsible for transmitting information gathered by sensor nodes from transmitter to receiver nodes using an agreed-upon routing algorithm. However, because of WSN's open, distributed, and dynamic nature, multi-hop routing is vulnerable to a diversity of attacks, cooperating with its safety and effectiveness. Traditional secure routing strategies, which are effectively set on the encryption algorithm and authentication methods, are unsuccessful for multi-hop decentralized WSNs.

Existing routing schemes are unable to distinguish malicious nodes in period-changing and dynamical WSN situations. Routing nodes in some routing algorithms are unable to distinguish between the truth and falsehood of routing evidence-

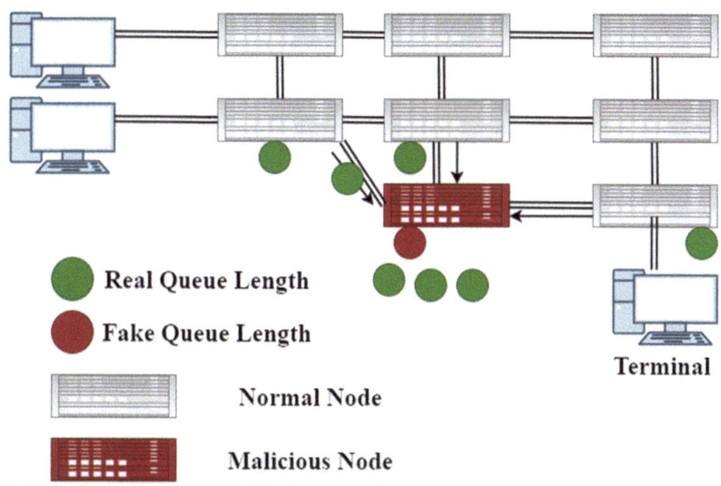

Fig. 17.2 The malicious node in a back-pressure routing algorithm (BPRA) [1]

free by other routing nodes. Figure 17.2 shows that an anomalous node can provide incorrect queue length information, increasing the likelihood of receiving a packet. This usually causes routing scheduling violations in other routing nodes. Because the actual-period alteration of routing evidence among two routing nodes is problematic to distinguish. Existing routing schemes have a hard time identifying such malicious nodes.

Once a suspicious node accepts packets of data from member nodes, rather than transmitting them to the next-hop member nodes, it throws away them immediately. This consequences in a data "black hole" in the system, thus the name "black hole attack," which is difficult to detect for routing nodes in WSNs. External attackers or inner appropriate nodes captured by outside attackers could be malicious nodes.

Trust organization has recently become a widely used method of guaranteeing routing network security. Its main strategy is for each node to keep and experiment with a trust model that uses neighbouring routing node's faith values to make routing decisions. Based on the trust value, this method enables the routing node to select a relatively reliable routing link. Nevertheless, it can only obtain the faith values of its neighbours, which is incompatible with a multi-hop distributed WSN.

Because of the aforementioned security concerns, a third intermediate is being considered as a solution to the routing nodes belief delinquency; however, the intermediary will be useless for multi-hop disseminated WSNs. In the meantime, malicious nodes are likely to attack and control a third-party trust organization centre, putting the system's security and fairness at risk. As a secure, autonomous, identity record system, the blockchain is well-suited to multi-hop decentralized sensor networks (SNs). A lot of studies have been done in the past several seasons on using blockchain to routing techniques. The blockchain is a distributed ledger keeping up with the latest by numerous nodes that is mainly supposed to address trust and security issues.

Blockchain requires not only a lot of computing power, but also a lot of memory and communication. As a result, integrating blockchain into a sensor node is not just complicated but also difficult, unless cost-cutting techniques for control and communication are established. Finding strategies to reduce the amount of storage used and, most significantly, the amount of computation used is a complicated task.

The cryptographic complexities of blockchain technology restrict its implementation in sensor nodes due to the limited resources that are available in sensor nodes. To take advantage of blockchain, however, an effort is made to design a private blockchain to avoid information manipulation, data damage or fabrication, and possible threats, intrusions, and data abuse.

In this chapter, designers introduce a new trusted routing technique for WSN applications based on blockchain and reinforcement learning. Designers deploy blockchain techniques to develop a decentralized routing data management system in which all routing data is stored on the blockchain via blockchain-token transactions (BTTs). To improve the reliability of route discovery across routing hubs, the technique makes use of the decentralized, unadulterated, and transparent qualities of blockchain activities. The blockchain network's dynamic, dependable, and comprehensive routing information is learned using reinforcement learning. Each routing node's dynamically updated reinforcement learning model is formed by the dynamically updated reward value supplied by each state's activity. To create more reliable and efficient routing decisions and select more reliable and efficient routing links as auxiliary routing nodes.

17.2 Related Work

An old-fashioned confidential routing arrangement for enlightening routing security and reliability is reviewed in this section. Using blockchain technology, here we present the relevant research approaches for routing schemes. Finally, we look into the current state of reinforcement learning's application in routing systems.

17.2.1 Conventional Trustworthy Routing Systems

WSNs face a significant and difficult challenge in providing a trusted routing environment. There are numerous related studies that have been conducted in order to appliance a trustworthy routing structure. By expanding the popular ad hoc on-demand distance vector routing protocol, Li et al. created an innovative trust-based routing protocol. The protocol uses a trust model to recommend reliable routing nodes and to enhance the security of the routing environment.

The majority of those researches are primarily based totally on a "recognition system," which assesses the person of different nodes so as to decide which path to take. However, constructing a reputation table necessitates knowledge of routing node behaviour in the past, which is insufficient to ensure WSN security in real time. Meanwhile, each routing node's reputation table may be altered, making complete trustworthiness impossible to guarantee.

17.2.2 Blockchain Network-Based Routing Mechanisms

Certain individuals have recently incorporated the data integrity and transparent characteristics of distributed ledger technology with routing techniques for increasing the level of trust between routing nodes. The approach replaces conventional public-key-infrastructure (PKI) with a blockchain system. Allowing for a decentralized inter-domain routing system by eliminating the need for central authentication. Individual integrated factor communication tree (IFT) algorithms and their behavioural characteristics vary. There are two types of routing nodes in blockchain-based communication: abnormal and normal. Smart contracts are used to aid routing nodes in establishing a secure path to destination nodes. On the smart contract, the source node acknowledges each hop's routing arrival and records malicious routing nodes with harmful behaviour. No more packets will transit through a known malicious node after then. A hostile node possessing Bitcratic Revenue (BCTR) tokens, on the other hand, could falsely claim to have received the packets, putting the scheme at risk.

17.3 Routing Approaches with Reinforcement Learning Algorithms

It is challenging to employ dynamic routing information successfully in WSN routing infrastructures. An identification routing set of rules is wanted for such dynamic routing environments. In order to boom the recognition of routing techniques, Boyan et al. were the first to use the Q-learning algorithm in conjunction with packet routing to learn the routing environment dynamically and find the shortest path. By mining complicated, dynamically updated routing network information, reinforcement learning can be utilized to enhance routing scheduling algorithms [5, 6]. The conventional backpressure routing set of rules has confined routing records of neighbouring routing nodes. As seen in Fig. 17.3, the loop routing problem occurs, causing a substantial delay in the entire routing process.

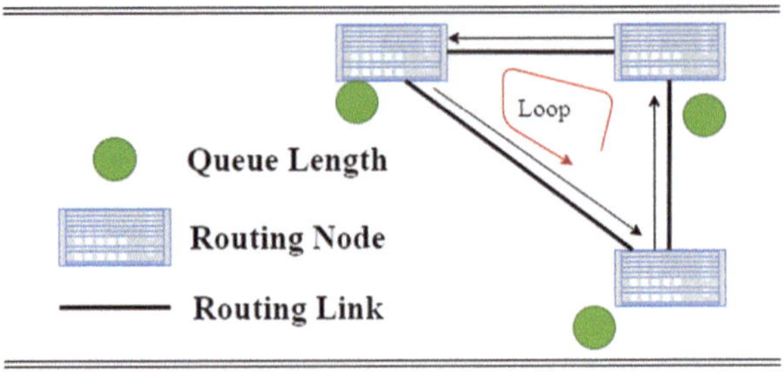

Fig. 17.3 Routing algorithms have a loop routing problem [1]

17.4 Blockchain and Reinforcement Learning Mechanisms to Improve Communication Network Routing Security and Efficiency in WSNs

A precise routing arrangement is much more beneficial for ensuring the routing effectiveness and security of WSNs. A large number of readings are being conducted to demonstrate trust between two wireless network routing nodes (WNRNs), a decentralization point (DP), cryptographic systems (CSs), trust management (TM), centralized routing decisions (CRDs), and other entities. Many other routing systems, on the other hand, are difficult to implement in real-world applications because dynamic differentiation between trusted and untrusted routing node behaviours is difficult. Meanwhile, it appears that preventing malicious node attacks effectively is proving to be a difficult task.

In terms of challenges, this research suggests that using blockchain and reinforcement learning approaches, a trustworthy routing technique can be used to improve the effectiveness and safety of routing for wireless sensor networks. On the blockchain technology, an actual-world wireless network routing (WNR) methodology is made available to obtain network routing information from network routing nodes, creating the data detectable and enormously problematic to change. The reinforcement learning process is employed to assist system routing nodes in dynamically selecting more accurate and competent channel routing links.

17.5 Blockchain Technology

Blockchain is a decentralized, publicly accessible databases that have the encrypted ledger. It is a technology that is accessible to everyone worldwide through a global database. A central authority owns traditional databases. The best example of a place where data is kept and accessed by a central server is a bank. Blockchain, on the other hand, does not belong to anyone. The distributed system is used by blockchain. Every node in the network stores updated data permanently and distributes it on a regular basis.

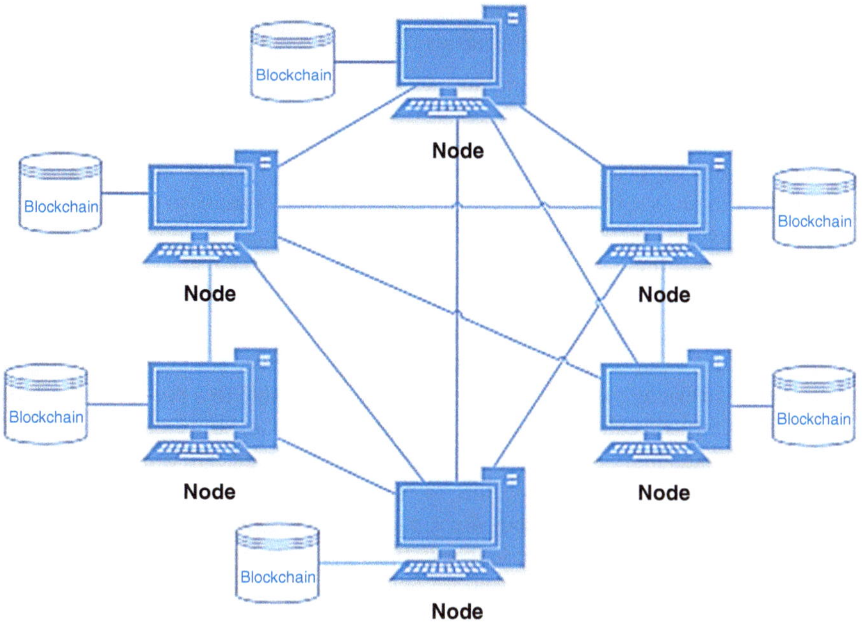

Fig. 17.4 Basic architecture of blockchain

Every node in the blockchain network has access to and may see already dispersed data. Nodes in the network will notice any changes to the data. Therefore, trying to trick the system by fabricating documents, transactions, and other information is practically impossible. Centralized systems are the norm. The single point of failure in this system is the main worry. Blockchain readily overcomes the single point of failure because it is a decentralized system. Figure 17.4 depicts the blockchain's fundamental design. Figure 17.2 indicates every node is connected with each other showing decentralized network. Same blockchain data is present on all the nodes. There are three different forms of blockchain, each with conditional transfer capabilities.

1. *Type 1:* In this type, a blockchain can only transfer digital currencies. Bitcoin is an illustration of this kind. Only cryptocurrency may be moved using the blockchain that is based on Bitcoin.
2. *Type 2:* These blockchains are capable of transferring digital currency, and some business logic can be created as needed. Logic can be implemented using a variety of languages, including Viper, Go, and Solidity. An illustration of this sort is Ethereum. Both business logic in the form of "smart contracts" and cryptocurrency in the form of "ether" can be generated on the Ethereum blockchain.
3. *Type 3:* This blockchain type disallows the transfer of digital currency. Only business logic can be used in this. The best example of this kind is hyper

ledger Fabric. Golang allows for the development of "Chain code," which is theoretically similar to a smart contract.

Based on access privileges again blockchain is separated into three sorts

1. *Public Blockchain*: No participant has established access credentials or preferences in this form of blockchain. Everyone on the network has "read/write" access. Bitcoin is a blockchain of this type that allows users to choose whether to exchange or just utilize the blockchain. Participants can join the consensus process or send transactions (Proof of Work, Proof of Stake, etc.). With this kind of blockchain, maximum transparency is possible. Participants are frequently rewarded for confirming transactions of this kind.
2. *Private Blockchain:* In this form, "read" access is enabled for all users or may be somewhat restricted; but, "write" access is only available to a pre-selected group of users. Only trusted peers will transact in some businesses, such as banking. Private blockchains are an option in certain circumstances. Although they are not compensated for certifying this form of transaction, participants will gain access to vital data because they are a part of the ledger. These include Ripple, Monax, Multichain, and Hyperledger Fabric.
3. *Permissioned/Consortium Blockchain:* With this form, a central authority grants "read/write" access and "consensus management" powers. Some of the concepts from both types are followed in this type, rather than strictly adhering to the fundamentals of public or private blockchains. This blockchain type can be used by a consortium of institutions, such as a hospital system. The examples of this class include R3, Corda.

As a new module, the blockchain makes a significant contribution to the security of the WSNs through its features. Should provide information integrity in a blockchain and traceability. Between the chains or peers, the blockchain creates consensus. A data rule is an expression that determines the set of legal data that can be stored within a data object. Use data rules to ensure that only values compliant with the data rules are allowed within a data object. Data rules will form the basis for correcting or removing data if you decide to cleanse the data. The consensus process is carried out independently using data status rules, contributing to the system's full independence. In completely truthful blockchain technology, there are five main components, i.e. distribution, encryption, immutability, tokenization, and decentralization as shown in Fig. 17.5.

Distribution The contributions to blockchain are completely separated from each other and are connected to the network. A comprehensive version of the Ledger that changes new transactions even though those who appear are managed to maintain a complete node within each contributing factor. In a production process, a distribution network is responsible for the data of warehouses and transport systems that receive inventory levels of goods and deliver them to the customer after any event.

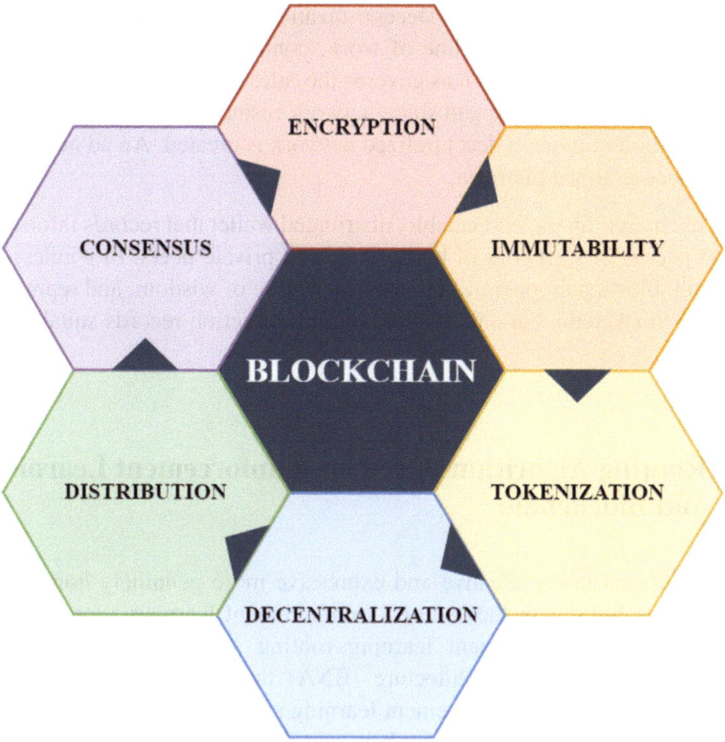

Fig. 17.5 Illustration of blockchain technology [3]

Encryption Public and private keys are being used by blockchain technologies to safely and semi-anonymously record all the information within blocks. Contributions will regulate their authenticity as well as other private details and will communicate only what they're doing in a transaction.

Immutability Comprehensive operations in the blockchain are cryptographically agreed to sign, time-stamped, and added to something like the ledger by forming as well as following a logical order as well as a sequence. Unless the participants agree on the need to do so, records in the blockchain technology cannot be corrupted or otherwise modified.

Tokenization A protected exchange of value in emerging blockchain technologies involves transactions as well as other interactions. Valuation comes as a result of tokens in blockchain, but it can represent anything from financial instruments to data to tangible infrastructure. The coupons as well approve contributions, a foundational driver of the business plan of the blockchain, to control their personal data.

Decentralization In this, due to a competition mechanism, both wireless network information and the guidelines for how the wireless network works are kept by nodes on the distributed platform. Decentralization implies that no central authority, as in the wireless atmosphere line of work, controls all the wireless networking nodes or indeed the data, as well as governs the rules. In its simplest form, when two or more nodes are connected and share network resources without going through a separate server computer, a decentralized network is created. An ad hoc connection may be a decentralized platform.

A blockchain is an understandable, distributed wallet that records information on a peer-to-peer (P2P) network of both public and private peers. In wireless-ad-hoc-networking, blockchain operates as a central source of wisdom, and representatives of a public blockchain can only display these transaction records suitable for each other.

17.6 Routing Algorithm Based on Reinforcement Learning and Blockchain

Routing multicast nodes creative and expressive more genuinely honest and successful routing links with the use of a reinforcement learning approach. Develop a comprehensive reinforcement learning routing scheduling algorithm for the blockchain-based network architecture (BNA) that is categorized as a routing algorithm that uses both reinforcement learning and blockchain technology. To aid the routing node in selecting the next best routing node, the reinforcement learning technique is employed.

Because of the dynamic getting-to-know nature of the nodes, the routing set of rules primarily based totally on reinforcement getting-to-know method lets in for routing option. The blockchain approach is used to record each hop of routing information in this case. The link is irregular or the transmission rate is extremely low if the hop is looped. The technique effectively decreases the possibility of packets passing through the connection. Simultaneously, the routing set of rules can dynamically find out some distance greater reliable and well-deliberate routing hyperlinks on the way to help routing nodes in making an awful lot of higher routing decisions. In this, an authenticated server node keeps each token transaction on the blockchain technology record; such that the quality of the transaction recorder's origin is backed up on all network routing nodes and server nodes.

By communicating trusted peers with municipal stakeholders, the framework allows transparent, safe, and reliable transaction infrastructures that speed up digital services. Uses blockchain technology to authenticate device identity, protect sensor data used in transportation, and manage access to city facilities with the assistance of blockchain approach for advanced security against cyber-attacks and fraud. In blockchain, the block structure is as shown in Fig. 17.6.

Similar to how a linked list is generated, a blockchain is a mechanism that links blocks to form a chain. A block header and content are the two main components of each block. The block header contains information about managing blocks and

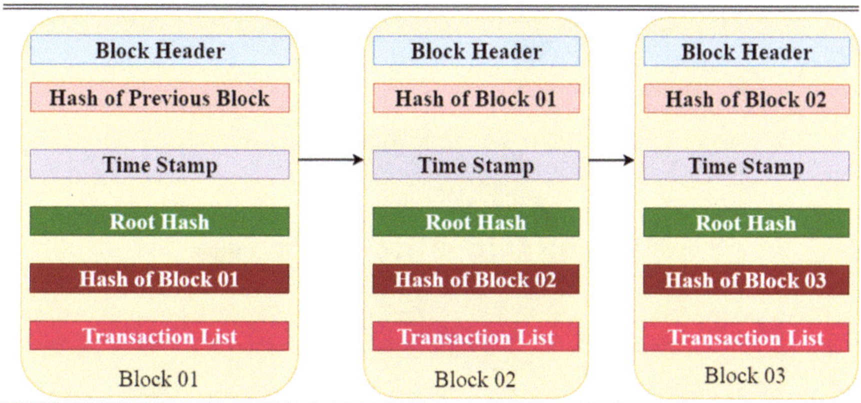

Fig. 17.6 Block structure in blockchain technology [3]

chains. On the other hand, the block body contains a list of transactions. Because the blockchain is a decentralized platform, nodes connect via a P2P network and validate transactions with their own digital signatures. In addition, some nodes maintain a ledger that contains a list of agreed blocks. Proof-of-Work (PoW) and Proof-of-Stake (PoS) are two consensus algorithms commonly used in blockchain to keep the data on the ledger in sync. Proof of work (PoW) is a decentralized consensus mechanism that requires members of a network to expend effort solving an arbitrary mathematical puzzle to prevent anybody from gaming the system. Proof of work is used widely in cryptocurrency mining, for validating transactions and mining new tokens.

The block header and block content are the two most significant components of each block. The significant sectors of the block header are shown in Fig. 17.7. To generate a root hash value from each block's transactions and check the integrity of the transactions within the block, blockchain technology employs a Merkle tree. A Merkle tree divides blocks transactions into pairs, with each transaction's hash recorded in a parent node. One step up the tree, the hashes of the parent nodes are preserved and arranged in pairs. This technique is repeated until you reach the root node.

A Merkle tree is a type of hash tree. Figure 17.8 shows the Markle Tree structure. Each leaf (node) of a Merkle tree is labelled with the cryptographic hash of the data block. Non-leaf nodes (also known as branches, interior nodes, or inodes) are tagged with the cryptographic hash of their child nodes label. A hash tree can be used to efficiently and securely verify the contents of a big data structure. Hash trees combine hash lists and hash chains to form hash trees. A list of transactions is contained in the block body. Digital signatures are used to verify each transaction.

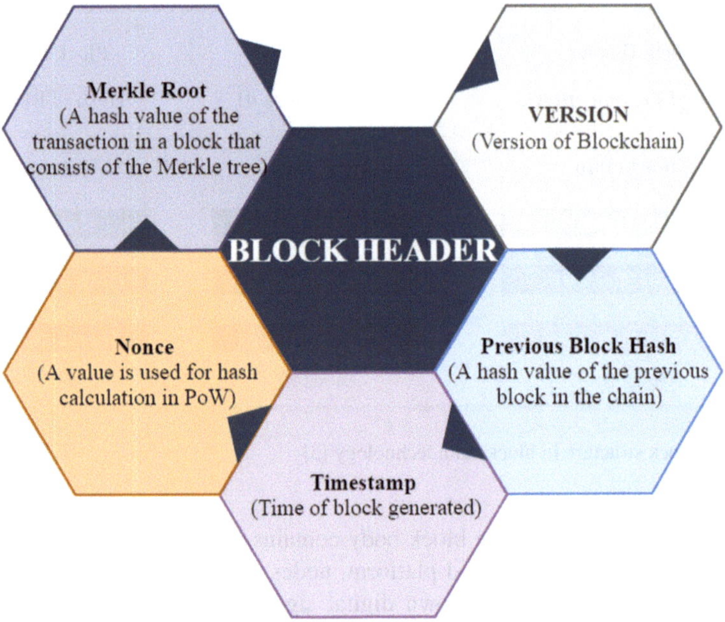

Fig. 17.7 The block header's most important elements [4]

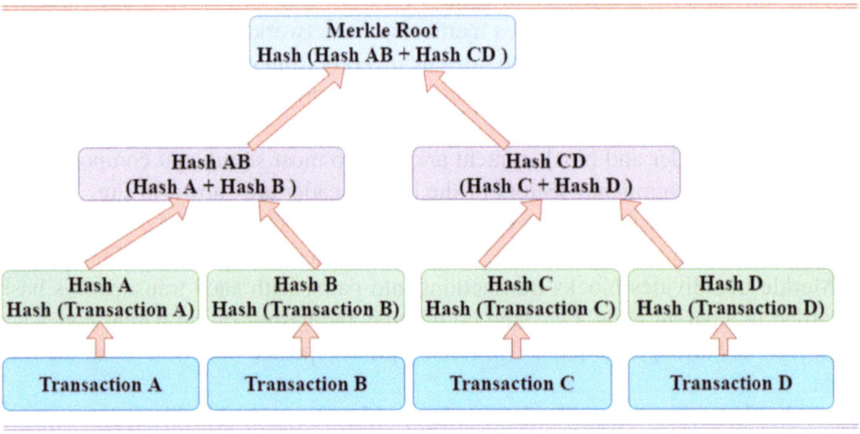

Fig. 17.8 The Markle tree [4]

17.7 Blockchain Network Procedure

In order to function the blockchain-primarily based totally routing community structure successfully, the routing community's statistics need to be transmitted to the blockchain community. Smart contract is the program logic where the conditions of the agreement between the buyer and seller are directly encoded into lines of

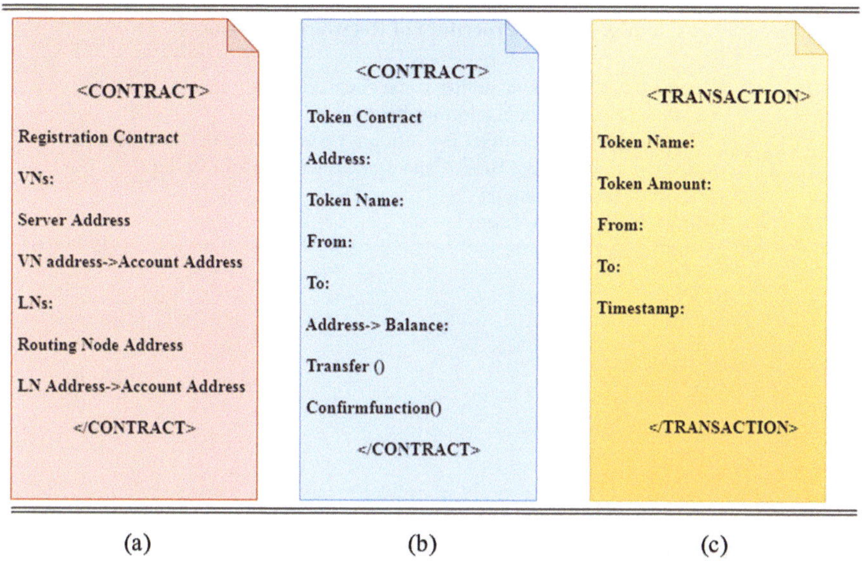

Fig. 17.9 Formats for smart contracts (SCs) and blockchain transactions (BTs). (**a**) Registration contract. (**b**) Token contract. (**c**) Token transaction

code, making it a self-executing contract. The agreements and underlying code are spread throughout a decentralized blockchain network. Transactions are traceable and irreversible, and the code regulates their execution. Without the need for a centralized authority, a legal system, or an external enforcement mechanism, smart contracts enable trusted transactions and agreements to be made between dispersed, anonymous parties. We can use such smart contracts to store the routing information. Like blockchain transactions like registration-contracts (RCs), token-contracts (TCs), and token-transactions (TTs), as shown in Fig. 17.9. The server nodes verify all of these contents before publishing them to the blockchain.

The nodes of the authenticated server manipulate all smart contracts. The results of the execution are then returned to the blockchain network. All routing nodes' identities are recorded in the registration contract. To make it easier to query the entire network node, there are server nodes. The first algorithm explains the technique for creating the registration contract. Mapping maps are variables built into registration contracts that track how blockchain addresses are translated into physical addresses. Allocation status is a built-in variable that indicates whether a node is registered. When a new node wants to join a subscription contract, it must first invoke the contract.

It then enters its data. Subsequently, the data is inputted, and the node proceeds to authenticate their identity by providing their physical address (PA). The subscription contract immediately registers its blockchain address *(ba)*. Whether *(ba)* is present in the state mapping is determined by the registration contract. The registration contract correctly stores the map (ba->pa) in the *(map)* mapping and the map (ba

Algorithm 1 Procedure of Registering a Node

```
Mapping map:blockchain.address->physical.address;
Mapping state:blockchain.address->0 or 1;
While true do /* The contract is waiting for a contract caller to trigger */
  Input: Contract Callers Blockchain address ba; Contract Callers
         Physical Address pa
  Output: Registration Result r;
r<-Null;
If sate (ba)=1 than
r<-Failure;
else
map (ba)=pa;
state (ba)=1;
r<-success;
end if;
end while
```

Fig. 17.10 Algorithm 1: procedure for registering a node

->1) in the (state) array when *(state = 0)*. If *(status = 1)* indicates that the node is registered, this action will result in failure. The code's logic links the node's blockchain address to the identity, and once registered, the information cannot be modified. Figure 17.10 shows the process for registering a node (Algorithm 1).

In this section, users extend reinforcement learning-based routing planning by using comprehensive, dynamic, and reliable routing information provided by the recommended blockchain platform. The correct blockchain-based reinforcement learning (RLBC) algorithm is summarized in Algorithm 2 as shown in Fig. 17.11.

In computer science, reinforcement learning (RL) techniques are really a subfield of machine learning worried with the appropriateness of software agents to act decisively in an atmosphere to achieve maximum the concept of aggregation reward. As shown in Fig. 17.12, the working principle of reinforcement learning (RL) techniques. One of the most significant techniques of the three basic machine learning modes, alongside supervised learning and unsupervised learning, is the reinforcement learning technique. The RL model is used to dynamically select the proper route discovery and avoid routing connections with malicious users. The RL-based routing scheduling algorithm (RSA) uses the entire global, innovative, and confidential routing information obtained from the suggested public blockchain.

Reinforcement learning and a blockchain-based routing algorithm are used to solve this problem. Increase the overall efficacy and safety of WSN routing. In addition to obtaining routing information from blockchain routing nodes, a realistic wireless network routing concept is provided. This makes routing information traceable and makes it impossible to make necessary changes to data from a wireless network (WN).

Algorithm 2 Reinforcement Learning & Blockchain based (RBC) Routing Algorithm

Input: Environment E; Action Space A; Initial Sate x_0 ; Reward Discount R_1 ; Learning Rate R_2; ;
Output: Policy P_i ;

```
Q.(x, a)=0, p(x,a)=1/lA(x)l ;
x=x0;
for T=1,2,......,do;
a=pi^p (x);
r= reward by routing action a;
x^1 = next state by routing action a;
a^1 = pi (x^1) ;
Qt (x, a)= Qt (x,a) + R2 (R1 + R1 Qt (x^1,a^1)) - Qt (x, a));
pi (x) = arg max_ak qt (x, ak) . Qt (x, ak);
x = x^1 ;
end for
```

Fig. 17.11 Algorithm 2: reinforcement learning and blockchain-based routing algorithm (RLBC)

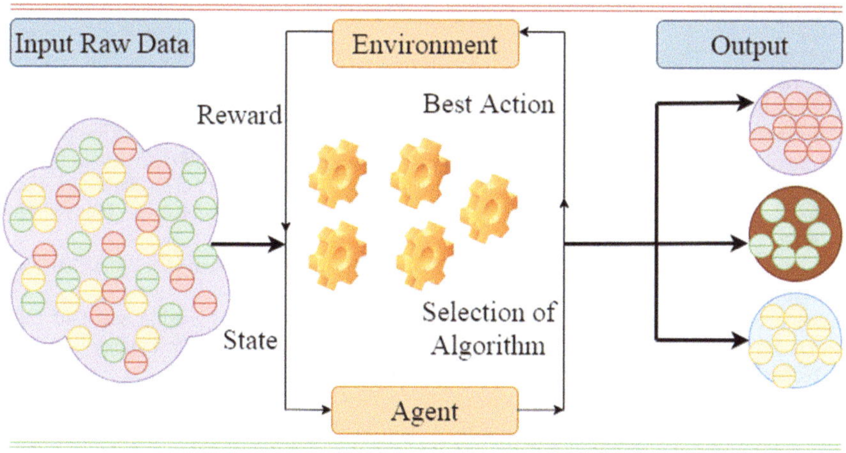

Fig. 17.12 Illustration of reinforcement learning technology [2]

Seismic stations, civil, battlefield surveillance, health surveillance, habitat monitoring, home automation, and traffic control are some of the applications for wireless sensor networks. WSNs of satellites could be built for emergency applications such as disaster management, rescue operations, and broadband communications in areas where no infrastructure is available.

17.8 Conclusions

In this chapter, the authors premeditated a trust routing approach based on blockchain and reinforcement learning. To improve the performance of the routing network in WSNs by creating a secure routing environment. The blockchain network is a decentralized system that provides a viable method for information management and a platform for reinforcement learning routing scheduling. The routing packets are represented by the blockchain token. Each routing transaction is released to the blockchain network after the validation node authenticates it. Making each routing transaction recorder traceable and impervious to change. Routing nodes on the blockchain network can acquire dynamic and reliable routing information. The authors also supplementary describe a thorough reinforcement learning algorithm to regulate the optimal routing path and avoid routing links on malicious nodes.

References

1. J. Yang, S. He, Y. Xu, L. Chen and J. Ren, "A Trusted Routing Scheme Using Blockchain and Reinforcement Learning for Wireless Sensor Networks", *Sensors*, vol. 19, no. 4, p. 970, 2019.
2. B. Jang, M. Kim, G. Harerimana and J. Kim, "Q-Learning Algorithms: A Comprehensive Classification and Applications", *IEEE Access*, vol. 7, pp. 133653–133667, 2019.
3. D. Nguyen, P. Pathirana, M. Ding and A. Seneviratne, "Blockchain for 5G and beyond networks: A state of the art survey", *Journal of Network and Computer Applications*, vol. 166, p. 102693, 2020.
4. A. Baldominos and Y. Saez, "Coin.AI: A Proof-of-Useful-Work Scheme for Blockchain-Based Distributed Deep Learning", *Entropy*, vol. 21, no. 8, p. 723, 2019.
5. Sable Nilesh Popat*, Y. P. Singh, "Efficient Research on the Relationship Standard Mining Calculations in Data Mining" in Journal of Advances in Science and Technology | Science & Technology, Vol. 14, Issue No. 2, September-2017, ISSN 2230-9659.
6. Sable Nilesh Popat*, Y. P. Singh, "Analysis and Study on the Classifier Based Data Mining Methods" in Journal of Advances in Science and Technology | Science & Technology, Vol. 14, Issue No. 2, September-2017, ISSN 2230-9659.

Chapter 18
Secure Data Hiding in Binary Images Using Run-Length Pairs

Gyankamal Chhajed and Bindu Garg

18.1 Introduction

As a result of the wide range of possible uses of digital documents, such as legal documents, certificates, digital books, and engineering drawings, there is great interest in their use. Further, sensitive documents such as faxes, insurance documents, and personal documents are digitized and stored. Securing the authenticity and integrity of digital documents is becoming increasingly relevant. Authenticating and detecting tempering and forgeries are therefore primary concerns. In order to mitigate these concerns, data hiding and watermarking have proven to be promising approaches. The use of Internet is extensively done for digital data transfer over large distance that is use of Internet is reducing efforts and takes negligible time for data transfer, but as use of Internet for data transfer increased chance of data being hacked. So security of data transfer from Internet is major issue and focused research topic. Even though there are number of methods for secure data transfer like cryptography but they have their own pitfalls as it is self-declarative about existence of some important information. So message hiding in images does not disclose carrying of any information secretly. Passing secret message through color images is easier as they have 256 color ranges for each RGB color band as opposed to only two colors available in binary images. Its challenging to hide data in binary

G. Chhajed (✉)
Research Scholar, BV(DU)COE, Pune, Maharashtra, India

Computer Engineering, VPKBIET, Baramati, Maharashtra, India
e-mail: gjchhajed@gmail.com

B. Garg
Computer Science & Engineering, BV(DU)COE, Pune, Maharashtra, India
e-mail: brgarg@bvucoep.edu.in

A. Neustein et al. (eds.), *AI, IoT, Big Data and Cloud Computing for Industry 4.0*, Signals and Communication Technology, https://doi.org/10.1007/978-3-031-29713-7_18

image maintaining visual distortion with high capacity. There are many spatial and frequency domain techniques. One of the ways is to use black and white pixel run length (RL) for hiding data.

18.2 Literature Survey

Data hiding, extraction, and image recovery can be easily done using image encryption and support vector machine model (SVM) [1]. For this purpose, image blocks are used where in each block, the one-bit message can be embedded. Extraction can be carried out using support vector machine model (SVM) as it is trained. By adapting image block pattern, a small-block-based technique is proposed wherein information is hidden in [2]. This technique claims the highest data hiding rate. This technique applies the principle of Webb's law and also uses blind data extraction. The computational complexity in data hiding and extraction algorithm is very low. In [3], to check the effect of embedding and also to identify that the data is hidden a binary steganalysis scheme has been proposed. The distortion for 4 classes of patterns is calculated considering 1-shape pattern as the reference. Further, a 2-dimensional feature set consisting of only 2 patterns is computed. To reduce visual distortion, the combination theory has been applied. In order to increase hiding capacity and to reduce visual artifacts, a special matrix has been used. Each block is changed one pixel at a time and [log2 $(m \times n + 1)$] bits are hidden into a block of size m x n. It allows for a higher payload embedding under the same noise conditions [4]. In [5], block classification and local complexity calculations have been used. In this scheme, each selected block is used to embed three bits data at most only two pixels are modified. In order to hide secret data, the complex region of the image is used so that it becomes difficult to detect the hidden data and also avoids grabbing attention from attackers. The texture property of binary images is considered in [6]. A similarity matching method based on texture is proposed as an efficient way to hide binary image data. In [7], new information security method is proposed. As compared to generally used LSB algorithm, the new proposed technique targets more unnoticeable and secure transmission of data that is secretly hidden. In [8], a novel scheme for authentication of binary image has been proposed, that has a small distortion in the cover image. In [9], in the data-embedding algorithm, hamming codes-based scheme has been used, where the ELSSM (Edge Line Segment Similarity Measure) algorithm is used for selecting and flipping pixels. A new multi-class steganalysis method was proposed in [10]. The capability of the proposed method is that it can identify the technique used for hiding data. The data hiding in binary text documents is proposed in [11] where the 8-connected boundary of character is used for embedding. By adding or deleting the center foreground pixel in a pair, a five-pixel boundary with a fixed set of pairs of patterns is utilized to embed data. After data hiding, an image quality preserving scheme is presented in [12]. The imperceptibility after embedding is determined by comparing modified bits with their adjacent bits. The discrete cosine transform (DCT) domain is used in [13] to check for the possibility of embedding a watermark in binary images. Using the relationship within the block to identify "flippable" pixels in [14–16] results in minimal visual distortion. A method for detecting and hiding data in binary images to maintain visual quality is proposed in [17]. Several proposed schemes are also discussed that locate updated regions. An approach for preserving pixels' connectivity is presented in [18]. In a moving window, "flippability" is centered at pixel. Yang and Kot [19] presents a two-layer blind binary image authentication scheme, in which the first layer enables the authentication, while the second layer identifies tampering locations. A scheme using morphological transforms is presented in [20]. The run-length (RL) histogram is monitored and modified to embed data using a reversible data hiding scheme. The influence of RL couples was used to conceal data from which black RL and white RL were formed [21]. An examination and analysis of current watermarking and steganography technologies presented in [22] is taken into account in the context of image processing in the spatial and transform domains. The regression analysis model is used to accomplish the efficient de-fuzzification operation [23]. Data bits pattern and image block pixel pattern matching is compared by scanning image in different ways and best way is selected for embedding [24]. Embedding and extraction Decision trees is used to select image part to hide and extract group of 4 bits of data in a image block of size 3X3 [25]. The 4 pixel LRUD pattern of 3X3 block is matched to hide 4 bits of message group [26]. The block diagonal partition pattern and connectivity in image block of size 3X3 is checked to embed data [27, 28].

18.3 Gap Analysis

The fact that any pixel change in a binary image for the purpose of hiding data is easily detectable and hence its challenging. Binary images have only two colors, so if a black pixel is changed to white or a white pixel to black, the image changes quite distinctly. Existing schemes, approaches, and methodologies have disadvantages such as limited data hiding capability, visual distortion caused by embedding, resilience, and data security. These gaps can be filled; if the original image is used in maximum possible way for data hiding, the capacity will be increased; if the pattern of the image in its original form is used to hide data as much as possible, visual distortion can be minimized; and if data is encrypted before hiding, it will be secure because it will be indecipherable.

18.4 Proposed Technique

In proposed system, the security of hidden data and the technique of data hiding are focused. Three levels of security are provided by using encryption with key, representing sparse matrix in single column form and arithmetic data compression. Secret message is encrypted by a key which is known to both the sides. Then it is represented in two-dimensional sparse matrix which is represented in one column form. The arithmetic compression algorithm is used to compress the secret message represented in above form. This encrypted data and compressed data is then embedded into the cover image using run-length pairs. In order to generate alternative black RL and white RL sequences, the binary image is scanned from left to right and from top to bottom. One RL pair is generated when a black RL is combined with its next white RL, resulting in a sequence of RL pairs. The RL having highest frequency are chosen for embedding data. The threshold for black run-length T and run-length pairs T1 are set for data embedding. At the receiver side the encrypted message is extracted from and it is decompressed in the reverse way to as done during embedding process. Then the data is decrypted by using key to get secret data.

18.4.1 Compressed Data Preparation

The input text message which is to be hidden is converted into bit stream.

This bit-stream is encrypted by key at first level of security. Then it is represented in two-dimensional matrixes. Since this matrix is sparse, the location where '1' bit is present in the matrix that locations are only considered and it is transformed from two-dimensional to one-dimensional matrix. This transformation provides security at second level. This transformed compressed data is further compressed by arithmetic compression algorithm by providing security at third level. This process of compressed data preparation minimizes the bits to be embedded and hence increases the hiding capacity.

18.4.2 Information Hiding Process

Figure 18.1 depicts the process of hiding information. Binary images are represented by the numbers 0 and 1 corresponding to the white pixel and black pixel, respectively. Identification of black-and-white RL histograms and formation of the RL pairs of the binary image are described in Tables 18.1, 18.2, and 18.3. The

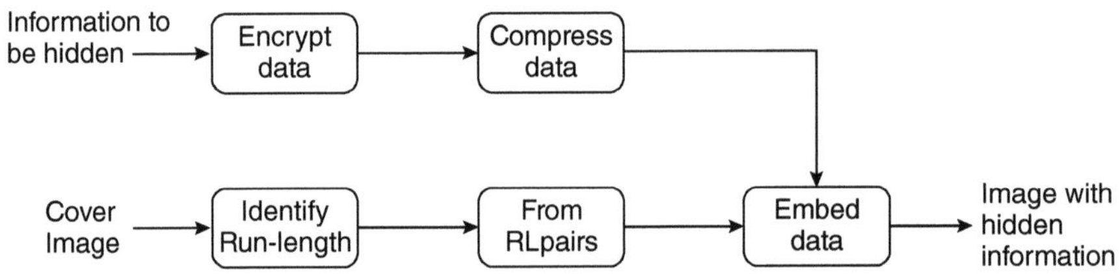

Fig. 18.1 Data embedding process

Table 18.1 Image representation as bits '1' and '0' corresponding to binary image and binary image with hidden data

Image	Col→	1	2	3	4	5	6	7	8	9	10	11	12	13	14	15
Original Binary Image	Row 1	1	0	0	1	0	0	0	1	0	0	1	1	0	0	0
	Row 2	1	0	0	0	1	1	1	0	0	0	1	1	1	0	0
Binary Image with hidden data	Row 1	1	1	0	1	0	0	0	1	1	0	1	1	0	0	0
	Row 2	1	0	0	0	1	1	1	0	0	0	1	1	1	0	0

Table 18.2 Original image run-length pairs and histogram

Original binary image RL pairs	Run-length (RL)	Black RL histogram	White RL histogram
[1, 2], [1–3], [1, 2], [2, 3],	1	4	0
[1, 3], [3], [2, 3]	2	1	3
	3	2	4

Table 18.3 Binary image with hidden data and corresponding histogram

Binary image after embedding 1010 bits with T = 1 and T1 = 4	Run-length (RL)	Black RL histogram	White RL histogram
[2-1], [1–3], [2-1], [2, 3],	1	2	2
[1, 2], [3], [3-2]	2	3	2
	3	2	3

information to be hidden is encrypted with shared key and represented in encrypted and compressed as described in earlier section. The threshold for black run length RL and length of RL pairs are defined and used for hiding information. Consider embedding data in black RL equal to 1, i.e., assign T to 1. Black pixels having run length one, i.e., T, "0" information bit is embedded and to embed "1" bit of information black pixels having run length two i.e., T + 1 is used. Apart from this, threshold for length of RL pairs is also set to embed and extract correct information (Fig. 18.2).

Assign a parameter T equal to 1, i.e., embed data into black RL equal to 1. In order to embed data, a pair of histograms is created. The sum of the black RL and the white RL in this RL pair remains unchanged. In this example pairs of length 3 and 4 are considered for hiding information. Since h(1) = 4, four bits can be embedded and assumed four bits are as follows: 1,0,1,0. Then the pixel at row 1 and column 2 is changed from 0 to 1 in order to change the black RL from 1 to 2. No change is made while embedding 0. Thus 4 bits are embedded by above method.

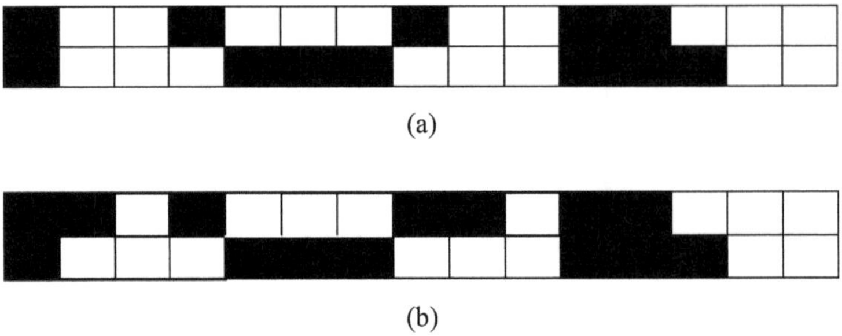

Fig. 18.2 Image part before embedding (a) and after embedding '1010' (b)

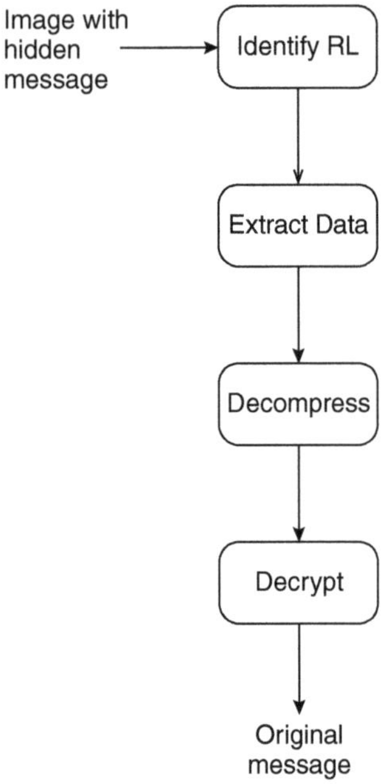

Fig. 18.3 Information extraction process

18.4.3 Information Extraction

An alternative black-and-white RL sequence is formed by scanning the binary image from left to right and from top to bottom. An RL pair is formed by combining one black RL and its next white RL, resulting in a series of RL pairs. RL pairs of set threshold are used for information embedding.

The information extraction process is shown in Fig. 18.3. The same sequencing as used for information embedding in reverse is used for information extraction. As per the set thresholds information is extracted. As shown in Table 18.1, when a black

RL of length one is encountered, a bit "0" is extracted, when a black RL of length two is encountered, a bit "1" is extracted. In this way, hidden secret information is extracted. This extracted information is decompressed and decrypted to get original information.

18.5 Algorithms

18.5.1 Arithmetic Coding Algorithm

Using this algorithm, maximum compression of the bits is successfully achieved as compared to other compression techniques such as Huffman coding, etc. In contrast to Huffman coding, arithmetic coding does not compress each symbol with a discrete number of bits. For every source, it reaches almost the optimum compression in terms of Shannon's theorem and is well suited to adaptive models.

Algorithm:

Step 1. Calculate the probability of each of the symbols.
Step 2. Calculate the range_low and range_high of each of these

symbols.

Step 3. Calculate l and h using the formula:
l = low+*range_low(S);
h = high+*range_high(S);
Step 4. Generate a compressed code for the data.

18.5.2 Embedding Algorithm

In this algorithm, the concept of run length pairs is used to hide data. The maximum number of run length is identified and assigned as the threshold value. This threshold then decides, which pixels are to be used for the information embedding. Two threshold values are assigned in order to prevent distortion to the image and hide the data in the image with maximum efficiency.

Step 1. Scan the cover image from left to right and top to bottom.
Step 2. Calculate the run length of black and white pixels and
    Prepare histogram of black pixels run length.
    The maximum black pixel histogram black RL is considered as threshold value (T) and RL pair with T is decided as threshold T1
    //For hiding data bit '1':
    If the run length is T in RL pairs length <=T1 and data bit is '1' then
        Change the white pixel to black to make black
        RL T+1
    Else keep it unchanged
    //For hiding data bit '0':
    If the run length is T+1in RL pairs length <=T1 and data bit is '0' then
        Change the black pixel to white to make black
        RL T,
    Else keep it unchanged for bit '1'
Step 1. Save the cover image.
Step 2. Stop

18.5.3 Extraction Algorithm

Step 1. Scan the cover image from left to right and top to bottom.
Step 2. Find index of RL pairs <= T1 having value equal to T or T + 1

Step 3. After calculating index
    If the run length is T then extract data bit as "0"
    If the run length is T + 1 extract data bit as "1"
Step 4. Stop

18.6 Result Analysis

The technique is tested with set of standard images and sample figures are shown below, before and after hiding data Fig. 18.4 and results are analyzed in Figs. 18.5, 18.6, and 18.7, respectively. The PSNR and MSE are used to check the performance of technique.

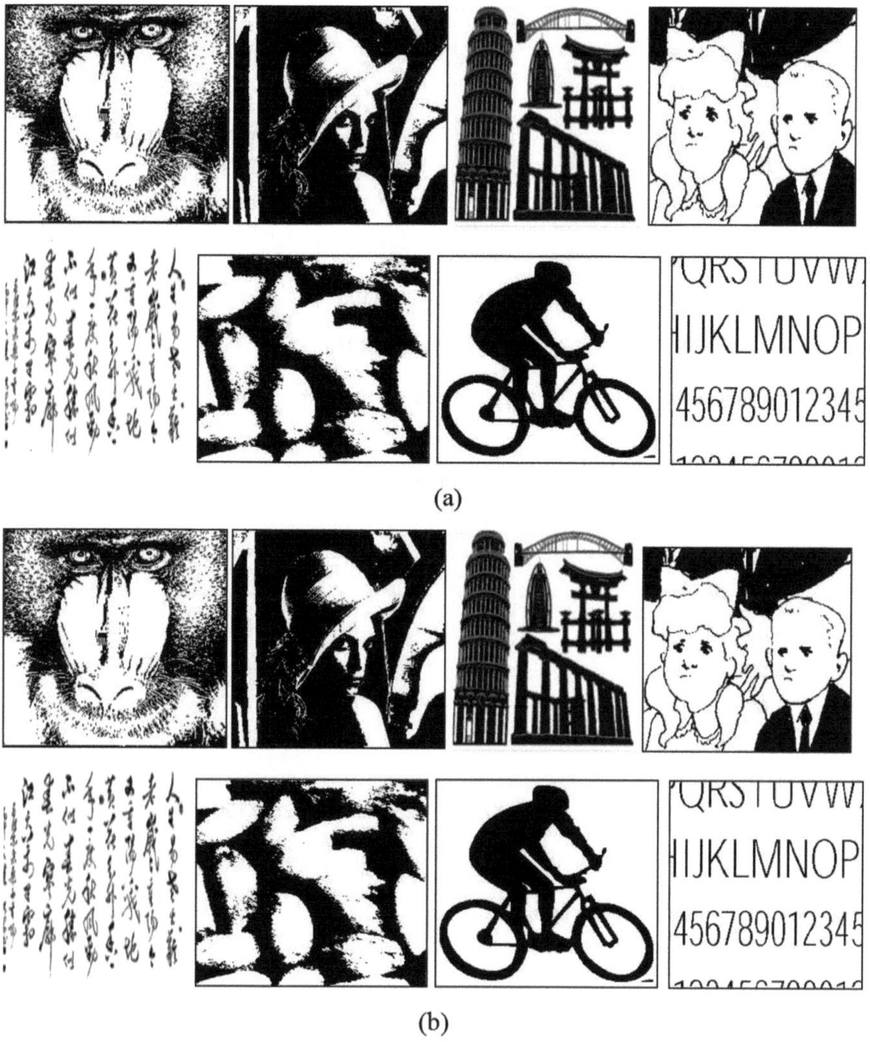

Fig. 18.4 Example tested images: (**a**) before message hiding and (**b**) after message hiding

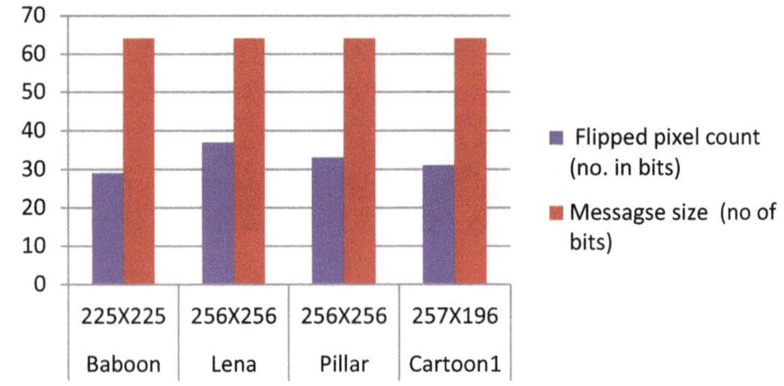

Fig. 18.5 Comparison of size of message embedded versus flipped pixels

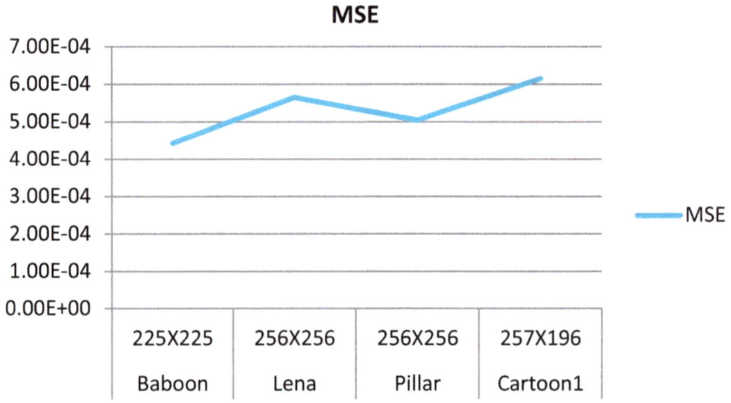

Fig. 18.6 Performance analysis using MSE

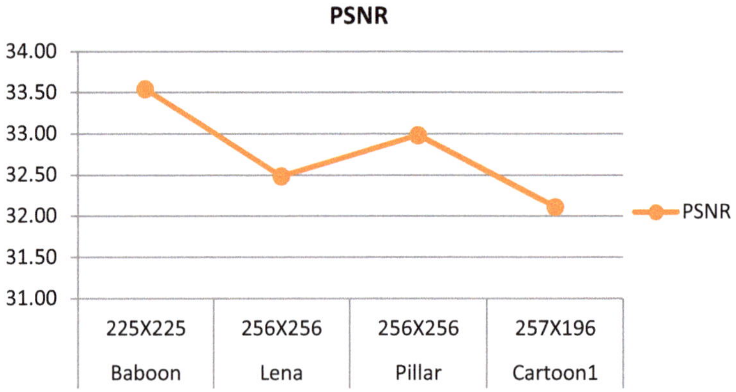

Fig. 18.7 Performance analysis using PSNR

Table 18.4 Result table

Sr. No.	Image	Image size	Total pixels	Flipped pixel count (no. in bits)	Ratio F/T	MSE in 10^{-4}	PSNR
1	Baboon	225 × 225	40000	29	0.45	4.43	33.54
2	Lena	256 × 256	88205	37	0.58	5.65	32.48
3	Pillar	256 × 256	65536	33	0.52	5.04	32.98
4	Cartoon1	257 × 196	83400	31	0.48	6.15	32.11
5	Image 1	256 × 256	88205	23	0.45	4.43	33.54
6	Image 2	256 × 256	88205	30	0.58	5.65	32.48
7	Image 3	256 × 256	88205	39	0.52	5.04	32.98
8	Image 4	256 × 256	88205	32	0.48	6.15	32.11

18.7 Discussion

We have tested the set of standard images using the MSE and PSNR performance matrix. Table 18.4 shows test results for images in Fig. 18.4a. The images with hidden data are shown in Fig. 18.4b. The table presents the information of image as size, embeddable blocks, flipped pixel count, flipped pixel count to size of hidden data, MSE, PSNR for the hidden data "success" and key "abc". Total encrypted hidden data has a length of 64 bits. The data is hidden sequentially in blocks satisfying the embeddability test. Result shows that about 50% of hidden data embedded without flipping pixel. The embeddability will be much better if the images have more number of RL pairs satisfying threshold set. The performance of any data hiding technique is highly depends on the pattern of cover image and sequence of bits in data to be hidden. The pattern of RL pairs in cover image and the number of RL pairs satisfying embeddability criteria matters for the performance of our technique. The number of flipped pixels is dependent on run length of black and white pixel pairs and message data bit pattern matching. The more the match, the less the visual distortion. The hiding capacity can be enhanced by setting other threshold values based on image pattern.

18.8 Conclusion

The various techniques available for data hiding have been analyzed and studied. This paper proposes a novel sparse matrix-based secured data encryption scheme for binary image data hiding using RL Histogram modification. Thus, using this technique, the hidden data can be encrypted and also can be embedded in the image with maximum efficiency and minimum distortion to the image and thus the hidden data from the image is extracted. In this proposed method, combination of encryption with data embedding is done using run-length pairs modification. This algorithm has reasonable imperceptibility and can be used for secret data transmission, according to experiments. This algorithm has a high correlation with subjective assessment, according to experiments. It can be applied in a wide range of situations range of applications involving visual distortion in digital binary document images, such as lossless compression and data hiding. Similar encryption strategies could also be used in contexts where pixels are modified in alternative ways via different channels or bit-patterns that affect only non-visible data.

References

1. V.M. Manikandan, V. Masilamani "Reversible data hiding scheme during encryption using machine learning," Elsevier ScienceDirect. Procedia Computer Science 133, 2018, 348–356
2. Wenjia Ding, Yulin Wang " Data Hiding in Binary Image with High Payload", Springer. Arabian Journal for Science and Engineering 43, 2018, 7737–7745 https://doi.org/10.1007/s13369-018-3130-5
3. BingwenFeng, JianWeng "Steganalysis of content adaptive binary image data hiding", Journal of Visual Communication and Image Representation volume 46, 2017, Elsevier, Pages 119–127 Elsevier Science Direct 2017 https://doi.org/10.1016/j.jvcir.2017.01.008
4. Nan-I Wu, Min-ShiangHwang : Development of data hiding schema based on combination theory for lowering the visual noise in binary image", Elsevier ScienceDirect 2017. https://doi.org/10.1016/j.displa.2017.07.009
5. Han-Zhou Wu, Hong-Xia Wang, Yun-Quing Shi, "Separable Reversible data hiding for encrypted palette images with color partitioning and flipping verification" IEEE 2016 DOI https://doi.org/10.1109/TCSVT.2016.2556585
6. Honey Mol M, Reji P.I "A secure binary data hiding and comparison technique", IEEE 2016. International Journal Of Scientific & Engineering Research, Volume 7, Issue 7, July-2016 810 ISSN 2229-5518
7. Thai-Son Nguyen, Chin-Chen Chang, "High capacity data hiding for binary image based on block classification"Springer 2015. DOI https://doi.org/10.1007/s11042-015-2768-1
8. S.Udhayavene, AathiraT.Dev "New data hiding technique in encrypted image: DK Lalgorithm", Elsevier ScienceDirect 2015 Procedia Computer Science 54 (2015) 790 – 798
9. Younho Lee, Heeyoul Kim, "A new data hiding scheme for binary image authentication with small image distortion" Springer 2011 Procedia Computer Science Information Sciences 179 (2009) 3866–3884
10. Kang LengChiew Josef Pieprzyk, "Binary Image Steganographic Techniques Classification Based on Multi-class Steganalysis" Springer ISPEC 2010, LNCS 6047, pp. 341–358, 2010 Springer-Verlag Berlin Heidelberg 2010
11. Q. Mei, E. K. Wong, and N. Memon, "Data hiding in binary text document," in Proc. SPIE, 2001, vol. 4314, pp. 369–375
12. Y. C.Tseng and H.-K. Pan, "Data hiding in 2-color images," IEEE Trans. Computer., 2002, vol. 51, no. 7, pp. 873–878
13. H. Lu, X. Shi, Y. Q. Shi, A. C. Kot, and L. Chen, "Watermark embedding in DC components of DCT for binary images," in Proc., IEEE Workshop on Multimedia Signal Processing, 2002, pp. 300–303
14. M. Wu and B. Liu, "Data hiding in binary images for authentication and annotation," IEEE Trans. Multimedia, 2004, vol. 6, no. 4, pp. 528–538
15. Chhajed G.J., V.S. Inamdar, V.Z. Attar " Steganography in Black and White Picture Images," 2008 Congress on Image and Signal Processing, CISP 2008, IEEE Computer Society, 978-0-7695-3119-9/08 DOI 10.1109/CISP.2008.626, pp. 141–144
16. Chhajed G.J., Shinde S.A. " Efficient Embedding in B&W picture Images," 2nd IEEE International conference ICIME 2010, 2010, 978-1-4244-5265-1/10, pp. 525–528

17. H. Y. Kim and R. L. de Queiroz, "Alteration-locating authentication watermarking for binary images," in Proc. Int. Workshop Digital Watermarking, 2004, Springer-Verlag Berlin Heidelberg 2005, pp. 125–136
18. H. Yang and A. C. Kot, "Pattern-based date hiding for binary images authentication by connectivity-preserving," IEEE Trans. Multimedia, 2007, vol. 9, no. 3, pp. 475–486
19. Huijuan Yang and Alex C. Kot, "Binary Image Authentication With Tampering Localization by Embedding Cryptographic Signature and Block Identifier," IEEE signal processing letters, 2006, vol. 13, no. 12, pp. 741–744
20. Huijuan Yang, Alex C. Kot, SusantoRahardja, "Orthogonal data embedding for binary images in Morphological Transform domain-A high capacity approach," IEEE transactions on Multimedia, 2008, vol. 10, no. 3, pp. 339–351
21. GuorongXuan, Yun Q. Shi, Peiqi Chai, Xuefeng Tong, JianzhongTeng, Jue Li, "Reversible Binary Image Data Hiding By Run-Length Histogram Modification," International Conference on Pattern Recognition, 2008, 978-1-4244-2175, IEEE., 1651–1653
22. Chhajed G.J., Ms. Krupali Deshmukh, Trupti Kulkarni, "Review on Binary image Steganography and Watermarking," in International journal in computer Science and Engineering (IJCSE), 2011, vol. 3, no.11, pp. 3545–3651, ISSN : 0975-3397
23. Bindu Garg, Rohit Garg, "Enhanced accuracy of fuzzy time series model using ordered weighted aggregation" Elsevier Applied Soft Computing 48 (2016) Elsevier, pp.265–280
24. G.J. Chhajed and B.R. Garg., "Novel Approach to Data Hiding in Binary Images Minimizing Distortion,". In: Smys S., Tavares J.M.R.S., Bestak R., Shi F. (eds) Computational Vision and Bio-Inspired Computing. Advances in Intelligent Systems and Computing, vol 1318. Springer, Singapore. 2021 pp 587–598 https://doi.org/10.1007/978-981-33-6862-0_46, pp.587–598
25. Gyankamal J. Chhajed and Bindu R. Garg. "Applying Decision Tree for Hiding Data in Binary Images for Secure and Secret Information Flow," Cybersecurity Measures for E-Government Frameworks, edited by Noor Zaman, et al., IGI Global, 2022, pp. 175–186. https://doi.org/10.4018/978-1-7998-9624-1.ch011
26. Gyankamal Chhajed and Bindu Garg," Data Hiding in Binary Images for Secret and Secure Communication Using Decision Tree". In: Haldorai, A., Ramu, A., Mohanram, S., Zgheib, R. (eds) 4th EAI International Conference on Big Data Innovation for Sustainable Cognitive Computing. EAI/Springer Innovations in Communication and Computing. Springer, Cham. 2022 pp. 69–84. https://doi.org/10.1007/978-3-031-07654-1_5
27. G. Chhajed and B. Garg, "Information Security by Hiding Data in Binary Images based on Block-Diagonal Partition Pattern", *2022 IEEE International Conference on Blockchain and Distributed Systems Security (ICBDS)*, 2022, pp. 1–6. https://doi.org/10.1109/ICBDS53701.2022.9935958
28. G.J. and B.R. Garg, "Hiding Data in Binary Images Using Block-Diagonal Partition Pattern", International Journal of Computing and Digital Systems (IJCDS), 2023, vol. 13, no. 1, pp. 953–963 (Scopus indexed). https://doi.org/10.12785/ijcds/130176

Chapter 19
Privacy-Enhancing Techniques for Gradients in Federated Machine Learning

Savita Lonare and R. Bhramaramba

19.1 Introduction

Since the last decade, Deep learning techniques have received massive attention due to their capability of prediction accuracy; with enormous data and powerful processors, deep learning is used in various fields. Nowadays, Machine Learning and Deep Learning techniques are used in business decisions and day-to-day life. It is involved heavily, affecting human beings' social behavior, habits, and decision-making. Even artificial intelligence (AI) applications have breached our daily lives. We use Apple voice assistance Siri, Google Assistant, Amazon Alexa voice AI, or Cortana which are very familiar voice assistants. Even the entertainment recommendation engine recommends videos on Netflix, YouTube, etc., as per your surfing behavior. Facebook also recommends friends on other videos by observing your actions and surfing history. Self-learning cars learn and improve by observing the driving behavior of the driver. In the medical industry, deep learning is used for disease prediction using patient's imaging data. Any susceptibilities in these AI systems can be a reason for false predictions and that will definitely result in negotiating its integrity and efficiency.

A traditional approach to applying the deep learning technique is lacking in managing user privacy as all data is collected at the central server before models are trained. Deep learning techniques require massive data, typically collected from the users. The user's privacy is the prime issue in the traditional approach as the

S. Lonare (✉)
Dr. D. Y. Patil Institute of Technology, Pune, India

Research Scholar at GITAM, Visakhapatnam, India

R. Bhramaramba
GITAM, Visakhapatnam, India
e-mail: bravi@gitam.edu

A. Neustein et al. (eds.), *AI, IoT, Big Data and Cloud Computing for Industry 4.0*, Signals and Communication Technology, https://doi.org/10.1007/978-3-031-29713-7_19

user's data may contain sensitive private information. For example, phone call details, chat history, or patients' medical reports. In such a case user or organization might not be keen to share the data with a third party or a machine learning server. Even an attacker can deliberately inject malicious data while the training process or intentionally manipulated the input parameter.

Due to this reason, distributed machine learning gained attention, especially Federated Learning techniques. Federated Learning (FL), proposed by Google in 2016, received widespread attention due to its distributed approach. FL allows the participant to collaboratively learn from a shared model while preserving the privacy of its local data. However, many security threats are still not attended to in FL.

Studies have proven that the shared model can be exploited [1, 2]. The research has shown that the adversary can also leak user privacy from the trained model. The attacker may passively observe the natural model computation or be an active participant in the FL, which knowingly introduces some parameters that affect the calculated weight. Another possibility is that the central aggregator is an adversary who modifies the aggregated model and then shares the polluted model with the participants. Privacy-Enhancing Techniques (PETs) need to be studied and applied in FL to solve these problems. However, there is no single solution to all privacy and security-related issues. Different PETs are used to achieve different aims. Some of them are listed below:

(i) Providing secure access to private data sets
(ii) Aggregated data analysis for the private data sets of several participants (organizations)
(iii) Secure data analysis by a trusted party
(iv) Decentralizing services which depend on user private data

The organizations may select different approaches for PETs which are best suited for them.

This work has used image construction as test-bed for showing FL vulnerabilities as image reconstruction allows one to quickly understand the extent of data that can be revealed by inverting the gradients shared for FL. Not only the textual data but the images such as identity cards, pan card, driving license, vehicle registration cards, medical imaging, and videos such as surveillance recording or any other video contents may leak much sensitive information of user.

19.2 Literature Review

Deep learning techniques have already proved their performance. It is successfully applied in self-driving vehicles [3], medical applications such as health monitoring [4], medical diagnosis [5], and handwriting recognition [6]. These problems used traditional deep learning techniques where huge data is primarily collected from the users and trained centrally. It leads to potential privacy leakage of users' private data. As discussed earlier, FL proposed by Google has turned out to be helpful to keep users' data at local sites only. As data need not be sent elsewhere, data remains private. Compared to traditional centralized deep learning techniques, FL also has some privacy leak threats [2, 7]. Some studies have proven that the privacy of users may be compromised through gradients [1]. Deep learning can even remember the training information [8]. The training data could be revealed to the adversary through the white box access or black-box access to the resultant model [9–11]. To provide privacy and security in FL, researchers have proposed many solutions. In [12], the authors have proposed a differential privacy (DP) scheme to protect user privacy and system robustness against attack. Here authors proposed an algorithm for local central DP against the backdoor attacks. In [13] authors proposed a secure weighted aggregation algorithm for FL. In this scheme, researchers used Homomorphic Encryption (HE) cryptography with a zero-knowledge proof-verification method. HE is again used in [14] to propose privacy-preserving regression.

The PETs provide the techniques to prevent exposure to participants' private data. It has been proven that the advisory may indirectly violate the participant's/user's privacy by gaining access to the personal data and inverting the gradients. Most of the solutions discussed above enlighten the techniques for the privacy of the user's locally trained model or private data. Another solution is to avoid malicious spoofing by the server. The key is to allow each user to authenticate the results returned by the aggregation server effectively.

The remaining chapter is organized as follows: Sect. 19.3 gives the architectural details of FL. Section 19.4 elaborates on experiments and results. Section 19.5 suggests security and privacy solutions for FL.

19.3 FL Architecture

Figure 19.1 shows the typical architecture of FL. It consists of two main components. The first component, i.e., a central server responsible for aggregating local models received from the participants/Clients and later sharing aggregated models, also called the global model, back to the participants. The second component is clients or participants responsible for training the local model with local private data and sharing the locally trained model with the server. In Fig. 19.1, step 1 denotes the initial communication from a central server to participants for sharing the initial global model. After receiving the initial global model, participants update the global model with locally available private data. This updated local model was later shared with the central server in step 2. On the server-side, all received local models from the participants are averaged using FedAverage Algorithm [15] to update the server-side global model. And this global model is now shared with participants.

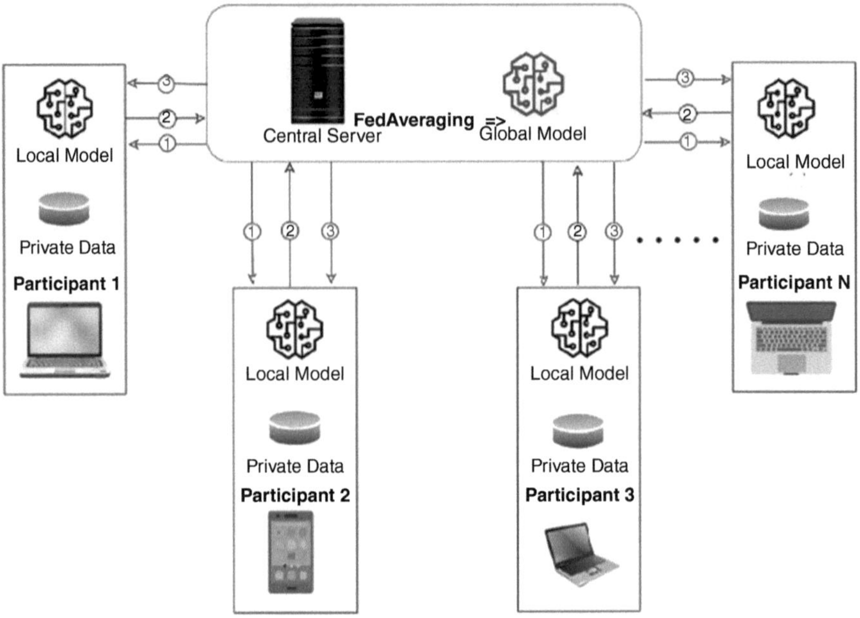

Fig. 19.1 Typical Architecture of FL

The ultimate aim of a machine learning model is to minimize loss L and optimize the parameters θ using training data consisting of feature $x_{i \text{ and}}$ respective label y_i to solve the following equation-

$$\min \sum_{k=0}^{n} L_{\theta}\left(x_{i,} y_i\right) \tag{19.1}$$

When the models are trained using the FL setting, the server solves Eq. 19.1 using multiple gradients received from various users. Participants/ Clients in FL share only the gradients instead of original data i.e.(x_i, y_i). The gradients were then averaged by server given by

$$\theta^{k+1} = \theta^{k} - \frac{1}{n}\sum_{i=1}^{n} \nabla_{\theta} L_{\theta} k\left(x_i, y_i\right) \tag{19.2}$$

19.4 Experiments and Results

This work proves that FL does not guarantee the privacy of the client's data, and the gradients can be inverted to recover original data. Here the server is considered honest but curious, interested in revealing the participant's private data.

Fig. 19.2 Image reconstruction from gradients output trained by Resnet152

Table 19.1 Peak signal-to-noise ratio (PSNR) for FL

One epoch, batch size of 2	Four images	Eight images
	PSNR – 15.66 db	PSNR – 15.95 db
Four epoch, batch size of 8	One image	Eight images
	PSNR – 17.05 db	PSNR – 18.45 db

The server cannot interfere with the FL process other than averaging the gradients. This work used the pre-trained image classification model Resnet18, trained on the ImageNet dataset. Imagenet dataset has 1000 different images. Instead of the recovery mechanism given in [16, 17], which uses optimization using the Euclidean distance formula, this work used the noise removal algorithm presented in [18].

Figure 19.2 shows the top row consisting of the ground truth, and the bottom row shows the reconstructed images. It is visible that the last image reconstruction has failed compared to the other three images. The experiments also show that reconstruction is highly dependent on the content of the images.

Table 19.1 shows the Peak Signal-To-Noise Ratio (PSNR) for the FL average on the CIFAR-10 validation data set. Experiments are done for reconstructing the different number of images and epochs. The experiment proves that as we increase epoch and batch size, the quality of recovered images improves.

There are very few techniques available to provide security to the gradient. Although there is no particular technique available for proving privacy and security to private data, one can go for any following PET. The next section provides a brief introduction to PET which may be used for privacy preservation.

19.5 Privacy-Enhancing Techniques for FL

19.5.1 Secure Multiparty Computation (SMC)

SMC was originally introduced to provide secure communication between multiple parties, where all parties are supposed to compute an output [19] jointly. SMC uses cryptographic techniques for secure and protected communication in multi-party communication systems. Recently SMC has been used in the FL framework for securely updating the weights received from participants. High computation efficiency is achieved in FL compared to traditional SMC as in FL. Only local weights need to be encrypted compared to the high volume of actual input data. High performance is why SMC is the most preferred technique in FL for privacy preservation. There is still a possibility that the private weights calculated by participants may leak information at the model aggregator server. To make SMC more robust, encryption can be combined with the asynchronous stochastic gradient descent to prevent participants' private data leakage on the server-side [20]. The work [21] combines two privacy-preserving schemes to lessen the risk of revealing a client's private data. The key technique they used is HE, and the other is differential privacy. Here authors have suggested applying differential privacy at the client level and encrypting the updated models. The scheme has proven against the honest but curious central server and other participants in the FL system. The advantages of using SMC in FL are that it is a commercially ready and proven technique for privacy preservation. As the clients never transfer the actual data outside their internal firewalls, no Trusted Third Parties (TTP) can view the private data. There is no need to remove or mask any data parameter to preserve privacy. It fulfills the requirements of the General Data Protection Regulation (GDPR) and sovereign data privacy agreement. It is proven that SMC achieves high accuracy [19–21]. Computation overhead and high communication costs are the limitations of SMC.

19.5.2 Differential Privacy Preservation (DPP)

DPP is an extensively used privacy preservation technique used in FL. The idea behind DP is to add noise to private data to preserve privacy of the data. The noise introduced in user data is very low to maintain the data quality. In FL, DPP avoids the inverse data retrieval by the advisory. In [21], the authors proposed a framework to use DPP for private Generative Adversarial networks (GAN). In work [22], the authors came up with a DPP scheme for the FL environment. The researchers also tried to improve the privacy of FL by combining DPP with the making of user data using the invisibility cloak algorithm. The disadvantage of this scheme is that it may reduce the overall system's performance.

19.5.3 Homomorphic Encryption (HE)

HE is a type of encryption that allows some computation on the ciphertext. HE allows calculations on ciphertext to generate results in encrypted form. The encrypted data need not be decrypted for applying measures using HE. The specialty of HE is that the operation results applied on the ciphertext exactly match the operation results on the plaintext. It is used in applications where computations need to be outsourced on sensitive data. HE can be used where the computational environment is not a trusted party and calculations are relatively simple. HE gives confidentiality of data and security against exposure to sensitive data. There are various versions of HE like Fully HE (FHE), Somewhat HE (SHE), Partially HE (PHE), etc. In [23], the authors proposed the HE scheme, BatchCrypt for Cross silo FL. Another work [24] also gives an FL scheme with HE to provide user data privacy.

19.5.4 Trusted Execution Environments (TEE)

In TEE, some area inside the central processor is secured to provide a trusted environment. The operating system or hypervisor cannot read the code and the data present in TEE. On the other hand, TEE can read the external memory. TEE might outsource the computations on the sensitive data to ensure that the server does not expose the data and code. TEE supports various types of analysis with lower costs as the calculations are performed on actual data. TEE is functional in handling data security and data exposure problem of private data. Like other existing cryptographic systems, Protecting secure keys is still a challenge in TEE. In [25], the authors experimented with secure coprocessors. Here authors validated trusted devices for the trusted execution environment. In [26], a collection of hardware modules is abstractly partitioned to create a secure computing area.

The PETs are not limited to the techniques discussed in this paper. But there are many areas where we may accelerate the research for the development of PETs. There is a need to promote research and development of innovations in PETs for FL techniques.

19.6 Conclusion and Future Scope

Privacy and security of data are the need in modern life. Although actual data is not shared in FL, PETs need to provide security to the local model and avoid information retrieval from the shared model. This paper shows how gradients can be inverted to recover the data. This paper explains how the different PETs may be used to improve the security level in FL. PETs are not considered an add-on feature; instead, it is one of the layers of information security culture. Various methods of PETs such as homomorphic encryption, differential privacy, secure multiparty computation, Private Aggregation of Teacher Ensembles (PATE), and the combination of these techniques need to explore possible applications in FL to make distributed machine learning more robust against privacy and security threats.

References

1. Milad Nasr, Reza Shokri, Amir Houmansadr, Comprehensive Privacy Analysis of Deep Learning [J], 2019 IEEE Symposium. Security. Privacy., pp. 739–753, 2020.
2. Hao Ren, Hongwei Li, Yuanshun Dai, Kan Yang, Xiaodong Lin, "Querying in Internet of Things with Privacy Preserving: Challenges, Solutions and Opportunities," IEEE Netw., vol. 32, no. 6, pp. 144–151, 2018, https://doi.org/10.1109/MNET.2018.1700374.
3. Ana I. Maqueda, Antonio Loquercio, Guillermo Gallego, Narciso Garcia, and D. Scaramuzza, "Event-Based Vision Meets Deep Learning on Steering Prediction for Self-Driving Cars," *Proc. IEEE Comput. Soc. Conf. Comput. Vis. Pattern Recognit.*, no. Dl, pp. 5419–5427, 2018, https://doi.org/10.1109/CVPR.2018.00568.
4. Qi Jianga, Jianfeng Maa, Chao Yanga, Xindi Maa, Jian Shenb, Shehzad Ashraf Chaudhry, "Efficient end-to-end authentication protocol for wearable health monitoring systems," Comput. Electr. Eng., vol. 63, pp. 182–195, 2017, https://doi.org/10.1016/j.compeleceng.2017.03.016.
5. Xiaoxia Liu, Hui Zhu, Rongxing Lu, Hui Li, "Efficient privacy-preserving online medical primary diagnosis scheme on naive bayesian classification," *Peer-to-Peer Netw. Appl.*, vol. 11, no. 2, pp. 334–347, 2018, https://doi.org/10.1007/s12083-016-0506-8.
6. Pastor-Pellicer, Joan, Castro-Bleda, María José, España-Boquera, Salvador, Zamora-Martínez, Francisco, "Handwriting recognition by using deep learning to extract meaningful features," *AI Commun.*, vol. 32, no. 2, pp. 101–112, 2019, https://doi.org/10.3233/AIC-170562.
7. Guowen Xu, Hongwei Li, Hao Ren, Kan Yang, Robert H. Deng, "Data Security Issues in Deep Learning: Attacks, Countermeasures, and Opportunities," *IEEE Commun. Mag.*, vol. 57, no. 11, pp. 116–122, 2019.
8. Congzheng Song,Thomas Ristenpart, Vitaly Shmatikov, "Machine Learning Models that Remember Too Much," *arXiv*, pp. 587–601, 2017.
9. Yixian Yang, Xinxin Niu, Lixiang Li, H. Peng, J. Ren, H. Qi, "General Theory of security and a study of hacker's behavior in big data era," *Peer-to-Peer Netw. Appl.*, vol. 11, no. 2, pp. 210–219, 2018,
10. Shenmin Zhang, Hongwei Li, Yuanshun Dai, Jin Li, Miao He, Rongxing Lu, "Verifiable Outsourcing Computation for Matrix Multiplication with Improved Efficiency and Applicability," *IEEE Internet Things J.*, vol. 5, no. 6, pp. 5076–5088, 2018.
11. M. Nasr, R. Shokri, and A. Houmansadr, "Comprehensive privacy analysis of deep learning: Passive and active white-box inference attacks against centralized and federated learning," *Proc. - IEEE Symposium Securit Privacy.*, vol. 2019-May, pp. 739–753, 2019
12. Mohammad Naseri,Jamie Hayes,Emiliano De Cristofaro, "Toward robustness and privacy in federated learning: Experimenting with local and central differential privacy," *arXiv*, 2020.
13. Jiale Guo,Ziyao Liu, Kwok-Yan Lam,Jun Zhao,Yiqiang Chen,Chaoping Xing, "Secure Weighted Aggregation in Federated Learning," October, 2020.

14. Irene Giacomelli, Somesh Jha, Marc Joye, C. David Page, Kyonghwan Yoon, "Privacy-preserving ridge regression with only linearly-homomorphic encryption," *Lecture Notes Computer Scicence* vol. 10892 LNCS, pp. 243–261, 2018, https://doi.org/10.1007/978-3-319-93387-0_13.
15. H. Brendan McMahan, Eider Moore, Daniel Ramage, Seth Hampson, and Blaise Aguera y Arcas, "Communication-efficient learning of deep networks from decentralized data," *Proc. 20th International Conference on Artificial Inteligence. Statistics. 2017*, vol. 54, 2017.
16. Zhibo Wang, Mengkai Song, Zhifei Zhang, Yang Song, Q. Wang, H. Qi, "Beyond Inferring Class Representatives: User-Level Privacy Leakage from Federated Learning," *Proc. - IEEE INFOCOM*, vol. 2019-April, pp. 2512–2520, 2019, https://doi.org/10.1109/INFOCOM.2019.8737416.
17. Bo Zhao, Konda Reddy Mopuri, Hakan Bilen, "iDLG: Improved Deep Leakage from Gradients," *arXiv*, pp. 1–5, 2020.
18. L. I. Rudin, S. Osher, and E. Fatemi, "Tv_Model.Pdf," *Phys. D*, vol. 60, no. 1–4, pp. 259–268, 1992,
19. Ran Canetti,Uri Feige,Oded Goldreich, Moni Naor, " Adaptively secure multi-party computation, In Proceedings of the twenty-eighth annual ACM symposium on Theory of Computing, pp. 639–648, 1996.
20. Le Trieu Phong, Yoshinori Aono, Takuya Hayashi, Lihua Wang, Shiho Moriai, "Privacy-Preserving Deep Learning via Additively Homomorphic Encryption," *IEEE Trans. Inf. Forensics Secur.*, vol. 13, no. 5, pp. 1333–1345, 2018, https://doi.org/10.1109/TIFS.2017.2787987.
21. L. Xie, K. Lin, S. Wang, F. Wang, and J. Zhou, "Differentially private generative adversarial network," *arXiv*, 2018.
22. S. Augenstein *et al.*, "Generative models for effective ml on private, decentralized datasets," *arXiv*, 2019.
23. Chengliang Zhang, Suyi Li, Junzhe Xia, Wei Wang, Feng Yan, Y. Liu, "BatchCrypt: Efficient homomorphic encryption for cross-silo federated learning," *Proc. 2020 USENIX Annu. Tech. Conf. ATC 2020*, pp. 493–506, 2020.
24. Ou Wei, Zeng Jianhuan, Guo Zijun,Yan Wanqin, Liu Dingwan,Fuentes Stelios, "A homomorphic-encryption-based vertical federated learning scheme for risk management," *Comput. Sci. Inf. Syst.*, vol. 17, no. 3, pp. 819–834, 2020, https://doi.org/10.2298/CSIS190923022O.
25. Sean Smith, Ron Perez, Steve Weingart, Vernon Austel, "Validating a High-Performance, Programmable Secure Coprocessor Secure Systems and Smart Cards," *Information System Security*, October, pp. 1–11, 1999.
26. T. Alves and D. Felton. "TrustZone: Integrated hardware and software security" Tech. In-Depth 3, 4 (2004), 18–24 "Trustzone," pp. 18-, 2004.

Savita Lonare received a master's degree in Computer Networks from SPPU, Pune, India. She is a research scholar at GITAM, Vishakhapatnam. She is working with Dr. D. Y. Patil Institute of Technology, Pune as an Assistant Professor. She has filed two patents in India and ten research publications. Her research interest includes Machine Learning and Deep Learning.

Dr. R. Bhramaramba received Ph.D. from India and is a Professor at GITAM, Vishakhapatnam. She has published over 50 research papers. Her research interests include Bioinformatics, Data Mining, Machine Learning.

Part V
Software Language Implementation, Linguistics, and Virtual Machines

Chapter 20
Multi-component Interoperability and Virtual Machines: Examples from Architecture, Engineering, Cyber-Physical Networks, and Geographic Information Systems

Nathaniel Christen and Amy Neustein

20.1 Introduction

A central theme of the following chapters will be computer programming languages, seen as practical technologies that have their own theoretical orientations and engineering requirements. One topic will be the relation between these artificial languages and natural (human) language. Another will be the relation between computer languages and various facets of "information representation"—for data modeling, designing data-sharing protocols, or database engineering. There is a connection between these themes insofar as data models constrain data structures, which supply the values on which programming languages operate. In other words, data models are an essential aspect of programming language "semantics."

As a preliminary to further discussion of these subjects, we will start with a review of computer languages as technical artifacts. Perhaps it is not quite accurate, in fact, to describe computer programming languages as "languages." This is partly because natural language evolves organically, whereas programming languages are *created* (usually "by committee"). They evolve in response to engineering challenges—in particular, those concerned with getting computers to "understand" high-level source code in lieu of physically realizable "machine" code. There are still some processing stages or semantic/syntactic formulations where one can make comparisons or even borrow formal models from one domain to the other, but in general, we should be careful about applying intuitions about natural language to programming languages too broadly.

For example, people do not use source code to "talk to" or even generically "communicate with" computers. Rather, computer code acts as the material through

N. Christen · A. Neustein (✉)
Linguistic Technology Systems, Fort Lee, NJ, USA
e-mail: amy.neustein@verizon.net

A. Neustein et al. (eds.), *AI, IoT, Big Data and Cloud Computing for Industry 4.0*, Signals and Communication Technology, https://doi.org/10.1007/978-3-031-29713-7_20

which programmers implement applications, which are tools pressed into service to manipulate computational states. More specifically, applications in general do not have default behaviors—instead, they are essentially computational environments, which do not initiate actions until guided by a human user.[1] Once the user *interacts* with an application, the application should have functionality to respond to that user's input. Typically, this response will be processed, and then the application will return to a passive state, awaiting further user actions. The passive–active–passive cycle is often called an "event loop."

Users typically interact with applications because they want to change something. At a minimum, they want to change what they *see*, while the program is running, and/or to modify some data somewhere. They might, for example, seek to save, upload, or download a file. For conventional computer systems (personal desktop/laptops, tablets, smart phones, and even *en situ* touchscreens built into industrial hardware and appliances), almost all information which a user acquires from applications is *visual* and *two-dimensional* (although a 2D screen might show a simulation of 3D graphics).[2] In this sense, the behavior which a user expects from their application is intimately connected with what they "see."

Fundamentally, then, the goal of an application's code base when responding to user input is to modify the on-screen display that users consume visually—even when the primary response is some side effect, such as saving a file, the display should have a change indicating that the operation was successful (or, if not, alerting the user that there was a problem).

These points might seem obvious, but they are worth emphasizing for any discussion where we examine the "semantics" of computer code and application data. Since the emergence of the modern Internet and World Wide Web, a lot of emphasis in computer science has concentrated on data *semantics*, or on how digital artifacts can model/represent empirical things in the world. The goal of viewing and/or analyzing data structures as a proxy for manipulating actual "things" underlies diverse branches of Information Technology, from object-oriented pro-

[1] Granted, there is a class of programs that run behind-the-scenes following a preprogrammed sequence, constructed with the expectation that human users will not interact with the program during its execution. Even here, though, we can argue that the choice to initiate the program in the first place constitutes a human action, so such programs do have at least *one* point where they are dependent on users' input. The choice of *when* to run a program will (potentially) influence its behavior, so this choice (which depends on a user's discretion) acts as a kind of parameter which users put in place to guide the program's actions. This holds even (or, indeed, especially) when users arrange for a program to run automatically (typically when the computer first starts up), such that they do not manually interact with that program *each time* it executes. Instructing the operating system to launch a program automatically is merely a convenient way for users to ensure that the program runs at their desired moments. Given these points, it is reasonable to say that *all* programs depend on *some* input from users, even if such input is restricted to a decision about when the program should (or should not) execute.

[2] To this, we can occasionally add audio feedback, though—setting aside interfaces explicitly designed for the hearing-impaired—audio is rarely an important data source except when playing an audio or audio/video file.

gramming to database engineering to the Semantic Web. But digital resources only model real-world phenomena at one step removed: when we talk about computational representations of external (either concrete or abstract) things, we are actually referring to capabilities for creating *visual* displays that convey information about those objects to human users, according to signifying conventions that users understand. If, say, a user wishes to learn the current score of a World Cup match, the application will perform correctly if it changes its visible state to show this data in a standardized pattern, such as a number (indicating goals scored) next to an Argentina logo (perhaps an icon of the Argentine flag) and an analogous number next to a France logo/flag.

There is nothing in the application's state in this example which actually "means" anything about Argentina, France, football, goals, or anything else as users understand it. Any "semantics" here is merely the result of adequate engineering. If the application behaves properly, the number next to the France logo will match the number of goals France has scored so far. We cannot say that the number visible on-screen is actually a "signifier" for the France goal total, because even a well-designed application could be incorrect (consider Internet latency, making it impossible for the software to update the displayed goal total in exactly "real-time"). We can only say that a properly implemented application will be a useful source of information about the soccer game because it is engineered to satisfy the requirement that, on most occasions, its visual interface will match the state we associate with correct information about the game. The "meaning" of data displayed by an application depends on an engineered correlation between a data source and "ground truth" facts; but such a correlation is something that has to be artifactually constructed by programmers and database administrators. This kind of semantics is very different from natural language word–object associations (that evolve organically via entrenchment among a given dialect community).

As a result, when considering how to formalize and evaluate programing language "semantics," we should not work in a paradigm where data structures are innate proxies for real-world concepts, the way that linguistic signifiers are communally recognized expressions of their correlated signifieds. Instead, data structures offer *tools* which engineers leverage to assemble pieces of information that on aggregate simulate or describe external objects with varying degrees of detail. Different contexts allow for different data structures that are molded to empirical phenomena, more or less selectively. To provide the score for a cricket match, for example, unlike a football score, one would need to represent the wickets taken for each side as well as their runs scored; thus a two-valued structure (for football) would be replaced by a four-valued one (or, equivalently, two number pairs). Of course, much more info could be modeled for any sporting event (time elapsed, for many sports, or the number of remaining balls in a T20 innings, or where the game takes place, or the stats on each player, and so on).

How much of this data is actually curated by any given piece of software depends on decisions made while the software is designed. Data structures can indirectly represent empirical concepts because they provide programmers with a tableau of representational tactics—datatypes and models, numerical encodings, units of measurement, valid ranges and other quantitative constraints, etc.—which allow data structures to track real-world objects by selectively sampling some of their properties. Even then, these structures are not intrinsically representations *of* their intended targets. Rather, applications may convert data structures into visual displays that stand in for objects by representing their properties textually or graphically (the former being, for instance, a printed number asserting goals/runs scored; the latter being a visually conveyed magnitude, such as a bar's height or color's hue to present weather temperatures).

Applications, in short, *simulate* semantics by curating data structures that are engineered to track real-world concepts *and* subsequently translating these structures into user-visible displays. A competent programmer must ensure that software performs these tasks correctly. Computer code manipulates applications' internal data structures and external displays. In this sense, computers do not "understand" code the way that people understand language. An application does not reach a desired state given user input by treating code as a kind of message or communiqué that it has to interpret. Instead, computers are engineered (via compilers and runtimes) so that source code can be employed as an indirect tool to predictably reconfigure applications' internal and external state such that the latter correctly conveys certain kinds of information.

Needless to say, the machinery for engineering the proper synergy between external facts, internal data structures, and visible displays is vastly different than the cognitive architectonics of people hearing (or reading) language. This is one reason why we implied that the expression "programming *language*" is a misnomer. Nonetheless, we believe that in some sense (natural) linguistics can benefit from analyzing programming languages, precisely because of their mechanical and artificial nature.

Specifically, much of the subtlety of natural language derives from interpretation and disambiguation. Almost all words have multiple meanings, and almost all sentences can be parsed in different ways. Addressees must therefore infer which construal of parse and word use is most likely to match the speaker's intent. There is nothing analogous to such hermeneutics in programming: software language implementations do not "guess" at meanings; they can only rely on precisely formulated rules to deduce which operations are designated by specific segments of source code.

Consequently, programming languages are much blunter and less expressive than human languages. But by discarding issues of interpretation and "most likely" meanings, computer code enables us to study how structural manipulations can be encoded in symbol systems that at least superficially *resemble* human language.[3] The chains of computation that extend from source code as written to visible and information-carrying display changes in applications' user interface are guided by

[3] In the sense that source code is composed with the same letters and punctuation marks that are the basis of natural writing systems; moreover, many symbol names and keywords in computer code are chosen to suggest words in natural languages (especially English).

formally describable systems that potentially have some correlates to how humans "process" language—not in the sense of how we interpret the meaning of sentences initially, but rather in terms of how we follow up. According to the *pragmatics* of speech acts, the "meaning" of a sentence lies in how cooperative listeners should respond (not just linguistically, but also potentially via concrete actions). This is analogous, at some level, to how software applications' semantics are determined by how they respond to user actions, making "correct" changes (i.e., those users anticipated and intended) to application state.

We have said that an application's external state is primarily manifested in its visual displays. This is true for conventional software, although we can envision technology evolving so that the 2D digital screen is not the only medium through which programs can show information. Perhaps 3D printing or even holograms will break through the limits of two dimensions. Virtual reality environments might increasingly incorporate kinesthetic, haptic/tactile, audio, and even olfactory qualia to complement visible scenes [24, 34, 36, 41, 42, 47, 55, 57]. And, of course, robotics introduces an entirely different set of dimensions for external configurations: a robot's external state is not only defined by what is visible on a display screen (if it has one), but on the position of the robot relative to its surrounding environment and all the degrees of freedom for its gears, joints, wheels, and other moving parts. In this way, robots can be programmed to adopt postures that physically alter their surroundings, by lifting and carrying objects for example. Such mechanical configurations are analogous to the purely visual configurations evinced by software whose effects though are mostly localized to a two-dimensional touchscreen or monitor. Nonetheless, the same internal architecture is in effect: robots are *caused* to take on their useful positions because internal data structures are engineered to guide external configurations according to desired outcomes, which is the same principle as old-school software wherein the "desired" outcome can be fully expressed through the visual content present at any moment on a computer screen.

The engineered causation between internal data and external configuration, then—be this visual (for most computational environments) or mechanical (for robots) or something more multi-sensory (for virtual reality)—is in any case the essential detail for analyzing computer technology: that is, for improving the science of Information Technology in ways that could make software more reliable. Although a lot of computer programming is done in an experimental, trial-and-error fashion—consider the ecosystem of bug reports, bug fixes, security patches, and other incremental improvements that are part of any application's life cycle—there is still room for examining software development (and programming languages) from a "theoretical" perspective. Note that ideas about effective computer code and application development have evolved noticeably during the "personal computing" era: conventions for conveying information through interactive visible displays are refined over time, yielding software that is progressively more intuitive and informative. Meanwhile, technologies for implementing computer languages themselves—compilers to translate source code into machine language and runtime environments to serve as containers where programs execute—have become more sophisticated and allow programming languages themselves to change over time, which in turn (in principle) makes software both less costly to implement and more trustworthy and secure.

Although physical improvements in the underlying hardware—ever-more-powerful silicon chips, say—have obviously fueled the power of modern computers (even quotidian ones owned by the general public), the evolution of programming-language and software-development methodology has not been driven by hard science *per se*, unlike advances in chip design, solid-state hard drives, or multi-core processors; but rather by relatively abstract considerations around the nature of parsers, compilers, data structures, virtual machines, and other accoutrements in the ecosystem of language implementation. In effect, software language engineering is almost philosophical, in that progress can be made through relatively abstract investigations (without empirical or physical substrata), but not in the sense of academic speculation without concrete payoffs: better computer languages means better software.

This is therefore the practical backstop surrounding chapters in Part 5 of this book, which will focus on issues of language implementation and virtual machines (VMs). Almost all modern programming languages invoke VMs at some stage in code processing, even if this is done prior to an application being executed (one way to gather compiler-related data about an application is to "virtually" simulate running code, so virtual machines can come into play during compilation even if the compiler's target is low-level machine code).[4] All computer code is eventually translated to "machine language"; i.e., to instructions that can be *physically* realized within a computer's central processing unit (CPU). A "virtual" machine, by contrast, is not a physical device that performs calculations, but rather a software system that emulates the behavior of such devices, to some approximation.[5] As a result, VMs are not limited to the data structures and instruction sets that can be physically realized in a CPU (or analogous settings, such as a graphics processing unit or a quantum processor): because VMs are software, not hardware, they can take on almost any set of kernel operations and built-in datatypes that can be programmed within the software through which the VMs are themselves implemented. In this sense, VMs are a valuable environment for analyzing programming language design and implementation because they are a flexible and adaptable environment for parsing, compiling, and executing high-level computer code.

For reasons sketched in this introduction, here we consider VMs especially in the context of how applications link internal data structures to external (mostly

[4] We are thinking of something like LLVM (low-level virtual machine) as a foundation for the C++ Clang compiler [33]. Even if it is not entirely accurate to speak of LLVM "virtually executing" C++ code during compilation, this is not a bad way to visualize compiler details in a relatively informal context. For example, type-checking the outputs of a procedure once specific types are assigned to its input parameters can be construed as an approximate "running" of the procedure where we abstract from the *specific* input values and consider only their types.

[5] Virtual machines typically model a wider range of operations than can be computed *in silico* directly; but such higher-level operations themselves are then destructured into sequences of lower-level machine language steps.

visual display) configurations—which we have argued is *the* central concern of software engineering in general. That is, we can use VMs to analyze systematically the processes through which *source code* formalizes digital machinery such that *user actions* are responded to by manipulating *internal data* resulting in changes to *visible displays*. Each of the scaffoldings connecting these four principal elements (source code, user actions, internal data structures, and visible displays) is governed by overlapping sets of conventions, specifications, and computational requirements; modeling the full space of their interoperation via virtual machines is a good way to make theoretical sense of the complex interworkings that come into play once an application is up and running.

20.2 Hypergraph Data Modeling

We claimed in the introduction that a software application's primary role is to await user actions and then, in response, produce some visual change (in the on-screen display) that presents information to the user (presumably information relevant to their original action), along with possible side effects (like saving a file). On this account, studies of applications' behavior would reasonably focus on the starting and ending points of these action/reaction cycles: how do we articulate the full space of possibilities for users to initiate actions within the application? For mouse-and-keyboard desktop setups, an action like "left click" on the mouse can be performed at any point in an application's window (or windows), and mouse events could click one of several buttons, or turn the scroll wheel, and they can be modified by pressing certain keys. In some contexts, likewise, users enter data by typing on a keyboard.

Applications have to systematically model each of these possibilities to create data structures noting all relevant details about user actions, which in turn are needed for responding correctly (left-click and right-click usually have different "meanings," for example). At the other end of the user-action/application-response cycle, applications need to subdivide their display window into parcels of screen space so that data updates are located in useful visual contexts (continuing the above example, the France score should be placed near the French flag/logo, and likewise for Argentina). This, of course, means that the computer screen is not just a "picture." It is an organized system analyzed through data models and types. And such analyses may potentially be extended to robots and/or virtual reality, which are not just 2D screens but possess 3D physical or visual-immersive (and maybe synesthetic) configurations.

Well-designed applications have multiple features and are easy or intuitive to use: this means that users have a variety of options for leveraging applications' capabilities, and that it is not difficult for users to learn or infer the steps needed to prompt their application toward their desired course of action. Applications can generally become "better"—more featureful and intuitive—by being flexible and subject to continuous improvement: acquiring more functionality over time and updating their interface-based users' experience (they may find different features easier or harder to access). For example, specific operations might be exposed in different ways, in that the user has multiple options for initiating certain responses (e.g., the effects of clicking on a certain button might be duplicated by pressing a specific key on the keyboard, or by activating the mouse's scroll wheel—consider scrolling down, hitting an on-screen "down" button, or pressing the down-arrow key).

Many applications also allow plugins or extensions to adjust the visual displays in ways not envisioned by the program's original designers.[6] This flexibility helps ensure they can improve and become more user-friendly, but more specifically these are engineering goals: there are steps which developers might take to maximize applications' ability to adapt and evolve.

Aside from a generic desire to build "quality" software, engineering applications for future adaptation also prompts developers to design software in rigorous, well-documented ways. Applications can be thought of, rather metaphorically, as virtual/digital "piping" linking a tableau of potential user actions (mouse, keyboard, etc.) to a space of possible screen configurations. A software engineer's goal is to build up this intermediate "plumbing" in an orderly rather than haphazard manner, so that future programmers have a clear picture of how the application's functionality can be augmented (without affecting its current behavior, unless a conscious decision is made to alter the application's pragmatics based on user feedback). Users initiate actions so as to invoke application features: to organize responses systematically, it is important to have an efficient structural breakdown of all the functionality which the application makes available, by analogy to how books might be organized on the shelves of a library. But this structuring principle also promotes extensibility—once a system is in place to enumerate functionality, it is easier to add new features because there is greater transparency in how new features can "slot in" alongside their predecessors. Similar points apply to the visual display of information: the more systematically we model how datatypes recognized by an application are translated to visual form, the more readily we can build new display units—such as isolated windows or dialog boxes—to display *new* datatypes, formulated in the course of providing new functionality.

Planning for continuous adaptation therefore implies that applications should deliberatively model the functionality they *expose* as features available to users: improvements can either provide alternative (perhaps more user-friendly) ways to invoke these same features, or to create new features that fit comfortably with the existing application by honoring its protocols and user-pragmatic conventions. As such, an important dimension of software design is formulating the protocols and concrete implementations through which applications "expose" functionality.

---

[6] We adopt the convention that a "program" is a self-contained software component, something that can be executed, and an "application" is one kind of program: one with a visible user interface. Most programs are applications, on these definitions (when discussing technical VM matters, sometimes, however, we can use the word "program" in a different sense, as a self-contained sequence of VM operations).

Consider the case of image processing (which will be the focus of Chap. 22). Computer Vision software will typically have many built-in functions for transforming and/or extracting data from images (or videos). Exposing these capabilities means allowing users to initiate image-processing workflows. The result of any workflow might be a modified version of the original image, or some extracted data structure that can be outlined or visualized. Because workflows are often chained together, "exposing" functions also involves software invoking procedures dynamically, so that although a user will manually initiate an overarching functionality, there may be many behind-the-scenes intermediary stages which users would not directly perceive. In this kind of environment, applications will typically offer an API (Application Programmable Interface) and/or an ABI (Application Binary Interface, which is more low level) such that important features can be accessed both directly through user actions *and* indirectly as intermediate processes launched in response to some higher-level user action.[7] Moreover, users need some way to view or study the results of Computer Vision workflows, so applications should expose functionality for altering their visual display in accord with the details of specific pipelines (e.g., presenting a new image in a new window). In short, applications should expose both analytic/processing functionality and display/visualization functionality.

To "expose" functionality is to offer multiple ways that operations may be accessed: by users directly; by engines that step through workflows; by application "macros," which are (for some programs) programmable sequences of user actions that typically are performed together (linking multiple actions into one group, initiated with one single action); by "updates" that can add new features without software needing to be re-installed; by "personalization" features that allow programs to enter different states depending on the identity of their current user; by configuration files; and by "scripting" environments where code written in a programming language (typically one with an interpreter that can be embedded in standalone software, such as LISP or PYTHON) can modify or fine-tune application state and performance. Supporting such personalization and fine-tuning is itself a feature that makes software more user-friendly, but (for reasons we mentioned vis-à-vis rigorous design) also prompts developers to architect software in well-organized ways, with a logical model of user actions, visual displays, and all the coding that connects the one to the other.

An important dimension of application development, accordingly, is that of modeling and formalizing how applications expose functionality and user interface details. Virtual machines, in turn, are a useful tool for analyzing the features exposed by applications (and how they are implementationally tied to the application as a whole). This is the primary scenario where we will consider VM designs as theoretical artifacts in their own right.

---

[7] We mention ABIs here in the sense that application procedures might be exposed to be called directly with raw data values, in contrast to what we will later call "meta-procedures" whose inputs and outputs are padded with extra layers of indirection.

20.2.1 Hypergraphs as General-Purpose Data Models

Ultimately, all of a program's functionality is achieved via procedures which are provided by the software code base (perhaps along with third-party libraries that get re-used, sometimes as high-level source code and sometimes as compiled assets). As such, one way to "expose" functionality is simply to allow external code (that is, code that is not explicitly developed as part of an application's core code base) to call procedures that *are* in the code base. The vast majority of such procedures, however, are "internal," intermediate computations that would not be meaningful or appropriate for external access. Accordingly, applications need a more overarching model of exposed features to ensure that external code utilizes internal capabilities in an orderly fashion.

In general, functionality is exposed to outside components by constructing a specific procedure whose role is to be an "entry point" for externally invoked features. That is, we assume that any exposed "functionality" requires multiple procedures to be realized, and one entry point is necessary to invoke a larger package of procedures that collectively implement an operation that we can *conceptually* regard as "one" function. For example, if an exposed capability (in a Computer Vision program) is running a superpixel segmentation on a 2D image via color-watershed, there will be many intermediate procedures necessary to complete that computation, but we can reason about the segmentation as one identifiable step in an image-processing pipeline. It is a *conceptual* functional unit even if it is not "one" function, in the sense of one source-code procedure. We propose to use the term "meta-procedure" to describe functions that are, in this sense, conceptually singular but implementationally multiple.

Application-level capabilities often require certain details to be specified—continuing the superpixel example, such an operation would need to know which image to target, and potentially would require certain threshold parameters (superpixels are relatively small image regions of similar color, but "how" small and similar depends on input values that might vary from one run to another). The "output" of such a segmentation would be a data structure (typically a "labeling" image, which adds an extra number to each pixel—alongside its existing color channels, such as red, green, and blue—identifying the superpixel to which it belongs); for visualization, sometimes superpixel segmentations will also be displayed by outlining the superpixels with a colored boundary (visually distinct from the underlying image). Thus a superpixel algorithm has inputs and outputs analogous to a single procedure. However, insofar as such an algorithm is not one procedure but rather (in our terminology) a "meta" procedure, these inputs and outputs are not delivered simply as binary data values (memory addresses or CPU registers) but rather must be marshaled and encoded between the software component that invokes the meta-procedure and the application that exposes it.

Unlike the execution sequence *within* one component, where one procedure can call other procedures directly, for such "meta" procedures there is a series of steps wherein input parameters are built up via a common protocol which both components can understand, and output results likewise encoded in reverse. The procedure/meta-procedure contrast is analogous to the difference between a face-to-face conversation between two parties and a long-distance negotiation where diplomats (or lawyers, etc.) exchange offers and proposals and counter-offers according to a fixed set of rules. Unlike in-person dialog, such indirect communication is less open-ended, and relies on trustworthy intermediaries to convey messages and information without distorting their meaning.

What diplomats, lawyers, and mediators may be to human communication, their analogues in Information Technology would be data-sharing protocols and APIs. Usually, software components interacting via these indirect, carefully mediated pathways can potentially have divergent implementations—they may be written in different languages, adhere to different coding paradigms, and so forth. Therefore, data has to be transferred from one component to another via neutral formats that are amenable to both sides: data in one context should be encapsulated in a neutral package and decoded (without distortion) on the other environment and vice versa. External-programming interfaces therefore depend on data representation systems that allow for information to be restructured according to the needs of different computing environments, while maintaining structural integrity (i.e., decoding the data back to its original form should produce an exact replica of the data prior to its originally being encoded).

Ensuring data integrity across divergent programming environments is a complex task. When being passed between components data is necessarily *serialized*, or converted from its innate binary/in-memory form to a textual or numeric encoding, and then reconstituted into a new binary package that should be "equivalent to" the original (we can define "equivalence" here in terms of bi-directionality: re-encoding the data and sending back from target to source, then re-decoding, should yield a copy of the original; or, more rigorously, modifying a smaller part of the received data and then back-sending should yield a duplicate of the original *except for* the specific change).

Data serialization formats must be designed to ensure that information is not lost in the encoding/decoding process. For example, a quantity that originally has floating-point type (so it *could* have non-integer values) might get encoded as an integer if, in fact, on a specific occasion it has no fractional part. Then, on decoding, imagine a value interpreted *as* an integer (rather than a **float** that happens to be equal to an integer). Future calculations could then become corrupted (in some contexts, arithmetic operations performed on an integer paired with a **float** will cause the floating-point value to be truncated to an integer, yielding incorrect results for algorithms expecting two **floats**), and/or erroneous data sent back to its source. Similar incompatibilities may exist between expected units of measurement (metric versus imperial, for instance) or value ranges (it is not obvious from a single number what are the sensible range of values which the number could take on, if modified: magnitudes for individual pixel colors, for instance, in most image formats are restricted to the range 0–255; percentages are often limited to 0–100; angles in degree to 0–359, and so forth). Failure to anticipate the full set of metadata conventions and standards that are prerequisite for multiple software components to act on shared data is a common source of data corruption (in the terms proposed by Srinivasa et. al. [52, page 4], "globally inconsistent state").

For these reasons, data-sharing protocols have to be designed around detailed and "expressive" representations, meaning that it is possible to systematically describe all facets of data profiles (types, ranges, units, scales, metadata) that could potentially be sources of ambiguity and encoding/decoding errors. To be sure, a lot of data sharing happens through relatively simple formats (such as JSON—JavaScript Object Notation—or CSV, Comma-Separated Values) that lack these higher-level guidelines, but such formats typically work in contexts where networking conventions are clearly defined—effectively relying on programmers' discipline and informal documentation to take the place of stipulations formally encoded in data protocols. In more "critical" contexts that should be less susceptible to programming errors—or for more open-ended protocols where networking between disparate end points should not depend on programmers honoring informal specifications—developers would tend to prefer more rigorous representations, such as XML (with validators and predefined tag and attribute sets), or RDF (Resource Description Framework), associated with the Semantic Web (in conjunction with explicit Ontologies).

In this chapter, we will discuss options for *hypergraph* representations (which in some sense generalize RDF, itself considered to be "graph-oriented," but, unlike hypergraphs, not internally multi-scale). We claim that hypergraphs are more effective at permitting rigorous data profiles to be described and confirmed (i.e., we can formulate hypergraph data paradigms that are more expressive still than, for instance, XML and RDF) while retaining the necessary attributes of formal verifiability and transparency. Indeed, hypergraph models in various forms have been proposed as extensions to both RDF (for example, in the transition from graph database engines to hypergraph engines such as HyperGraphDB, Grakn, or AtomSpace) and XML (in the context of "concurrent" markup and related supersets of XML trees, which tend to focus on allowing tags to overlap one another instead of being strictly nested, as they are in XML and its variants, notably HTML).

Although data (meta-) models are an important theoretical topic (particularly when the discussion turns to optimization [2, 27, 29] so that we are considering not only *whether* queries or validations can be completed, but how to do so efficiently), here we are equally concerned with practical software concerns: in particular, the coupling between *representing* information and *exposing* application functionality. The most common reason we seek to (digitally) encode data is to send it between separate components (including ones designed with little or no explicit inter-connections, except for joint participation in data-sharing capabilities that have to be implemented *ex post facto*). Metamodels are most valuable when they permit inter-operative capabilities in an additive manner—i.e., applications should be able to expand the scope of other applications with which they might network. Representational systems that engender computational artifacts (parsers, for syntax; and validators, for semantics, let's say) that can be implemented, embedded, and leveraged by applications—as their capabilities are refined to support new protocols—are most effective when the amount of effort consumed by the requisite programming is minimized. Data-sharing protocols should thus be engineered to pass through application integration phases as quickly as possible, but likewise applications should be architected to accommodate new data-sharing initiatives without substantial re-coding.

20.2.2 Examples: Building Information Management and Medical Imaging

Generically, we can describe a metamodel as relatively "expressive" to the degree that data structures and their concomitant semantic paradigms can be transparently encoded according to the modeling system's rules. Expressivity might imply a level of redundancy, or at least superficial redundancy when viewed from the sole perspective of data encoding. For example, the so-called Industry Foundation Classes—widely used in AEC (Architecture, Engineering, and Construction) technology and BIM (Building Information Management)—employ two different sorts of data annotations (called "attributes" and "properties") that both are roughly analogous to XML "attributes." The IFC attribute/property distinction might seem locally superfluous in that asserting a given data point via a property rather than an attribute (or vice versa) does not appear to convey greater information—the information is borne by the field's value, not by its property-or-attribute classification. However, this distinction is not semantically vacuous; its significance emerges in the larger scale of schema standardization and validation. Data fields asserting properties of objects (in the sense of real-world physical materials incorporated into building designs) are classified as either *attributes* or *properties*, with the distinction being that attributes are fixed within schemata for objects of any given kind, whereas property sets can be introduced in objects' context more flexibly. In type-theoretic terms, attributes are immutable parameters in types' schema, such that objects of the same type have the same attributes, whereas properties are not bound to types with equal force (usage patterns might dictate, for example, that common type instances share property sets *in some context*, but these are not essential maxims of the type itself).

In the realm of bioinformatics, a similar distinction is made in the context of DICOM (Digital Imaging and Communications in Medicine), wherein "tags" (which have descriptive labels but also two-part numeric codes) are encoded using a system that distinguishes "Standard Data Elements" from "Private Data Elements." The latter are assigned a numeric pair whose "Group" (first) Number is odd. PACS (Picture Archiving System) viewers [11, 56] (software that recognizes the DICOM format) will ignore Private Data Elements by default, so the Standard/Private distinction is intrinsic to protocols through which DICOM data are processed. This distinction is not functionally identical to IFC attributes/properties, but reveals similar motivations insofar as data specifications are also guidelines for implementing software that manages data structures conformant to a given standard. Standardization projects' tasks include defining requirements on software as well as data representation, and it is logistically significant to distinguish standardized parameters that software *should* recognize as a compliance criterion from non-standard kinds of values that particular users or groups of users working within a given protocol may want to introduce internally (but software may legitimately ignore).

The BIM attribute/property and DICOM Private/Standard distinctions are suggestive illustrations of challenges encountered when applying generic data modeling principles to specific domains. Most general-purpose representation frameworks (consider XML and RDF, for instance) lack a mechanism to directly express disjunction in parameters' level of standardization, as concretely found in these examples (not to imply that such details could not be represented indirectly, e.g., via meta-attributes or DTD stipulations, but the point is that one is thereby leveraging the affordances of a given representation scheme to accommodate semantic distinctions not specifically anticipated by that scheme, rather than organically encoding the semantics to begin with). Also, note that—for fields indexed by character strings—descriptive labels are designed to be meaningful for human users/readers, but in most modeling approaches labels do not have an "internal structure" recognized by the framework (at least if we consider, say, XML namespaces as separate labels from element names). The DICOM pattern wherein two-part numeric codes are employed for something akin to namespace/element separation, and then the use of an odd/even to signal Private/Standard tag rules, is also an idiosyncratic formulation that does not have a natural correlate in multi-domain modeling protocols.

Data representation theories that are grounded solely in "logical" reasoning, or mathematical representation, can miss such real-world complications. If we start from symbolic logic (or from a mathematical picture of graphs or trees as formal systems), we might be inclined to recognize *properties* (essentially metadata on graph sites, which is how properties work in conventional property-graph database engines), or *attributes* in the XML sense (metadata on tree nodes), but our theoretical framework could neglect to consider the possibility that a data-sharing protocol might need two *different* property/attribute mechanisms. The semantics of a property/attribute distinction (in contexts such as IFC, or DICOM public/private) derives not from logical models but from real-world implementational concerns. A metamodel that supports this larger semantic context is therefore sufficiently expressive to properly encode IFC or DICOM data; a less expressive model (one that collapses attributes and properties into a generic notion of "fields," say) would fall short, at least without compensating by leveraging its own formal resources (e.g., classifying fields as attributes or properties via metadata). Here we can appeal again to application-level concerns: the paucity of less expressive systems would become evident insofar as reifications, indirections, and other ad hoc solutions to representational blind spots can render application integration more complex and time-consuming.

Metamodels are more expressive to the degree that they offer a greater range of representational parameters with which structural and semantic conventions might be communicated. For example, JSON can be deemed more expressive than primitive CSV-style records because JSON distinguishes arrays (which are structurally akin to lists) from "objects" (i.e., associative arrays, lists indexed by field names rather than numbers). Likewise, XML is more expressive than JSON insofar as elements' data can be asserted either via attributes or via nested elements (for sake of argument, treating XML as a *de facto* superset of JSON, albeit with different syntax, wherein arrays correspond to sequences of similarly tagged child nodes). Associative arrays in XML might be coded *either* as attribute key–value pairs on *one* node *or* as sibling nodes with unique tag names. This "redundancy" allows for conventions to cohere (in a given data-sharing protocol) stipulating when one or another of these alternatives should be adopted, thereby supplying an extra layer of semantic detail that is not present in JSON.

In the case of property graphs and/or hypergraphs, data fields can be associated with an "object" (in the sense of an integral data structure) via properties (attributes on a node) or via node-to-hypernode relations (sometimes called "projections"), a duality reminiscent of property/attribute in IFC. Graphs, of course, have the further stipulation that any two nodes (or hypernodes) may be linked by edges (themselves equipped with labels, label namespaces, controlled vocabularies, and potentially constraints/axioms enforced by "Ontologies"). Formats such as XML and JSON, which are more "syntactically" oriented, tend to be hierarchical, in that any specific value in a JSON object or array may itself be another object/array (rather than atomic value) and likewise XML elements may have other elements as children. By contrast, formats such as RDF and property graphs, which are more "semantically" focused, encode relations via edges across nodes (rather than hierarchical nesting). Hypergraphs combine both styles of representation, with hypernodes containing nodes hierarchically *and also* edges between two (or two or more) nodes and/or hypernodes (the precise rules as to which constructions are possible will vary from one system to another, but these are reasonable approximations).

Any representational system, as these examples point out, provides a range of parameters that could be pressed into service when formalizing a protocol for encoding specific kinds of data via structures conformant to the specific system. More expressive systems have a wider arsenal of parameters; for instance, along the lines of the above gloss, property-hypergraphs (models that either add hyperedges to property graphs or, equivalently, add properties to hypergraphs) are more expressive than either property graphs or document-style hierarchical trees alone, because properties-on-nodes (as in XML attributes), nested hierarchies (hypernodes containing child nodes akin to child XML elements), and inter-node connections (as in Semantic Web-labeled edges) are all potential representational devices in the property-hypergraph context. One conclusion to be made here is that property-hypergraphs form a flexible general-purpose metamodel, but the relevant point for the moment is that expressivity can be "measured" via the range of distinct representational parameters afforded by the system, at least intuitively.

This intuition could be made more precise—insofar as we have deferred any rigorous definition for "representational parameters." That is to say, for a reasonably systematic analysis of metamodels, we should specify building blocks of constructions recognized through any metamodel. The overall concept may be clear enough—in general, the parameters of a modeling system are the full set of structural elements that might potentially be employed in fully describing any given structure covered by the system—but one would like a still tighter definition. For this chapter, we approach this problem from the perspective of virtual machines.

20.2.3 Virtual Machines in the Context of Data Metamodels and Database Engineering

The correlation between virtual machines and representational paradigms should be clear: suppose we take any structure instantiating a particular metamodel. Presumably, such a structure may be assembled over multiple stages. Insofar as node–hypernode inclusion is a constructional parameter, for instance, then a structure can be modified by including a node within the scope of a hypernode. Similarly, insofar as inter-node relations (via directed edges or hyperedges) are significant constructions, then a structure could be modified by adding an edge between existing nodes. In short, any structure can be derived from modifications on precursor structures. The full set of structure-modifying operations available for a given representational paradigm might then be enumerated as (at least one part of) the opset of a hypothetical (or realized) virtual machine. As such, VMs provide a potential formalizing environment for analyzing data metamodels.

Conversely, data models can serve as a prompt for motivating the proper scope of a virtual machine. That is, VMs may be designed subject to requirements that they permit the accumulation of any data structure conformant to a given metamodel. Similarly, VMs could be characterized in terms of how they support various "calling conventions," in the sense of protocols through which computational procedures delegate to other procedures (supplying inputs, reading outputs, spawning concurrent execution paths, and so forth). This section's chapters will focus on VMs based on hypergraph data models, employing such structures both from the perspective of data representations and interlocking procedures (i.e., describing procedures in terms of sequences of calls to other procedures).

Virtual machines are useful tools for studying software interoperability insofar as both data representations and communications protocols could be modeled in terms of VMs (at least as abstract summaries of working code; or, more ambitiously, application-networking frameworks might provide actual VMs through which applications can route their networking logic, analogous to query languages as host-language-agnostic conduits for database access). As emphasized earlier, applications typically seek to "expose" functionality to external components. In order to do so rigorously, it is necessary to stipulate preconditions on how functionality being externally invoked should be designated—for instance, if a Computer Vision application has multiple algorithms available for many processing tasks, external code needs to specify which implementation is being requested—and how input parameters (and then output results) should be encoded. Applications can use VMs as a kind of neutral environment where third-party code builds up representations of exposed-functionality invocations incrementally. Indeed, many programs support scripting via languages such as PYTHON; it is plausible to generalize this idea to VM platforms amenable to multiple scripting languages, so long as they are compiled to the relevant "byte" code. Alternatively, even if applications do not expose VMs to third parties directly, they could use VM-backed processes to process API requests: since VM-targeted code may be updated without recompiling the application, the API could then adapt to new networking situations.

The prior discussion centered on data sharing between disparate *software* components in general, but in this context VM models overlap with constructions pertaining to interoperability between applications and *database engines* in particular. Consider the general setup of database systems: applications store data structures for future reuse by passing some (suitably encoded) serialization of the relevant information to a database, which arranges the data into a layout optimized for storage and retrieval/querying. Typically, the database will attempt not merely to preserve the presented data structure for future reconstruction but will index or destructure it in such a manner that such specific data can be selected as matching a future query.

At any moment in time, an application will be working with one or more data structures that might be called "live"; they are directly implicated in the state of the application at that moment. The current user (or a different user) might, accordingly, be interested in reconstructing that application state (or some relevant subset thereof) at a future point in time—at which point database queries are typically necessary, because the relevant information is no longer in live memory. To be sure, a database does not necessarily store "application state" as such—although developers can potentially create "application state objects" and persist those so that users may resume prior sessions—but the typical rationale for persistent data storage is supporting users' desire to resume prior work, return to previously viewed files, and so forth.

As a concrete example, suppose we are considering an application that (at least as one of its features, and in the context of particular GUI windows or components accessed through the software) supports photo tagging and annotation. Imagine an image database allocated for documenting ecological and/or infrastructure damage from natural or man-made events, such as the 2022 Russian invasion of Ukraine, and perhaps setting the stage for reconstruction plans. A typical image for such an application might be a photograph of damaged buildings or facilities, or plots of land on which real estate will be rebuilt. Potentially, images would depict remnants of prior buildings that were destroyed and/or could be annotated with data relevant to reconstruction designs (e.g., in recreating damaged neighborhoods a certain number of residential units might be targeted for each parcel of land or each block subject to redevelopment, perhaps optimized by models measuring the ideal urban density for that specific neighborhood based on environmental criteria, utility grids, public transit, and so forth).

Let us explore this case study. A characteristic session for such a software component would involve users viewing individual images, previewing image series depicted via thumbnails, searching for images (based on criteria such as street address, city/district name, or perhaps GIS coordinates), switching between 2D images and 3D street views, and—once in the context of a specific picture—viewing annotations and other associated data in tabular or otherwise structured forms (e.g., tables or key–value pairs displayed through independent GUI windows floating above the graphics viewport).

Assuming such an application works with relatively large image collections (enough to be impractical for users simply to browse images in a filesystem folder, say), functionality for finding and tracking images would need to be based on systematic query capabilities, i.e., some form of image database: image file paths, feature sets for retrieval/similarity searches, metadata and formatting details, etc., hosted in a database so that images can be selected inside large series by variegated query strategies.[8]

Suppose an image depicts a city block where damaged buildings have to be replaced. Data associated with that image could include estimates of the number of people who lived in this location prior to the Russia/Ukraine war; the number of residential units slated for redevelopment; cost estimates; links to public transit info, street views, data concerning utilities grid, etc. Plausibly, such data would be held in a database and loaded alongside the image, or in response to user actions signaling an interest in the relevant data points. The database, in effect, is a means to an end, whereas from a user's perspective the important detail is that the application can enter a state where multiple data points are visible side by side: users might view a photograph in one window juxtaposed with a window or windows showing civic/residential data.

Continuing this specific (hypothetical) example, the envisaged scenario has image-viewport components serving as entry points for a diversity of civic/architectural information. Presumably, the specific kinds of data available will vary from one image to another, and users would signal through interactive GUI features their interest in accessing certain branches of the available data over others. That is, assume there is not a fixed metadata/associated information package that automatically accompanies each image, but instead that data and images are linked on a dynamically changing case-by-case basis. That setup would call for coding strategies that work to organize the available information and user-interaction pragmatics coherently. The resulting software-design choices would, moreover, propagate to GUI and database designs as well. Once some aggregate of data is identified as a coherent unit, this structural decision must be accounted for at the GUI level (insofar as users request *that specific* data from application states where they are viewing an image carrying the appropriate information) and the database integration level ("that specific" data has to be queried from the larger database context when needed). In other words, the interaction between GUI, database, and application logic is more complex than if each image were given a fixed set of data

[8] Images themselves (i.e., their pixel data) might also stored as "blobs"—binary large objects—but the term "image database" generically covers cases as well where images files are hosted on a filesystem with only their metadata stored in the database directly.

fields isolated from user pragmatics. This shows how database theory intersects with disciplines (e.g., Software Language Engineering) with practical benefits vis-à-vis efficient application design/deployment.

20.2.4 Database Engineering and Type Theory

Suppose again that one information bundle that could be associated with (some) photographs outlines redevelopment plans: data points such as the number of residential units targeted, renderings of building designs, contractors, contact information, and so forth. Presumably, analogous data would only be applicable to images showing blocks or plots where relevant plans are in the works; and moreover, such images would link to other forms of data as well (about, say, utilities grids). As part of the empirical background, in effect, one might conclude that various facts pertaining to individual reconstruction projects/contracts can both be aggregated (as interconnected data points) and isolated from other information potentially relevant to a viewed image. In general, software design depends on sensitivity to how information spaces, practically speaking, might be carved and organized. The decision to isolate (say) construction-project data as integral units would be made against that kind of design/decision context. Having this topic reified as a specific "category" of data affects GUI programming and event handling (because user-visible components must be implemented enabling users to access information in that category, which in turn yields such signals as context-menu activations and the need for appropriate event handlers) as well as database interop (insofar as data must be packaged in persistable forms). Subsequently, representations of integral construction-project data (in the form of, e.g., a cluster of interrelated datatypes) would be manifest as software artifacts threaded through a code base.

The specific patterns of data organization and programming language types engineered for an application reflect practical concerns and aspirations to optimize user experience; these patterns do not necessarily map neatly to native database architectures. Neither relational databases (built up from tables with fixed tuples of single-value columns) nor conventional graph databases (whose representations are confined to one layer of labeled edges) generically match the multivariate and multi-level structure of real-world information typically managed at the application level. This is why application-level datatypes generally need to be restructured and transformed when routed between applications and database back ends.

Software engineers are responsible for ensuring a proper synchronicity between application and database state, although (as intimated above) this does not (by and large) entail application state being directly persisted in a back end. Instead, back-end updates are a property of application state at certain moments; insofar as users have performed edits or in general made changes resulting a mismatch between the data as currently seen by the user and what is stored in the database, the latter has to commit such changes for perpetuity. Update-worthy application state then needs to be distributed over (potentially) multiple database "sites" that are collectively implicated in an update.

For the sake of argument, consider an update (or part thereof) consolidated into a single (application level) datatype instance, e.g., one object of a given C++ class. Quite possibly, there is no one-to-one correspondence between application objects and database values or records, particularly if the classes in questions contain many data fields and/or multiple one-to-many relationships and values that are more involved than simple strings or numbers (e.g., pointers to other objects). The structural mismatch between application data and database records is evident in technologies such as object-relation mapping (ORM)—or "Object Triple Mapping" for the Semantic Web—marshaling data for relational databases or triplestores, respectively, [5, 13, 32, 35, 40]. Engines with more flexible representation paradigms, needing less reconstruction of application data exported to a back end, can be advantageous precisely because data-persistence capabilities end up consuming less "boilerplate" code [6, 9, 21, 23, 54]. Hypergraph databases are a case in point: the kind of complexity in datatypes' internal organization (multiple multivariate fields for a single object, for instance) that gives rise to ORM/OTM-style transforms maps organically to hypernode or hyperedge constructions.

Modifying a database entails conveying a package of updates between applications and the database engine. The structure of this communiqué in turn reflects database architecture. An SQL INSERT statement, for example, derives its form from the layout of table-based data models: adding a record entails naming the table to which it will belong, which brings on board the specific list of fields (columns) whose values have to be accounted for in the query. Our impression that relational algebra is a relatively crude or inflexible metamodel derives, it would seem, in large part from the quantity of bridge code needed to implement database updates through query commands such as INSERT that are restricted to the relational architecture. To the extent that hypergraph engines (for example) are more flexible in principle, this advantage only becomes concretely evident to the degree that hypergraph databases support a query system such that updates (and analogous modifications to a database instance) are initiated with relatively less boilerplate code.

As we alluded to earlier, hypergraph data models have comparatively greater parameters available for representing information. In particular, this implies that we have greater flexibility in formulating queries to modify database instances. It is worth mentioning at this point that "queries" need not involve instructions passed to a database engine in the form of character strings (e.g., SQL code). Indeed, forcing applications to build query code on the fly is often an antipattern (spurring boilerplate code bloat)—better solutions involve query "factories" that assemble queries via procedure/method calls in a host programming language (perhaps through an embedded domain-specific language, as one finds with LINQ vis-à-vis C#). On the other hand, in the best case scenario, a database would *also* support a query language that could be executed as text strings outside of an application context (and without relying on a host programming language), for examining the contents of a database in situations removed from application-based access (debugging, analytics, general admin functionality, and so forth). In other words, ideally, engines will encompass a query engine that works with query structures compiled from *either* application-code factories *or* standalone query code, which is one rationale for embracing a query evaluation virtual machine.

The specifics of database updates—sticking again for sake of exposition to single type instances—depends of course on types' internal organization. For typically atomic types (like 1-, 2-, 4-, or 8-byte integers), updates might only entail replacing one value with another, but more complex values ("objects," in effect[9]) with multiple and/or multi-value data-field updates may include adding or removing a value from a collections-type instance as well as altering such a value, and so on. For many compound types, the collection of fields will span more than one "collections" value (e.g., vectors, stacks, queues, deques, and unordered sets/multi-sets, plus map arrays that are pair lists with possible nonduplication restrictions on the first element) subject to add/remove operations, in contrast to full-on value-to-value replacement. Some collections have modification restrictions (e.g., values can only be added or removed from one or both ends of a list, or, as in typical "sets," all added values must be unique) or enforce constraints such as monotone increase and decrease.[10] All of these are potential paths toward legal mappings of type instances between states which initialized typed values can take on, given their internal organization.[11]

The mechanisms for aggregating multiple values into individual type instances tend, accordingly, to be more complex for general-purpose programming languages such as C++ (see unions, pointers, arrays, multiple inheritance, enumerations and their base types, etc.) compared to databases (see SQL or RDF types); this is a proximate cause of complications in persisting application-level data. Conversely, databases with more refined type systems can (at least potentially) absorb application data more conveniently. For this to work in practice, however, the engine needs a query system that duly leverages type-system expressivity. Properly recognizing application-level intra-type organization depends on representing (and then exposing to a query interface) the full spectrum of morphisms through which a single type instance might be updated.

For complex types, an effective query system would have update operations that are more targeted than just replacing one value with another *tout court*; instead,

---

[9] Taking the perspective that in some systems objects are formally defined as aggregate structures (rather than atomic data points) and, even if not precisely stipulated, object/atomic value distinctions tend to align *de facto* with object classes against, say, built-in types (cf. C++).

[10] Consider a collections type that has only one insertion operation but will automatically place new values at either the beginning or end of the list to preserve increase direction; here, new values have to be either greater or less than all prior values. Or, automatically sorted lists need only one insertion operation because the insert procedure would deduce the proper insertion point.

[11] Nor is this discussion complete; we could also mention reclassifying "union"-type values (whose instances can be one of several types) or various types depending on binary arithmetic (cf. tagged/"decorated" pointers, or enumerations whose instances can be members of nominal-value lists *or* bitwise combinations thereof, or unions where multiple type tags are valid by virtue of shared binary encoding, in effect using one tagged value to update another member of the union—a simple case being integer/bitset unions where setting the integer automatically sets or clears corresponding fields in the bitset).

updates may only involve one or some subset of all data fields and (for multi-value fields) could involve insertions/deletes in a collections context rather than a direct value change. The degree to which a database engine seamlessly interoperates with applications depends on the latter constructing data packages (without undo effort) that signal the *kinds* of updates requested alongside the relevant new values (notating, e.g., collections-context changes as distinct from single-value morphisms). Intuitively, flexible architectures (e.g., hypergraphs) accelerate the requisite implementations, although actually supporting a query interface satisfying such ambitions depends on a confluence of factors (e.g., underlying database architecture but also query representation and query evaluation protocols and the tools to compile code/text or procedurally generated queries to internally represented structures suitable for evaluation).

Although this section's discussion has focused on updating a single type instance, complex intra-type layout implies a diverse set of query-based update options. An UPDATE statement in SQL, say, only (directly) supports one particular "genre" of updates (value-to-value edits in one or more discrete columns). Given application state that can be expressed in terms of modifications to existing persisted values, keeping the back end in sync entails accounting for the changes embodied in the new application state by presenting the back-end engine with (in general) a series of multiple updates. The more flexibly updates can be encoded, the less development time needs to be expended figuring out how to translate application-state changes to updates the engine can process. This is one reason why a diversity of data modeling parameters can streamline application integration: a flexible tableau of recognized constructions implies that applications may describe updates with relatively less boilerplate destructuring. Consider the multitude of "sites" where data associated with a hypernode could be asserted. In the context of property hypergraphs, at least (and some systems may have more elaborate constructions as well), properties *on* a hypernode, nodes *in* the hypernode, and edges *from* a hypernode to its peers. This articulation of site varieties is, self-evidently, also a list of update forms. Having multiple protocols for describing updates allows different forms of updates to be recognized with distinct semantics. For example—consider again the attribute/property distinction in IFC—updates to *properties* (which would in general be ad hoc annotations on a hypernode less strictly regulated than nodes encompassed *in* hypernodes) might be subject to different validators than updates performed via node insertion or (potentially ontology-constrained) edge insertion.

In effect, multi-parametric modeling tableaus in the database *architecture* propagate to multi-dimensional options for encoding updates, which in turn allow for multiple co-existing update protocols each with their own semantics. Applications can then choose which protocol most efficiently describes any particular change in application state.

Of course, these points vis-à-vis changing *existing* database values have analogs in the context of inserting *new* values. Ideally, applications will have flexibility in how to encode data structure for insertion into a back end via database queries. This is not only a matter of notating all information that should be persisted, but also providing cues to the engine about how the new data should interact with other records (e.g., using primary keys or globally unique identifiers to secure inter-record links analogous to—perhaps translating—live-memory pointers) and subsequent find/select queries. What are the criteria through which a database record, once deposited, should be retrieved again in the future? Most database systems will give nodes/records unique global identifiers, but the whole point of typical search queries is that records should be located based on known data in contexts where an application does *not* know the requisite **gid**.

Consider again the case study of image-curation software that could be used in a redevelopment/urban planning context, such that photos include depictions of future building sites. As suggested earlier, data associated with each picture might reference construction-project data (e.g., the number of units slated to be built, or the identity of the firm contracted to oversee the project), alongside generic information (image format, dimensions, color depth, GIS coordinates) not specifically endemic to AEC. One consideration when designing such an application would be how users would find images that they had not seen before, or had not revisited for an extended period of time (so that the presumptive image **gid** is not cached in recent history). It would make sense to track images by longitude and latitude, for example, assuming that users would know such data. Perhaps street address (accounting for the possibility that large-scale redevelopment might alter the street grid so that pre-war addresses, say, become obsolete; one might still maintain a mapping from such addresses to GIS coordinates so they remain useful for queries); or (less granularly) by district or city. We can similarly envisage scenarios where architectural details are mentioned ("find images of sites within a 10-kilometer radius featuring planned buildings over 4 stories tall").

For queries along these lines to work, back ends need to structure type instances such that (potentially large collections of) values can be filtered into subsets meeting specific criteria: GIS coordinates restricted to a given locale, housing matching a given contractor name, and so forth. When exporting info to a back end, the relevant details are not only specific values comprised by new data but also which fields may serve as eventual query parameters, and how they should be indexed. Such "selectability" criteria should be encoded alongside persisted data structures themselves, and convenient application integration entails supporting protocols for noting pathways for query retrieval and doing so with minimal boilerplate code (for analogous reasons as with update queries).

Of course, selection/retrieval and update protocols tend to be defined on types rather than the type instance level. To the degree that certain data fields are indexed and queryable (for retrieving instances when their global id's are not at hand *a priori*) decisions as to which fields to thereby expose tend to be made for each type—as part of the type's design and contract—rather than negotiated on a case-by-case basis per instance (though note that properties in property graphs are possible exceptions; indeed, this is one rationale for property graphs in the first place). Similarly, type-level modeling tends to define which update protocols to invoke for different state changes. Accordingly, an expressive query interface may allow stipulations regarding updates and searches to be described as attributes of *types* as well as single instances (e.g., records and/or hypernodes) and to be deferred from instances to type contracts (analogously, inferred in instance contexts by virtue of type attributions). An obvious corollary here is that query systems should recognize type descriptions as well as encodings of type instances.[12] Formally, this results in a technical merger between database engineering and mathematical type theory [31, page 21], [12, 16, 20, 28, 48], and [50].

20.3 GIS Databases and Digital Cartography

This discussion has just touched on themes related to flexibly typed database systems: the idea being that expressive database engines support *strongly typed* data (in the sense that all records have type attributions that offer guarantees about their internal structure, such as a specific set of defined fields) but also allow new types to be introduced into an active database. A good case study of where such design is apropos is that of maintaining geospatial databases that track information associated with geographical coordinates—for example, roads, bridges, buildings, waterways, transportation infrastructure, or land plots, but also any kind of scientific data (environmental, sociodemographic, etc.) that can be overlaid on a digital map. GIS systems are typically divided into two distinct data categories: there are "basemaps" that represent the most common or generic types of GIS data (roads, buildings, and the like), which are then augmented with supplemental "data layers" that are narrower and more domain-specific (such as ecological data). Digital cartography renders *base-map* visuals by subdividing geographic areas into relatively small segments (within a 256×256-pixel area, for example) translating geospatial (typically latitude/longitude) coordinates into image coordinates. Digital maps are composed at multiple scales of resolution; in a common format, so-called XYZ tiles, the lowest zoom level produces a "map" of the whole world (basically an outline of the continents), whereas the highest level (around 20, or slightly less depending on the environment) is sufficiently detailed that maps show the contours of individual buildings and street intersections. In XYZ, each successive zoom level doubles the underlying magnification, so that each tile at one level splits into four tiles at the next level.

Software known as "tile renderers" and "tile servers" [38, 61] map latitude and longitude coordinates (or other GIS units, such as those of Mercator projections) to tile coordinates that include zoom level as one axis (the "z" in XYZ). An XYZ triple designates one geographic area at a specific zoom level; individual latitude/longitude and Mercator points may then be identified via fractions of the tile width and height. Via such mapping, locations of objects in a GIS database are

[12] By implication, query languages would then require, as a subset, "type-expression" languages where types' attributes, internal organization, and back-end protocols are duly notated (so that type information can be "loaded into" the system and consulted as the engine resolves how to accommodate insertions, updates, and filtering for specific instances).

converted into pixel coordinates, so that features such as roads and building can be marked visually. Tiles are rendered according to graphical rules called "styles," which stipulate details such as the colors that should be used to outline buildings or roadways, or the icons for points of interest (transit stops, historical sites, special-purpose buildings, or areas such as schools, hospitals, parks, government offices, and so forth). Tile rendering requires some computational algorithms because streets/highways and other noteworthy spots need to be named or labeled, and the positioning of these identifiers (unlike the fixed latitude–longitude of the labeled locations) is indeterminate, so cartographic engines try to calculate how to position street names (for example) so that this text does not intersect with other map data (such as roadway lines).

The end result of a tile-rendering pipeline is a miniature map in the form of an *image*, often via photo formats such as PNG or JPEG. A GIS application will form maps that users see by composing together multiple tiles. This provides the "basemap" layer, representing a general-purpose view of a given geographic region. On top of these basemaps, applications will usually superimpose domain-specific data points (sometimes called "attribute" data) relevant in specific contexts: data layers might show locations where Ukrainian buildings or infrastructure was damaged in the 2022 war; or distributions of COVID-19 cases; or environmental hazard sites; or tax lots and zoning codes for architecture, real estate, urban planning, and property management; or sightings of specific plant or animal species, and so on. In effect, whenever scientists or researchers have data sets indexed by geographic coordinates—epidemiological, ecological, sociodemographic—such information can potentially become a data layer that is visualized by superimposing special-purpose icons or colorations on a generic basemap. These extra layers can be added at different processing stages: for instance, they might be consumed by the tile servers themselves, as a post-processing step immediately after rendering base-map tiles; or they could be superimposed on basemaps by GIS front ends, immediately prior to showing maps to users (in this case applications would typically download tiles and data layers separately, and merge them together at the last moment).

We provide this digital cartography overview mostly to mention certain details concerning GIS data systems. Although there is a core dimension of familiar mappable elements (roadways, buildings, and so forth) with common presentation guidelines (formalized via tile-rendering styles), the space of information that could be added onto basemaps is open-ended. Any data set with latitude/longitude coordinates (or fields convertible to such coordinates, such as a street address) can be visualized against the backdrop of a GIS basemap.

If we consider GIS data to encompass common base-map info *together with* special-purpose supplemental layers, then GIS databases are a good example of "strong but flexible" typing—it is impossible for a GIS system to anticipate *a priori* the full range of data profiles evinced by objects visualized in a map context, within supplemental data layers. Every mappable object has latitude/longitude coordinates (or a collection thereof, giving a geometric outline), but in addition objects will encompass other data points as well, relevant to their specific domain (civil engineering, etc.) but opaque to mapping applications themselves. In this sense, GIS systems manage non-GIS data by tracking opaque objects without consuming them directly—it is up to individual applications to download data within supplemental layers and reconstruct the relevant domain-specific info, so that users can employ street maps as entry points for accessing information that is then visualized through other features in the application. For example, urban planners taking on the responsibility of rebuilding Ukrainian cities might start with street maps showing sites of destroyed buildings and infrastructure, and then (typically by clicking on a map) transition to windows showing structured data about individual locations, such as the number of families formerly living in a damaged residential complex, or details about utilities and water/gas/electricity grids, or estimates of environmental hazards (chemical leakages, say, or unexploded munitions) resulting from the conflict.

The point here is that objects containing such domain-specific data need to be packaged and stored in a GIS database even if they are not GIS data proper, and then correctly reconstructed and presented to users in conjunction with digital-cartography front ends. This data management workflow is roughly analogous to the case of routing data packages across third-party networks, discussed above (in ways we will summarize more completely at the end of this section).

20.3.1 Geospatial Data and GUI Events

Before that, though, a final topic relevant to the current discussion relates to front-end implementations for GIS software.

Most of the time, users consult GIS maps with the intent to find and get information about specific addresses or locations, i.e., to switch from a cartographic overview to a more detailed display of information specific to a given spot on the map (perhaps a restaurant's menu, or departure times at a bus stop, or a business's hours of operation ...). Presumably, their GIS software (which can include domain-specific applications that have digital-map windows as one front-end format among others) has the capacity to render such location-specific data in a format relevant to the user. When accessing street maps via a Web browser, for example, the simplest way to access non-GIS data is to open third-party websites in a separate browser tab. Consider users intending to search a library's card catalog: finding the library on a map, they might then rely on the map generating popups with location info, hopefully including a link to the library's website—following that link would allow the user to query the desired catalog.

Outside of a browser, on the other hand, we can imagine a city's overarching library system providing an app that includes cartographic displays as one window where users can seek individual library locations, giving them the option of clicking on locations associated with a map icon; but then presenting a catalog-search window as a GUI component of the same application, rather than a separate website. This would be an example of an application that implements both digital-map windows and other forms of displays side by side, routing the user interface between them based on user actions.

Whether via multiple intra-software GUIs or through external websites, the common denominator in both scenarios is that users interact with GIS maps as a way of requesting information about specific places. Supporting GIS front ends for these purposes requires, at a minimum, the ability to map geographic locations to site-specific information. For example, a library-system app would need to know which latitude/longitude coordinates refer to which library branches, so that users clicking on an icon at a given geographic coordinate would be routed to the appropriate card catalog. In terms of front-end development, it is likewise essential for the software to know how on-screen coordinates map to latitude/longitude pairs.

This last step may seem obvious, but most users probably do not notice the computational details involved in deciphering their actual GIS locations. When a user sees a map, they intuitively understand that each point in the display corresponds to a specific GIS position (with identifiable latitude and longitude coordinates). Clicking a mouse when the cursor hovers over a given site—or tapping on a touchscreen—is thereby understood to be a gesture that signals the user's interest in selecting the location thereby singled out as a focus point.

From the application's point of view, however, the user action is not intrinsically endowed with geospatial axes, but instead is localized on the screen (they have clicked on a given pixel location relative the overall dimensions of their screen/monitor). There are a variety of mathematical quantities to be tracked when converting screen coordinates into latitude/longitude, including the current zoom level and the specific geographic rectangle that is visible in the map's viewport (following how the user has "panned," viz., orthogonally scrolled the display, so that the click location is quantified as a certain distance right and down from the window's top-left corner, which corresponds to moving a certain distance west and south). Moreover, because map tiles are acquired through coordinate systems such as XYZ rather than through latitude/longitude rectangles directly, applications will typically need to map on-screen coordinates to (e.g.) XYZ triples as an intermediate step, and then on to latitude/longitude as desired (in the case of XYZ, there might actually be still a further coordinate transform involving Mercator projections).

Although the equations to effectuate the various inter-coordinate calculations just outlined are not prohibitively difficult to implement, an application's ability to properly respond to front-end user gestures depends on having complete information about the digital map which the user sees at any point in time, prior to his or her interactions (e.g., current pan and zoom state). Depending on the map source used, such information may or may not be available. For example, a GIS front end could be constructed through a Web-view window, pointed at a street map URL (e.g., GoogleMaps). Via this arrangement, however, there is no way for the host application to get detailed pan/zoom info from the embedded map display, so the host software has no way to translate user actions (such as mouse clicks). The application could certainly register the fact that users have clicked on a given screen coordinate, but they will not be able to convert screen coordinates to latitude/longitude without knowing the current map state, and this is not data that systems such as GoogleMaps make computationally available.

For these reasons, relatively full-featured GIS applications have to manage map-rendering displays on their own (rather than deferring these operations to an embedded service), which involves connecting to a raw map-tile service (a tile server) rather than to a higher-level component such as Google Maps, and internally managing issues such as local tile storage and scheduling updates. This is why open-source tile servers such as Open Street Maps are an important contribution to the software development ecosystem, because they remove many barriers that would otherwise exist to applications implementing fully functional GIS front ends as one facet of their user interface.

On the other hand, even well-intentioned projects such as Open Street Maps have certain limitations (including data usage restrictions), so large-scale projects have an incentive to implement their own tile-server back ends rather than rely on third parties. Consider the case study of Ukraine reconstruction: assuming multiple organizations and international funders commit to significant reconstruction projects for Ukraine—and that groups agree to collaboratively engineer and utilize digital resources for sharing information and ideas that could be useful for reconstruction projects—it would make sense for the shared resources thereby deployed to include a dedicated tile server specific to digital maps of Ukrainian territory, including both maps designed to show street/urban details and to diagram natural/geological features. Apart from giving software open-ended access to raw cartographic images and GIS data, a special-purpose tile server could choose rendering styles and conventions (for instance, duplicating tiles with street names, place descriptions, and other textual info in multiple languages and alphabets: Ukrainian, English, German, etc.) collectively designed to maximize maps' usefulness for the specific purpose of a multi-national rebuilding effort. This could ensure that (e.g.) users unfamiliar with the Cyrillic alphabet would be able to see versions with transliterated place names and that base coloring for features such as roadways and building outlines would not conflict with the display of non-GIS data (e.g., tracking patterns of refugee displacement and return).

When analyzing front-end event processing, a further detail involves distinguishing between user gestures targeted at visible base-map features and those focused on supplemental data layers. With respect to base maps, although structured information such as details about individuals roads and buildings is implicit in tile images—because GIS databases are consulted in order to render tiles in the first place—the actual tiles which GIS front-end components receive from tile servers are just images, which function from a software development point of view more or less like photographs. Although a front-end event handler can (with suitable state info) translate event coordinates (screen locations) to latitude/longitude, the application would not (at least from map tiles themselves) be able to ascertain whether the user's selected latitude/longitude point corresponds to a particular building, a roadway, a street intersection, transit point, or any other kind of GIS feature. This further information must be obtained from a remote source (such as an address-lookup service) and/or a local database which applications maintain internally. In the case of Ukraine reconstruction, for example, a dedicated tile server could include information packets co-extensive with tiles identifying street-map landmarks keyed to geospatial locations, so that client software would be able to translate screen coordinates to points of interest.

These comments apply to details visualized through basemaps. The situation is different, however, with respect to supplemental data layers. In this case, typically, front-end applications acquire base tiles and non-GIS data separately and then overlay the latter on the former prior to rendering map windows for users to see on-screen. Because the client application thereby *itself* correlates non-GIS data objects with map locations, the correspondence between these objects and latitude/longitude coordinates is already present in the system (it does not need to be externally queried from an address locator, for instance). At the same time, there are still computational details which applications must reckon with in order to properly route user events. For example, implementations can produce the visual effect of non-GIS overlays visibly superimposed on base-map background in different ways.

By way of illustration, consider the options for a GIS front end coded via QT (the most popular cross-platform GUI framework for C++ and, via bindings, for most scripting languages). There are at least two possibilities: in one case, data layers could be superimposed on basemaps by overriding the functionality for "painting" basemaps as images, such that icons, diagrammatic elements (arrows, lines, and so forth), or color effects (e.g., semi-transparent colors to represent something like population density) are "painted" on top of base-map images. In other words, non-GIS layers are painted onto GIS tiles via techniques similar to how those tiles are rendered with GIS info proper (roadway/building outlines, etc.) in the first place. The result of that operation would be a composite image that, perhaps via a **QImage** subclass, gets treated by the QT displays as equivalent to ordinary images (with no internal structure). In this case, all user events would share the same profile, as signals relative to image coordinates, and the application must determine whether a mouse click (or similar gesture) is targeting a point in the base-map image or an overlay painted over it.

On the other hand, an alternative technique would render the composite of basemaps and overlays through **QGraphicsScene**, where the basemap is treated as a raw image, but overlay data are introduced as scene objects (for example, icons representing specific points where non-GIS data is accessible), that the application tracks and holds in memory as data structures fully separate from base-map images. In this case, clicks on scene-object icons would be registered via different event signals and a different application-response pipeline than clicks on base-map locations, so programmers have to engineer two different event-handing systems, one for basemaps and one for supplemental layers. These two options ("painting" non-GIS layers versus introducing them as graphics-scene objects) have respective trade-offs: different functionality is more or less complex in one scenario versus the other, and vice versa.

This specific example—implementing GIS displays with supplemental data via QT windows—is a good case study in how applications should systematically prepare and document their event-handling logic. Failure to clarify how a QT application is approaching the task of rendering layers on GIS basemaps, and on how these design choices propagate to event-handling requirements, could make such applications difficult to maintain once the original developers move on to other projects.

20.3.2 Representing Functional Organization

Supplemental GIS data layers are good examples of a computational theme that recurs in multiple contexts: the co-existence of spatial and function design or organization, which software seeks to model, simulate, or visualize. The domain of "space" can vary; in the GIS context, *spatial* is actually *geospatial*, where "space" is construed as features and locations on the Earth's surface (plus, sometimes, height above or depth beneath the ground). When mapping utilities, transit systems, traffic flows, and so forth, we are dealing with multi-part networks that are spread out over a geographic area, and geospatial relations are a fundamental dimension to the functional interrelation between disparate parts. But such systems are also comprised of non-geospatial (or not entirely geospatial) relationships: transfer points, in the case of transportation; highway exists and roadway capacity, in the case of traffic simulations; aquifers, water mains, and treatment centers, in the case of utilities (specifically water systems), etc. It is a spatial fact, for instance, that two roadways or train lines cross at a given point, but whether these points permit exits/transfers is a functional relationship—geospatial line data alone do not report whether or not drivers can turn off one road onto another, or riders switch from one line to another, at their contact points.

Other domains have other concepts of space (not necessarily *geo*spatial). Architectural and building systems assemble into the 3D form of a house or factory, but they also capture the machinery of modern-day construction, with electricity, heating/cooling, ventilation, fire safety, and so forth—a wall is not just a spatial border between rooms but a stopgap against a spreading fire, a barrier to sound, a perimeter enclosing rooms that have specific entry ways and dimensions. Or, in robotics, image-processing techniques parse visual input to create a spatial map of a robot's environment (here "space" is both the internal pixel expanse of an image and the projected ambient space where a robot moves), but robots also need to understand their surroundings in functional terms [3, 4, 8, 14, 15, 22, 30, 44, 51]—understanding external objects as artifacts to be moved, obstacles to be avoided, or affordances enabling the robot's own movements (walking steps, say). In all of these contexts, spatial and functional systems are overlaid one on another, where spatial relations constrain and substantiate functional relations, but additional parameters and axioms are needed to describe the functional systems in detail—simply mapping out their component's spatial networks is not sufficient to define functional–organizational properties (again, say, intersections between highways or train lines may or may not be a traffic link or transfer point).

Formats such as Industry Foundation Classes have evolved alongside "green tech" and "smart grid" technology, or in general an increasing emphasis on employing computer models to promote sustainable, eco-friendly engineering; and these use cases exemplify the co-engineering of spatially and functionally articulated data structures. For example, architects have learned that carefully selecting building plans and materials can lead to more efficient designs—ideally, "net zero" carbon footprint—by limiting a building's reliance on electricity (and environmentally harmful chemicals, such as freon) for, e.g., heating and cooling [45, 59]. Organizations such as the World Resources Institute have promoted icons and visual markers that may be introduced in graphic media to indicate buildings' energy-efficient measures: double or triple window panes, evaporative or radiative cooling, natural lighting and ventilation, and so on [7, page 22]. Such icons might be positioned on 2D images—photographs, artist's renderings, floor plans—or also on digital maps; consider street maps annotated with information showing how a particular community is planning to reach carbon emissions goals. As a visual cue summarizing green tech designs, such icons may also become interactive GUI features, where users could click on the icon to learn about the building materials chosen (specific kinds of windows or HVAC installations, say) and how they reduce carbon footprint.

As a case in point, warm weather properties can minimize the use of air conditioners by orienting exteriors to create shaded areas and position windows away from sunlight, arranging ventilation systems to produce cooling air flows (which also has a drying effect, reducing interior moisture that could yield health hazards), selecting materials that naturally absorb heat—and potentially harness solar and/or wind to offset energy use. Computer models come into play because concepts such as "net zero" emissions are not abstract goals; they are concrete targets that may be measured and predicted. Software can estimate a property's energy efficiency while its architectural plans are still being finalized, and CyberPhysical sensors might track actual energy use post-occupancy (which in turn evaluates the accuracy of the initial simulations) [46, 53].

In this sense, green technology has become an important component of computer-aided design in domains such as architecture and civil engineering. In addition to their roles as construction blueprints and visual tools to assess the aesthetics and livability of property designs, building plans thereby gain an additional dimension as data structures defining the functional organization of physical constructions, which in turn are subject to analysis by special-purpose software components that simulate air flow, air quality, energy use, and related metrics that collectively quantify building efficiency [18, 37, 58]. Although spatial relations among architectural elements are obviously significant factors in these simulations, such analyses require additional data based on mathematical models of (e.g.) air currents or solar-thermal heat gain [26]. Because CAD applications (for conventional AEC) are not "scientific computing" platforms in most contexts (apart from measuring buildings' structural integrity, perhaps), such "green tech" analyses require external software or plugins taking CAD data as a starting point but then incorporating additional, domain-specific parameters.

Similar principles apply to non-geospatial attributes within GIS data layers. Functional networks overlaid on GIS terrain can be analyzed much like structured systems within architectural space [1, 39, 60]. As a concrete example, consider hydrology and "hydroinformatics" [10, 17, 19, 25, 43, 49]. Indeed, we will continue the earlier case study of Ukrainian reconstruction. It has been well-documented that the 2022 war crippled the country's urban infrastructure in many places, so that utilities such as water distribution systems have to be rebuilt alongside residential and industrial units. The multi-faceted scale of reconstruction priorities actually gives preliminary designs greater flexibility because engineers are not limited to repairing utilities to meet the needs of a fixed residential population, or conversely accepting constraints on new construction due to limitations of existing urban infrastructure. Under the assumption that refugees might want to return to their prior cities or neighborhoods but not necessarily their exact previous address, planners have leeway to explore how to reconstruct both residential housing and utility systems to optimize factors such as clean water and inexpensive electricity: density of habitation can be structured to minimize the costs associated with providing energy, water, gas, waste disposal, etc.

In the case of water systems, rebuilding plans would reasonably start from existing aquifers and water sources (such as lakes and reservoirs) that are geological artifacts, not man-made ones. Once these sources are tapped, however, water distribution networks have some level of flexibility, allowing for engineering choices: what kind of pipes to use, where to place occupied buildings and hydrological infrastructure (such as holding tanks), how to recycle wastewater (which points to the functional interrelationship between water and sewage systems), how to provide irrigation (which likewise connects water distribution to agricultural optimization), how to protect water sources from contamination due to floods, existing hazard sites, and so forth. Such assessments depend on structural models of water utilities as networked systems, where mathematical models can compute the quantity of water that may run through specific network paths (to minimize the risk of buildings' losing access to water when one part of the system is damaged, for example), or to estimate water demand (based on the types of properties, considering the different needs of industrial, commercial, or residential locations), at properties which are water systems' end points, or to simulate how floods or environmental hazards might damage and/or contaminate the system. Computationally, such network models would normally represent water distribution via graphs, where (for instance) nodes represent physical sites (aquifers, buildings, treatment sites) and edges represent infrastructure such as sewage drains or pipes, which might then be annotated with numeric data (for instance, a water main or secondary pipe's maximum capacity).

Graphs representing water systems in this sense could be geospatially visualized because most of these structural elements have fixed geographic locations—i.e., the nodes in hydrological graphs may be assigned latitude/longitude coordinates, and therefore the entire network can be laid out alongside street grids or geological feature maps. At the same time, functional analysis of water networks would need to consider many engineering specifications apart from GIS position alone (e.g., the capacity and manufacturing details of pipes and storage tanks). This would be an example of non-GIS data that might be curated within the scope of a geospatial database but consumed by domain-specific client applications, rather than digital cartography renderers themselves.

Consider the following scenario: a company, or non-profit, that specializes in hydrology is tapped to produce proposals or assessments of reconstructed water systems considered for damaged Ukrainian cities. That work is encapsulated in structural models that simulate the design of new water networks, with system details mapped out via overlays on Ukrainian street maps. Simultaneously, architectural firms are at work formulating plans for new residential (and commercial/manufacturing) sites to replace infrastructure irreparably damaged during the war. When designing an apartment complex to house a specific number of families, for example, architects could then estimate the tenants' anticipated water use and refer to the hydrological company's simulations to check that their proposed networks will provide adequate water for the new building, modeled as an end node within the overall graph. From a computational perspective, we can see this situation as a kind of multi-part analytic framework where the software employed by the hydrologists and by the architects should interoperate with sufficient rigor to allow the water network models to answer questions arising in the architectural context. For example, the architect's software might need to obtain quantitative parameters specific to the water distribution proposals and use such data as inputs to simulations of future residents' water consumption.

We may envision variations of this scenario, too: environmental engineers could similarly consult water system proposals but run simulations of the network's performance, not in the context of residential demand, but rather for the effects of drought and floods, or the risk that water in the system could be tainted by environmental contaminants or flood runoff. Or, agricultural engineers may simulate the performance of water supply when these systems are tapped to provide irrigation, which in turn would vary by season and in conjunction with choices related to crop variety and cultivation methods (whose computational models would be derived from agricultural science).

Because this chapter is focused on computational methods rather than botany or hydrology *per se*, we will not examine the specifics of the different scientific domains in play for these various examples. Instead, we want to focus on the application design concerns that these case studies illustrate. We have here a hypothetical water distribution network model serving as a core data source which multiple applications, each with their own specific domains (architecture, ecology, farming), will tap for parameters and calculations specific to their area of focus. All of this software has to interoperate according to data-sharing protocols that ensure each application's access to its requisite information despite their differences in scientific domain. Accommodating multi-faceted information networks along these lines is a more complicated task than engineering multi-part services spanning applications with largely overlapping functionality (e.g., passing Electronic Health Records between biomedical facilities, or supporting electronic payments/transforms among e-commerce and/or financial services applications).

Furthermore, in addition to the data-sharing issues just identified, the water network example also points to the role of GIS data: although the most important network data are not intrinsically geospatial, it is nonetheless true that many details of a water system can be visualized in a GIS environment, and geospatial relations are among the parameters that would be considered during simulations (for example, the geographic distance between a water source and building location would presumably be a factor in predicting uninterrupted water supply based on, e.g., network redundancy). Visualizing water systems against a street-map background is a good framework for human users to picture the network in overview (as opposed to schematic diagrams, for example). We can thus imagine that front ends for the various applications accessing the water network data might use digital maps as an entry point. Civil engineering software, for example, could give users the option of selecting a specific building by starting with a street map, zooming in to a desired location as necessary. In the context of hydrology-related simulations, this front-end functionality might be augmented by rendering water distribution data within the maps that are presented to users, e.g., showing water routes drawn as GIS "ways" analogous to highways or transit lines, and providing icons marking individual building's connections to water systems as entry points for requesting data about an individual property's water use/access.

In this environment, then, software simulating water networks would be depositing data in an overall GIS environment wherein specific data packages can be associated with geospatial locations, even if the encapsulated data are not intended for use by GIS software directly. Geospatial location then serves as a kind of index through which disparate applications could obtain non-GIS information so that they may then unpack and plug in to their own analytic capabilities. Here, GIS databases act as way stations for data packages routed between disparate applications. Perhaps an analogy to old-fashioned (non-electronic) mail is appropriate: the post office does not examine the contents of someone's letter, but it implements a deposition, storage, and delivery service allowing packages to be sent from one point in the country to another.

Notice GIS systems thereby potentially serve a purpose that is a variation on cross-application data sharing considered earlier in the chapter. Before, we mentioned the use of APIs allowing an application to receive third-party data from an external source, "exposing" specific capabilities that permit the host application to correctly interpret and process such external information. That prior discussion made the implicit assumption that applications in such an arrangement would be inter-communicating directly: an originating component explicitly connects to a target application via the latter's API. However, the case of non-geospatial data in a GIS database points to how cross-application interop can be more delayed and circuitous. Hydrological software could deposit data about water networks through GIS systems that are subsequently consumed by third-party applications (for architecture, agriculture, and so forth) without either application having direct contact: all data would be routed and stored through GIS components (which, in general, would not consider non-geospatial data directly but at least would keep track of its association with geographical coordinates and GIS objects proper, such as buildings or geological features).

Having outlined the structure of this kind of mediated data sharing—relying on intermediate components such as a GIS database to allow cross-application networking without direct inter-application connections—the next question would be to address the computational requirements of data intended to be shared through an indirect, multi-party environment and of the components wherein data packages are constructed, then stored, and then acquired and re-used. The following chapter will present arguments that hypergraph data models are particularly well-suited for this kind of multi-part protocol, both with respect to formalizing data models and with respect to analyzing computer code through which the functionality to (respectively) construct, store, and acquire the respective data packages is implemented.

20.4 Conclusion

This chapter has briefly touched on several disparate themes in the realm of software engineering, themes that will be further examined, to varying degrees, over the next three chapters. In particular, domains such as geographic information systems and building information management present use cases for multi-application data sharing through intermediaries such as a GIS database; for the juxtaposition of spatial (e.g., geospatial, image space, or 3D/CAD designs, or some combination thereof) and functional structures, with spatial and functional facets analyzed and visualized side by side; and for application front-end event-handling pipelines, discussed here in the context of users interacting with digital GIS maps. Each of these contexts presents engineering complications that software components need to manage—these are the sort of implementational tangles that can, in the absence of well-constructed code libraries and design patterns, inhibit our ability to engineer applications in a timely and cost-effective manner.

Situations such as Ukraine reconstruction point to how software development inertia has real-world consequences: ideally, a variety of special-purpose applications will come on board soon after an envisaged Russian cease fire, helping the international community to rebuild Ukrainian cities and facilities in short order. The more user-friendly and feature-provisioned this custom software, the more effectively civil engineers can leverage the sheer scale of needed reconstruction to implement "smart city" technologies.

For each of these three themes (decentralized/mediated data sharing, spatial/functional data juxtaposition, and event processing), we may identify specific concerns that add layers of complexity and engineering requirements to the respective classes of applications. In terms of data sharing, programmers should identify how to package domain-specific data in formats that can be stored in a domain-neutral setting (like a GIS database) and then reconstituted by other applications with sufficiently overlapping capabilities, a problem that involves both how future applications will query for the specific information they need and how they will build in-memory structures (in their own coding environment) that are properly aligned with and behaviorally replicate the objects originally deposited. Programmers can certainly use object-oriented techniques (for example) to encapsulate the properties and behaviors of specific object classes, but it requires extra layers of design to build object systems that are sufficiently semantically explicit to orchestrate cross-application interoperability. In the context of event-handling, applications should engineer a centralized protocol for documenting event signals and response routines. It is one thing to code event-handling procedures (e.g., via QT signals and slots); situations such as GIS front ends demonstrate however that applications require a further dimension of structural coherence, insofar as a code base needs not only to provide procedures responding to user events, but to allow the total system of event signals and responders to be documented, queried, and augmented in an orderly manner. Frameworks such as QT provision support for the base-level implementations, but other paradigms are necessary to architect the overall system in this kind of logical manner.

And, finally, with respect to spatial/functional organization, applications need to construct data packages encapsulating system-functional details that can be keyed to locations such that spatial coordinates serve as indices and handles to extra data layers (again, "space" here can have different or multiple meanings: geographic extent, image space, the virtual space of CAD designs, etc.). At one level, associations between spatial points and data objects might be covered by a straightforward key–value mapping, with locations as keys and object identifiers as values. But the larger problems of integrating spatial and functional data are more complex: how should GUI displays targeted at spatial arrangement (GIS maps in the geospatial context, say, or floorplan blueprints in architectural CAD) be extended to include icons or signal generators such that users can access data (or access functionality leveraging data) handled by functional-analytic plugins/components co-existing with the spatial displays? How should underlying "spatial" data stores (such as GIS databases) support the depositing, querying, and retrieval of objects that are largely opaque to the stores themselves (as in non-GIS data integrated with GIS databases)?

To appreciate the full scope of these issues, we should consider the problem domain from the perspective of a programmer intending to support analytics concerning functional organization in an environment where data are predominantly organized in a spatial context (for some notion of "space"). In GIS, for instance, imagine the perspective of a developer building an application tracking data structures (consider water systems, as in the above examples) where functional units have geospatial coordinates but also many parameters that do not measure GIS magnitudes or attributes. Ultimately, such data must be encapsulated into persistable units and linked to GIS locations, so that peer applications can pull the relevant objects from a database via (say) latitude/longitude. How would engineers know that the full range of parameters is properly reconstructed by external applications so that their computations or simulations are consistent with the originating applications' models?

It would be one thing if external software were simply cloning the original program's full data sets, with some guarantee that the later computing environment is sufficiently analogous to the former that both copies of the data will be behaviorally equivalent. However, working through a GIS intermediary occludes this synchronization: we assume that later applications are obtaining data piecemeal from specific GIS locations or areas, rather than getting a full-scale replica of an original information space (in the context of large-scale data sets, such as a hydrological survey of an entire country, cloning the data *tout court* could well be impossible; even for smaller-scale data sets, however, working with partial samples rather than full copies of original data might be more consistent with the later application's requirements). How can originating applications guarantee data integrity and analytic accuracy under the assumption that external software will run analyses over parts rather than full copies of their data sets? How can these applications identify the "operational closure" of any given sample within their data sets such that these integrity validations are guaranteed?

Current IFC standards are a good illustration of how these issues play out in practice. Seen as a design model, CAD systems compose IFC elements as building blocks that piece together to define an integrated data space through which architectural plans may be visualized, in various formats (blueprints, 3D models, "artist rendering" style 2D images). At the same time, the structural network evinced by interlinked IFC data can also be queried as a foundation for simulations involving buildings as functionally organized systems (with respect to energy consumption, fire safety, internal temperature dynamics, and so forth). Structured chains of IFC objects thereby serve as a multi-faceted foundation from which, on the one hand, spatially and visually oriented resources might be constructed (such as floor plans) and, on the other hand, data needed for functional simulations can be initialized. Such integration between spatial and functional structures is driven by the co-existence of parameters within IFC objects that address concerns related to buildings' functional-systematic profiles (energy efficiency, for example) as well as those related to CAD model space. These facets of the IFC data model could potentially serve as an inspiration for analogous standardization of information classes within other domains where spatial and functional networks co-exist, such as GIS or—in the sense of visual object-recognition alongside structured environment models—robotics and computer vision.

Setting aside the specific details of IFC, however, the issues of spatial/functional juxtaposition, event-handling, and mediated multi-application data sharing point to software engineering challenges that—one could reasonably claim—point toward next-generation software development environments. Existing programming tools and design patterns arguably do a relatively poor job of addressing such concerns effectively, measured by how readily developers have tools at hand to implement code that manages the relevant complexity in a rigorous manner (so that applications can be deployed more quickly and maintained for a longer or more productive life span). These are the kind of practical challenges that call for new ideas in software language engineering, to circle back to the discussion at the start of this chapter. In the next chapter, we will focus in specifically on one such paradigm, namely hypergraph models of computing processes, query evaluation, and data interoperability.

References

1. Joginder Ahlawat and Saurabh Kumar Sah, "Optimal Drinking Water Distribution System Designing using Network Analysis and Geospatial Technology". https://globaljournals.org/GJHSS_Volume17/5-Optimal-Drinking-Water-Distribution.pdf
2. Renzo Angles, *et. al.*, "Foundations of Modern Query Languages for Graph Databases". http://renzoangles.net/files/csur2017.pdf
3. Paola Ardón, *et. al.*, "Building Affordance Relations for Robotic Agents—A Review". https://www.ijcai.org/proceedings/2021/0590.pdf
4. Paola Ardón, *et. al.*, "Towards Robust Grasps: Using the Environment Semantics for Robotic Object Affordances". https://paolaardon.github.io/files/papers/18ardon2018towards.pdf
5. Jeroen Bach, "Theory and Experimental Evaluation of Object- Relational Mapping Optimization Techniques: How to ORM and how not to ORM". Thesis, University of Amsterdam, 2010. https://homepages.cwi.nl/~jurgenv/theses/JeroenBach.pdf
6. Jeffrey M. Barnes, "Object-Relational Mapping as a Persistence Mechanism for Object-Oriented Applications". https://digitalcommons.macalester.edu/cgi/viewcontent.cgi?article=1006&context=mathcs_honors
7. Renilde Becqué, "Accelerating Building Decarbonization: Eight attainable policy pathways to net zero for all". https://wrirosscities.org/sites/default/files/19_WP_ZCB_final.pdf
8. Daniel Beßler, *et. al.*, "A Formal Model of Affordances for Flexible Robotic Task Execution" https://ai.uni-bremen.de/papers/bessler20affordances.pdf
9. Jeang-Kuo Chen and Wei-Zhe Lee, "An Introduction of NoSQL Databases Based on Their Categories and Application Industries". https://mdpi-res.com/d_attachment/algorithms/algorithms-12-00106/article_deploy/algorithms-12-00106-v2.pdf?version=1559011753
10. Junde Chen, *et. al.*, "Intelligent Monitoring Method of Water Quality Based on Image Processing and RVFL-GMDH Model". https://ietresearch.onlinelibrary.wiley.com/doi/full/10.1049/iet-ipr.2020.0254
11. David A. Clunie, "DICOM Structured Reporting". http://www.dclunie.com/pixelmed/DICOMSR.book.pdf
12. Pierre-Évariste Dagand, "A Cosmology of Datatypes: Reusability and Dependent Types". Dissertation, University of Strathclyde, 2013 https://citeseerx.ist.psu.edu/document?type=pdf&doi=199771 0cea4b67b9e53df96d6dee7f2c3e5c5556
13. Martin Fowler, *Patterns of Enterprise Application Architecture*. Addison-Wesley, 2002 https://www.martinfowler.com/books/eaa.html
14. Sourav Garg, *et. al.*, "Semantics for Robotic Mapping, Perception and Interaction: A Survey" https://www.nowpublishers.com/article/DownloadSummary/ROB-059
15. Leni K. Le Gof, *et. al.*, "Building an Affordances Map with Interactive Perception" https://arxiv.org/pdf/1903.04413.pdf
16. Jan Hidders, "Typing Graph-Manipulation Operations" http://adrem.uantwerpen.be/bibrem/pubs/hidders.3.pdf
17. Thomas Hies, *et. al.*, "Enhanced Water-Level Detection by Image Processing" https://www.researchgate.net/publication/262337135_Enhanced_water-level_detection_by_image_processing
18. Rafael Horn, *et. al.*, "The BIM2LCA Approach: An Industry Foundation Classes (IFC)-Based Interface to Integrate Life Cycle Assessment in Integral Planning" https://www.mdpi.com/2071-1050/12/16/6558/pdf-vor
19. Shaun Howell, *et. al.*, "Integrating Building and Urban Semantics to Empower Smart Water Solutions" https://www.sciencedirect.com/science/article/pii/S0926580517301759
20. Bart Jacobs, *Categorical Logic and Type Theory*. Elsevier, 2001.
21. Christopher Ireland and David Bowers, "Exposing the Myth: Object-Relational Impedance Mismatch is a Wicked Problem" https://oro.open.ac.uk/43318/1/download.php_articleid%3Ddbkda_2015_2_10_50020

22. Peter Kaiser, *et. al.*, "Extracting Whole-Body Affordances from Multimodal Exploration". https://h2t.anthropomatik.kit.edu/pdf/Kaiser2014a.pdf
23. Samiya Khan, *et. al.*, "Storage Solutions for Big Data Systems: A Qualitative Study and Comparison". https://arxiv.org/pdf/1904.11498.pdf
24. Julian Keil, *et. al.*, "Preparing the HoloLens for user studies: an Augmented Reality Interface for the Spatial Adjustment of Holographic Objects in 3D Indoor Environments". https://d-nb.info/1208303023/34
25. Praveen Kumar, *et. al.*, *Hydroinformatics: Data Integrative Approaches in Computation, Analysis, and Modeling*. Taylor & Francis, 2019. https://vdoc.pub/download/hydroinformatics-data-integrative-approaches-in-computation-analysis-and-modeling-1thfcnva63kg
26. M. B. Luther, *et. al.*, "A simple-to-use calculator for determining the total solar heat gain of a glazing system". https://anzasca.net/wp-content/uploads/2014/08/p511.pdf
27. Divya Mahajan, *et. al.*, "Improving the Energy Efficiency of Relational and NoSQL Databases via Query Optimizations". https://www.sciencedirect.com/science/article/pii/S2210537918301112
28. Mike Marti, "Translating Between Graph Database Query Languages". Thesis, ETH Zürich, 2022. https://www.research-collection.ethz.ch/bitstream/handle/20.500.11850/575738/1/Marti_Mike.pdf
29. Wim Martens, *et. al.*, "Representing Paths in Graph Database Pattern Matching". https://arxiv.org/pdf/2207.13541.pdf
30. Troy McMahon, *et. al.*, "Affordance Wayfields for Task and Motion Planning". https://ocj.name/papers/tamcm_iros2018.pdf
31. Bálint Molnár and András Benczúr, "The Application of Directed Hyper-Graphs for Analysis of Models of Information Systems". https://www.mdpi.com/2227-7390/10/5/759
32. Martin Ledvinka and Petr Křemen, "Formalizing Object-ontological Mapping Using F-logic". https://kbss.felk.cvut.cz/reports/2019/19ruleml-report.pdf
33. Chris Lattner and Vikram Adve, "LLVM: A Compilation Framework for Lifelong Program Analysis & Transformation". https://llvm.org/pubs/2004-01-30-CGO-LLVM.pdf
34. Tommi Laukkanen, *et. al.*, "Virtual technologies in supporting sustainable consumption: From a single-sensory stimulus to a multi-sensory experience". https://e-tarjome.com/storage/panel/fileuploads/2022-06-27/1656317001_e16754.pdf
35. Martin Ledvinka and Petr Křemen, "A Comparison of Object-Triple Mapping Frameworks". https://www.semantic-web-journal.net/system/files/swj1910.pdf
36. Daniel Martin, *et. al.*, "Multimodality in VR: A survey". https://arxiv.org/pdf/2101.07906.pdf
37. James O'Donnell, *et. al.*, "SimModel: A domain data model for whole building energy simulation". https://simulationresearch.lbl.gov/publications/simmodel-domain-data-model-whole
38. Anders Olofsson, "Behavior-driven Tile Caching in Web GIS Applications". Chalmers University, Thesis, 2013. https://publications.lib.chalmers.se/records/fulltext/217024/217024.pdf
39. Jiman Park and Byungyun Yang, "GIS-Enabled Digital Twin System for Sustainable Evaluation of Carbon Emissions: A Case Study of Jeonju City, South Korea". https://www.mdpi.com/2071-1050/12/21/9186
40. Radoslav Radev, "Extending Object-Relational Mapping with Change Data Capture". http://www.cs.ubbcluj.ro/~studia-i/contents/2015-1/04-Radev.pdf
41. Ismo Rakkolainen, *et. al.*, "State of the Art in Extended Reality—Multimodal Interaction". https://humanoptimizedxr.org/wp/wp-content/uploads/2021/06/HUMOR-State-of-the-Art-in-Extended-Reality-Multimodal-Interaction-1.pdf
42. Ismo Rakkolainen, *et. al.*, "Technologies for Multimodal Interaction in Extended Reality—A Scoping Review". https://www.mdpi.com/2414-4088/5/12/81
43. Mark Randall, *et. al.*, "Geographic Object Based Image Analysis of WorldView-3 Imagery for Urban Hydrologic Modelling at the Catchment Scale". https://www.mdpi.com/2073-4441/11/6/1133
44. Michael Rescorla, "Chrysippus's Dog as a Case Study in Non-Linguistic Cognition". https://philosophy.ucla.edu/wp-content/uploads/2016/08/Chrysippus27s-Dog.pdf

45. Danielle Richardson, "Sustainable Urban Planning: Turning the Concrete Jungle into Green Buildings". https://research.library.fordham.edu/cgi/viewcontent.cgi?article=1125&context=environ_2015
46. J. Rio, *et. al.*, "Expansion of IFC model with structural sensors". https://www.researchgate.net/publication/308164615_Expansion_of_IFC_model_with_structural_sensors
47. Filippo Sanfilippo, *et. al.*, "A Perspective Review on Integrating VR/AR with Haptics into STEM Education for Multi-Sensory Learning". https://www.mdpi.com/2218-6581/11/2/41
48. Patrick Schultz, *et. al.*, "Algebraic Databases". https://arxiv.org/pdf/1602.03501.pdf
49. Ehsan Shafiee, *et. al.*, "Parallel Evolutionary Algorithm for Designing Water Distribution Networks to Minimize Background Leakage". https://ascelibrary.org/doi/full/10.1061/%28ASCE%29WR.1943-5452.0000601
50. Joshua Shinavier, *et. al.*, "Algebraic Property Graphs". https://arxiv.org/pdf/1909.04881.pdf
51. Hyun Oh Song, *et. al.*, "Learning to Detect Visual Grasp Affordance". https://www1.icsi.berkeley.edu/pubs/vision/visualgraspaffordance.pdf
52. Srinath Srinivasa, *et. al.*, "LogicFence: A Framework for Enforcing Global Integrity Constraints at Runtime". https://ieeexplore.ieee.org/document/4041608
53. Michael Theiler and Kay Smarsly, "IFC Monitor—An IFC schema extension for modeling structural health monitoring systems". https://www.sciencedirect.com/science/article/pii/S1474034617306134
54. Valter Uotila, *et. al.*, "MultiCategory: Multi-model Query Processing Meets Category Theory and Functional Programming". https://arxiv.org/pdf/2109.00929.pdf
55. D.W.F. van Krevelen and R. Poelman, "A Survey of Augmented Reality Technologies, Applications and Limitations". https://ijvr.eu/article/view/2767/8825
56. Johan Van Soest, *et al.*, "Towards a semantic PACS: Using Semantic Web technology to represent imaging data". *Studies in Health Technology and Informatics*, Volume 205 (2014), pages 166–179, https://www.ncbi.nlm.nih.gov/pmc/articles/PMC5119276/
57. Jose D. Velazco-Garcia, *et. al.*, "A Modular and Scalable Computational Framework for Interactive Immersion into Imaging Data with a Holographic Augmented Reality Interface". https://pubmed.ncbi.nlm.nih.gov/33045556/
58. R. Verstraeten, *et. al.*, "IFC-based calculation of the Flemish Energy Performance Standard". https://biblio.ugent.be/publication/482379/file/6814885.pdf
59. Natalie C. Wheating, *et. al.*, "Embodied carbon: A framework for prioritizing and reducing emissions in the building industry". https://repository.usfca.edu/cgi/viewcontent.cgi?article=1574&context=capstone
60. Linlin Zhao, *et. al.*, "An Integrated BIM-GIS Method for Planning of Water Distribution System". https://www.mdpi.com/504658
61. Xiaodong Zhou, *et. al.*, "A PDF Tile Model for Geographic Map Data". https://www.mdpi.com/2220-9964/8/9/373

Chapter 21
Virtual Machines and Hypergraph Data/Code Models: Graph-Theoretic Representations of Lambda-Style Calculi

Nathaniel Christen and Amy Neustein

21.1 Introduction

This chapter will examine interconnections between three of the focal subjects for Part 5 of this book: virtual machines, hypergraph data models and metamodels, and the common structures (among many dissimilarities that were reviewed in Chap. 20) between human and computer languages. We will continue last chapter's discussion of data model and programming language "semantics." We argued in the previous introduction that digital artifacts represent real-world objects only indirectly (we can contrast this with word–object associations in natural language). Applications "signify" empirical facts because they can be engineered to maintain a consistent (even if occasionally fallible) correlation between application-state and particular real-world propositional situations of interest (recall the case of apps showing a frequently updated World Cup score).

To adopt terms familiar in the Philosophy of Science, we would argue that signification in this environment is (at most) an *emergent property* of an overall computational system. Whereas a photograph of a key moment in the World Cup final—assuming it is authentic—represents an event with some immediacy, a computer program that someone might consult to track the game's score can only be deemed an indicator for the relevant propositional content (something like, France and Argentina are now tied 3-3) insofar as the program fulfills its design requirements. Semantics, in this realm, is *implemented.* We have to assess the inner workings of a digital artifact to confirm that the phenomena we as users experience as "signifiers" (e.g., two numbers next to France and Argentina flags/logos, respectively) actually play that role.

N. Christen · A. Neustein (✉)
Linguistic Technology Systems, Fort Lee, NJ, USA
e-mail: amy.neustein@verizon.net

A. Neustein et al. (eds.), *AI, IoT, Big Data and Cloud Computing for Industry 4.0*, Signals and Communication Technology, https://doi.org/10.1007/978-3-031-29713-7_21

Because signification is emergent in this sense, the semantics of application state can only be rigorously defined on a holistic scale: it is a product of many parts within the application acting and interacting correctly. Consider the symbolizing device of printing a number next to the two country's flags: as users, we perceive that each displayed number correlates with the respective country's goal total, and therefore stands toward that total in something like a signifier-signified relationship. But such correlation only *happens* because many operations are completed behind the scenes: repeatedly (we may assume) connecting to an Internet service (itself presumably monitored and updated by someone who witnesses the game directly) to receive a data package including the current score, decoding/deserializing that data into live memory, mapping the fields in the resulting live objects/values to specific screen areas, and mapping the numeric scores into printable numerals whose glyphs are arranged to be viewed in those boxes on the screen. All of these operations involve data structures modeled, shared, and manipulated, and chains of procedures responsible for one step or another in the overall data-management process.

To understand the holistic semantics of computer programs as emergent systems, then, we need to model the "local" or behind-the-scenes semantics of individual data structures, and the procedures that manipulate them. This is a semantics that emerges as datatype instances are initialized and then passed from one procedure to another (potentially modified along the way) and, in some cases, mapped to a visual/iconic form. It is a semantics of holistic systems that take inputs in the form of user actions and/or received data (e.g., from an Internet connection) and output (mostly) visual artifacts on-screen. It is a *procedural* semantics because chains of operations are typically needed to progress from the inputs to the outputs.

To model this semantics, then, if we seek to analyze computer software by analogy to studying human language, we need to represent procedure calls: how tokens in source code map to specific functions that can be executed as subroutines, interrupting the program flow at the call site and then returning, and how procedures modify and manipulate data structures according to constraints of datatypes and design patterns (for example, a type can never hold a value incompatible with its potential instance set: an unsigned integer, say, can never be negative). Programming language semantics, in short, is built up in layers, with inter-procedure conventions establishing the working substructure through which (via GUI design and responsiveness to user actions) applications take on states and present information to users, which have emergent significatory effects.

Formal models of data structures and procedure calls are therefore intrinsic to the semantics of artificial computer languages, with no obvious analogues in "natural" language. In later chapters, we will consider to what degree extralinguistic "cognitive processing" plays a somewhat analogous role in human language understanding, but for the moment we can register this as a foundational contrast between these two genres of languages. Whereas the building blocks of human dialects are parts of speech, morphosyntactic markings, anaphoric/deictic resolution, and word senses within a lexical continuum, the analogous building blocks in source and compiled computer code are procedures, their parameters, symbol tokens iconifying specific values (tracked and potentially modified) over time, numeric encodings of data structures (not only quantitative ones, but also text, enumerations, visual graphics, and multimedia), and text or character-string representations of values likewise (such as numerals written into source code to represent numbers).

Virtual machines (VMs) are a useful theoretical tool for studying these "semantic" building blocks. To be sure, VMs have many applications, and analyses of computer languages inspired by natural linguistics are probably not the most important of these. Still, a "linguistic" perspective is a useful intuitive guideline for considering how VMs relate to computer code and programming language runtimes.

There are several different use cases for VMs. We might recognize the following:

Implementing procedures Virtual machines in this sense play the role of execution engines mimicking the behavior of physical processors (see Chap. 20, footnote 4). Here VMs actually run procedures and produce values and/or side effects as an alternative to compiling code to machine language.

Simulating procedures Instead of actually executing procedures, VMs may simulate how procedures *would* behave if they were called with specific values, so as to obtain information for compilation, diagnostics, testing, quality assurance, and so forth.

Encapsulating calls to specific kinds of procedures As discussed in Chap. 20, applications and code bases often "expose" certain functionality to be called by external/third-party components. A typical case is when relatively complex steps must be taken to instantiate environments where certain kinds of functionality are available. This chapter and the next will consider examples in the context of cyber-physical systems and computer vision. In the context of code libraries for interfacing with cyber-physical devices, or image-processing algorithms, respectively, such libraries are often designed with the goal of "hiding" complications related to image/data format and mathematical models. These libraries would be engineered to encapsulate relevant functions so that they may be invoked by external code, without complex domain-specific details being directly addressed by the calling code.

Programmers could therefore invoke operations for (say) computer vision or cyber-physical networks without needing detailed understanding of these respective domains. Virtual machines may add an extra layer of indirection by including encapsulated domain-specific procedures alongside lower-level instructions as part of their "operation set." In this way, VMs can play a role analogous to (or permitting the implementation of) scripting interfaces to domain-specific modules (consider, say, a PYTHON interface to a predominant computer vision library, OPENCV: many programmers find it easier to invoke OPENCV routines in PYTHON than to call these functions directly via C++).

Synthesizing or interoperating between divergent programming languages and computing environments Because—as touched on in Chap. 20—VMs are more flexible than machine language and physical CPUs, they can provide a neutral platform that is agnostic to language-implementation details. For example, the philosophical differences between divergent programming styles (functional vs. object-oriented, say) are reinforced by what is optimized by compilers and runtimes. Languages encourage code writers to adopt certain design patterns, and in turn they are optimized so that code that adheres to such patterns runs more quickly or consumes less memory (and other computing resources, such as CPU time), or is subject to more stringent evaluation (compile-time error detection, for example), or all of the above. Because different languages optimize in different ways, it is hard to find a neutral medium where code written in two different source dialects might be compiled to a single byte code, say, or where data structures created in one language can be utilized by procedures in a different language. Virtual machines, compared with CPUs, are typically slower but more flexible: because they do not seek the most optimized implementation, they may serve as a hybrid environment through which multiple languages may interoperate.

This chapter and the next will discuss some of these potential features/applications of VMs in a little more detail. The plan is first to summarize a specific model for VMs based on procedure-call semantics; subsequent discussion will consider the role of VMs in larger ecosystems, such as those related to image processing.

21.2 Virtual Machines and Hypergraph Code Models

As a preliminary to analyses later in this section, we will make a few comments that might situate this discussion in the context of topics covered elsewhere in this book. Specifically, we intend to motivate the discussion by appealing to technologies related to industrial computing, cyber-physical systems (CPS), and computer vision.[1]

One potential use case for virtual machines is encapsulating cyber-physical networks: we assume that CPS devices are sensors and/or actuators managed remotely by more-conventional computers. In general, sensors measure physical quantities (air quality in a room, say), whereas actuators have tangible physical effects (optimizing air quality by adjusting vents, say). Neither kind of device typically has extensive computational capabilities; instead, data are routed from the devices to a central location that uses software to process sensor data and convey instructions to actuators. These networks might, in turn, cover multiple intermediate points (consider CPS data warehoused in the cloud); but, for exposition, we can focus on the essential end points—on one hand the devices themselves

[1] Throughout this book section, we speak of "Virtual Machines" specifically in the sense of digital artifacts that emulate central processing units or other physically realized computing engines. Many discussions particularly in the context of industrial technology refer more broadly to virtual machines that simulate engineered physical artifacts—as "digital twins" [3, 4, 8, 33, 44, 52, 55, 67]—or to virtual operating systems. Discussions about VMs in the former sense, however, can carry over to this broader usage because VM execution environments are often intrinsic parts of digital twins and/or virtual operating systems.

and on the other software applications through which humans monitor and, if desired, manipulate CPS technology. Even if the actual devices do not perform calculations (they may not have the means to execute computer code), sensors and actuators *are* capable of converting some physical quantity into digital signals which computers themselves can interpret. Programmers developing CPS applications will rarely interact with devices independently; instead, the "physical" links with CPS networks[2] will typically be handled by low-level driver programs, which could expose capabilities to access sensor readings through, for instance, C language functions.

Developers might employ virtual machines in the CPS arena to streamline this process—instead of programmers needing to write code that interfaces with C drivers directly, the low-level function calls can be managed via virtual machines that interoperate with higher-level applications in a language-agnostic manner. Recent research in data semantics and virtual simulations of CPS networks and devices can benefit the engineering of relatively general-purpose middleware tools that route data and instructions back and forth between application-style environments and CPS hardware [15, 58, 61, 64, 65]. An analogy might be made to database systems: although there are many different data persistence engines, a handful of specific architectures (relational/SQL plus the predominant NoSQL frameworks) have emerged as templates that cover the interactions of applications with almost all databases they are likely to encounter. Interop protocols can therefore target a core group of database models, yielding results applicable to a large range of actual technologies. Analogously, consolidating the extensive variety of CPS sensors/actuators into a core set of data profiles would support generic middleware components that work against many specific CPS networks.

In this sense, CPS VMs can be analogous to database query languages, in their design and rationales: just as query formats give engineers the option of communicating with a database from a diversity of application environments, the protocols for accessing CPS data could similarly be encapsulated in VM operations. Moreover, query engines (as noted in Chap. 20) can ideally function either as standalone languages of their own (processing query code outside an application context, e.g., on a command line for debugging and maintenance) or via query factories programmable from general-purpose languages. The situation is comparable for CPS, and one could pursue the same duality in how CPS network capabilities are exposed (either through host languages or special extra-application code).

If a VM is indeed devoted to specific cyber-physical networks (or groups thereof), then a logical first step when implementing the VM would be to articulate the collection of procedures (through driver libraries, say) to be encapsulated. More likely, a more generic VM would be deployed in different contexts, each of which (by targeting a specific group of CPS devices) would work against particular groupings of driver code, on a case-by-case basis. The overall VM would thereby introduce capabilities to interface with some collection of kernel functions exposed by driver

[2] Which of course can be physical only in an ethereal sense, e.g., passing data via wireless services.

libraries, with the actual integration to those procedures finalized during deployment. In any case, though, the general pattern is that any particular deployment of a VM would be "seeded" with some preliminary set of functions (i.e., built-in VM operations, some perhaps added on in deployment).

Someone might reasonably point out that interoperability between heterogeneous components—whether the differences are just in programming environments (for disparate pieces of software) or more substantial, as with connections between applications and Internet of Things endpoints—typically falls under the rubric of data sharing protocols and APIs, rather than VMs proper. That is, CPS driver libraries could certainly expose access to devices through an API, but this would only indirectly be an issue for VM (via, say, programmers using a scripting language to facilitate API calls). While this is true in principle, we briefly argued in the previous chapter that VMs can serve as a logical extension to APIs: there are rationales for exposing functionality via connection points handled through VMs, either to ensure interop protocols' adaptability or to present an interface that can be hooked in multiple ways—via scripts, query-like languages, application code building up requests incrementally, etc. This potential VM use case fits well with the theoretical perspective on VMs developed in the current chapter. Here, we will consider VMs as systems organized around a collection of kernel functions, which in turn form the basis of procedures implemented through the VM that call such functions in sequence. While some kernel operations may be generic in the sense that they simply provide expected features of coding environments—declaring variables' symbolic names, types, and scopes, registering procedure inputs, binding outputs to symbols or input positions on enclosing procedure calls, and so forth—there is no reason why kernel operations could not encapsulate access to domain-specific capabilities, such that VMs mimic features of APIs.

In this scenario, VMs furnish functionality imitating "kernel" driver or operating system calls, initializing input arguments with values passed from applications, and sending results back. The VM accordingly bridges application-level and low-level kernel functions, which involves not only encapsulating access to low-level functionality but also negotiating the discrepancy between relatively high-level and low-level programming environments. Most application programming languages, for example, recognize constructs (e.g., object orientation, or functional programming via lambda closures/call continuations, plus exceptions, mutable references, and so on) outside the scope of low-level code (even C). Effective "bridge" protocols allow applications to access CPS data (as one category of low-level resources) within familiar higher-level paradigms. For example, programmers already working in an object-oriented environment might reasonably find it intuitive to apply object models to CPS elements, such as individual devices. Driver code most likely would not recognize object-oriented calling conventions directly, so a VM could be pressed into service to translate application-level descriptions of procedure calls and their input parameters into simpler binary packages which drivers can handle.

In this sense, a VM fitting these requirements would simultaneously model "high level" calling conventions and translate procedure-call requests between higher and lower levels. We make this point in the context of CPS, but the idea would hold for many cases wherein VMs encapsulate access to some integral set of low-level functions (at least low level relative to applications that would benefit from the requisite functionality). Consider image analysis: procedures exposed via computer vision libraries are not directly comparable to low-level driver code—they may themselves be implemented in high-level languages such as C++—nevertheless, libraries such as OpenCV or ITK tend to demand a rather complex scaffolding set up to enable analytic procedures being called. We could imagine circumstances where wrapping complex image-processing pipelines in a simpler API would be convenient for code libraries managing image series. For instance, a database exposing image-processing operations through query expressions would ideally support query evaluation covering computer vision capabilities (going beyond obvious information one may want to query from an image, such as dimensions and color depth).

Indeed, database queries, image processing, and CPS network administration each represent plausible use cases wherein VMs can serve as adapters between application-style coding environments and procedure collections that are too specialized or low level to fit comfortably in application development norms. These cases might overlap, of course; databases can warehouse CPS data such that queries monitoring real-time device state would logically coexist with queries against database content (tracking device info), or image databases can support queries dependent on computer vision.

The database query perspective points to an interesting heuristic analogy, insofar as packaging procedure calls from an application environment to a lower-level context resembles the task of packaging application-level datatypes and updates into records and fields natively recognized by a persistence engine. Similar to how live-memory data structures (objects, say, in object-oriented environments where the relevant data are interpreted in light of objects' polymorphic type, which could be subclasses of their declared type) are destructured for insertion into database sites, high-level calling conventions (again, object orientation provides useful examples insofar as **this/self** values are represented apart from other input parameters) need to be restructured into the (generally simpler) forms suitable for low-level functions (e.g., moving **this/self** to be an ordinary parameter, which may require truncating to base-class binary layouts, and some level of type erasure).

Continuing this analogy, flexible database architectures bridge application-level datatypes with persistent data structures/records so that application code can work with back-end values according to the norms of application-context programming; equivalently, VMs might wrap low-level functions such that they fit the profile of high-level procedures called according to high-level conventions (with objects, exceptions, and so forth). A flexible VM would play this bridge role in multiple contexts, perhaps allowing native functions to be registered as kernel operations and striving to be usable from multiple host languages: that is, from one VM, we can envision encapsulating access to a variety of procedure collections (examples we have cited here include functions exposed vis-à-vis database queries, image processing, and CPS networks) and routing descriptions of procedure calls from a variety of programming languages (which might have object-oriented or functional characteristics, or some combination). To the degree that such generality is desired, VMs should anticipate integration with diverse application-level languages preferring different calling conventions and procedural contracts.

One maxim for VM design, then, at least in this sort of context, is to prioritize capabilities to model and carry out procedure calls described through diverse calling conventions, rather than narrowing in on specific calling conventions that derive from a preferred programming model (functional or object-oriented, say). Object-oriented conventions such as method overrides/polymorphism and exceptions/exception handling might be natively expressed via the VM, but likewise functional idioms such as lazy evaluation and overloading based on type state. To clarify the last example: procedures can potentially be given different implementations by virtue of values' type state at the moment of call (essentially a more granular classification than type attribution itself), often mixing type state with value destructuring. A canonical example would be procedures operating on list-style collections, which in one overload would accept only *non-empty* collections where the last (or first) element is passed separately, alongside a structure representing all other elements (this convention is ubiquitous in recursive algorithms, since the "tail" can then be passed to the same procedure recursively, becoming destructured by the calling mechanism into its own head-plus-tail pair; of course, procedures implemented via this strategy also need an overload taking an empty list, which serves as a halting point).

In short, an ambitious VM can ideally work with a broad set of calling styles embraced by diverse programming styles, e.g., object methods, exceptions, lazy evaluation, type states, and parameter destructuring. To this list, we might add dependent types and functional-reactive idioms (e.g., the so-called signal/slot conventions, reviewed in greater detail in Chap. 24).

Since not all VMs actually aspire to multi-language support to the open-ended degree implied here, theories informing VM implementation do not necessarily analyze data models or design patterns targeted specifically at language "agnosticity" (so to speak); a more common scenario is that theoretical perspectives emanate from coding paradigms that give rise to distinct flavors of programming languages, potentially at some level proselytizing for favored paradigms rather than aiming for broad generality (in the sense that VMs *for functional languages*, say, reflect a general assumption that functional methodology is in the general case a better coding style than alternatives). By contrast, we are interested here in describing theoretical VM models that remain nonjudgmental as to which conventions are better in which context (or take the view that multiple coding styles each have their own use case so should be supported as such). A rigorous VM "model" should have multiple dimensions (addressing types and data encoding, for example), but of course an essential concern is modeling procedure calls and calling conventions, so we will pursue this topic as a starting point for a (relatively informal) system to encapsulate VM details in a schematic fashion.

21.2.1 Applicative Structures and Mathematical Foundations

Theoretical (and applied) computer science often approaches procedures from the viewpoint of mathematical functions, essentially mapping that transform inputs to outputs. Codifying the principles of "functionhood" in general forms the central project of analyses that bridge math and computers (e.g., lambda calculus). In terms of mathematical "foundations," these various formulations (including lambda calculus and its derivatives and, for instance, Combinatory Logic) entail strategies for clarifying what are sometimes called "applicative structures," or generic patterns involving the *application* of a function/procedure to one or more arguments/parameters [28, e.g., Chapter 14], [57, Chapter 18] [9, 12, 17, 22, 23, 25, 29, 39, 42, 63]. Since this is such a basic phenomenon in the mathematical realm, it is understandable that some researchers in "philosophy" of mathematics and its foundations would focus on applicative structures (perhaps indirectly via, say, lambda calculus)—even though mathematical expressions are written down according to a wide variety of conventions (consider formulae for integration, for ratios, for polynomials, and so forth), we can image a system that translates mathematical terms to a more systematic logical representation (which would presumably resemble something like Lisp code).

The concerns evoked by applicative structures are not only orthographic, however, because there is also a *semantic* dimension in the sense of spaces of functional values qua semantic entities. The problem of semantically characterizing *a* function or procedure (e.g., as a calculational process, or a—maybe time/context-dependent—input-to-output mapping, or the like) is one matter, but whatever our foundational semantic theory in this sense it is natural to extend it via functional composition (analogous to how linguistic semantics involves the semantics of nouns and verbs, but more thoroughly also the compositional principles of verbs together with nouns yielding sentences/propositions). The set of possible applicative structures "generated" by some collection of functions (or function symbols) is analogous to the set of discrete functional values that can be defined by the combination of multiple functions, wherein the results of one function applied to its arguments become in turn (one of the) parameters of a different (or, recursively, the same) function.

Systematically, an "applicative system" would be a set of function symbols alongside "variable" symbols with a rule that applicative *structures* include expressions of the form $fx_1x_2...$, where xs are variables, and that for any applicative structure the modification formed by replacing a variable symbol with another applicative structure is also an applicative structure. Defined in terms of the closure of such substitution operations, applicative systems can be seen to follow a generative pattern very similar to labeled trees—each node is either a symbol node or a branch with its set of child nodes—with the added detail that labels are partitioned into two sets (function and variable symbols, respectively) such that leftmost child nodes always have function labels and other non-branch nodes always have variable labels.

The analytic purpose of applicative structures can potentially be served by systems employed in other branches of mathematics—e.g., "term algebras," in formal logic, or n-ary trees in computer science, or an extension of the "free magma" groupoid concept to (sets of) n-ary (not just binary) operations. For instance, applicative structures might be seen as term algebras if we consider only function terms in the context of first-order logic, ignoring predicates, constants, or relations (constants can be treated as nullary functions, and predicates/relations modeled via functions whose codomain is the set of two boolean constants *true* and *false*). Applicative structures may also be defined in terms of freely generated strings in a language with balanced parentheses—essentially a "Dyck" language (after Walther von Dyck) intersected with a suitably restricted function/variable language [32, 35, 37, 43].

In effect, applicative structures can be defined by restricting the forms generated by first-order logic, or by regular languages according to the Chomsky–Schützenberger representation theorem (in formal linguistics) [20, 31, 40, 66]. Different such formulations could be preferred depending on which seems like an intuitive basis in the specific context being analyzed. However, although any language roughly equivalent to n-ary trees may be intuitively simple, it is worth defining the precise structures one is working with axiomatically to ensure that there is a strict isomorphism between representations in different contexts, as opposed to spaces that overlap but do not exactly align (for example, balanced-parentheses strings allow for nullary function applications that are not necessarily substitutable with constant values; it is unclear how to map that specific possibility to n-ary trees). In any case, our strategy here is to define applicative structures recursively in a manner that works whether we approach them syntactically or semantically, and to analyze their properties mostly through a representation based on matrices, which in turn could potentially be used to verify isomorphisms between these structures and alternative encodings (here we carry out such an analysis in a graph-theoretic context, though we propose similar methods might apply to, e.g., term algebras).

We will make a couple of technical points about applicative structures here, not so much because the mathematics is particularly sophisticated or consequential but so as to establish a baseline of comparison for the graph-theoretic encoding of (generalizations of) applicative structures we will discuss subsequently.

Definition *Applicative Systems*: For any fixed set of *function symbols* and *variable symbols*, an *applicative structure* is any element of a set that is the closure of the set of primitive expressions (consisting of a function symbol followed by a number of variable symbols that matches its arity, assuming we assign a specific arity to each function, or else to any number of variable symbols) under substitution operations wherein an applicative structure is inserted in an enclosing structure taking the place of a variable. Applicative systems are then sets of function and variable symbols (and possibly declarations of function arity) together the full set of possible applicative structures generated on their basis. For generality, we can allow functions `f` to have dynamic range arity (multiple integers n such that `f` followed by n variables is a valid expression); in this sense, systems that do not recognize arity limitations at all would implicitly allow arbitrary range arity for all function symbols.[3] Note that the generative rules allow for zero-arity functions, whose semantics would be procedures that yield results even without inputs (nested structures can be wholly comprised of one single function symbol).

Definition A given symbol may appear multiple times in an applicative structure; the above definitions were formulated with the idea that "the same" symbol in different positions is, according to the generative rules, a different symbol, but for clarity we can say that one symbol can have multiple *tokens* in a given structure. Each symbol token has a *nesting level* such that the nesting level of a function symbol matches that of variable symbols following it (that are not themselves part of a further nested structure), and replacing a variable with a nested structure forces the function symbol at the left of the latter structure to have a nesting level one greater than the replaced variable. By the construction/generation process, there will always be exactly one function symbol with least nesting level, which we can stipulate to be zero. Symbol tokens also have *positional indices* defined such that function symbols are assigned index zero, and variable symbols following them (incrementing across but skipping over nested structures) are assigned successively greater index numbers. Tokens are uniquely identified by a positional-index list whose length is determined by (viz., one greater than) the token's nesting level, notating the token's own positional index, and also those that would be assigned to its parent nodes (treating the structure as a tree) were they tokens rather than nested structures. Positional-index lists induce an ordering on all tokens in an applicative structure (comparing the first number in two respective lists, then the second as a tie breaker, and so forth; akin to ordering leaf nodes on a tree where left-ward and higher nodes are prior to those below and/or to their right). Applicative structures are *non-recursive* if each function symbol has only one token. It is helpful to further define non-*root*-recursive structures as those where the function symbol whose token has zero nesting level appears only once.

Definition Applicative structures can be grouped into equivalence classes wherein any structure in the class would map onto a peer structure under a permutation of the applicative system's variable symbol set. Since we can give an ordering to the variable set, assume each variable is assigned a number code, which in turn allows us to define a *canonical presentation* of an applicative structure by permuting its variable symbols so that tokens that are prior (in the ordering based on positional indices just described) are assigned symbols with lesser codes, choosing each new symbol to be the one with lowest code value not yet used. Each applicative-structure

[3] A *partial* applicative structure would be one that belongs to an applicative system not subject to arity restrictions but which is excluded when arities are recognized, due to one or more function symbols lacking a sufficient number of following symbols or nested structures; partial applicative structures in this sense can be semantically interpreted as designating "meta-functions" that, when the variables are replaced by fixed values, reduce to actual functions whose parameters are the missing symbols implied by insufficient arity thresholds. Of course, this discussion assumes we also have a notion of "fixed values" being assigned to functions as parameters.

equivalence class is thereby represented by one structure with such canonical presentation, and we can restrict attention to these structures in particular. Note that the generative rules for applicative structures have to be separated into two groups because (in the general case) we have to avoid unrelated symbols with the same "label" colliding according to the generative step whereby a nested structure is substituted for a variable. For a nested structure with its own collection of variables, the structure may need to be rewritten (cf. lambda-calculus alpha conversion) as an equivalent structure (but with symbol permutation)—not necessarily (indeed not usually) one in canonical presentation—so that each nested symbol is different from all symbols in the enclosing structure (unless the symbol representation is explicit, i.e., the point is to model a function application where the same input value is copied in multiple places).

Representing function arguments via "variable" symbols (which are understood to be substituted with values from some domain in the context of *calling* functions) is consistent with lambda calculus, and with the idea that applicative structures describe valid strings in a language codifying notions of function application. However, it is possible to develop alternative representations of the same structures, such as De Bruijn indices (due to Nicolas de Bruijn) that in effect model argument places by numeric indices rather than symbols [5].

We propose to use a similar convention based on the idea of forming symbols that have numeric values but also "colors." Specifically, consider, first, characterizations of the set of functions that may be identified on the basis of one function symbol—say, `f`, which assume to have a fixed arity a, say, three. Then `f` itself could be notated `f123` (using the color red to signify function symbols and blue numbers for variable symbols). Given `f`, we implicitly also have a family of related functions that are identical to `f` but operate on different domains, taking argument lists longer than `f`'s arity but ignoring the extra values: we propose denoting these via `f1234`, `f12345`, etc., using *gray*-color numbers for discarded extra elements. Conversely, we can also form functions with arity *less* than `f`'s by repeating numbers for the arguments, e.g., `f121` or `f122` where note that numbers occurring multiple times are shaded in light blue. We will refer to this as "red-blue-gray" notation—schematically, equivalent to notation with symbols (each colored number, i.e., color/number pair, being a distinct symbol) except that the colors allow grouping into ordered sets that may be subject to further constraints. Likewise, the function symbols themselves can be replaced by numbers encoding any ordering of the list of available function symbols that form the core of our applicative system.

Suppose we start with some function-symbol list—for instance, earlier in this chapter, we alluded to collections of kernel functions that "seed" a virtual machine wrapping access to CPS networks. Intuitively, it would seem to be a natural property of any computational or mathematical (or even linguistic) environment for which a notation of `f` itself would be meaningful—e.g., denoting a possible computation, or a mathematical function that yields a value when parameters are fully specified—that so too would be variants of `f` with greater or lesser arity as notated (here) through red–blue–gray strings. In other words, if we take some semantic predicate—

say, *describes a computation*—then we may want to consider the full set of functions meeting this criterion that can be generated from some initial "seed" collection. Given `f` as a seed, that is, one wants to elucidate the set of functions (expressible solely via `f`) that describe a computation, for instance (or replace "describe a computation" with some other predicate). We seem to have:

Lemma 1 *Assuming* `f` *designates a function with arity a, there is a unique list of red–blue–gray strings that each describes a function, mutually structurally distinct, enumerating all functions that can be described on the basis of* `f` *alone (including* `f` *itself). (Note that we are considering functions generated only by extending or contracting* `f`*'s arity; a different enumeration would involve recursive descriptions where a call involving* `f` *yields a value that is input to another call involving* `f`*.)*

Proof Consider strings with exactly a (not necessarily distinct) variable symbols. Without losing generality, we can assume that symbols (blue-colored numbers) with lesser numeric value appear to the left in their first token. Let $a' < a$ be the number of *distinct* variable symbols in a given string, i.e., the arity of the function being described as a variant of `f` (as compared to the arity of `f` itself). Then any a-length permutation of a' numbers (in the range $1 \ldots a'$)—restricted to cases where the first occurrence of smaller numbers is always before the first occurrence of any larger number—describes a structurally unique variant of `f`. Analogously (distinguishing blue and light blue), the darker-blue-only substrings are increasing by one starting at one, and light-blue symbols can be freely interspersed except that any light blue n has to follow the corresponding darker n. For each of these strings, consider the strings themselves and then modifications that add the gray-color numbers starting with $a' + 1$, and so forth, where the gray-colored numbers can only occur in an arithmetically (-by-one) increasing list. Each string generated in this fashion represents a structurally distinct variation on `f`, suggesting that the full set of strings generated accordingly embodies all `f` variations, at least those that could be notated by a language that encodes function calls through symbols representing input parameters.

Definition We will call a red–blue–gray string "lambda feasible" if it satisfies the restrictions employed above vis-à-vis darker-blue and gray numbers.

Generalizing this analysis to something like lambda calculus requires notating function composition, by substituting nested applicative structures for variable symbols. For this context, we propose extending "red-blue-gray" notation to include "black" numbers that stand in for nested structures. Call a "red-blue-gray-black matrix" a set of rows formed from red-, blue-, gray-, or black-colored numbers (unlike normal mathematical matrices, the rows need not have equal size, though if desired we can always introduce a "dummy" symbol, e.g., a black zero, to pad shorter rows on the left, yielding something that looks like a normal matrix; likewise, colored numbers can always be mapped onto disjoint integer subsets, so we can if desired treat red–blue–gray–black matrices as isomorphic to normal integer matrices).

Lemma 2 *For any fixed function symbol and (ordered) variable symbol set, there is exactly one (countably infinite) set of finite applicative structures in canonical presentation that can be generated from those symbols.*

Proof We will proceed by representing each applicative structure via a red–blue–gray–black matrix. The simplest applicative structures have no nesting, just a function symbol followed by a list of variable symbols. Since we are attending strictly to canonical presentation, we can permute the latter list so that lower-numbered symbols appear to the left, and distinct symbol numbers are allocated incrementally. Symbol numbers in this context function similarly to "De Bruijn indices" (in a presentation of lambda calculus formalized through de Bruijn's nonstandard notation) mentioned earlier. Lemma 1 argued that we can describe all variations on an `f` *without* composition as the set of "lambda-feasible" red–blue–gray strings with `f` (which can be denoted by a red number as well as a symbol) as the sole function notated. According to the construction rules generating applicative structures, we then have to consider substituting nested structures for variable symbols. Using red–blue–gray–black matrices, encode such nesting by inserting a black-colored number instead of a blue one, selecting black numbers according to the rule that lower numbers occur before (to the left of and above) higher ones, and that the black-number values map to the index of subsequent rows in the matrix; each row, then, is its own red–blue–gray–black string. Re-encode blue numbers in nested structures by adding to their value the smallest offset possible without conflicting with blue numbers in earlier rows. Call red–blue–gray–black matrices subject to these restrictions (on the black numbers as well as lambda-feasible restrictions for each row) "lambda-feasible" matrices. Based on their construction, we claim the set of such feasible matrices is isomorphic to the set of applicative structures, so that the lambda-feasible red–blue–gray–black matrices offer a systematic enumeration of all applicative structures.

There is a certain amount of mathematical bookkeeping implicit in the above presentation, which might obscure the fact that applicative structures are very basic; they should indeed be seen as rudimentary to the point of being rather uninteresting in themselves. More involved systems, however, emerge by relaxing certain conditions; for example, we might consider allowing self-referential applicative structures that express infinite recursion. The "red–blue–gray–black" matrix form points to one plausible avenue for modeling this process, because negative matrix entries could be allowed to "refer back" to prior rows, notating the idea of instruction sequences in a computing machine looping back to prior instructions. Other classical developments (often phrased within lambda calculus rather than applicative structures *per se*) involve encoding natural numbers via repeated iterations of a single "successor" function (one being the successor to zero, etc.), such that—again appealing to matrix notion—there is a sequence of matrices encoding the sequence of natural numbers. Any application of functions *to* natural numbers can then be encoded alternately as a *substitution* of these special matrices for the relevant variables.

Additional applications of applicative-structure theory turn on the notion that applicative structures model function-composition semantics. To the degree that we can (with suitable semantics) treat functions as *values*—as points in an ambient function "space"—then functions may be composed in various ways to generate new such values. In the simplest case (functions of arity one), at least without recursion, composition can be analogous to an algebraic operation—from `f` and `g` get $f \circ g$ (and $g \circ f$, which is generally different). Once one or more functions have arity two or greater, however, we have a set of multiple composition options—$fx(gx)$, $fx(gy)$, $f(gx)y)$, $f(gx)(gx)$, $f(gx)(gy)$, for example, are all potential compositions of a two-arity `f` with unary `g`, listing only non-root-recursive structures where `f` takes priority over `g` (the full list as such is larger, and, if we allow unrestricted recursion, infinite). We can therefore describe each form of composition via applicative structures, yielding a more complete description of function values' semantic terrain than is designated via individual function symbols themselves (this discussion leaves unaddressed the question of whether there are function values that *cannot* be encoded via applicative structures).

Depending on one's perspective, applicative structures can be seen as either essentially semantic or syntactic phenomena. Syntactically, we can treat these structures as characterizing valid strings in a language comprised of function and variable symbols, and one single grouping construct (via nested structures, which syntactically take the form of sub-terms that can be grouped into quasi-atomic units, substituting for the actual atoms, viz., variables). Certain syntactic formations, however, also seem to have semantic interpretations: as already observed, the $f \circ g$ (and $g \circ f$) compositions correspond to what are implicitly semantic relations, that is, the composition of two functions to yield a new function. Moreover, mappings *between* applicative structures sometimes appear to express semantic relations: $g \circ f$ is the compositional *inverse to* $f \circ g$, and the inverse of one (say, binary) function fxy could be notated as fyx—in short, we can extend applicative systems to treat certain non-canonical structures as notations for variational forms of functions derived from their "base" forms by substituting which argument is placed in which position, the simplest example being $fxy \rightarrow fyx$ (note the similar point mentioned earlier, that partial applicative structures can be read as designating "meta"-functions). The fact that some applicative structures thereby have semantic interpretations—even if we consider applicative systems as essential syntactic constructions (enumerations of valid expressions in a certain simple class of formal languages)—has led some researchers to consider applicative structures (particularly in the guise of Combinatory Logic) within fields as diverse as linguistics and psychology. Combinatory logic essentially uses a set of combinator symbols (external to both function and variable symbols) and string-reduction rules to enumerate all applicative structures generated by a system's intrinsic symbol lists [18, page 5], [14, 27, 56]: for any structure formed from symbols f_1, ..., x_1, ... (fs and xs being functions and variables, respectively), there is a combinator **C** such that the string $\mathbf{C} f_1 \cdots x_1 \cdots$ reduces to the relevant structure (where **C** can itself be a string of other, more primitive combinators). In this context, combinators play an enumerative role analogous to (what we are calling) red–blue–gray–black matrices (although we personally find combinators' reduction rules feel more ad hoc than the state-machine-like interpretation one can give to a red–blue–gray–black matrix; we'll leave the details to a footnote).[4]

The linguistics-based interest in combinators evidently reflects the idea that certain applicative structures (and intra-structure morphisms) codify compositions or modifications that express semantic operators, and, in general, certain applicative structures embody syntactic constructions that are sufficiently common or entrenched as to emerge as semantic conventions, not just syntactic forms (the underlying principle, articulated for example in construction grammar, being that semantic constructions originate at least in some cases from recurring syntactic formations, such that *syntax per se*—novel phrases, say, which rely on grammar rather than idiomatics to signal the intended composition—represents forms that are *not* sufficiently entrenched as to have "automatic" meanings, e.g., those of single words, but instead need to be parsed) [1, 13, 19, 26, 41, 50, 51]. In this context, combinators are at least intuitive emblems suggesting how syntactic entrenchment yields semantic conventions [6, page 6], [7, Chapter 9], [16, 49, 59]. More generally, we can also observe in the linguistic context that the overall space of applicative structures (over all words in a sentence, say) can be partitioned into subspaces (semantic entrenchment being one example, but we could also analyze different applicative structures as having more or less linguistic coherence, based on criteria such as verb-to-noun relations).

Apart from *syntactic* interpretations, applicative structure might also be seen from semantic angles in more narrowly mathematical contexts, for instance when we consider properties *of* functions, such as those derived from algorithm theory. As an example, if `f` denotes a function that may be evaluated through a calculation implemented with guaranteed termination, then composing `f` with another function with the same property yields (or describes) a different function that also by guarantee terminates (simply allow `g`, say, to terminate, and then feed its value to `f`). More generally, we can take the applicative system over a set of functions

[4] An intuitive way to picture combinatory logic is as follows: disregard the interpretation of function/variable symbols as functions and arguments, respectively, and consider merely symbol strings as expressions in a formal language. Extend this language with combinator symbols that are associated with reduction rules that impose a partial order on the set of permissible strings (such strings are differentiated by a grouping operation, potentially nested, as well as symbol lists). A canonical example is the **S** combinator such that $\mathbf{S}fgx \succ\!\!- fx(gx)$ (where $\succ\!\!-$notates reducing to) or the **B** that appears to connote composition (seen as semantic): $\mathbf{B}fgx \succ\!\!- f(gx)$ (or the famous "Y" combinator, which can be described by the rule $\mathbf{Y}g \succ\!\!- g\mathbf{Y}g$). Note that "reduction" here, however, at least in the mathematical presentation of combinatory logic, does not have an explicit semantic interpretation, but merely notes that some strings come before others in the $\succ\!\!-$partial order; moreover, strings in the language contain free-form admixtures of combinators, functions, and variables, which have no apparent semantic interpretation (the general premise being that only "maximally" reduced strings should be approached semantically). We find these details to render combinatory logic proper somewhat counter-intuitive, at least on its own terms (to be fair, though, although some papers systematize combinatory logic in isolation, it is more common to adopt combinators as merely notational conveniences for certain constructions in lambda calculus).

sharing properties such as guaranteed termination to be a larger set with the same property. In this sense, we can regard applicative structures as part of the backbone for analyzing different kinds of function spaces according to algorithm-theoretic properties or profiles (e.g., computability, termination, different complexity classes, and the like). These potential applications, alongside those mentioned above in the context of recursive function theory and infinite recursion, as well as encoding number theory/arithmetic, constitute some of the extensions to underlying applicative-structure theory that have some mathematical significance.

Our concerns here are less mathematical, so we will extend applicative constructions in a different direction, grounded in graph theory. We will suggest that applicative structures have natural correlates among (directed) graphs, and that this setting is in some ways more intuitive and less cumbersome than the above presentation appealing to notions such as "substitution" and function/variable "symbols."

Outside of mathematics proper, there is still some value in enumerating the full collection of applicative structures (generated by a preliminary function list). If we treat these structures as syntactic phenomena, then intuitively a plausible *semantics* for notions of function application would consider formally distinct applicative structures as semantically distinct, entities. The mechanisms for enumerating distinct structures may be less important than the mere fact of having a well-defined algorithm for producing and screening for all valid constructions. We use (what we call) red–blue–gray–black matrices to provide these criteria: intuitively, a reasonable semantics for applicative structures would recognize every formation described by a distinct such matrix as semantically distinct, and recognize the set of lambda-feasible red–blue–gray–black matrices as a complete listing of all valid strings in a language that describes functions (for some meaning of "function") derived from kernel "seed" functions by arity expansion (via unused arguments), projection (arity reduction through repeated parameters), and composition. Later in the section, we will discuss more specific semantics, but the enumeration with red–blue–gray–black matrices yields an initial picture of the space over which such a semantics should quantify.

21.2.2 Hypergraph Models of Calling Conventions

The definition of function application modeled via applicative structures (and similarly the lambda calculus) should be deemed insufficiently expressive for analyses related to modern programming languages. To be sure, in the above discussion, we were evasive about functions' "semantics," such that an incomplete account of "application" *per se* is arguably warranted by generality. When turning to programming languages in particular, however, we have more detailed semantic notions to work with; for instance, *functions* can be considered as computational *processes* that follow some *procedure* so as to (in general) derive an output in the presence of given inputs. Such an overview is still mostly intuitive and informal, but it begins concretizing the notion of "function" in *procedures* that, presumably, in the general case execute series of smaller steps in sequence (by contrast, say, to regarding functions set-theoretically as mathematical relations between inputs and outputs, each function being a subset of the total space of mappings conceivable between its domain and codomain).

Given the functions-as-procedures paradigm, a first step is to broaden the notion of applicative structure appropriately to accommodate how programming languages (via which procedures are implemented) recognize multiple *forms* of inputs and outputs (objects vs. ordinary parameters, say, or thrown exceptions as opposed to ordinary return values). To designate these variations in forms of inputs/outputs, we will describe procedure parameters as being grouped into "channels," which can be interpreted as gathering a node's adjacency set into (potentially) two or more partitions. We will introduce some nonstandard terminology, hopefully justified by the virtual-machine model that "drops out" of the theory we are sketching here, as a practical use case. The central underlying notion is to designate certain graphs as "syntagmatic" insofar as they model procedure calls with (in general) multiple parameters grouped into variegated "channels" (an explanation for the choice of the terminology *syntagm* and *syntagmatic*, usually encountered in linguistics, is offered in Chapter 6 of [45]).

Definition A *channelized in-neighborhood* (we can drop the "in-" when it is clear that out-neighborhoods are not relevant) of a directed graph is the in-neighborhood of one vertex v where the edges incident to v are ordered and grouped into (typically) one or more *channels*. A "channel" in this context can be considered, most generally, as any collection of edges (focusing on the edges themselves, not their incident nodes; in the case of labeled graphs channels would then be, in effect, sets of label tokens) but define a *syntagmatic channel* as a channel all of whose directed edges share the same target node; by default *channel* and *syntagmatic channel* can be used interchangeably. The central vertex v might be labeled with a string taken from a prior collection of names (intuitively, names of procedures insofar as graphs diagram procedure calls). Channels may also have "descriptive" labels (which serve to associate channels with channel "kinds").

Definition A *channel system* on a directed graph is a set of criteria constraining the legal channelized neighborhoods around any vertex. Example restrictions would be stipulations that vertices cannot have multiple channels with the same kind, or (in some context) restricted to specific possible channel kinds, or that channels need some fixed number or range of edges (e.g., a channel embodying procedural *outputs* may be restricted to have at most one edge). More detailed restrictions can apply to multiple channels in interaction. Consider modeling the alternative between normal procedure *outputs* and *exceptions*: throwing an exception precludes returning from a procedure normally, and vice versa. In terms of channels, this means that a non-empty channel representing (normal) outputs precludes a non-empty channel representing exceptions, and vice versa (more precisely, as clarified below, the two channels cannot simultaneously have edges out-incident to nodes with *non-void state*).

Definition A "lambda-restricted" channel system is one specific system that it is convenient to name ahead of time, intended to mimic a minimal lambda calculus. See the following lemma.

Instead of an open-ended notion of "applicative structures," we can speak of applicative systems *constructed relative to* channel systems. This more general sense of applicative structures builds up incrementally from channelized neighborhoods conformant to the relevant channel system. By way of illustration, consider a graph-theoretic encoding of "classical" applicative structures (the form embodied in lambda calculus) as defined earlier.

Lemma 3 *Any "classical" applicative structure can be modeled via channel systems with the following characteristics: there are two channel kinds—inputs and outputs—and output channels must have exactly one edge. Call channel systems subject to these limitations "lambda-restricted" as anticipated above.*

Proof Proceeding by induction on the generative rules for (the kind of) applicative structures defined above. The simplest case is one function symbol followed by some number of variable symbols. Express this via a neighborhood with one "procedure" node (standing for a procedure to be called) and a collection of "argument-nodes" with edges pointing in to the procedure node (all these edges grouped into one channel). Next, we expand outward to encode nesting (substituting structures for single variables). Consider two channelized neighborhoods N_1 and N_2 restricted to the channel system described in the lemma statement (inputs plus exactly one output). We can form a connection from N_2 to N_1 by inserting an edge between the vertex in N_2 that is adjacent (in the outgoing sense) to the single output-channel edge in N_2 and one of the input-channel edges in N_1. That is, the out-adjacent vertex in N_2 will now have two out edges, one targeting the function node in N_2 and one targeting an argument node in N_1. We can see by induction an equivalence between this graph-theoretic model and the original closure definition for applicative structures. Indeed, introducing these connecting edges is structurally analogous to "substituting" nested structures for variables. We will clarify that claim with a further definition.

Definition The *syntagmatic neighborhood set* of a vertex is an expansion of channelized in-neighborhoods by connecting a different neighborhood to an initial neighborhood via an edge between a "peripheral" node in each neighborhood (i.e., excluding the vertex in-adjacent to other nodes in the neighborhood). Because each channelized neighborhood could have its own syntagmatic neighborhood set, these structures can model nesting. Stipulate that any syntagmatic neighborhood set proper has exactly one channelized neighborhood that does *not* have a connection to another neighborhood, i.e., all of its vertices have at most one out edge. Call the central node of this neighborhood central for the neighborhood set overall, and stipulate that syntagmatic neighborhood sets must be *connected* in the sense that the central node of each component channelized neighborhood must be connected to the central node (via paths allowing either in- or out-adjacency).

Definition An *input ring* around a syntagmatic neighborhood set is a collection of labeled nodes (each with distinct labels) that are not otherwise parts of the construction. Consider the ring nodes to be connected with (non-procedural) nodes via a "supplication" edge (intuitively, to represent the idea that the node is associated with a value based on its supplier node's label); two nodes adjacent to the same ring node via such edges have "shared supplication."

Lemma 4 *Syntagmatic neighborhood sets within a channel system restricted to output/input kinds (and max-one output channels) can be unambiguously encoded via matrices equivalent to "red-blue-gray-black" matrices represented earlier for applicative structures in canonical presentation.*

Proof Label procedure nodes (i.e., center nodes) with *red* numbers based on their string labels (assigning like numbers to like strings). Edges in channelized neighborhoods are ordered, so form matrix rows by notating the procedure number followed by *blue* numbers assigned to argument nodes; if an argument node shares supplication with a different node already numerically labeled, adopt that number, otherwise adopt the least available numeric label. This applies only to argument nodes that are not connected to other nodes across syntagmatic neighborhood-set connections. In the latter case, label the "nested" channelized neighborhood with *black* numbers (starting at one for the central node overall and incrementing as needed) and, for argument nodes connected to other neighborhoods, insert a negative integer whose absolute value is the external neighborhood's index number. Create a matrix row for each channelized neighborhood, in order of their index numbers. The resulting matrix will structurally mimic red–blue–gray–black matrices in the context of applicative structures outlined above.

Theorem 1 *Applicative structures as presented earlier and syntagmatic neighborhood sets within a lambda-restricted channel system are isomorphic to the same set of lambda-feasible red–blue–gray–black matrices, assuming they share the same set of function symbols/procedure labels.*

Proof Lemmas (1) and (4) detail the construction of red–blue–gray–black matrices from applicative structures and syntagmatic neighborhood sets, respectively. We claim that comparing the two constructions documents that each step in the construction would modify the resulting matrices in equivalent ways, and so the set of matrices generated from applicative structures will be identical to that generated from syntagmatic neighborhood sets—both are precisely the lambda-feasible matrices according to the earlier definition of lambda feasibility that precludes infinite recursion.

In other words, lambda-restricted syntagmatic neighborhood sets are a graph-theoretic encoding of applicative structures in the classical lambda-calculus sense.

We have formally reviewed "classical" applicative structures, however, primarily to demonstrate that within the context of channel systems these structures represent only one (relatively restricted) model of function/procedure application, characterized by a simplified (input/output) channel semantics (plus at most one output per function). The constructions for channelized neighborhoods and syntagmatic neighborhood sets can be naturally extended to more expressive channel systems, yielding (so we claim) graph representations that are more consistent with actual programming languages (whereas lambda calculus models a kind of abstract programming language for purposes of mathematical treatment).

There are at least two significant extensions that can be made when transitioning from "lambda-restricted" channel systems to ones that are more free form: first, channels can have multiple kinds (beyond just input and output), and, second, it is possible for the output to one procedure to be a *function value* that in turn is applied to other arguments. The latter scenario implies that an output node (in one neighborhood) can be linked (via a directed edge) to the *central* node of a different neighborhood, not just to one of its argument nodes (in this case, the central node would not be labeled with a string denoting a function name but rather would receive a function value from that output edge).

Definition *Syntagmatic Graphs*: Consider syntagmatic neighborhood sets as above, with the following generalizations: each component syntagmatic neighborhood set may have channels of multiple kinds (though we can maintain, as an optional restriction, the stipulation that each neighborhood has at most one channel of each kind), and the central/procedural node of a neighborhood may have an incoming edge that passes a function value assigned to that node. This latter possibility can be accommodated by defining a special "function value" channel kind that may be occupied by at most one edge; the non-central node incident to that edge can be considered a special form of argument node, one that has an incoming edge from another neighborhood (representing a passed function value), but instead of this argument node being an *input* parameter to the central node's procedure, it is a function value representing the actual procedure which that node iconifies. This formation would thereby model situations where nodes encapsulate numerous possible procedures and the actual procedure designated is dynamically calculated (presumably just prior to the function call that is notated via the syntagmatic graph).

Intuitively, each syntagmatic graph describes a structure connecting one or more procedure calls, interconnected by parameter passing—outputs from one procedure become inputs to another. Working in a more flexible "channel system," however, the connections between such calls can have more subtle relationships than output-to-input chaining. For example, multiple nodes in a syntagmatic graph may represent the same "variable" (cf. "shared supplication" nodes from earlier) that in turn might be modified by one procedure, so an input given a new value by a procedure and then passed to another procedure is a form of inter-procedure precedence—the effects of the first are consequential to the second—even if the two are not linked by an explicit output-to-input hand-off. These are the kinds of scenarios that a procedure-call semantics reflecting actual programming languages (as opposed more strictly mathematical function theories) should address; they will be the focus of Sect. 21.3.

21.3 Semantic Interpretation of Syntagmatic Graphs

Assuming we remain within the context of applicative structures proper, the point of these formations is to represent the idea of functions (in some sense) that take input values. Notating this process via variable symbols allows function composition (and arity-reducing projections) to be described; thus we have "substitution," replacing variable symbols with concrete values. When all free variables in an expression describing (potentially nested) function applications are thus substituted, we have sufficient information to evaluate the function, or—in the sense of lambda-calculus "beta" reduction—reduce (or "collapse") the aggregate expression to a single resulting value. This is the central dynamic figured by applicative structures in their simpler, classical sense—input parameters yield applications that "reduce" the inputs to a single output value that may in turn be input to other functions. Chaining inputs and outputs in this sense engenders a model of computations as graphs, where edges encode how values travel between applications, with multiple inputs potentially each being outputs from multiple precursor function applications.

The semantic interpretation we adopt here for "syntagmatic" graphs is noticeably different than this model (we have analyzed the specific differences in [45, page 150]). Rather than taking input/output as a fundamental divide within function parameters, we consider more general channel systems here. Note that in contrast to (more typical) graph representations of computation where outputs are marked by directed edges *away from* procedure nodes, in syntagmatic graphs (hereafter SGs) all edges point *into* procedure nodes, including those embodying "outputs" ([45, section 9.2] has some comments about why this can make sense). More elaborate rationales for the notational and interpretive differences between SGs and other procedure models are tangential to the current chapter, so we will focus instead on outlining the semantic interpretation itself.[5]

Definition Call a *marking* of a syntagmatic graph to be an association between some of its nodes and a collection of typed values, against some type system. Markings might be context-dependent; that is, a graph could be subject to multiple markings concurrently, each restricted to one context. We will say that values are *bound to* nodes, but indirectly, as explained in the following.

Definition Syntagmatic graph nodes can be associated with *types* and *states*. For the present, we will not rigorously define "types," but we will comment that types are introduced vis-à-vis nodes *through* states. For any type, each possible instance of that type is a potential *state* for syntagmatic graph nodes, the state of being "occupied" by that specific value. However, not every node state need correspond to a type. In particular, nodes can have a state corresponding to a "void" or lack-of-value.

[5] Some of this terminology is derived from Petri nets [2, 24, 34, 36, 38, 46, 48, 53].

That is, we approach the notion of "void" as a *state* possibly evinced by nodes, rather than through type systems themselves; we do not need a "bottom" or "nothing" type with some special value selected (essentially by fiat) to represent non-initialization or pre-initialization. We find it more semantically coherent to express such "nothing" states as the state of *having no value*, rather than as the presence of a construed "nothing" value (and nothing type which it instantiates).[6]

Against this background, the semantics of procedure calls can be expressed via before-and-after states for each argument node incident to a procedure node. The procedure's semantics is, as such, the cumulative state changes, or *change in marking state*, effectuated by the procedure. In a general case, some of these changes will be void (no-value state) nodes transitioning to having a value; these would typically correspond to "output" nodes in classical applicative models. However—accommodating scenarios such as mutable references—nodes with the state of binding to some value could migrate to the state of binding to a *different* value. In general, notions such as input-versus-output are expressed in this system "semantically" as a manifestation of state changes, rather than "syntactically" as edge direction or arrows assigning different directions to inputs versus outputs.

Roughly as an analog to "beta" reduction, we propose the term *digamma* reduction to express marking-state changes due to a procedure call. A digamma reduction is the cumulative state change in all nodes affected by the procedure (or, seen syntactically, all nodes adjacent to the procedure node). For multi-(channelized) neighborhood SGs, digamma reduction happens in multiple stages, each contextualized to a single neighborhood. Note that we like the term "digamma reduction" partly as an oblique reference to "sigma" calculus (an object-oriented extension to lambda calculus, and the Greek digamma numeric symbol looks like an enlarged lower-case sigma) and partly because "gamma" is a common symbol for graphs, so "digamma" suggests "two graphs," or a computational interaction between one graph assembling a procedure call and one graph implementing it.

Semantically, then, SGs model digamma reductions *in sequence*, each localized to individual channelized neighborhoods. This idea can then be extended to sequences of SGs in turn. We are getting closer to representing actual computer code; one further step might then be incorporating something like "stack frames." Assume

[6] Note that in type theory we can also define "bottom" types via constructions that preclude bottom from ever being instantiated at all (e.g., by stipulating that bottom is a subtype of any other type, i.e., the subtype relation between that and any type is defined to be true). A type system may have an uninhabitable type insofar as predicates on intertype relations (such as subtype-of) can be non-paradoxically defined on the full collection of types, including uninhabited ones. A bottom type does then have a reasonable correspondence with uninitialized code symbols because this is the only type that correctly characterizes the state of such symbols prior to their being assigned a value (without a value, there is no inhabited type where we can say the symbol's value is an instance of that type). Still, the purpose of assigning a type to symbols in practice is to constrain how symbols can be used (when passed to procedures, in particular), and for these purposes, it is arguably misleading to use type attributions *per se* to model value lifetimes. For this reason, it seems better to capture the fact of pre-initialization via a *state* characterizing the symbol itself rather than a *type* attributed to its value.

that SGs are represented *in sequence* and moreover that such sequences occur in the context of a *symbol scope*, or an environment where string labels can serve as "carriers" for typed values; we will sometimes refer to "SG-scopes," scopes contextualizing SG sequences wherein symbols designate carriers that (when in an initialized state) hold concrete values. In particular, assume that carriers can reveal "carrier states" reciprocating the states defined on nodes earlier: for any type, there is a spectrum of states equivalent to each type instance, but there are also states involving *lack* of typed values (the generic example of such a state we will call *pre-initialized*). The states available for carriers could be called "type-based" in that they are organized around typed values but include non-type states as well. SG nodes may then take on states by associating with carriers such that carrier state propagates to the nodes. Nodes are in "shared supplication" if their carriers are synchronized to be perpetually in the same state (this can be considered a refinement of the earlier definition). Symbols in SG scopes might derive from multiple sources: assume that scopes' symbol list can be grouped into *declared* symbols internal to the scope, *argument* symbols bound to procedures' signatures, and (potentially) *global*, or perhaps "ambient" symbols shared among multiple scopes (we will use "ambient" for the general case and "global" for analogs to "global variables" in typical programming language).

Definition *Tripartite Scope*: We will call SG scopes *tripartite* in that they are mappings from symbol names to carriers (or groups of state-synchronized carriers) with type-based states, and symbols can be classified as *declared*, *argument*, or *ambient*. The semantics of these groupings will be clarified below.

Definition We will say an "SG-described procedure" (or just "procedure" when the SG topic is obvious) is an SG-sequence unfolding in a tripartite scope where argument symbols are bound to values held by carriers in a *different* procedure (or, recursively, the same procedure in a different context, viz., different symbol bindings). Such interactions between SG-described procedures model the semantics of procedure calls. Of course, procedure calls in turn are modeled by channelized neighborhoods in individual SGs. Specifically, channelized neighborhoods embody procedure calls wherein peripheral nodes' carrier states in the *calling* neighborhood become the basis for initializing carriers in the called procedure; specifically, the carriers bound to argument symbols in the latter's tripartite scope. The switch in execution context from the calling to the called graph is analogous to binding symbols to values in process calculi, or beta-substitution in lambda calculus, except that carriers' connections (via their argument nodes) to procedure nodes are organized via channels, which can affect procedure-call semantics. For example, it may be stipulated that carriers in channels modeling "outputs" are understood to be in pre-initialized state for the duration of called procedures' execution, and attempting to utilize (e.g., read values from) such carriers *within* the procedure is a logical error. There are various ways of enforcing this kind of restriction, of course, but channels offer a convenient representation of the relevant constraints viewed under a semantic interpretation.

We now have a semantic interpretation of channelized neighborhoods: each such neighborhood embodies a procedure call, which entails binding call-site carriers to procedures' argument symbols and then carrying through other procedure calls notated via SGs associated with the *called* procedures. When they are finished, the original carriers will (potentially) be in a modified state, so the called procedure effectuates a "digamma reduction" on carriers around the calling site. State changes then propagate across channelized neighborhoods insofar as values from output-like channels in one neighborhood are handed off to input-like channels in later-executed neighborhoods. Each SG has a "top-level" neighborhood that is the last to "execute" (viz., compel a procedure call) or, from a different perspective, the root of a directed acyclic graph showing carrier hand-offs and shared supplication across channelized neighborhoods. Note that output channels on the top-level neighborhood cannot be connected to other neighborhoods in the form of procedure outputs become inputs to a subsequent procedure (there is none)—so either top-level procedure calls are useful solely for their side effects or the carriers in their output-like channels are bound to symbols in the current scope. In the general case, we will say that an SG is *anchored* by one or more symbols if those symbols (strictly, the carriers associated with them) acquire values from the SG's top-level neighborhood (particularly from the channels there that obey output-like semantics).

The mechanism just described constitutes how *declared* symbols in an SG tripartite scope can acquire values; they become initialized from procedure calls enacted *in* the procedure, rather than from arguments passed *to* the procedure (or from an ambient space of values visible from multiple scopes). Once initialized, declared symbols can then supply values for procedures likewise called *inside* a procedure—that is, an argument symbol in one procedure can be derived from a declared symbol in a calling procedure. It is reasonable to assume in the typical case that all procedure symbols (whether declared, argument, or ambient) hold values that originate from a *declared* symbol somewhere, so the anchoring points where declared symbols are first initialized represent precisely the points in any holistic collection of SG-described procedures where new values are "introduced" into the system.

This last point has implications for how we theorize *typed values* as well. Implicitly, here we have assumed that we are operating in a strongly typed system where types and values are interconnected: any value is *typed* (an instantiation of a type) and types exist by virtue of their possible values. We assume that in the general case types are not equivalent to their *extensions*, that is, to any fixed set of values. There are indeed some types whose extensions are fully "enumerable," so we can specify exactly how many possible instances a type has: the type corresponding to signed one-byte integers, for example, denotes precisely the set of numbers at least -128 and at most 127 (the analogous range for *unsigned* bytes is 0–255). However, for types such as *lists* of integers, the extension cannot be specified a priori, in the sense that it is impossible to know from a specific computing environment (at a specific moment in time) whether or not a particular value (here, a particular list of integers) that logically fits the type's criteria can in fact be represented. The size of a list that may be instantiated is limited by factors such as the computer's available memory; in theory, the extent of a collection type like "list of integers" is infinitely large because there is no rule to group all potential such lists into a finite set (if there were, take the longest list, an an integer to the end, yielding a new list that has no reason not to be included in the first place). We assume moreover that we are working in a *non-constructive* type environment where we cannot consider types' extensions as a logical construction abstracted from computational feasibility in concrete computing environments. We can discuss types' *hypothetical* extension, which may indeed be infinite, but we assume we need some other mechanism to manage uncertainties regarding types *actual* extension (in some contexts, such as lazy-evaluated lists or ranges, working directly with a logical model abstracted from actual realizability is appropriate, but in contrast to functional programming theory, for example, we do not take such "constructive" formalizations as an essential aspect to the relevant type systems).[7]

So as to give a rigorous semantics to *non-constructive* types, instead, we leverage the interconnections between types, symbols, and SG nodes insofar as we are working in an SG context. Recall the heuristic that all symbols acquire values originating from declared symbols, which in turn have a specific initialization point (although some symbols might be initialized by non-constant reference, passed into a procedure for initialization rather than being bound to a procedure result, the values they are bound *to* in this case will itself be determined by an anchoring initialization, either in that procedure itself or some other procedure called in turn; so, even if not all declared symbols receive their values via anchoring, all declared symbols receive values that *originate* from an anchoring somewhere). In other words, when we say that a symbol has some typed value, we mean that a symbol's value is derived via initialization from an anchoring point, where a declared symbol's value was set via digamma reduction at an anchoring point—the symbol was associated with a node that transitioned from a pre-initialized to a type-bound state as a result of binding to nodes in an output-like channel. Symbols may be "re-anchored," initialized to new values due to anchoring points, but (assume we work in a fixed-type environment) symbols and carriers retain the same types once they are initialized, so long as they remain in an initialized state. Types themselves, then, can be construed in terms of continuities in carrier state: once a carrier is initialized to be in a state determined by a type instance, it will remain in states determined by the same type so long as it is initialized at all; and all initializations can be traced to anchoring points, together with synchronization among states of distinct carriers. We do not need to theorize types' logical extensions in general because we are not concerned with the totality of a type's possible values, only with the states directly associated with the type at specific anchoring-initialization sites.

Many programming languages distinguish between *constructors*, which are intrinsic to a type's definition, from other procedures that return values of a given

[7] See [45, Chapter 4], for a more extensive analysis of non-constructive type systems and why assuming that types are non-constructive in general (rather than the opposite) is practically appropriate.

type. To the degree that this distinction is in effect, any value can be traced back specifically to a *constructor* for its type. To be sure, a value might be bound to the output of a *non*-constructor procedure, but in that case this return value will have been formed initially by a constructor proper called in that procedure, or perhaps in a procedure it in turn calls, iteratively. In a channel system we can leverage a comparable distinction by defining a special kind of channel, which functions like a normal output except that procedures whose signature includes such a channel are considered to be constructors, or analogous to constructors (in [10] we used the term "co-constructor" to suggest allowing procedures being analyzed as if they were constructors even in programming contexts where, for technical reasons, they would not be classified as such according to a programming language whose code is being analyzed). For generality, we will call procedures with constructor channels "co-constructors" with the assumption that different environments may be more or less lenient in conditions where co-constructors may be declared (only in code specifically defining a type, say, in contexts where a narrower policy is warranted, but without imposing restrictions along these lines a priori).

The significance of co-constructor channels is that they allow us to pinpoint the specific anchorings where new typed values are assigned ab initio. Assuming an SG system uses co-constructors, for any carrier in a type-bound state we can be sure that the value traces back to a digamma reduction on a carrier specifically situated in a co-constructor channel. As such, our semantics for types may not give a set-theoretic account of types' extensions, but we do have a mechanism for locating the origination point for all typed values. In particular, values are acquired via carrier state changes localized to co-constructor channels; such a change only occurs when a procedure has completed, with an initialized value in that channel available to be bound to its eventual symbol. Any value therefore witnesses the fact that a procedure with a co-constructor channel completed and triggered a state change accordingly. If a type system is designed to recognize types as (associated with) certain contracts, such that a type being instantiated confirms certain properties about the value thereby manifest, we can recognize such contracts in an SG-style system so long as co-constructors adhere to type-specific contracts in the course of depositing values in a co-constructor channel.

Once a carrier with a given type is initialized, its value can of course be handed off to other carriers assigned the same type (or potentially a supertype thereof, or some other related type wherein casting is possible). Channel systems and SG representations do not preclude polymorphism—specifically, the labels assigned to procedure nodes need not uniquely identify one single procedure available to be called, but may instead provide a premise from which to select one of numerous available procedures, each with the same name—so type resolution may come into effect: if there are multiple candidate procedures, the one whose signature best matches the procedure call's argument types is selected. In the context of channels, such disambiguation extends to recognize different channel kinds, which become part of a function's signature. In effect, each node in a channelized neighborhood has a type (based on its carrier state) that might be matched against candidate procedures' signatures. Types therefore are consequential primarily in the context of overload resolution. The issue here is not types' extension *per se* (two types having the same extension would have no bearing on their appropriateness for matching call-site neighborhoods to procedure signatures; and co-extensive types are not interconvertible unless such casts are implicitly declared according to the same conventions as casts between types with different extensions). However, since successfully matching call sites to signatures implies that all affected carriers' types are suitably aligned, whatever guarantees on carrier states that may be implicit in type-bound states become transferred from the calling to the called procedure, insofar as the formers' carrier states carry over to the latter's argument nodes. Therefore, carrier states associated with being initialized according to a given type propagate from co-constructor anchorings across all subsequent procedure calls, with type-specific guarantees propagated alongside.

In effect, types' semantics in such a framework is based on the premise that once there is an *originating* initialization via digamma reduction in a co-constructor channel, then there is one specific carrier whose state is thereby changed, and subsequently (presumably) a series of further procedure calls where that originating carrier's state becomes synchronized with subsequent carriers in subsequent procedure calls. Each type is essentially a premise warranting the propagation of this shared carrier state: the type's presence as a condition on procedure signatures or on declared symbols indicates that a carrier whose state is synchronized in accordance with such propagation would be in a valid state according to the expectations of the procedure that uses the corresponding value. For the sake of discussion, we will refer to this approach to the issue of type semantics as a *propagation semantics*, in contrast to a *set-theoretic* semantics that would construe types' semantic interpretations in terms of their extensions, or a *constructive* semantics that would read type semantics through the logic of constructive patterns that could give rise to values of a given type. The essential point about propagation semantics is that type contracts would be checked at origination points (to the degree that they are in effect) and presumed to have been enforced whenever there is an originating digamma reduction in the first place; and that subsequently any use of a type as a polymorphism-disambiguating device should proceed under the assumption that any carrier state inherited as synchronized with the state of an originating carrier in this sense should be deemed a valid state for the called procedure's purposes.

This discussion regarding type semantics has passed over some non-trivial details, such as issues of type casting and inheritance, which we will briefly address in the next subsection.

21.3.1 Distinguishing Non-constructive from Extensional Type Semantics

A plausible objection to the above presentation is that it is burdened with special terminology and representational conventions to define constructions that are not radically different from traditional notions of, say, stack frames and lambda abstraction. Beyond just expanding from inputs and outputs to more flexible "channel" systems, we have constructed a semantic interpretation for these systems based on formulations such as type-based states and channelized neighborhoods, without clarifying the theoretical merits of this overall semantic presentation. Our contention is that the full semantic details (not just, say, extending applicative structures to richer channel systems) become valuable in the context of virtual machines. We will attempt to warrant this claim by addressing VM implementation, in particular.

We suggest the idea of digamma reduction and propagation-from-constructors to replace extensional and/or constructive semantics for type theory. Implicitly, then, we propose that (what we here call) "propagation" semantics can displace two other paradigms for type theory, *extensional* and *constructive*. We will try to defend the idea first that this is a reasonable three-fold division in options for type "semantics" and second that there are formal or theoretical merits to the "propagation" approach.

An extensional semantics would essentially proceed by grounding the semantics of any specific type in the set of its possible inhabitants. To attribute a type to a value is thereby to assert that the value belongs to the type's extension set. This paradigm has certain benefits, to be sure, such as providing an elegant approach to functions defined on a type that have different behaviors for different parts of the extension. As a canonical example, a procedure operating on list-like types may require one implementation when the lists are empty and a different one when they are non-empty. Writing functions with distinct code bodies for different extensional alternatives (another example would be types with possible "null" values, such as **Maybe** in Haskell or pointers in C/C++) is a widespread practice in functional programming. Extensional semantics gives a concise reading to this coding style: insofar as types semantically are effectively equivalent to their extensions, any function defined over a covering collection of extension parts is equivalent to a function defined over the type as a whole.

On the other hand, we can model extension partitions and the prospect of quantifying over them as a feature for procedural implementation without actually equating types with extensions; we merely need to posit that types can be *associated* with extensions in such a way that extension partitions are well-founded. Even where types have nondeterministic extensions, such associations work so long as code employing their partitions tolerates similar nondeterminism at the partition level. For example, the nondeterministic facet of list-like types lies with non-empty types because there is no way to know how large instances can be (in practice), so this uncertainty translates to the part of their extension including non-empty instances. But code working with such types has to accept uncertainty anyhow (for instance, by recognizing that extending a list's size might fail), so the proper accommodations need merely to be accepted in code narrowed to an extension part that would be present in code without extension alternation.

A larger issue is perhaps *how* extension-partition alternatives should work; in particular, whether mapping a value to one partition or another is a compile-time or runtime operation. The code-over-alternatives paradigm is not only a notational convenience; it manifests benefits of functional languages' type system that allows questions like whether lists are empty to be resolved at compile time (at least more often than other languages). If we have a function written so that the code branches differently for empty and non-empty lists (or, say, for valid and null pointers), the relevant programming language might treat this as an alternative to be resolved at compile time or as something deferred to runtime (so that the alternatives exist more or less like if-then execution paths), or some combination. Ideally, it seems, a language engine would compile-time optimize when possible but allow decisions to be deferred to runtime when necessary (rather than insisting on compile-time uncertainty renders code non-compilable at all). By analogy, a compiler might reasonably stipulate that type-attribution ambiguity (as when procedures have two equally valid overloads, say) is a logic error such that offending code should not compile (unless types are deliberately chosen for pre-runtime ambiguity as with "variant" types, such as **boost::any**) but ambiguity in how to map values to types' extensional *partitions* should not necessarily foreclose compilation in the same manner.

These, however, are issues in the context of tracking compile- and runtime information about values, and quality of a language's support for managing value-partition mapping is an orthogonal matter to the semantic interpretation of partitions. In general, languages can take advantage of extension-based reasoning without embracing the theory that types are semantically derived from their extensions.

Moreover, we can similarly distinguish types' semantics from construction sequences, in that constructive type theory essentially works by using construction patterns as a model for type extensions, so that the extension embodied by that model fixes the type's semantics. Every non-empty list, say, may be derived from a smaller-by-one list by appending a specific value to its end; so for any list there is a specific construction sequence that terminates at that list, and we can take the set of those sequences as a model for the list type's extension. Extension partitions can then be defined over distinct construction patterns. In the case of lists, non-empty list derivation via smaller-by-one instances is analogous to marking values that can be constructed according to a specific pattern (involving a different list and an **append** operation), so the non-empty part of the extension is precisely the set of instances constructible via that pattern. This is the rationale for functional programming languages branching over construction patterns.

Constructing lists from their smaller-by-one kin actually leverages two facets of "constructive" types: first, that for any given value of a given type, we can identify a constructor that produces that value; and, second, that construction sequences are isomorphic to the type's extension and therefore "model" the type, in the manner that sets endowed with specific operations model "theories" (i.e., axiom combinations) in symbolic logic. Unless we consider these notions only as logical abstractions, however, in concrete programming they are often wrong, or at least misleading. If a procedure is supplied a pointer to a list (any list-like type), there is no way to determine a construction sequence for that value short of examining each item in the list one at a time—especially if the pointer is derived from a binary stream supplied by some external component, along the lines of cross-application interop discussed in Chap. 20. There is not even a guarantee that we can learn the *length* of the list without similarly scanning the whole value. The presence of a **size**()-like procedure depends on the public interface, and there are scenarios (such as **stacks** and **queues**) where the type may only provide **push/pop** (or **enqueue/dequeue**) access to the collection's state; with no size-related (public) functions apart from a test for **empty**(). In these cases, functional programming idioms related to treating each value as a destructured head/tail pairing have no optimization value compared to procedural if...then branching (although a language could still support these idioms as "syntactic sugar").

Of course, collections types *could* be engineered so that destructuring becomes convenient. If resizable lists are stored via multiple inverted arrays (so that the list head always has smaller memory addresses than its preceding elements), then obtaining a reference to the smaller-by-one "tail" becomes just pointer arithmetic (we might use multiple array allocations rather than one single array to ensure that elements never have to be moved in memory, which would invalidate pointers/iterators, if the list's size grows beyond the memory initially requested for it). In general, then, the idea of "constructive" types—seen not as a theoretical issue of type semantics but as a tool modeled in source code—depends on how types themselves are engineered, their memory management protocols and public interface. Similar comments could be made about extension partitions. Rather than a logical paradigm, we can treat *extensional* and (or and/or) *constructive* semantics as *design patterns* that may guide *some* types' implementations (in the sense of code libraries that realize types via providing implementations of their requisite procedures).

Again, though, the fact that some types might be associated with extension partitions defined via construction patterns need not signal the semantic equivalence of types and type extensions (or type extensions and construction sequences). While recognizing extensional reasoning as a useful tool for language implementation, we believe that it is more conceptually accurate to model types' semantics instead via "non-constructive" notions we discuss here, such as digamma reduction, co-constructor channels, and carrier-state propagation. In other words, we believe that this latter group of formulations better capture how people reason about computer code and design patterns *in the general case*, only switching to the alternative semantics when types' explicit designs warrant as much.

This argument, however, is localized to the question of defining semantics for type systems, which is arguably a philosophical issue more than a directly practical one. We intend to refine the discussion by considering how the semantic notions hereby sketched out could yield practical benefits. In this context, we will focus on the hypergraph structure of computer code (within the framework of "syntagmatic" graphs) and the use of channels as structuring vehicles.

21.3.2 Syntagmatic Graph Sequences as a Virtual Machine Protocol

We are representing syntagmatic graphs as hypergraphs, in two senses: partly, for one, because *channels* embody a grouping operation (gathering multiple edges, by analogous to hypernodes being sets of nodes—although channels are not identical to hyper*edges*, wherein *one* edge spans multiple nodes). Second, we also leave open the possibility of SG nodes having internal structure (i.e., becoming hypernodes). However, such hypergraph representation is mostly heuristic; technically, we consider hypergraphs to be essentially a visual shorthand for *virtual machine* constructions. Here we appeal to comments we made in Chap. 20: a hypergraph (or, in essence, any structured representation) iconifies the series of steps needed to construct it in its precise state. For example, focusing on channelized neighborhoods, the elements of such structures—procedure nodes, peripheral nodes, and channels—correlate with operations setting out these details sequentially (identifying a procedure node, e.g., by label; identifying a channel kind to serve as a context for subsequent peripheral nodes; initializing the latter nodes; switching to a different channel kind as needed; and so on). This outline could be expanded to syntagmatic graphs in general by aggregating multiple channelized neighborhoods and then asserting where carrier hand-offs (via cross-neighborhood edges) occur; to SG sequences by constructing multiple SGs along with their respective anchorings; and finally to procedure descriptions by enumerating declared, argument, and ambient symbols in a tripartite scope. From this perspective, SG procedures can be seen as compact representations summarizing sequences of VM operations, or perhaps as configurations that guide the construction of VM procedures by defining an end goal toward which VM operations approach incrementally.

Ordinary VMs, of course, depend on such concepts as stack frames and procedure calls; one of the fundamental operations (or sequences thereof) typical VMs carry out involves pushing input values onto a stack and then redirecting to the address of a procedure that uses them, at least insofar as the VM in question emulates assembly code to provide a runtime for higher-level programming languages. Modeling procedure calls via hypergraphs, and in particular *channels* (in the sense we propose here) adds representational parameters to this model, by analogy to how hypergraphs present a richer expressive tableau than other metamodels for encoding data structures. The benefits of added metamodeling detailing the procedure-call context actually help illustrate why expressivity pays dividends with respect to data structures (e.g., in query evaluation) as well.

The primary rationale for VMs is to execute code, of course (whether scripts, queries, or some other programming category that does not feat neatly into one or the other, such as workflows, or one or both ends of remote-service functionality). Secondarily, VM compilation could support static code analysis even if the intermediate code is not actually run. Code analysis factors into execution as well, at least to the degree that a VM can internally support runtime checks and guarantees—insert debugging breaks where certain conditions are met, preventing code from running in certain circumstances, allowing procedures to make contract-like assumptions (stronger than type checks alone could enforce), and so on. For example, stipulations that values should fit within a fixed range (narrower than theoretically possible via their types) could potentially be verified or implemented by weaving "gatekeeping" code into VM operation sequences. In short, VMs acquire more flexible capabilities—they present opportunities to implement requirements engineering style features, even if only via add-ons—to the degree that they present op-sets and data models configured for static and dynamic/runtime analysis. These observations motivate our proposals to augment the modeling parameters for VMs (in procedure-call contexts) via formulations such as channels and carrier state.

To be sure, a suite of static-analysis capabilities is endemic to traditional (e.g., stack-based) VMs as well (reachability, control paths, initialization guarantees, etc.). We would suggest, however, that future generations of VM technology will seek a broader scope vis-à-vis contexts where code analysis is applied, considering GUI programming, cyber-physical systems, multi-modal front ends, AI integration, and other capabilities that may be unified under the rubric of "Industry 4.0." Future VM architecture may be optimized for the *intersection* of such user experience and multi-modality concerns with traditional static analysis and requirements engineering.

Consider the case of channel systems. According to the SG model we have outlined here, procedure calls are represented via channelized neighborhoods, and in such neighborhoods all edges incident to a central ("procedure") node lie within a channel. In VM translation, this configuration implies that procedural stack frames are split up into multiple channels, so that whenever an argument is "pushed" as a parameter for some near-future call, this action occurs in the context of a specific channel kind. As a result, argument-push operations may be analyzed in terms of semantic protocols specific to the currently active channel. For example, VMs might want to add extra information in the context of channels representing "message receivers" in object-oriented contexts (which we have earlier alluded to as "sigma" channels, essentially the **this** or **self** of languages like C++, JAVA, RUST, etc.), particularly when such channels contain more than one node (insofar as most languages providing kernel functions accessed from a VM do not allow multiple **this** objects: the demo code for this part's chapters has examples of how to emulate multi-sigma calls in C++, but in general it would presumably take extra effort to map multi-sigma VM calls to the underlying native functions). Moreover, channel-specific semantics can be implemented on an extensible basis, allowing the VMs to be augmented with special-purpose channel kinds that supply their own functionality for managing (in effect) stack frames when such channels are active.

Furthermore, channels can act as a grouping mechanism tieing together procedure arguments that are logically interconnected (to a greater degree than merely co-existing as arguments). An example would be unit/scale-decorated types: suppose a function calculates a formula that requires two arguments with the same dimension and measurement units (kilometers, say) plus a third argument that is just a scalar; evidently, the first two arguments are mutually constrained in a fashion distinct from the third. Or consider a calculation with a scalar plus two mathematical vectors that should have the same length—the latter condition might be modeled with a dependent-type construction on the second vector, or some runtime check for the two vectors together, in either case demarcating the vectors as paired arguments distinct from the scalar. One strategy to satisfy these runtime use cases would be to implement channel semantics where special guarantees (finer than type checking alone) are enforced, while the channels are being populated.[8]

Doubtless, some of the features enabled by channels could be achieved via other means.[9] Channels, however, present a convenient interface for organizing the range of functionality entailed by the relevant channel semantics. Insofar as every argument node in an SG occurs in the context of a channel, before any such nodes are set in place the VM would first construct a channel of the relevant kind; that is, there is a specific operation to "open" a channel given its kind. This operation is therefore a VM site (for static or dynamic analysis) that can be targeted by extension code enforcing or examining channel semantics. Once channels are opened, operations exist to indicate the type and value source for nodes added to the channel. Because these latter operations always occur in the context of a specific channel kind, they can be filtered or re-implemented based on the channel kind in effect, so extensions could modify the treatment of certain channels while preserving the underlying VM implementation in most cases. For example, a VM extension could modify exception-handling protocols by re-implementing operations for inserting nodes into (or, on the call side, initializing carriers in) special "exception" channels, which are partitioned from "ordinary" output channels (given the obvious behavioral contrast between exiting via exceptions versus normal returns). Procedures typically involve multiple channels, so there are VM operations for "closing" one channel and opening another of a different kind, which results in subsequent operations occurring in the context of a different semantics. Explicitly introducing channels as part of the underlying VM machinery allows variegated channel-semantic protocols to be implemented in an organized and extensible manner.

Last paragraph's discussion regarding (for instance) special-purpose channel semantics perhaps does not obviously connect the idiosyncratic constructions endemic to SG representation with concerns in the latter sense, so we will try to present a case more concretely. Consider first the issue of "reactive programming" in GUI contexts. Analysis of procedures as a series of calls to other procedures—what we are calling SG sequence descriptions—fails to directly address how call sequences in this sense fit into over application execution (we touched on this theme at the start of Chap. 20, as well). In general, an application is not a one-

[8] In this case, the one-channel-kind-per-neighborhood restriction might be relaxed, since mutual connections could exist between (say) some input parameters and not others; or one could adopt something like "subchannels" that semantically refine the channels around them.

[9] For example, aggregating input parameters as mentioned last paragraph might be achieved alternatively via explicit dependent typing or via smashing tuples into single (aggregate) arguments. However, dependent types are notoriously difficult to implement in the context of software language engineering, and the latter alternative could be syntactically unwieldy (if done explicitly in source code) or complex in its own right to implement (if done behind the scenes).

dimensional *program* that simply executes some sequence of operations and then terminates. Instead, applications construct a graphical and runtime environment and then wait for user-initiated actions, responding accordingly, and resetting to a passive state awaiting further user actions. Users signal their intention for applications to take specific steps via interactions that can generically be called *gestures* (e.g., typing something via a keypad, or clicking somewhere via a mouse). Gestures in turn are presented to application code as *signals* that are *handled* by implemented procedures. The overall programming model entailed by composing applications as collections of procedures poised to handle signals—rather than fixed operation sequences designed in advanced—is generically called *reactive* or (with some additional technical specifications) *functional-reactive* programming [11, 21, 30, 47, 54, 60, 62]. Such a programming model introduces a variety of issues for VM implementation, at least for VMs that model/analyze or execute reactive procedures.

In particular, the central idea of reactive programming is that certain procedures are called (or initiated) in response to signals (typically those due to user gestures, though certain signals may be prompted by non-user changes to the current environment that might be relevant for applications, such as a sudden loss or gain of Internet connectivity).[10] Procedures in this sense are not called *from other procedures*, so the normal analysis of procedure calls in terms of stack frames being transferred from one site to another has to be modified. A canonical approach to reactive programming involves *signals* and *slots*, with the idea that instead of one procedure directly calling calling another, procedures instead emit "signals" that are *connected* to other procedures, which in such contexts become "slots." Unlike hard-coded procedure calls, signal-to-slot connections may be dynamically altered, created, or suspended. Such a mechanism depends on a centralized routing component to observe when a signal has been emitted (for instance, added on to an "event queue") and transfer control to one or more slots registered as connected to that signal. Insofar as a centralized processor in this sense is active, it can also (in typical application-runtime frameworks) receive signals originating *outside* the application, i.e., resulting from external conditions apart from one *procedure* emitting signals. Typically, such external signals would result from user-generated events, such as clicking a mouse button or moving the mouse.

In effect, application code based on signals and slots includes some procedures that are called because they are registered as "slots" whose signatures match signals that may arise from *outside* the application proper. These procedures serve as "entry points" where operations specific to the application originate. In other words, applications are environments that, after an initial setup, wait in suspension until external signals initiate a chain of actions in response. The details of such signals

[10] More precisely—because there are not necessarily signals that actually notify applications about, say, network-connection changes—a dropped connection would implicitly *cause* signals to be emitted. For example, an object representing a Web request could emit a network-error signal rather than a usual "finished" signal.

cannot be known ahead of time—for example, one cannot say which user gestures will occur (whether the user first types something, or moves the mouse, or clicks the mouse, etc.) nor their specific details (which keyboard keys are pressed, which mouse buttons are clicked, where on-screen the mouse cursor is located in the latter event, etc.). Application code therefore needs to prepare for multiple forms of external signals, and to analyze their properties to respond correctly (in accord with user intentions and/or design requirements). A left-mouse click with the cursor hovering over one GUI element typically signifies a different user intent than a right-mouse click over a different element, for example.

The unique characteristics of reactive programming have numerous consequences for VM design. First, note that applications typically implement many procedures so as to possess requisite capabilities to handle user actions. When and whether a given procedure is called depends on user pragmatics; for instance, a procedure involved in saving a file (at least a file users know about, as opposed to, e.g., a database-related file storing configuration information) would only be called in circumstances where users signal their desire to save files they are currently working on. A well-organized code base will specify which procedures could potentially be called as the *initial* handler responding to external signals. This information makes it possible to factor in external signaling conditions during code analysis. For example, suppose file saving is disabled for some reason (e.g., the current user does not have permission to modify the local file system). In that case, procedures that are *only* called within a call chain leading from external signals specific to saving files would become effectively unreachable. Notions such as reachability and execution paths have to be evaluated in the context of procedures registered as external signal handlers.

Consider a scenario where an application, being upgraded, is redesigned to support two different file-handling models: one for local filesystems and one for cloud storage. Certain procedures (e.g., one to check whether the file has been modified since the time of last save) may be relevant for both scenarios; others would only be active in contexts where cloud saving is possible (an open Internet connection, say) or, respectively, local file system access. Introducing new procedures to manage the cloud storage case alters the application's inter-procedural "connectivity," potentially requiring analyses to be updated with respect to conditions wherein a certain procedure might be called (or might be unreachable). The fact that such information *about* applications is subject to change (insofar as applications are continually refined or redesigned) indicates that information *about* application code should be managed in a systematic fashion. This might be done through VM representations directly, or at least source code and/or documentation *available* to VM compilers can draw metadata from such sources. In other words, we can assume that applications are engineered in coding environments where information (including, for example, which procedures play the role of external signal handlers) is detailed with enough rigor to be read by VM compilers and/or runtimes. For example, VM code could then internally incorporate representations related to reactive control flow and GUI objects that iconify pragmas for initiating application actions from users' points of view.

Analogous to "co-constructor channels" as special-purpose sites notating the origination of typed *values*, a channel-based VM could similarly support special-purpose *input* channels that carry external signal data (we might call these "reactive" channels). In the same way that "co-constructors" are declared by providing co-constructor channels, external signal handlers could then be identified through the presence of reactive channels. This makes it easy to identify all execution points where external signals could trigger procedure chains: simply observe for whenever a reactive channel is "opened" during the course of building a channelized neighborhood.

Functional-Reactive Programming (FRP) has engendered various strategies and analyses for extending type systems to incorporate signal/slot signature (how should signals and slots be assigned functional types, because they are different from ordinary procedures)? Channel systems offer one solution to this problem because channel semantics and carrier state are available as modeling parameters orthogonal to types themselves. For instance, slot procedures and the signals they specifically respond to would not need to be notated with special type variants (like "World" environments or "discrete time" or other constructions endemic to FRP); instead, we could stipulate that idiosyncratic *channel kinds* designate specific procedures as signals or slots; carriers *in* those channels would hold normal typed values.

External signals (as initiators of procedure chains that disrupt applications' passive "event-loop" states) offer one example of how reactive programming and VM design intersect, but there are related scenarios that could also be mentioned. For example, applications' GUIs are typically seen as a 2D visual extent populated with viewable objects that are simultaneously spaces to represent pieces of information to users (e.g., text, via readable characters; or numbers, via printed characters or indicators like dials and sliders; or graphics, via image or 3D displays) and origination points for user gestures (e.g., scrolling on a slider to increase/decrease a value). Our above comments touched on the latter capabilities, but interpreting user gestures depends on the information currently presented through the relevant GUI control. As such, it is helpful to track systematically how GUIs are populated with data. A useful maxim is that each class representing distinct GUI controls should be paired with a separate class representing the data that is visible within such controls. This is straightforward if a control (or in general a GUI "gadget") indicates one simple quantity (consider a slider that adjusts the zoom level for viewing an image), but even more complex GUI areas whose display is spread over multiple subcontrols can be associated with a multi-field datatype (these sorts of correlations are analyzed further in Chap. 24). Two datatypes are then closely interconnected: one represents some data aggregate from a computational perspective (ensuring that all fields are properly initialized, packaging the data for persistence or serialization, and so forth), while the other translates the same information into visual indicators for user interaction. To the degree that GUI and application-level datatypes are closely aligned, VM code can be annotated to mark and leverage such alignment.

In general—continuing these GUI-related examples—interconnections between GUI components, user actions, and the datatypes shown and affected by either tend to appear in multiple contexts. Consider (as above) an image zoom level, indicated (and adjusted) by a slider control. It is not uncommon for zooming (in and out) also to be initiated by small arrows adjacent to a slider, or by context-menu options on image displays themselves, or by keystroke sequences such as control plus either plus or minus (holding the "control" key and tapping the "plus" or "minus" keys to increase or decrease zoom). Presumably, altering zoom levels via these other gestures should cause the zoom slider to be adjusted proportionately. There are, then, potentially five different gestures that might affect zoom levels, each associated with GUI controls in different ways: directly interacting with either a slider or arrow buttons, or via a context menu (within an image display), plus via keyboard actions. Such interrelationships should be tracked with some degree of rigor to maintain a consistent "User Experience."

Continuing this example, a common pattern in GUI programming is to implement "tool tips," or floating text blurbs that appear when users hover over GUI controls, explaining the purpose and pragmatics associated with the control itself. Insofar as multiple controls can be used for image zoom, each should be provided tool-tip text accordingly; it would be useful to confirm that policies along these links are sustained in an application code base, ideally via static code analysis. Important gesture/feature connections need to be programmed in multiple contexts, apart from the underlying signal handlers themselves, including tool tips, help-menu information, documentation, application history and undo/redo capabilities, notifications (e.g., small labels sometimes displayed near the bottom of application windows clarifying recent actions, such as a string confirming that an image was zoom to a particular percentage), and unit or integration testing.

Or, consider again issues with application upgrades. In the example we suggested vis-à-vis cloud versus filesystem saves, there may be multiple controls and datatypes associated with the cloud functionality, including windows where users register credentials to access a cloud service and strings giving a remote path for files on cloud servers—each of the GUI elements presenting the corresponding strings (path names, user names, passwords, etc.) are logically interconnected by their common utility in the guise of cloud file backup. Meanwhile, the "image zoom" case can likewise be extended to hypothetical "upgrade" examples: consider the fact that some image displays use modified *mouse gestures* for zooms, such as cursor up/down with a key pressed, often the shift key (these pragmas are borrowed from 3D displays, which try to fit the 6 or 12 degrees of freedom in 3D graphics to conventional mouse and keyboard gestures—rotations; zoom; and translations, i.e., moving parallel to x, y, or z axes; each of which can occur within the model or within the "camera"—using the keyboard to compensate for mouse-move gestures having only *two* degrees of freedom). Consider an application that decides to support this latter 3D-style zoom gesture added on to prior functionality; this decision would propagate to concerns such as those itemized at the end of last paragraph (test suites, documentation, help menus, undo/redo, etc.). Applications that tend to be intuitive and responsive from a user experience point of view—where it is easy for users to understand how to initiate their desired actions by interacting with the software, and to learn the requisite steps when they do not know the pragmas ahead to time—

consistently model the full network of interconnections between application-level capabilities, GUI display elements, and user pragmatics.

It is also worth pointing out that such goals overlap with database engineering. Controls to set zoom levels may be factored in to a database profile insofar as the optimal (or most recent) zoom level for viewing an image might be stored as one metadata point in an image database. Similarly, a file's cloud-hosted save path would be a relevant piece of information to track in a database for which that file is an external resource object. Cyber-physical networks belong in the discussion as well: CPS devices, for example, tend to have quantitative profiles (data ranges and measurement units) that would be relevant both for database persistence and for GUI admin controls. When a GUI indicator models CPS sensor readings or actuator settings (a thermostat's temperature level, say), the relevant value range and units (e.g., Fahrenheit or Celsius) should be explicated in GUI code (obviously, an indicator needs to know how each value to be displayed compares to valid minima/maxima, and should identify and clarify for the user, perhaps supporting alternation between, Fahrenheit/Celsius, or metric/imperial, and so forth).

21.4 Conclusion

User experience (UX) sometimes appears to be treated as if it were a stylistic and subjective dimension to software engineering more than a technical or mathematical problem, but effective user interface design—and its interconnections with such themes as database engineering and CPS networks—hopefully shows that rigorous UX engineering requires highly structured models of software and GUI components, as well as application-level datatypes, with their interconnections and value synchronizations. There are various ways to enforce disciplined engineering standards to sustain quality UX, but at least one set of tools for this purpose can derive from VM resources—whether in the form of script-like layers interposed between GUI elements and application procedures, or runtimes for unit testing and prototyping, or representations of application code for static analysis (or some combination).

We contend that trends such as Industry 4.0 foretell an increasing convergence between technologies related to user experience and user interface, cyber-physical systems, 3D graphics, and requirements engineering. The forces behind such dynamics reflect how digitization and computer analytics has the power to increasingly shape industries such as manufacturing, architecture, green tech, and scientific research. Modern graphics cards can fluidly display interactive 3D models of industrial parts (on a small scale) to building renderings and urban designs (on a large scale) and everything in between, allowing users to visualize planned or manufactured objects/spaces for design, training, or simulations. Digital "twins," moreover, serve as presentations or anchoring points for data aggregates specifying objects' or designs' requirements and specifications. The digital twin concept implicitly posits relationships between an object's or design's 3D spatial form and its functional purpose, material/mechanical properties, fault tolerance, energy consumptions, and similar usability and quality-assurance metrics. Stakeholders might use software both to form visual images of materials and spaces as physical environments and as functionally organized systems, with software being able to migrate between more geometric (e.g., 3D rendering) and more data-oriented (e.g., specs table) modes. Given these application capabilities, digital platforms can simulate and analyze physical and/or functional systems with an unprecedented degree of computational accuracy and, simultaneously, user interaction, allowing software ecosystems to be a larger presence in industrial/architectural (and etc.) planning, construction, and maintenance.

We do not sketch this account from the point of view of advocacy, but instead to point out that maximizing the benefits of existing digitization capabilities depends on software that systematically and interactively unifies visualization and data curation/analytic aspects (e.g., 3D graphics alongside specs presentations), which in turn calls for advanced database and application development capabilities—at least if software fitting these characterizations is to be developed in a cost-effective, extensible/adaptable, and widely available manner. We have emphasized how virtual machine engineering can fit into this overall sea change in application development requirements and standards, with a more holistic attention to such details as database query processing and GUI design than were, arguably, emphasized by earlier VM paradigms.

References

1. David Adger, "Constructions and Grammatical Explanation". https://www.qmul.ac.uk/sllf/media/sllf-new/department-of-linguistics/26-QMOPAL-Adger.pdf
2. Nadeem Akhtar, *et. al.*, "Hierarchical Coloured Petri-Net Based Multi-Agent System for Flood Monitoring, Prediction, and Rescue (FMPR)". https://ieeexplore.ieee.org/stamp/stamp.jsp?arnumber=8926354
3. Kyoungho An, *et. al.*, "Cloud Computing for Cyber Physical Systems: Reliability and Security Challenges and Solutions". http://www.dre.vanderbilt.edu/~gokhale/WWW/papers/CC4CPS13_FT.pdf
4. Anees Ara, *et. al.*, "A Secure Service Provisioning Framework for Cyber-Physical Cloud Computing Systems". https://arxiv.org/abs/1611.00374
5. Stefan Berghofer and Christian Urban, "A Head-to-Head Comparison of de Bruijn Indices and Names". https://www.sciencedirect.com/science/article/pii/S1571066107002319
6. Cem Bozşahin, "Combinatory Logic and Natural Language Parsing". https://users.metu.edu.tr/bozsahin/bozsahin-tjeecs.pdf
7. _____, "Combinatory Linguistics". https://library.oapen.org/bitstream/handle/20.500.12657/24661/1005450.pdf
8. Yi Cai, *et. al.*, "Sensor Data and Information Fusion to Construct Digital-Twins Virtual Machine Tools for Cyber-Physical Manufacturing". https://www.researchgate.net/publication/318291645_Sensor_Data_and_Information_Fusion_to_Construct_Digital-twins_Virtual_Machine_Tools_for_Cyber-physical_Manufacturing/link/596140e8aca2728c11e0b5fe/download
9. Felice Cardone and J. Roger Hindley, "History of Lambda-Calculus and Combinatory Logic". http://www.users.waitrose.com/~hindley/SomePapers_PDFs/2006CarHin,HistlamRp.pdf

10. Nathaniel Christen, "Hypergraph-Based Type Theory for Software Development in a Cyber-Physical Context". In *Advances in Ubiquitous Computing: Cyber-Physical Systems, Smart Cities, and Ecological Monitoring*, Amy Neustein, ed., Elsevier, 2020, Chapter 3. https://raw.githubusercontent.com/ScignScape/dog-cognition-research/main/documents/NathanielChristen-Hypergraph.ngml.pdf
11. Guerric Chupin and Henrik Nilsson, "Functional Reactive Programming, Restated". https://nottingham-repository.worktribe.com/preview/4922128/ppdp2019.pdf
12. Liron Cohen, *et. al.*, "The Effects of Effects on Constructivism". https://www.cs.bgu.ac.il/~cliron/pubs/Effects.pdf
13. Evie Coussé, *et. al.*, "Grammaticalization Meets Construction Grammar: Opportunities, challenges and potential incompatibilities". http://eviecousse.be/publications/cousse-ea-forthc-grammaticalizationmeetscxg.pdf
14. René David, *et. al.*, "A Direct Proof of the Confluence of Combinatory Strong Reduction". https://arxiv.org/abs/0905.2545
15. Parastoo Delgoshaei, *et. al.*, "A Semantic Framework for Modeling and Simulation of Cyber-Physical Systems". https://user.eng.umd.edu/~austin/reports.d/IARIA2014-PD-MA-AP-Journal-Paper.pdf
16. Jean-Pierre Desclés, "Reasoning in Natural Language in Using Combinatory Logic and Topology An Example with Aspect and Temporal Relations". https://www.semanticscholar.org/paper/Reasoning-in-Natural-Language-in-using-Combinatory-Descl%C3%A9s/e3b140a490dbce5a1be7abd35b18d22bab0ae52e
17. Pietro Di Gianantonio and Furio Honsell, "An Abstract Notion of Application". https://users.dimi.uniud.it/~pietro.digianantonio/papers/copy_pdf/ana.pdf
18. Alessandra Di Pierro, *et. al.*, "On Reversible Combinatory Logic". https://www.sciencedirect.com/science/article/pii/S1571066106000843
19. Peter Ford Dominey, *et. al.*, "Dynamic Construction Grammar and Steps Towards the Narrative Construction of Meaning". https://users.cs.fiu.edu/~markaf/doc/w14.dominey.2017.procccgnlu.1.163_archival.pdf
20. Manfred Droste and Heiko Vogler, "The Chomsky-Schützenberger Theorem for Quantitative Context-Free Languages". https://arxiv.org/abs/1208.3942
21. Bernd Finkbeiner, *et. al.*, "Synthesizing Functional Reactive Programs". https://arxiv.org/pdf/1905.09825.pdf
22. Jean H. Gallier, "Realizability, Covers, and Sheaves: Application to the Simply-Typed Lambda-Calculus". https://repository.upenn.edu/cgi/viewcontent.cgi?article=1280&context=cis_reports
23. Silvia Ghilezan and Simona Kašterović, "Semantics for Combinatory Logic With Intersection Types". https://www.frontiersin.org/articles/10.3389/fcomp.2022.792570/full
24. Robert Gold, "Petri Nets in Software Engineering". https://www.thi.de/fileadmin/daten/Working_Papers/thi_workingpaper_05_gold.pdf
25. Masahito Hasegawa, "The Internal Operads of Combinatory Algebras". https://www.kurims.kyoto-u.ac.jp/~hassei/papers/2022_mfps2022_pre.pdf
26. Martin Hilpert, "Construction Grammar and its Application to English". http://members.unine.ch/martin.hilpert/CxG%20sample.pdf
27. J. Roger Hindley, "Axioms for Strong Reduction in Combinatory Logic". https://www.cambridge.org/core/journals/journal-of-symbolic-logic/article/abs/axioms-for-strong-reduction-in-combinatory-logic/9B7CBE781BAE43FF5148ABE720D4517A
28. J. Roger Hindley and Jonathan P. Seldin, "Lambda-Calculus and Combinators, an Introduction". Cambridge University Press, 2008. https://www.cin.ufpe.br/~djo/files/Lambda-Calculus%20and%20Combinators.pdf
29. Furio Honsell and Donald Sannella, "Pre-logical Relations". https://homepages.inf.ed.ac.uk/dts/pub/prelogrel-report.pdf
30. Paul Hudak, *et al.*, "Arrows, Robots, and Functional Reactive Programming". https://citeseerx.ist.psu.edu/document?repid=rep1&type=pdf&doi=e523884ac0c118b6e42261c6c1eaaa8fde7d57f7

31. Mans Hulden, "Parsing CFGs and PCFGs with a Chomsky-Schützenberger representation". https://dingo.sbs.arizona.edu/~mhulden/hulden_ltc_2009.pdf
32. Adam Husted Kjelstrøm and Andreas Pavolgiannis, "The Decidability and Complexity of Interleaved Bidirected Dyck Reachability". https://cs.au.dk/~pavlogiannis/publications/papers/popl22.pdf
33. Petr Janda, *et. al.*, "Implementation of the Digital Twin Methodology". https://www.daaam.info/Downloads/Pdfs/proceedings/proceedings_2019/072.pdf
34. Radek Kočí, *et al.*, "Object Oriented Petri Nets—Modelling Techniques Case Study". *Second UKSIM European Symposium on Computer Modeling and Simulation*, 2008. https://ijssst.info/Vol-10/No-3/paper4.pdf
35. Konstantinos Kogkalidis and Orestis Melkonian, "D^3 as a 2-MCFL". https://omelkonian.github.io/data/publications/d3.pdf
36. Michael Köhler and Heiko Rölke, "Properties of Object Petri Nets". *International Conference on Application and Theory of Petri Nets*, 2004, pages 278–297. https://link.springer.com/chapter/10.1007/978-3-540-27793-4_16
37. Yuanbo Li, *et. al.*, "Fast Graph Simplification for Interleaved Dyck-Reachability". https://research.cs.wisc.edu/wpis/papers/pldi20-graph-simplification.pdf
38. Fei Liu, *et al.*, "Coloured Petri Nets for Multilevel, Multiscale and Multidimensional Modelling of Biological Systems". *Briefings in Bioinformatics*, Volume 20, Issue 3, (2019), pages 877–886. https://academic.oup.com/bib/article/20/3/877/4590142
39. Samuele Maschio, "On Posetal and Complete Partial Applicative Structures". https://arxiv.org/pdf/2211.11326.pdf
40. Paul-André Melliès and Noam Zeilberger, "Parsing as a Lifting Problem and the Chomsky-Schützenberger Representation Theorem". https://hal.archives-ouvertes.fr/hal-03702762/document
41. Laura A Michaelis, "Construction Grammar". https://spot.colorado.edu/~michaeli/documents/Michaelis_CG_ELL_offprint.pdf
42. John C. Mitchell and Eugenio Moggi, "Kripke-Style Models for Typed Lambda Calculus". https://person.dibris.unige.it/moggi-eugenio/ftp/kripke.pdf
43. Michael Moortgat, "A Note on Multidimensional Dyck Languages". https://link.springer.com/chapter/10.1007/978-3-642-54789-8_16
44. Tiago Mück, *et. al.*, "CHIPS-AHOy: A Predictable Holistic Cyber-Physical Hypervisor for MPSoCs". https://par.nsf.gov/servlets/purl/10119096
45. Amy Neustein and Nathaniel Christen, *Innovative Data Integration and Conceptual Space Modeling for COVID, Cancer, and Cardiac Care*. Elsevier, 2022.
46. Sergey P. Orlov, *et. al.*, "Application of Hierarchical Colored Petri Nets for Technological Facilities' Maintenance Process Evaluation". https://www.mdpi.com/2076-3417/11/11/5100
47. Ivan Perez and Henrik Nilsson, "Bridging the GUI Gap with Reactive Values and Relations". https://ivanperez.io/papers/2015-HaskellSymposium-Perez-Nilsson-BridgingGUIGapReactiveValues.pdf
48. Robert G. Pettit IV and Hassan Gomaa, "Modeling State-Dependent Objects using Colored Petri Nets". http://csis.pace.edu/~marchese/CS865/Papers/pettit-final.pdf
49. Steven T. Piantadosi, "The Computational Origin of Representation". https://www.ncbi.nlm.nih.gov/pmc/articles/PMC8300595/
50. Giulia Rambelli, *et. al.*, "Distributional Semantics Meets Construction Grammar: Towards a Unified Usage-Based Model of Grammar and Meaning". https://hal.archives-ouvertes.fr/hal-02146537/document
51. Ivan A. Sag, "Sign-Based Construction Grammar: An Informal Synopsis". https://web.stanford.edu/group/cslipublications/cslipublications/pdf/BoasSag-sag-chapter.pdf
52. Borja Bordel Sánchez, *et. al.*, "Enhancing Process Control in Industry 4.0 Scenarios using Cyber-Physical Systems". http://isyou.info/jowua/papers/jowua-v7n4-3.pdf
53. Anil Sawhney, "Petri Net Based Simulation of Construction Schedules". https://dl.acm.org/doi/pdf/10.1145/268437.268747

54. Christopher Schuster and Cormac Flanagan, "Reactive Programming with Reactive Variables". https://users.soe.ucsc.edu/~cormac/papers/16crow.pdf
55. Vivek Kumar Sehga, *et. al.*, "A Comparative Study of Cyber Physical Cloud, Cloud of Sensors and Internet of Things: Their Ideology, Similarities and Differences". https://ieeexplore.ieee.org/document/6779411
56. Jonathan P. Seldin, "The Search for a Reduction in Combinatory Logic Equivalent to $\lambda\beta$-reduction". https://core.ac.uk/download/pdf/82407499.pdf
57. Sebastian Shaumyan, "A Semiotic Theory of Language". https://publish.iupress.indiana.edu/projects/a-semiotic-theory-of-language
58. Murray Sinclair, *et. al.*, "The Identification of Knowledge Gaps in the Technologies of Cyber-Physical Systems with Recommendations for Closing These Gaps". https://incose.onlinelibrary.wiley.com/doi/10.1002/sys.21464
59. Mark Steedman, *et. al.*, "Plans, Affordances, and Combinatory Grammar". https://homepages.inf.ed.ac.uk/steedman/papers/affordances/pelletier.pdf
60. Kohei Suzuki, *et. al.*, "CFRP: A Functional Reactive Programming Language for Small-Scale Embedded Systems". https://www.worldscientific.com/doi/pdf/10.1142/9789813234079_0001
61. Szymon Szomiński, *et. al.*, "Development of a Cyber-Physical System for Mobile Robot Control using Erlang". https://annals-csis.org/Volume_1/pliks/246.pdf
62. Sam Van den Vonder, *et. al.*, "Tackling the Awkward Squad for Reactive Programming: The Actor-Reactor Model". http://soft.vub.ac.be/~svdvonde/papers/ecoop2020-tackling-the-awkward-squad-the-actor-reactor-model.pdf
63. Silvio Valentinia and Matteo Vialeb, "A Binary Modal Logic for the Intersection Types of Lambda-Calculus". https://www.sciencedirect.com/science/article/pii/S0890540103000890
64. Alexander Vodyaho, *et. al.*, "Model Based Approach to Cyber–Physical Systems Status Monitoring". https://www.mdpi.com/2073-431X/9/2/47/htm
65. Richard West and Gabriel Parmer, "A Software Architecture for Next-Generation Cyber-Physical Systems". https://www.cs.bu.edu/~richwest/papers/west_cps.pdf
66. Ryo Yoshinaka, *et. al.*, "Chomsky-Schützenberger-Type Characterization of Multiple Context-Free Languages". http://isw3.naist.jp/IS/TechReport/report/2010003.pdf
67. Yu Zhang, *et. al.*, "High Fidelity Virtualization of Cyber-Physical Systems". http://web.cecs.pdx.edu/~xie/pubs/IJMSSC13.pdf

Chapter 22
GUI Integration and Virtual Machine Constructions for Image Processing: Phenomenological and Database Engineering Insights into Computer Vision

Nathaniel Christen and Amy Neustein

22.1 Introduction

This chapter will use image processing as a domain to examine issues related to virtual machines (VMs). We will use image-processing algorithms as a case study in the sorts of workflow modeling and quantitative operations that might reasonably be incorporated into VM designs, albeit a VM specialized for a specific computational domain. For the sake of discussion, this chapter will base some examples on a novel, rather idiosyncratic image format that we have called "XCSD" (the acronym stands for "Extensible Channel System" and "Subdivision Indexing," to be clarified later). This novel format is optimized for certain image-processing operations related to approximate segmentation and color-based (more than shape-based) background/foreground separations. We will orient the exposition around hypothetical VMs for which XCSD supplies the "native" image format, and which might be used with image databases (e.g., for query evaluation) storing content in this format.

In the case of image processing, there are literally hundreds of algorithms that might be run to transform, or extract information from, a 2D image (not to mention image series, videos, or 3D point clouds), so mapping each algorithm to kernel VM instructions is impractical. Obviously, even an image-oriented VM will have to provide some foreign-function interface where various algorithms could be registered and accessible through VM code; the VM runtime could accordingly be provided with foreign-function pointers through the application that hosts and initializes the VM itself. On the other hand, there might be some operations related to image processing that are so fundamental that they should be recognized by

N. Christen · A. Neustein (✉)
Linguistic Technology Systems, Fort Lee, NJ, USA
e-mail: amy.neustein@verizon.net

A. Neustein et al. (eds.), *AI, IoT, Big Data and Cloud Computing for Industry 4.0*, Signals and Communication Technology, https://doi.org/10.1007/978-3-031-29713-7_22

a domain-specific VM automatically, without needing an extra step of loading "foreign" functions (providing the relevant capabilities).

Preliminary to discussing algorithms directly related to images, we will review several themes arising in that context but applicable more broadly, within the overarching subject matter of reasonable scope and operation sets for (relatively high level) VMs. This initial discussion will highlight type-theoretic issues and primitive mathematical functionality that go beyond the ubiquitous integers, floats, and fundamental arithmetic capabilities that one would expect from any VM (however broad or narrow in scope, and high level or low level in design). Whether or not these specific types and operations are appropriate for a specific VM, given its goals and motivations, these examples will hopefully suggest how the question of what are legitimately "fundamental" or "low-level" types and calculations is open-ended.

Numerous constructions usually associated with high-level languages that compile to VMs or intermediate representations (perhaps only in a temporary sense in the course of being compiled to actual machine code) may prove to be worthy of direct support and implementation at the VM level—aside simply from convenience (insofar as certain high-level operations do not need to be repeatedly translated down to multiple VM instructions), the specific conditions and rationales for the relevant high-level types can intersect with VM opsets in ways that justify accommodating them directly. Again, this chapter will illustrate such points with concrete examples. Subsequently, we will situate the types and operations reviewed in the preliminary discussion in the context of image processing, so that first analysis will extend into computer vision as a case study for (what one might call) "domain-specific" VM design.

22.2 Type-Theoretic Constructions at the Virtual Machine Level

The types that lay at the foundation of almost any computer system's type hierarchy are integer and floating-point values with different byte lengths—often 1-, 2-, 4-, and 8- byte integers (distinguishing signed and unsigned types) and 4 or 8 byte floating-point numbers (decimal approximations). It is less common for relatively low-level environments, such as conventional virtual machines, to model *pairs* of integers or decimals as "base" types, assuming that such values would more likely be implemented as types in a high-level language (e.g., **std::pair** in C++). However, number pairs are such an essential component of image-processing operations that a VM specifically targeting this domain would have reason to recognize pairs as equally fundamental to single values; since any pixel or location in an image needs two independent numbers to be identified, and since image locations are the building blocks around which computer vision algorithms are constructed, such two-number pairs are if anything more "foundational" than single values.

For similar reasons, one could equally argue that several *different* number-pair types are foundational. One consideration for image processing is that different computer vision libraries utilize different conventions about how positions within an image are to be notated. A detail that varies across platforms concerns where location axes have their origin point: one could have the top-left corner correspond to the point (0, 0)—with axes increasingly down and to the right—or alternatively the origin at the bottom-right (actually, any of the corners) or even the image center.

Another question is which of the two axis values—the horizontal or vertical coordinates—are notated first in a number pair. These are not particularly complex details, of course, but care must be taken to ensure that analytic procedures are called with the proper location encoding, i.e., that there is alignment between libraries and calling procedures with respect to location–coordinate conventions.

Ensuring such alignment could be promoted via strong typing: assuming that number-pair types are further narrowed to reflect coordinate conventions, only locations encoded according to the target expectations can be passed to the corresponding procedures. Such policies might include how coordinates are named: e.g., one common vocabulary for pairs, as in tables or spreadsheets, is to identify horizontal "lines" as *rows,* and vertical as *columns.* Moreover, row/column pairs usually notate the *row* coordinate first. Therefore, constraining a pair to be specifically *row/column* documents that the expected interpretation of the first coordinate is vertical (i.e., a vertical distance from the origin) and the second is horizontal. Conversely, a different popular convention (reflecting mathematical norms) is to label horizontal values as **x** and vertical as **y**. In this case, normally the **x** value comes *first*; which illustrates possible confusion, because a number pair would be interpreted differently depending on whether it is understood to be row/column (vertical then horizontal) or x/y (horizontal then vertical).

To add further confusion, some procedures accept inputs in terms of *lengths*, e.g., width and height, where similar procedures would accept location–coordinates. So, without full contextual information, it is ambiguous whether a pair designates a coordinate or a pair of length measures. Consider a procedure to calculate the average color or dominant color within some region bound, where the inputs must define a rectangle spanning the focus of the computation. That rectangle might be described via a pair of locations—top-left and bottom-right corners, for instance—or alternatively via *one* coordinate pair and one length pair, e.g., top-left *corner* plus *width* and *height.* In other words, the second number pair might be a *location* (x/y or row/column) or might be *dimensions* (width/height). Confusion can be avoided by having a width/height type distinct from location types, analogous to distinguishing row/column from x/y location–coordinates.

At the same time, note also that some procedures might take number pairs that do not model geometric quantities such as position or length, where terms such as "width" or "column" could be misleading—in the generic C++ **std::pair**, for example, the first value is called "first" and the second "second." For such "non-geometric" pairs, one might indeed prefer still another pair type with coordinates labeled as first/second, say. Indeed, code for the XCSD format (for example) supports numerous variations on pair types, including options based on start/end, top/bottom, left/right, horizontal/vertical, and above/below (e.g., if a pair's C++ type is, say, **tl2**—meaning top/left, two-byte values—the first coordinate would be accessed by a **.top** member and the second by **.left**).

Even were a virtual machine to natively model number pairs, theorists preferring relatively minimalist VM design might think to provide one such pair type (or perhaps several distinguished only by byte length), anticipating that higher-level distinctions between conceptually distinct, but structurally isomorphic types (like row/column, x/y, width/height, and first/second) would be made by source code, not VM byte code. Ensuring that strong pair types are never mismatched would then be the responsibility of the compiler emitting VM code, not the VM runtime. Consider, however, that pair labeling could reasonably be an issue within special contexts, such as query evaluation: an image database might, for example, allow for queries that invoke the rectangle-focused procedure mentioned above (along the lines of "find images whose central third rectangle color-averages" to nearly some given shade). Assuming type-level distinctions such as x/y versus width/height are recognized by the query *language*, then query *evaluators* would of course at some stage ensure that an x/y coordinate is not passed to a rectangle constructor that expects width/height dimensions, say.

Moreover, the "compilers" that would bridge *query* code to VM instructions might be significantly different than those working on code from other genres, such as scripts (with notions of types, scopes, and variables closer in spirit to compiled programming languages). These circumstances could argue for stronger type distinctions being recognized even at the VM level, at least to some degree of detail, to help guarantee that type-enforced conventions are applied in similar ways across all environments that use the VM, notwithstanding their distinct compilation models (database queries compared with scripts, for instance).

From this perspective, then, assume that a VM is indeed designed to model *several* number-pair types. This decision immediately has implications for the *operations* supported through the VM, because some operations seem intrinsic to managing paired quantities; for instance, transpose (switching the first and second values) and arithmetic operations (which might involve a single value—altering both numbers by the same amount—or a second number pair, applying the relevant operation between the first and seconds pairwise). Here is a more complete list of some operations that may be used often when working with number pairs and would be candidates for native encoding in a VM context:

(Strictly) Ascending, descending, spaceship As boolean operators, return true if the second value is larger (respectively, smaller) than (or equal to, for the non-"strictly" options) the first. The "spaceship" version (borrowing the terms from many high-level programming languages) would return 1, 0, or -1 depending on whether the second value is greater, equal, or less than the first.

Zeros mask, nonzeros mask, nonnegative mask, spaceship mask Return pairs whose components are one or zero depending on whether the two input pair values are zero (respectively, are not zero; and, are at least zero). Similarly, "spaceship mask" is a plausible name for a function that applies a unary spaceship operator to each coordinate in a pair: yielding a new pair, each member either 1, -1, or 0, depending on whether the respective number in the original pair is positive, negative, or zero.

Area, ratio Product of the two numbers (or perhaps absolute value of the product); respectively, ratio of the second to the first (or perhaps a version that transposes descending pairs before taking the ratio, so the result is always at least one, or vice versa, so the result is at *most* one).

Binary merge Combine the two numbers into a single value, with the former numbers occupying different bit positions (this operation could have several variants wherein one of the two inputs is bit-shifted by different amounts).

Maximum, minimum, difference, positive-integer difference, gap Greater or lesser of the two numbers; or, return the second subtracted from the first (the "positive integer" variant would result always in a nonnegative output, mapping negative differences to zero), or "gap" between the two numbers as an absolute value (i.e., maximum minus minimum).

Duplicate, "end at" Given one number, construct a pair by setting the second value to have a desired gap greater than the first (or, allowing this parameter to default to zero, form a pair with two equal numbers); alternatively, modify the second number in an already-existing pair to conform to such a gap.

Linear intersections Given two pairs, form a new pair that takes one value from each pair, analogous to the $-|$ or $|-$ operators in diagram languages such as **tikz** (in effect, define a third point as the middle vertex of a right triangle whose hypotenuse is the given two points, thus the intersection of two lines, one horizontal and one vertical, including the two input points; there can be two versions for such a function, corresponding to the two extra corners of a rectangle if the given two points are one diagonal, or equivalently which input has the horizontal line and which the vertical). As a variant form, take one pair and then two other pairs (defining a line) and calculate the nearest distance of the former to the latter.

As arithmetic base Treat the two numbers as components of a two-digit number in some base (provided as a parameter): multiply the first number by that input and add the second. This operation can be extended to value-triples recognized as distinct data types analogous to pairs, where the first number would be multiplied by the square of the input.

Manhattan, Chebyshev, and other distances For cases where pairs represent 2D locations on an integer grid, there are several possible metrics that could be used to represent lengths (via *integers*) between grid points—of course, one could compute real-valued Euclidean distances (approximated by floating-point numbers), but we will focus instead on formulae expressed solely through integers (i.e., distances use the same type as point components themselves). Familiar choices are the "Manhattan" metric, which calculates distances by counting orthogonal steps leading from a start to an end point), or the "Chebyshev" metric, which counts the minimum number of orthogonal *or* diagonal steps, along grid intersections (a "grid-diagonal" would be the shortest diagonal between two grid points, in effect two single orthogonal steps at right angles). Supposing $\{x, y\}$ and $\{a, b\}$ are two locations, the Manhattan and Chebyshev equations would be $|x - a| + |y - b|$ and $\mathtt{max}(|x - a|, |y - b|)$, respectively. Also, the XCSD format is built around a combination of these two distance measures (analyzed in the next section). One benefit of a Manhattan/Chebyshev "hybrid" (at least in the form discussed below) is an almost-Euclidean precision with respect to ordering distances, while retaining only integer-based expressions: specifically, two location pairs on an integer grid will only have the same Manhattan/Chebyshev hybrid distance if they have the same Euclidean distance (and vice versa). In any case, each of these (Manhattan, Chebyshev, and the hybrid) are comparison operations that would be appropriate for any image processing or geometry library.

Integer-based line operations The previous item reviewed several distance metrics appropriate for discrete-geometric systems. We can similarly incorporate other computations appropriate for geometric work restricted to integer-based mathematics. For example, consider the problem of calculations pertaining to the Bresenham's line between two points (i.e., the points drawn for graphics displays approximating a smooth Euclidean line, accommodating the restriction to a rectangular pixel grid) [5, 38]. Given two end points and a single intermediate coordinate, one might want to calculate the intersection of the formers' Bresenham's line with the horizontal or vertical axis translation through the latter, or then go on to select some set of points adjacent to that intersection on either side (this last step would need information about the line specifically; it could not be ascertained indirectly via Euclidean calculations).[1]

Floor, ceiling, fit in range Modifies a pair's components so they are at least (respectively, at most) some value. Similarly, given two pairs, modify the first pair (as needed) to satisfy range restrictions expressed by the second pair. For example, alter each number (of the first pair) so that it is no less than the first coordinate of the second pair, and no greater than its second coordinate.

Binary insert Treating the first number as a bitset, encode the second number through a designated bit sequence within it (as a unary operator, simply shift the second number left by some count of bits, applying binary-or of the result to the first; as a two-argument function, the second input could specify an *end* bit, so in effect the second number would be truncated of leftmost bits as needed and then copied into the first using a bitmask-guarded "**or**" operation that only alters bits in the specified range).

[1] An example of where this sort of problem might arise would be implementing mouseover effects where we want to highlight some region of a Bresenham's line, as a visual cue for cursor location or the image point where some action would take effect.

Most of the aforementioned operations are used in the demo code associated with this chapter (largely for calculations related to the XCSD image format), so this code can furnish examples of these operations employed in concrete situations.[2]

One benefit of providing a relatively thorough suite of operators defined for number pairs is encouraging programmers to think of algorithms in terms of operation sequences on number pairs as integral units, not as sequences applied to the two pair elements in isolation. Usually, pairs designate some logically connected entity (e.g., a geometric point), and operations have an interpretation where it is conceptually accurate to consider the pair as one value in a two-coordinate space, rather than two separate numbers. As an example, XCSD has a function called "quadrant_code_against" that is defined (within a C++ macro) as a series of pair operators:

```
ul quadrant_code_against(const ty##size##s& rhs) const
{
 return (*this - rhs).spaceship_mask()
  .plus((*this - rhs).zeros_mask())
  .floor(0).times({2, 1}).inner_sum();
}
```

That code may not be more readable than alternatives that work with each coordinate separately, but it fits within a paradigm where we are encouraged to think of pairs as self-contained mathematical entities whenever possible.

Although the names chosen for some of these operations make more sense in some types' contexts than others ("area" as a label for the inner-product absolute value fits width/height better than row/column, for example), it seems reasonable to provide operators for all versions of structurally similar types absent compelling reasons to make specific exceptions. Moreover, distinctions such as row/column versus width/height (and x/y, first/second, top/left, etc.) are orthogonal to signed/unsigned, integer/float, and byte lengths for the components. Accordingly, we can derive a fairly large number of "primitive" pair types, reflecting different combinations of coordinate names, signed/unsigned, 1/2/4/8 bytes, and floating-point-based types (albeit some operations applicable to integers, such as ones based on bitsets or binary encodings, would not carry over).

On the principle that most operations available for one pair type should be available for others, this results in quite a few duplicate operations differing only by input types. Assuming that high-level coding techniques such as templates are not available for VM implementations, such a scenario seems to demand brute-force coding of multiple function versions (perhaps aided by code generators). It is not difficult to maintain a function-pointer table with hundreds of entries if needed, so supporting this diversity of "primitive" operations does not impose a major overhead for a VM runtime. However, once code generators (and perhaps more complex grammars) are involved, the development environment for *implementing*

---

[2] See https://github.com/scignscape/dog-cognition-research. This repository was constructed specifically for Chap. 23, but we are using it to host content for other chapters belonging to the current book section (Part 5) as well.

a VM runtime would become relatively more complex and would need to be set up properly.

Since we are not constrained by existing VM or intermediate representations, there is room for creativity in how opsets are designed, in terms of their textual labels (or even their numeric opcodes) from the point of view of people reading "virtual assembly" code for the VM, as well as runtime engineering (opcodes might have bit flags carrying information about their respective operators, for example), with the idea of organizing groups of operations differing only by (e.g.) input/return types. Likewise, a well-designed VM infrastructure can allow designers of high-level languages targeting the VM some creativity as well. The pair-operator code for XCSD mentioned above (which expands via macros to C++, not a VM, but could migrate to other contexts) is constrained by C++ syntax and operator overloading, but other VM-specific languages would be free to, for instance, create new binary operators for pairs. We alluded above to one example: the **tikz** −| and |− could be used as syntactic atoms in a high-level language, compiling to line-intersection functions (analogous to the <=> spaceship recently being added to the C++ operator set, in current specifications for that language). Similar comments could be made for other operators that seem common enough that they might warrant operator-like syntax (for readability) rather than function-call notation—examples might be min/max, absolute value (applying the "**abs**" operator to both components), floor/ceiling, ranges, and so forth.

Also, a comparable custom-syntax scenario might apply to conventional arithmetic (and boolean/binary) operators in contexts where it is necessary to distinguish an operation between a *pair* and a *single* value (e.g., multiplying both coordinates by the same scale), or conversely another pair (e.g., vector multiplication[3]), or (alternatively again) *internally* within one pair. For instance, there are three important versions of, say, addition in a number-pair context: given $\{x, y\}$, there is $\{x + s, y + s\}$ (for some scalar s), or $\{x + a, y + b\}$ for pair $\{a, b\}$, or $x + y$ (the "inner" sum, a scalar result).

These alternatives would similarly exist for almost all "primitive" numeric operators for most coding languages, which would include arithmetic functions and also boolean calculations in both the true-false sense (e.g., treating positive or nonzero values as "**true**" and zeros as "**false**") and bitwise manipulation, plus binary operations such as left- and right-shift. A high-level language might notate these variations via compound tokens combining one part or character representing the operation itself and another character (or characters) indicating a specific variant (as a hypothetical example, a bare **+** might represent addition *between* pairs, while **+.** represents "inner sum" and **.+** encodes addition with a duplicated scalar).

Apart from overloading functions for many structurally similar types, there are further dimensions to the overlap between VM operations, types, and code management, some of which are indirectly posed by the previous examples. Consider the process of casting a pair from signed to unsigned. Plausibly, this might

3 In the sense of Hadamard product; cf. Julia's **.*** operator.

be achieved in several ways: first, take absolute values; second, map negatives to zero (essentially using a "floor" operator with parameter zero); or, third, simply reinterpret any negative value as the binary encoding of a positive value (so -1, say, would become the maximum integer of the corresponding unsigned type). Each of these could potentially be expressed as operators between pairs as well as type-cast logic; as such, one consideration when formulating a VM's underlying type system is to define operators that provide the calculations for type casts (in case someone wants to achieve type-cast-like effects via a function call, or perhaps to enable scenarios where the VM could be configured to perform type casts in different ways by specifying a desired operator).

More generally, it is worth noting how type casts intersect with operator application: consider an "inner difference" function that subtracts the second pair value from a first. Assuming we sustain the principle that unsigned pairs should by default perform operations yielding other unsigned pairs/values, we might want the *nonnegative* version of inner difference, which in turn implies casting a signed result to unsigned. This is analogous to a "nonnegative-inner-difference" procedure that applies subtraction but with a floor of zero—however, such a variation might *also* be useful in a signed (including a floating point) environment, or as the nullary version of a unary operator that takes an inner difference raised (as needed) to a floor that might be something *other* than zero, passed as a parameter.

22.2.1 Issues with Overflow/Underflow and Loop Termination

Similar ideas apply in the context of addition with possible overflow: should we consider addition involving one-byte integers to be capped at 255, or to "wrap around" past zero? The latter option does not make much sense in the case of colors, say (it would almost certainly be an error if an algorithm considered "pure black" zero to be one unit greater than "pure white" 255), and moreover could be a notorious source of infinite-recursion bugs (cf. tests for <=255, which might never revert to false, depending on the type involved). Types related to colors (in conventional one-byte-per-component encodings) are good examples of types for which basic arithmetic operations should almost always be guarded against overflow (addition maxed at 255 without circling around to zero, and subtraction floored at zero without negative overflow back to 255). Conversely, in some contexts, integer-maximum wraparound effects are exploited deliberately; analogously, casting an integer to a single byte (typically in the context of merging multiple values via bitmasks) performs a function analogous to zeroing out bits with index 9 and greater.

The overarching theme here reflects different conceptual roles played by numeric types. In the case of one-byte integers, for example, in some contexts, the full 0–255 range represents meaningful values (e.g., color), whereas elsewhere only a subset of this range is actually in effect. A ready example is enumerations (a.k.a. enums) that might use **unsigned chars** as the base or "carrier" type but only define, say, ten values.

Most instances of the underlying type therefore do not match any enumerated value; the **char** (i.e., 8-bit) type is employed simply because most systems do not have types smaller than one byte (one cannot take the memory address of "half a byte," for instance). At the same time, the numeric details of how enums are encoded can sometimes become consequential even for application-level code. Consider the relatively common situation where an enum type—together with some other data—is encoded into a single (4 byte, say) integer where the enum is assigned a specific bit range (if there are ten enumerated labels, say, the bitmask must allow four bits for the enum). A programmer would immediately know to left-shift the **enum**'s value (cast to an integer as needed) by the requisite number of bits but could easily forget that the left-shift applied to **char** could result in *another* **char** and therefore truncate bits on the left; the **enum** instead must be cast to *int* first (or whatever type will encode the shifted bits) even if the original base type is, say **unsigned char**.

Similarly, consider using one-byte integers (say) in contexts where the underlying mathematics reflects modular arithmetic. An example is encoding directions (such as for labeling adjacency steps for inter-pixel color comparisons): a common option is to encode the eight options from top-left, top, top-right, etc., through bottom-right, via integers 0–8. On the other hand, these directions can also be derived via coordinate subtraction, spanning the eight pairs $(-1, -1)$ through $(1, 1)$ (excluding $(0, 0)$), so procedures may need to interconvert the pair-based and mod-8-based encodings. Still another possibility is to label directions such as "northwest" or "top-left" and so forth via a distinct enumeration class, where the enumeration labels would be mapped to corresponding numbers in the 0–8 range, but descriptive terms such as "left" might be more readable than numeric codes.

A variation on this example would be that of notating directions between image locations on a larger scale via 16 direction possibilities, incorporating those between strict orthogonal, horizontal, and vertical. Actually, the XCSD code (mentioned earlier) in some contexts recognizes 24 different directions (to accommodate variations in even/odd pixel or region counts, where the center in an image partition might correspond to a line of regions or to the gap between two regions). The XCSD format utilizes such direction codes in several calculations (mostly related to the proximity of regions vis-à-vis an image center) where arithmetic operations are performed on the underling numeric value (e.g., mapping a direction to its nearest strict diagonal counter-clockwise). Insofar as numeric codes represent directions, then "rotations" (via modular addition or subtraction) and modular wraparounds become mathematically significant.

Modular integer ranges can potentially be supported directly via a strong type system allowing programmers to explicitly define a (distinct) type for integers modulo 24, say. Alternatively, one may adopt something like "modular guarded" arithmetic (and binary, e.g., shift) operations, where each operation could be given an extra parameter to serve as a modular-arithmetic context (note that equally relevant for image analysis and other geometric contexts can be variations on modular arithmetic where zero is sometimes replaced by the modulus, or results are *subtracted from* the modulus). Some of these arithmetic details might just be encoded relatively inelegantly by if-then branches and the like ("if(result >24) ..."

and so forth), but we have more flexibility on the VM side to identify various forms of modular guards (perhaps via runtime flags that can be attached to ordinary integers) and of modular arithmetic in general (e.g., zero-to-modulus swaps).

The main point here is to identify coders' rationales for employing specific types and to recognize situations where algorithms need to satisfy conditions more precise than the types alone express (an integer encodes mod-24 values, an enumeration is mapped to numeric quantities for geometric computations, such as mapping named directions under rotations/reflections, and so forth). Whereas these conceptual details might be merely implicit (or consigned to comments) in high-level source code, VMs can treat such conditions more rigorously by annotating types, or flagging runtime values, or providing variants on common operations (e.g., a modular variant on arithmetic functions).

Consider, again, integer-overflow situations: if a programmer is looping *up to* an integer maximum (e.g., 255), they might select a type for the looping variable to *avoid* overflows; so if a number i, say, is to represent one color component and be used for a loop, the algorithm could declare i as a *short* (or a *signed short* to avoid negative wraparound) since 256 (and -1) then becomes expressible values, preserving the loop termination. At this point, however, we have a *short* encoding a color component whose only *conceptually* meaningful values are in the **char** ranges; if the i were used in a color-related computation, it would need to be cast to its conceptually accurate **unsigned char** (aside from an extra coding step, consider how this could trip up code review or compiler-warning sensitive processes). A more fluent solution might be to keep i itself as one byte but cast it to **short** for the loop test, something like

```
/* u1 is unsigned char; u2 is unsigned short */
for(u1 i = 0, ok = true; ok;
    ok = ((u2) i) + 1 < 256, ++i)
```

... which admittedly is not very attractive as C++ code. This is an example of where a VM could provide extra functionality beyond the reach of compilers targeting smaller-scoped assembly languages: consider a variation on inequality operators that would combine a (temporary) type cast and boolean test into a single instruction. Having that form of test available in a VM opset might inspire language designers to support it with distinct syntax, so that a pattern such as "cast-then-compare"—which is more conceptually precise than using artificially larger types to avoid overflow/underflow conditions—is directly supported by high-level code, potentially making it clearer and more widely adopted.

In general, VM engineers have opportunities to identify computationally similar but conceptually distinct operations. Consider a basic increment (++) operation: this may occur in a context where the range of possible loop end values (e.g., the size of a list) is well below the maximum expressible integer of a relevant type; but it might also occur (as in color components) where we are looping up to that maximum value, which itself serves as the loop terminator (this in turn is the kind of situation where overflow errors due to "perpetual-truth" can occur—comparisons of **unsigned char** against 256, say, never fail). The former use of the increment occurs in a context where there is an a priori range setting outliers on the loop (from the first to the last index in a list, say), whereas in the second case the type itself (e.g., all possible values for one color component, in a format like RGB32) sets a range span. One can argue that the two uses of increment (and analogously decrement, especially in the context of unsigned types when looping down to zero) are conceptually different enough to warrant two different VM instructions, especially considering that extra care is appropriate for the second case (due to wraparound bugs).

Similar attentiveness to operations' "conceptual" significance applies to functions on number pairs as discussed above, especially in contexts where pairs represent logically integral values (such as locations within an image). When operations have a geometric interpretation (not just arithmetic), it is possible that they may be naturally presented in several different versions for different contexts, in the same way that modular arithmetic has multiple interpretations (including geometric ones related to directions and/or rotations). A VM instruction set might therefore recognize these distinctions, providing a more flexible compilation target and perhaps cueing higher-level language engineers to similarly acknowledge the relevant distinctions (e.g., via specialized syntax).

Moreover, operation variants and the types that they work on are conceptually interrelated, so engineers at both higher and lower levels should be attentive to type-system concerns interrelated with conceptually distinguished low-level operations. This chapter thus far has presented examples in contexts such as modulus, integer overflow, and number-pair types; another set of examples derives from enumerations and their encoding strategies, which we address next.

22.2.2 Different Variations on Enumeration Types

In the simplest sense, enumeration types represent sets of named values, which have numeric codes (for the sake of storing instances in memory) but whose numeric representation has no particular significance. This setup, however—where enums' actual encoding carries no semantic weight—is only accurate in some contexts; elsewhere, there are various possibilities for enums bound more tightly to their numeric basis. For example, nominals representing days of the week, or months of the year, have a fixed sequence, which in turn is reflected in their encoding. These are examples of types where both named labels and a limited range, modular-arithmetic representation have conceptual merit (consider the problem of "adding" some number of days or months, or iterating around a week, or year, cycle).

Or, consider a collection of named color values: each of these would be identified by a label but may also be mapped to a numerically encoded color space, so that quantitative inter-color operations could be implemented (measuring the distance between two named colors, for example, or mapping them to grayscale). Some operations on this enum type would therefore yield numeric results (such as color distance); other operations could map one enum to another by applying quantitative formulae (such as mapping chromatic hues to gray tones, or to brighter/darker shades). So the enum type would implement intra-type and out-of-type operations consistent with a mathematical foundation, at variance with the convention that enumerated values are string labels without quantitative interpretations.

Another case of consequential numeric encoding that can be illustrated via color values would be using enumerators as array indices: consider enums that label the components of color encodings according to different color spaces, such as RGB (the enum labels would obviously be something like {**red**, **green**, **blue**}) or HSV ({**hue**, **saturation**, **value**}). If colors (for these three-part models) are represented with three-valued arrays, then the enumerated component labels should map to 0, 1, or 2 so as to form array indices.

Also, some use cases for enums are to introduce a collection of related labels into a programming environment, without necessarily implying that there is a single semantic or conceptual interpretation that covers each of them. For instance, it is plausible that an enum employed in the context of managing calendars (not just raw dates but schedules, appointments, and so on) would have labels for days (Sunday–Saturday) but *also*, say, *weekday* and *weekend* (in contexts such as being open (all) weekdays, or happening "every weekend"). In this case, the numeric encoding for days of the week would operate on two levels: most labels would be related sequentially (Sunday, Monday, etc.), but "weekday" and "weekend" would be grouping labels instead, signifying collections (Monday to Friday and Saturday/Sunday, respectively).

A plausible implementation for such a type would be via bitmasks: assign the numbers 1–64 (increasing over powers of two) to the seven weekdays (i.e., each code has exactly one bit set, in positions 1–7) and then assign the codes 31 (the bitmask of all five weekdays, assuming the lowest value is "Monday") and 96 (the bitwise combination of Saturday and Sunday) to "weekday" and "weekend," respectively. Presumably, this kind of encoding would be most natural in a contexts where bitmasks could be employed to represent various combination of days (e.g., coding "Monday and Wednesday" as the bitwise merger of the Monday and Wednesday codes). Of course, this is only feasible in contexts where enumerations can be cast to integers (whether explicitly or implicitly).[4]

[4] Meanwhile, note that bitmask encoding would introduce a further calculation step in contexts where days of the week are sometimes also given by numbers. In response, some code might prefer to use sequential numbers and assign, say, 8 and 9 to "weekday" and "weekend." Any procedures inputting a days-enum would therefore need to check for these two idiosyncratic values and branch accordingly. This may be imperfect style from a viewpoint that code that has numerous conditional branches is harder to maintain and understand (compared to mostly "branchless" programming). On the other hand, using a "bitmask" encoding would entail extra procedures being implemented (for converting the 1–64 values into their 1–7 equivalents, perhaps via a "`ffs`," or "find first bit set" operation, almost equivalent to taking a base-2 logarithm). Moreover, relying on numeric values to ascertain details through the enum—"on Wednesday" is not a matter of testing one enum value against *Wednesday*'s for *equality*, but rather for nonzero bitwise-or, because "weekday" includes Wednesday (in this sense bitmasks make certain day-related logic very concise)—forces the enum codes to have specific values, which might feel dubious from a type-theoretic perspective according to which enums should be proper "sum-over-unit" types. In effect, two different intuitions about

Mainstream programming languages evince different patterns with respect to mapping enums to numeric values. In C++, for example, although we can specify the *byte length* of an **enum** type by requesting a particular underlying type (e.g., **enum class e : unsigned char** forces each enum value to be represented by only 8 bits), enums—at least "strong" **enum class** ones—do not automatically *cast* to integers. They can, however, be manually cast, or integer-style arithmetic semantics can be added *to* enums via operator overloading (in the XCSD code, certain `ENUM_FLAGS_OP` macros expand to implementations of operator overloads that make most arithmetic operations available to enum values, for enumerations that require them). With other languages (Java, for instance), utilizing enums' numeric values is more difficult, whereas elsewhere (e.g., C#) conversion to integers is automatic.

This variation between languages reflects inconsistency across underlying semantics: enumerated nominals have different degrees of conceptual connection to an underlying arithmetic space in different contexts. There are, indeed, at least six possible options (we are proposing non-standard terms for the final three items here, but hopefully the exposition will furnish rationales for the vocabulary chosen):

Raw nominals Collections of named labels ("nominals" in the statistics sense) that are to be assigned numeric codes simply for purpose of computational representation, but where the actual numbers have no special meaning (an exception could be allowed for a label used as a "default" or fallback when a specific value is missing, which may be coded with a zero to coincide with defaults in other settings); this is the case that, technically, best matches the type-theoretic understanding of enumerations (in terms of summing unit types—those with exactly one instance—so their sum is a collection of instances that have no internal structure, at least apart from the means to distinguish one from any other).

Cyclical or sequential enumerators This would cover cases where labels correspond to entities that have an implicit order ("ordinals" in statistics) so we can speak of one label coming "before" or "after" another in the sequence as conventionally understood. In some contexts, we could also recognize "differences" (as in, Friday is four days past Monday) and wraparound (after Sunday, we return to Monday, for instance), which in turn implies modular arithmetic.

Enumerators as *de facto* numeric constants A different use case applies to labels that have a specific numeric code because they are intended to give a recognizable name to an actual numeric value. Of course, in some cases important numeric values that can be assigned descriptive names are independently introduced as constants, which does not involve a specific enumerator list—for instance, C/C++ has macros such as `SHRT_MAX`, `INT_MAX`, `INT_MIN`, and the like that define maximum (respectively, minimum) values for different integer

"elegant" (and maintainable/understandable) programming style are in tension, as far as restricting branching *or* preferring enum types not bound to precise underlying values. Our general point is that these kinds of tensions tend to occur in contexts where enums span concepts that (albeit interrelated) have a variety of semantic interpretations, such as *Wednesday* versus *weekday*.

types; these are global constants that have some obvious logical interrelationship, but there does not appear to be any benefit in grouping them together as a distinct *type* whose specific purpose is to identify just these values. On the other hand, consider a situation where we want to create a list of `Latin1` codes for characters that have specific meanings in some formal context (e.g., the list of whitespace glyphs: space, tab, new line, form feed, carriage return). Here it is plausible that a procedure might accept a type that expresses a member of such a list (implementing actions for a code-generating grammar, say), while also each label needs to embody a specific constant value (we want the enumerator to translate directly to a corresponding ASCII and Unicode number), which in a sense combines the sorts of contexts where global constants would be preferred and those more amenable to proper enums. If indeed an enum is chosen to represent the constants as a *type*, the numeric values would of course have semantic significance (on the other hand, most quantitative operations between two labels would not make sense—there is no meaning to adding one ASCII code to another, say—so enums in this case should not necessarily be provisioned with a full suite of arithmetic operators).

Isolated bitmask enumerators Names that do not have intrinsic mathematical interpretations but which are assigned numeric codes built around powers of two, for the sake of notating combinations of such nominals via a single number. This kind of encoding is common for labels that represent "flags," with the possibility of merging multiple flag values or "activating" multiple flags at once. For a geometric-flavored example, suppose we want to specify alignment options relative to a bounding rectangle: the actual labels recognized by the enum might be *left*, *top*, *right*, *bottom*, and *center*, but these labels could be used in combinations such as "top-left," using the bitwise-or merger of codes for *top* and *left* to indicate alignment according to both of these directions. We call enums in this case "isolated" to suggest that *within the label list itself* each nominal has a numeric code with no intrinsic meaning; it is only for creating numeric representations of *combinations* (like **top** | **left** as a mathematical construction) that encodings' binary structure becomes significant.

"Fusional" bitmask enumerators We propose this terminology to express cases where some enumeration labels represent combinations of other labels. The difference between this case and the prior one is that here combinations are not only numeric *expressions* that can be interpreted as simultaneously declaring multiple nominals at once, but instead labels that are part of the enumerative collection itself can be assigned mathematical values that translate to sets of other labels. The weekday/weekend case would be one example, given a bitmask encoding such as proposed for the weekday example; or, suppose we extend the "alignment" example just cited such that **TopLeft** and other combinations are actual parts of the **enum**. Types in these scenarios have the ability to form labels for commonly used collections of other nominals; consider, say, the type **Qt::TextInteractionFlag** (part of the QT application development framework) that defines settings for how a GUI control displaying natural-language texts allows for user actions (selecting, copying, following hypertext links, and so forth). In addition to several independent/isolated labels, there are two bitmasks that are part of enumerator list: **TextBrowserInteraction** (which defines the most common fusion of interactions appropriate for controls playing the role of text browsers) and **TextEditorInteraction** (which, similarly, sets the defaults for text editors—which differ from browsers in that text may be modified within the control, while it is being displayed). Of course, other flag combinations may also be used; but providing labels for the *most common* combinations helps to clarify the specific conceptual status of those fusions specifically.

Free-form or arithmetized enumerators This case would cover situations where numeric values associated with enum labels are significant, but for multiple or intersecting reasons not covered by the above items. Relevant examples can include cases where mathematical functions are applied to enum values in some algorithmic contexts. Earlier we mentioned a 24-valued "direction" code employed internally by XCSD; were these codes to be expressed via an enum, they would need a specific mapping to integers so as feed the enums directly into certain XCSD-specific algorithms (related, for example, to the process of ordering image regions based on proximity to the center and, secondarily, by clockwise progression *around* the center). For technical reasons (the details are not especially relevant here), these directions are coded by *signed* values in the range -1 through 22; there are specific computational reasons for this range in particular, indicating that the numeric codes are intrinsically associated with corresponding enum labels (it is not a matter of assigning numbers essentially at random, or merely in an increasing sequence to honor ordering and/or cyclicality). Other "free form" examples would come from situations where codes may be assigned with an eye to bitwise fusions, but not all labels fit that binary pattern. The above "alignment" examples can have variations that fit this last case. In QT, the **Qt::AlignmentFlag** enum gathers—into one type—the labels for vertical (**AlignTop**, **AlignBottom**, **AlignVCenter**, and **AlignBaseline**) and horizontal (**AlignLeft**, **AlignRight**, **AlignHCenter**, **AlignJustify**); and also for *one* named combination (**AlignCenter**, which fuses **AlignVCenter** and **AlignHCenter**). In addition to these labels, the list includes a special flag that is only applicable for text in natural languages with a right-to-left writing system, and "mask" labels for convenience—specifically, a mask that extracts the *horizontal* and another for the *vertical* component of an alignment code; those masks would be used mathematically to split a single code into two labels (indicating that the labels need specific numeric values to make the binary operations work properly).

As the QT alignment example shows, sometimes enumeration types have labels serving different conceptual roles: the QT type essentially combines horizontal and vertical options, plus a special supplemental flag, and a bitmask for convenience, into a single label set. Proper numeric encoding is needed to orchestra the interplay of concepts and values involved here. More to the point, this example shows that expressing multiple concepts through a single label collection increases the likelihood that numeric values will be significant.

The various alternatives outlined here reflect a duality in the "conceptual" role of enums: in effect, enum types fulfill two distinct purposes—namely, the use of mnemonic or descriptive *labels* (instead of raw numbers), on the one hand, and (on the other) isolating a specific range or group of numbers as a distinct *type*. These two roles can be combined in different ways for different enumerations, which in turn raises questions about the proper type-theoretic protocols appropriate for enums. For example, how should procedures whose signatures indicate parameters of an enum type work with inputs whose numeric values do not match any nominal label? Should compilers reject code where there is no algorithm to prove that a certain input conforms to the specific list of values associated with enum labels? Or should it allow values that can be formed from those matched to labels via boolean **or** operations (merging their respective bits) but reject others? Or, perhaps, map anomalous values to one "default" label? Moreover, we can distinguish cases where a compiler could verify that a value will *fail* via either of those constraints, as opposed to cases where the details are uncertain—e.g., the enum might be obtained via serialization, or some other external source that may (or may not) endeavor to restrict values to "meaningful" options. These issues are also reflected by how enums are used by procedures that take them: if they are fed to **switch** statements, for example, then a **default** execution path could catch any values not matching named labels.

In short, the enumeration mechanism is actually a combination of multiple semantic patterns, and these differences tend to get reflected at the point where procedures take inputs of enum types (rather than their underlying numeric types). In the VM context, procedures could therefore indicate what "flavor" of enum type they expect, through some sort of flag or annotation. Instead of a single enum protocol, compilers (or a VM runtime) could then fine-tune enum handling based on procedures' specifications. For example, a procedure might indicate that it should *only* be called with an integer value that matches an enum value, or that (as a runtime feature) anomalous values get mapped to one specific label. Moreover, enum *types* could be equipped with different varieties of integer casts, distinguished between contexts in which conversions would be applied. In the most restrictive case, enums would never be cast to their underlying values except for a narrow set of low-level operations (such as serialization via binary packs); in the least restrictive, enums would be freely utilized as quantitative magnitudes, subject to a full suite of arithmetic, binary, and boolean operations. In between those extremes, enum types could be annotated to accept casts in certain circumstances but not others, or to support a proper subset of mathematical functions.

This discussion has set forth some of the conceptual issues associated with enumerations, and before that with number-pair types. The examples cited here serve primarily as case studies; ultimately, we intend to ground the analysis in VM engineering in particular. The overarching theme of this chapter's discussions thus far has been that VMs might work with type systems that have programming and design-pattern nuances that are not obvious in "formal" type theory. High-level coding patterns are not necessarily reflected, or reciprocated, at the VM tier; compilers might "compile down" source constructions to quasi-assembly languages.

But this is a design choice; it is certainly possible to plan VMs that directly represent many high-level design patterns and provide supportive compilation targets enabling compilers to support such patterns in the languages they target. As such, we will transition to a more VM-oriented (and, indeed, image-processing-oriented) focus in the following section.

22.3 Integrating Virtual Machines with Image-Processing Operations

This chapter so far has addressed topics, such as two-coordinate data types and enumerations, which are only tangentially related to image processing. The current section will focus on images more specifically. When considering VM "integration" with image processing, we refer primarily to setting up virtual machines with inherent functionality related to invoking computer vision algorithms. Presumably many VMs will implement a general-purpose foreign-function interface that would support calls to native-implemented graphics-related procedures (along with any other high-throughput domain, where highly optimized code in languages such as C or C++ is preferable for performance reasons). In the current context, we intend to focus however on VMs that are explicitly oriented toward image processing, where this specific problem area is a central design emphasis instead of one group of foreign-function capabilities among others.

This means that the VM model will incorporate features specifically relevant to images. For example—as already discussed last section—representations of image locations and intra-image lengths, in the form of distinct numeric-pair types, could be incorporated as "kernel" data types for the relevant VM. This is a straightforward example of how basic choices for VM design might be influenced by image-related use cases. We will present other examples below.

22.3.1 Exposing GUI Functionality

As outlined in Chap. 20, a central focus of VM functionality in these chapters involves applications "exposing" features for third-party components or plugins; within this context, an important area of functionality attends to GUI state. In computer vision, a straightforward example of GUI interop would be showing intermediate stages of an image-processing pipeline. The OPENCV library, for example, provides an **imshow** procedure that displays an image in its GUI window. Typically, this procedure is called in conjunction with specific OPENCV algorithms in order to visualize those algorithms' outputs, particularly while the pipeline is initially being formulated and fine-tuned.

For example, the algorithm in question might be a feature detector that looks for "keypoint" locations (that match a specific point in an image), straight lines (matching extended, relatively crisp edges, or visible lines), contours (matching closed curves, considered to enclose a 2D area), or "superpixels" (relatively small image regions of largely homogeneous interior color, also 2D). These alternatives correspond to zero-dimensional, one-dimensional, and two-dimensional features, respectively [14, 15, 37].

In most scenarios, feature detection is not intended to transform an image or to produce intrinsically visual data; instead, sets of keypoints, lines, contours, or segments/superpixels are subject to further statistical analysis. For instance, "landmark" keypoints may be used for image registration—aligning multiple images (MRI scans, say, or photos taken with a head-mounted camera that are stitched together to form 360° "panoramic" tours) by locating pre-identified keypoints common to all images in a series. Landmark keypoints may also be used for segmentation (by connecting certain keypoints to form polygons outlining or containing a region of interest).

Whether via this technique or others (e.g., superpixels or closed contours), segmentations are then typically utilized to isolate a desired foreground region and analyze its color and/or shape (metrics such as its color average, or the eccentricity of an ellipse containing the region) that in turn can yield object classification or other "semantic" interpretations of image data—consider software that records traffic patterns by isolating visuals of cars in the context of street cameras.

Image data generated via this kind of pipeline are typically entered into a database or in some other manner tracked analytically, so insofar as human users access these results it would be in the form of spreadsheets or other structured, largely textual formats. However, when initially formulating a computer vision pipeline engineers often need feedback concerning intermediate processing stages, so algorithms typically feature detectors that can be paired with operations to produce secondary images visually summarizing the algorithm's performance. Keypoints, feature lines, contours, and superpixels may be visualized by drawing markers at keypoint locations or highlighting lines and superpixel/contour boundaries with pronounced colors (distinct reds, yellows, light blues, and so forth, selected to avoid confusion with actual colors in the image). These graphics are then overlaid on the original image to represent an algorithm's results pictorially.

Such intermediate images are not analyzed themselves (they are not technically part of a pipeline) but provide performance/accuracy feedback.[5] In other contexts, intermediate graphics that *are* part of a workflow proper—such as blurred/simplified or grayscale images analyzed in lieu of an initial picture, so as to reduce noise or allow for restricted-channel analysis—might likewise be previewed during development, to help programmers conceptualize how the pipeline operates.

[5] Many segmentation and feature-detection algorithms, for example, have multiple parameters that can be adjusted, and programmers will often experiment with different settings in the context of a sample image for which the desired analysis is known ahead of time; so summary visuals signal how well the algorithm matches that goal.

In short, an intrinsic demand of computer vision components involves rendering intermediate images visually, even if those visuals are not analytically part of a workflow. Insofar as image-processing capabilities are part of a full-fledged desktop-style application, computer vision routines can leverage GUI features to show intermediate images, potentially with interactive capabilities allowing users to examine results in greater detail. For example, keypoints might be represented on images through boxes that receive mouse events independently from the larger picture (recall the **QGraphicsScene** comments from Chap. 20), providing for instance context-menu options specific to one keypoint. In OPENCV, each keypoint detected via SIFT (scale-invariant feature transform) and similar computations has multiple corresponding parameters reflecting their surrounding image context.[6] Keypoint attributes are an example of how visualizing computer vision workflows may involve interactive displays with more functionality than merely graphing images decorated with feature-depicting annotations (like highlights for feature points and superpixel boundaries).

This example also points to the potential complications involved in applications "exposing" GUI functionality for image processing. The case of *previewing* an intermediate-stage graphic may involve nothing more than a file path for the desired image. Once the presentation involves context menus and intra-image interactions, however, the interop between a GUI host and an image-analysis (semi-autonomous) component becomes more elaborate.[7] Context-menu options might furthermore be linked with procedures available in the *analytic* component, so the front end would need to register "callbacks" deferring functionality back to the client; there would be a bidirectional conduit for metaprocedure invocations (recalling terminology from Chap. 20) instead of a client requesting one host feature (as would be the case with showing an intermediate image non-interactively).

In short, both components would need to expose respective functionality according to a mutually implementable protocol. For reasons we have discussed, situations where components "expose capabilities" along these lines are a good example of leveraging virtual machines to model, fine-tune, or dynamically extend data-sharing protocols.

As a general rule, computer vision algorithms are largely self-contained, in that code needed to run the actual mathematical analyses—once an image for processing is loaded into memory—may be grouped together into a single library without often calling external procedures. On the other hand, image-processing workflows will typically need to defer to host applications for file paths to load images in the first place (or some other resource reference, such as handles into an image database) as

[6] These data structures allow keypoints occupying the same ground-truth location to be tracked among multiple images, e.g., for registration: we can compile keypoint sets for two images and look for points with similar statistical profiles, as if matching fingerprints; scale- and rotation-invariant keypoints in particular will have analogous profiles in multiple images whose contents overlap in some area of physical space.

[7] In a context such as examining keypoints, the front end would need access to a data set of keypoint attributes alongside the underlying image, plus instructions on how to populate context menus over keypoints and over generic image points, respectively.

well as (discussed in the prior paragraphs) for visualizing analytic results. Because of the sheer diversity of image analysis methods, applications should incorporate computer vision in a flexible manner, with the option of adding new kinds of analysis over time (partly because different methods are better suited for different image series, depending on their coloration, occlusion effects—and in general how crisply pictures show foreground/background boundaries—textural complexity, and similar variables that might affect the accuracy of a pipeline). In this sense, we can speak of "exposing" functionality in two directions; image-processing components share their analytic capabilities, while host applications expose GUI and filesystem routines, and so forth.

The prior discussion has considered interop from the host application's point of view, but we will now transition to image-processing components themselves. As a case study, we will consider the "XCSD" format that was formulated partly in conjunction with developing material for these chapters. This discussion will indicate the rationale for XCSD's approach to color processing, memory layout, and other implementation details, but this is not a technical documentation of XCSD intimating that this is necessarily a superior representation or paradigm. Instead, this chapter employs XCSD mostly as a case study for implementing new image-processing functionality in a pre-existing host application environment.

22.3.2 *Extending Host Applications with Image-Processing Workflows*

As is implicit in this discussion, the form of connections between applications and image-processing components that we will emphasize here involves hosts *calling* computer vision algorithms provided by external libraries, perhaps mediated through virtual machines. A different notion of "overlap" between VM and image-based engineering would be dedicated VMs for *implementing* computer vision algorithms, analogous to how graphics processing units provide computing environments tailored to concerns such as ray tracing and other graphics-rendering calculations. Designs for compiling image-processing code *to* special-purpose VMs, ones that are optimized for running such algorithms very efficiently, constitute a separate topic, which is outside the focus of this chapter (except indirectly since calling conventions within such algorithms might overlap with calling conventions for initiating image-processing workflows or workflow steps, the latter topic fitting more within the bounds of invoking computer vision capabilities externally from a VM context).

Image analysis functionality can be grouped into several areas, such as contour detection, region morphology, texture description, diffusion analysis, and methods related to color statistics (e.g., color histograms). For purposes of exposition, this section will discuss algorithms related to the XCSD format mentioned earlier. The basic focus of XCSD concerns how pixel data are stored in computer memory; in particular, pixels are lined up into groups at different "tiers" or nested levels of detail. The XCSD code accompanying this chapter supports three tiers, based on powers of three; so 3×3 pixel boxes are grouped together, as are 9×9 and 27×27. This structure is noticeably different from conventional "matrix"-based image formats, where one entire row of pixels (spanning the whole image width) would typically be coded in memory as one pixel run (sometimes called a "scanline"), followed by each successive row of pixels in turn. This representation is of little intrinsic value, however (apart from simply recording pixel data, that is, the *organization* of this value in memory adds no further benefit) unless one is working with sub-images spanning the whole distance from left to right margins. In XCSD, pixels are stored in memory such that 3×3 boxes represent one contiguous chunk and can be accessed via raw pointers (rather than copying from scanline-based data to temporary buffers); then, on higher tiers, the same applies to 9×9 and 27×27 groupings. Figure 22.1 shows several graphics that can be produced from XCSD images summarizing some aspects of the XCSD format and memory layout.

At the uppermost scale on this "tiered" system, the 27×27 boxes are called "tierboxes," which are aligned in memory from the center of the image outward. The XCSD code uses calculations based on the Manhattan–Chebyshev hybrid distance representation (mentioned earlier) to arrange tierboxes such that contiguous pixel runs correspond to meaningful image regions; specifically, diamonds, octagons, squares, and rectangles centered on the overall image center. This provides a rationale for tierbox memory layout, clarified below in more detail. *Within* each tierbox, pixel runs are stored on 9×9 and 3×3 scales. The goal is to increase the likelihood that a particular cluster of pixels of analytic interest can be accessed as a raw pixel run (via pointers to start and end memory) without needing multiple copy operations. Also, XCSD allots memory space for data associated specifically with boxes on each level, from 3×3 to 27×27. This allows algorithms to consider the image on coarser scales, avoiding features on the individual-pixel level that could be statistical "noise," as desired. Pre-computed values such as color averages can be recorded in the image data along side pixel-level information itself.

The multi-level representation just described gives rise to the notion of "subdivision" indexing, wherein individual pixels are accessed via their location in nested three-by-three groupings rather than absolute coordinates. Such indexing is useful in contexts where we may want to examine image features present at different levels of detail; shifting focus from the pixel tier to the 3×3, 9×9, or tierbox levels simply involves dropping a subdivision-indexing coordinate. Also, XCSD allots space for encoding image data of varying provenance (e.g., texture codes or region-of-interest masks) alongside channels, both at the pixel level and higher tiers (referred to as "extension" channels). The XCSD format can therefore record analytic data internally, such as masks on different tiers placing units on that tier within (or on the border of) one or more regions of interest.

The last two paragraphs have explained some features and rationales for XCSD and mentioned the origin of concepts such as "extensible" channel systems and subdivision indexing, reflected in the format's name. The purpose of this chapter is not to analyze XCSD itself in detail, but rather to adopt it for case studies in VM integration with image processing. We will do this through the lens of algorithms associated with XCSD, which hopefully serve as representative examples of how VMs can interface with algorithms related to image formats in general.

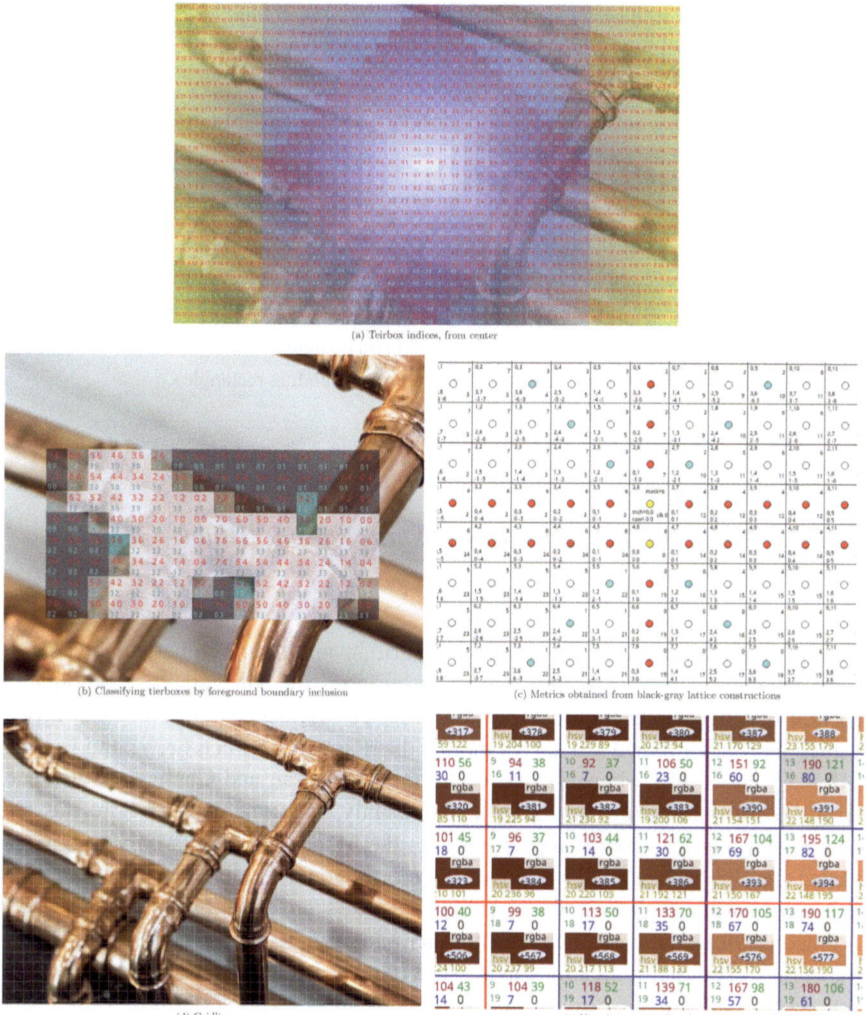

(a) Teirbox indices, from center

(b) Classifying tierboxes by foreground boundary inclusion

(c) Metrics obtained from black-gray lattice constructions

(d) Gridlines

(e) Visualizing internal tierbox data

Fig. 22.1 Visualizing components and quantities for the XCSD format. XCSD includes various tools for visualizing how images are stored internally according to the XCSD protocol, including: (**a**) seeing tierbox indices (data are stored from the center outward); (**b**) foreground and region-of-interest boundaries can be defined at the tierbox level; (**c**) reviewing internal metrics XCSD employs to locate pixel memory addresses (according to subdivision indexing); (**d**) viewing subdivision lines into (by default 27×27) boxes; (**e**) generating detailed high-zoom summaries of images showing data for each pixel and each of the subdivision scales

Table 22.1 Possible **MCH** cycles and cycle lengths

Center	Zeros	Lengths
Double-black	None	4 or 8 (depending on bounding rectangle)
	Diagonal	2 or 4 (depending on bounding rectangle)
	Orthogonal	4
	Both	1
Gray–black or black–gray	None	4 or 8 (depending on bounding rectangle)
	Diagonal	2, 4, or 6 (depending on bounding rectangle; complete is 6)
	Orthogonal	4
	Both	2
Double-gray	None	4 or 8 (depending on bounding rectangle)
	Diagonal	4 or 8 (depending on bounding rectangle)
	Orthogonal	4
	Both	4

22.3.3 Manhattan/Chebyshev Distances and "Black–Grey" Grids

Earlier we mentioned that XCSD uses a "hybrid" combining Manhattan and Chebyshev distance metrics. For the sake of discussion, we will call this combination an **MCH** distance representation (not a metric, *per se*, because the result is a two-valued pair rather than a scalar). Formally, the **MCH** pair between $\{x, y\}$ and $\{a, b\}$ is $\{M - C, 2C - M\}$, where M and C are the Manhattan and Chebyshev distances (see the formulas from last section).[8] Visually, the **MCH** merges how the two component metrics "count steps": whereas the Manhattan counts only orthogonal steps, the Chebyshev freely counts *either* orthogonal or diagonal steps, and the **MCH** represents a path *first* along a diagonal and then, "turning" by 135 degrees, orthogonal steps outward. The raw-number pairs can be supplemented with one of the eight directional indications as needed (between two points the diagonal-then-orthogonal path can have four initial directions and two possible further directions for the orthogonal component). Picture **MCH** pairs as the sum of a diagonal vector (sloping at 45 degrees from the relevant line) end to end with an orthogonal vector.

In order to document the relevant mathematical properties of **MCH**, we will introduce a couple of auxiliary constructions related to discrete geometry. The main results we are leading to are summarized in Table 22.1 with respect to finding locations equidistant from some point. The math involved here is barely above grade-school level, but there is some combinatorial trickiness in fully enumerating all possibilities for distance comparisons, so this presentation hopefully does not

[8] Actually, XCSD uses a slightly different pair we will call **MCH'** equaling $\{C, M - C\}$, but this text's version of **MCH** is more intuitive for exposition. They are readily interconvertible: **MCH'** = $\{innersum($**MCH**$), first($**MCH**$)\}$ and **MCH** = $\{second($**MCH'**$), innerdifference($**MCH'**$)\}$.

seem more technical than the subject warrants. In particular, it is significant to contrast locations (especially in image processing) that represent *pixels* themselves and that represent boundary points *between* pixels; this comment would then generalize to overall image regions. If a picture has odd pixel sizes both horizontally and vertically, the exact center is one pixel; otherwise, it is the gap *between* two pixels (for an even/odd mixture) or four (in the both-even case). This motivates the following modification to the notion of discrete geometry on an integer grid or lattice:

Definition Call a "black–gray" grid a lattice with lines grouped into two classes, *black* and *gray*, where each horizontal (respectively, vertical) black line is surrounded by two gray lines and vice versa (i.e., the two colors are interspersed). The "points" on such a grid would lie at intersections between lines, and given the color differences, there are four kinds of points: double-black (intersections of two black lines), gray–black, black–gray, and double-gray (reading the horizontal color first, then vertical). Unless stated otherwise, we will refer to **MCH** distances as comparisons between two intersection points on a black–gray grid. In this context, "points" discussed without further qualification mean "grid points," i.e., intersections between grid lines.

The purpose of this kind of grid is to address certain mereogeometric or mereotopological situations where distances can combine entities of differing dimensions. Consider a checkerboard pattern; imagine placing chess pieces either *inside* board squares (corresponding to double-black grid points), or along edges *between* squares (gray–black mixed) or at corners where four squares intersect (double-grays). The black–gray construction distorts actual geometry (if we picture gray and black as evenly spaced lines, these colors do not affect the dimensions of points or lines involved, so the mereogeometric interpretation is not visually implicit; but the point of black–gray grids is for discrete geometric properties, such as distance, which do not correspond to Euclidian space anyhow). We are particularly focused on the following:

Definition Call an **MCH** *cycle* on a black–gray grid to be a collection of double-black points equidistant, by some specific value, from a central point, within the context of a rectangle (possibly a square) centered on that point. If the stipulated center point is not double-black, calculate the shared **MCH** distance according to the double-black point adjacent to the center (either two points alongside, for black–gray or vice versa, or four points around, for double-gray) that is nearest to each candidate double-black point.

Note that two **MCH** pairs are "equal" if both components are equal (they may differ in the vector directions). If two **MCH** pairs are *not* equal, there is no way for them to represent the same *Euclidean* distance (assuming we treat the diagonal and orthogonal vectors as immersed on a Euclidean plane) so **MCH** provides a reasonable alternative to Euclidean distances with respect to ordering point pairs in terms of their respect distances, as mentioned in the last section, or grouping those representing the same Euclidean (and therefore **MCH**) distance.

Observation The MCH and Euclidean distances are similar with respect to ordering point pairs into lesser, equal, or greater internal distances. In particular, (i) for two MCH distances that internally sum to the same number of steps, the MCH and Euclidean orderings coincide. Also, (ii) strict equality between distances matches between both representations, and (iii) if one MCH coordinate is held constant, or if both either simultaneously increase or decrease, the changing coordinates will increase or decrease with Euclidean distance.

Proof Suppose we have a normal (one color) grid of evenly spaced orthogonal lines separated by a fixed unit, so we can assign them integer coordinates. One formula for expressing Euclidean distances in terms of MCH would be: (1) $d = \sqrt{x^2 + 2xy + 2y^2}$, where x and y are the MCH *orthogonal* and *diagonal* components. Assume the pair $\{x, y\}$ is modified by the scalars j, k, respectively, for $|j| < x$ and $|k| < y$, with d^2 correspondingly modified by h. The equation for $d^2 + h$ would be $(x + j)^2 + 2(y + k)^2 + 2(x + j)(y + k)$; subtracting the square of d as in (1) from this yields (2) $h = j^2 + 2k^2 + 2xj + 2yj + 4yk + 2xk + 2jk$. It is obvious that h is nonnegative when j and k both are. If both are nonpositive, note that $|2xj|$ would be $\geq j^2$, and likewise for $|4yk|$ and $2k^2$, given $|j| < x$, $|k| < y$, so the nonnegative terms (the squares) in (2) are more than matched by terms with negative or zero values (all the others), concluding that h in this case has to be negative or zero. These two cases (where j and k are not contravariant) confirm (iii)—whenever x and y are both made larger, or, respectively, both smaller, their corresponding Euclidean length will be larger or smaller, as expected. The remaining question concerns mixing positive j with negative k or vice versa. Suppose $\{x, y\}$ and $\{x + j, y + k\}$ represent the same total number of steps, i.e., they sum to the same amount, so $j + k = 0$. Then, plugging $k = -j$ into (2) yields (3) $h = j^2 - 2yj$. In this case, $|j| = |k| < y$, so the second term is greater in magnitude than the first, which forces h to be negative when j is positive; when j is positive, h of course will be as well because both terms in (3) then end up positive. This proves that k and h are covariant, which conforms to MCH ordering: a greater diagonal term, for the same total steps, is defined as a larger MCH value; since h has the same sign as k, the Euclidean distance will increase or decrease along with the MCH, demonstrating (i). For $j + k$ not zero, we can rewrite (2) to include $(j + k)$ terms, as in (4) $(j + k)^2 + 2(x + y)(j + k) + 2k(y + j)$. Setting $s = j + k$, (4) becomes $h = s^2 + 2s(x + y + k) + 2k(y - k)$. Let $a = x + y + k$ and apply the quadratic formula to solve for $-s$: (5) $a \pm \sqrt{a^2 - 2k(y - k) + h}$. If $h = 0$, we can move the bare a in (5) across the equals sign, and let $b = -a$ (note $b^2 = a^2$) to derive (6) $(b - s)^2 = b^2 - 2k(y - k)$. Expanding (6)'s left side and rearranging yield (7) $s^2 = 2bs - 2k(y - k)$. This is only possible if s and k are both zero because by stipulation s, b, k, and y are integers, and a $\sqrt{2}$ would fall out of the right-hand side upon taking the root when converting s^2 into (integer) s alone. Since $s = j + k$, $s = k = 0$ forces $j = 0$, meaning that $h = 0$ implies $j = k = 0$—and, also, the converse is obvious from (2)—proving that two MCH values can yield the same Euclidean distance only if they are identical, which confirms (ii).

In short, **MCH** is a reasonable replacement for Euclidean distance in many contexts because there are few situations where point pairs have greater **MCH** but less Euclidean distance, or vice versa (as shown by the above proof, the only scenario where the two representations differ in this sense is that an **MCH** pair that sums to fewer steps, but has greater diagonal component—i.e., $k > 0$ from (5) in the proof—may yield a greater Euclidean length than an alternative **MCH** pair with more steps but less diagonal). Calculations such as those related to **MCH** cycles are not affected by those discrepancies.

Observation An **MCH** cycle can have 1, 2, 4, or 8 points.

Proof Every point in an **MCH** cycle will have some pair $\{diagonal, orthogonal\}$ shared by all points, which will differ by direction. Since there are eight possible directions attributable to each pair, a cycle can have eight different points. We then have to identify cases where cycles will have *fewer* points. Note that if either or both coordinates in the **MCH** pair are zero, then the distinction between diagonal and/or orthogonal directions goes away, eliminating some points. The 1-point case corresponds to zeros for both diagonal *and* orthogonal components on a double-black center (so there is only one point, the center itself). A double zero on a black–gray or gray–black center represents a length-two cycle (because we consider the two double-black points around the center), while a double-zero **MCH** for a double-gray center engenders a length-four cycle. An **MCH** with one zero and one nonzero component can have four directions (since both a nonzero diagonal and an orthogonal with *no* diagonal supports four) and therefore represents four distinct points in a cycle (this is another length-four case). Moreover, for any **MCH** with nonzero *orthogonal*, it is possible that half of the points in a full cycle (within a sufficiently large rectangle) are excluded because they lie outside the bounds of the actual rectangle associated with the specific cycle.

Note that there are several factors influencing the length of an **MCH** cycle: the dimensions of the bounding rectangle (and the degree to which one side is longer than another, causing some cycles to "lose" points); whether the center is double-black, double-gray, or a combination; and the presence or absence of zeros in the desired **MCH** distance. An orthogonal zero, for example, results in four points along two diagonals without any additional orthogonal projections that could extend either horizontally or vertically; as a result, the cycle can have only four (not eight) points, but also it cannot be contracted due to width/height discrepancies in the bounding rectangle.

The reason why XCSD employs **MCH** cycles is to itemize tierboxes that are equidistant from the image center. As mentioned earlier, XCSD computes a memory layout according to which tierboxes closest to the center occupy lower memory addresses, with pixel data encoded in contiguous raw memory expanding outward from that center. This "expansion" is understood to progress through **MCH** cycles. The **MCH** format has an intrinsic ordering based first on the total length (adding orthogonal and then diagonal) and then, for paths with the same sum result, treating paths with fewer *diagonal* steps as shorter (diagonal steps are longer from a Euclidean perspective). Given any collection of points around a center, we can partition the set into XCSD cycles that have a natural ordering between one another; moreover, *inside* each cycle, the points are distributed clockwise or counter-clockwise (since the basis for separating points is alternative directions superimposed on a single directionless **MCH** pair). The end result is a coherent algorithm for ordering points subject to constraints that points nearer to the center should be placed before those further away.

For XCSD, the "points" are actually tierboxes (or gaps between them), but this is consistent with black–gray grids embodying mereogeometric relations through discrete/integer mathematics, without the grid being a faithful *representation* of the modeled space's actual (Euclidean) geometry. In short, the algorithms we have described in the black–gray context work also for the practical task in XCSD of deriving the proper memory layout for image tierboxes (and, by extension, individual pixel runs).

One rationale for memory layout oriented to an image center is to increase the likelihood that the pixel data for an image region of interest can be obtained simply via pointers to the start and end of a memory block (without needing multiple copies into a temporary buffer). Due to the nature of **MCH** cycles and ordering, a contiguous pixel run (assuming it is aligned to tierbox boundaries) would take on a diamond, octagon, or square shape around the image center (or potentially a rectangle, if the length of one side of the demarcated region matches the shorter image dimension). While many significant image regions of course will not be centered on the full image's center, there are still situations where focusing on pixels closer to the overall center is desired. For example, signals extracted via image analysis are more likely to be significant for labeling, classifying, or matching images when they emerge from an image's central region more than its periphery. The color spectrum or histogram weighted toward the center typically bears more significance than further from the center, for example, and similarly for prominent contour shapes, textures, diffusion processes, and so forth. And colors near the center are more likely to be characteristic foreground tones than ones closer to the edges.

Also, more simplistically, thumbnail or preview image summaries might need to strip away peripheral content, even if the image is also scaled down. This points to potential benefits of center-oriented memory layout. Copying pixels (even multiple pixel runs, e.g., one per scanline) is not a very time-consuming operation, especially in the context of full-scale image analysis, where executing computer vision algorithms could easily take much more time than setting up an initial pixel matrix. However, some image processing may occur in contexts such as image databases where performance delays may noticeably affect usability, in context where it could be necessary to work through hundreds or thousands of images in response to one user query/action. In these kinds of contexts, avoiding even the relatively minor step of pixel copying can improve performance, in situations where one wants to construct thumbnails for hundreds of pictures (matching a query, say), or perform basic analysis on a large image series.

For these reasons, XCSD is designed as a format particularly suited to hosting images in a database, where basic operations such as image compression, calculating dominant colors, and building image thumbnails may be highly optimized. It is also true that the **MCH** representation, which we have described here in the context of memory layout, can have other benefits. For some analytic algorithms, for example, having a near-Euclidean distance formula on an integer grid may be useful. Furthermore, memory organization based on **MCH** cycles includes (partly to calculate the layout ordering in the first place) numerous data points applicable to each tierbox, such as distance and orientation against the image center, information that may have some meaning in certain computer vision contexts. For example, the weight given to some color, texture, or diffusion scale within the bounds of a given tierbox—or perhaps as compared between tierboxes—might be affected by the tierboxes' distance from center, or from image central–orthogonal or diagonal axes, or angular distance from some analytically significant line (consider a trendline through control points suggesting a foreground region). Some contexts may prefer angular distances measured via integers (e.g., via **MCH**) rather than Euclidean floating-point approximations.

These points are suggestive of potential **MCH** use cases; our goal here is not to advocate for **MCH** *per se*, but simply to justify the claim that this is the *sort* of construction that we might want to engineer in an image-processing context. We therefore claim it is a reasonable example for discussions within image analysis contexts. That is, we propose to use **MCH**-cycle ordering and related algorithms as case studies for VM engineering that encapsulates computer vision capabilities (or an interface to them).

22.3.4 XCSD Operators as Representative Image-Processing Functions

The above discussion has hopefully motivated some details concerning algorithms specific to XCSD. In particular, the ordering of tierboxes around an image center is determined by **MCH** cycles for progressively larger **MCH** values. These and related calculations give rise to a series of functions giving basic quantitative information about tierboxes, such as:

Translating tierboxes from row/column coordinates to MCH pairs This results in a measurement of the basic distance of the tierbox from one central tierbox—or one of two or four central tierboxes, depending on even or odd tierbox row and column counts (the even/odd details correspond to gray/black, black/gray, or double-gray cases reviewed above for **MCH** cycles on black–gray grids).

Mapping index numbers for a desired center-originating ordering of tierboxes to MCH values Such an ordering can be seeded by expanding outward according to Chebyshev distance, forming equivalence classes of Chebyshev-equidistant tierboxes. Within each such class, **MCH** distances expand outward from rows and columns passing through the image center (or centers; again, tracking even–odd effects as with black–gray grids) toward diagonals. This algorithm has to be adjusted for the case where the Chebyshev distance is greater than half the shorter image dimension (in terms of tierbox counts) so that the **MCH** cycle is restricted to the image's longer axis.

Calculating tierbox positioning within each MCH cycle The choice of how to order tierboxes *within* one cycle seems mostly arbitrary. In XCSD, we chose to order counter-clockwise from the left (for images with landscape orientation) or top (for those with portrait orientation, i.e., height greater than width). Our reasoning was that if an image *were* to be chopped from only one side, it may be slightly more likely to cut from the right or from the bottom.[9] Shrinking an image by deleting tierboxes toward the end of contiguous memory requires no copying data, of course, which provides a rationale for placing more-likely-expendable tierboxes later in memory than earlier.

Identify tierbox position within its MCH cycle in terms of image directions and octants In general, an image can be divided into eight wedge-shape slices (i.e., octants), and members of a cycle for **MCH** without zeros will lie in one such octant (those *with* zeros may lie on the lines between octants, e.g., diagonals). This is one example of an algorithm utilizing the "direction" enumeration we mentioned in the last section. As explained there, XCSD's notion of direction has 24 options, corresponding to length-8 and length-4 cycles (the latter representing pure diagonals) and then distinct codes for pure orthogonals (**MCH**s with diagonal component zero) distinguishing even–odd permutations at the center point (as modeled by different gray/black grid combinations). Code needs to translate formulae expressing how different **MCH** pairs translate to one of these four direction codes (this furnishes an example of where enumerations need precise numeric values, mentioned last section, because these algorithms make some use of modulo and bitshift operations when "collapsing" the 24 directions to accommodate different **MCH** cycle lengths).

Map "subdivision indexing" to orthogonal coordinates Inside tierboxes, XCSD recognizes indexes across 3×3 and 9×9 tiers, mirroring the actual memory layout. For example, the center of each tierbox is located via code "555." Obviously, sometimes it is convenient to access points or box regions in a more conventional x/y or row–column format, so one requisite calculation is converting subdivision location codes into orthogonal distances against tierbox corners and those of the image overall.

Compute color averages within box regions of different tiers One rationale for the subdivision system is to establish data structures for 3×3 and 9×9 boxes (as well as tierboxes) that proxy the collection of pixels within them. If an image is compressed by a factor of three, of course, pixels "collapse inward," so 3×3 regions *become* pixels on the smaller scale. Given a desired color-averaging

[9] Given that our vision tends to be oriented left to right and top to bottom; in other words, we are perhaps inclined to perceive a foreground focus leading from the center top/leftward more than the opposite, increasingly the possibility that it will be peripheral content toward the bottom/right that is deemed expendable, if the image is chopped.

procedure, XCSD precomputes color means so that "compressed" images are represented within the larger image's memory, in case algorithms intend to work with the smaller-scale images directly, or to analyze on several scales at once. Apart from color averages, data from different sources (e.g., region masks) can be attached as an "extension channel" on multiple scales. For example, a numeric code designating one region of interest (in the sense that a given pixel or box region lies on that region's interior) can be notated as a channel for individual pixels, or boxes on the 3×3, 9×9, or 27×27 tiers, or any combination thereof. The same applies to codes representing textural patterns, annotations, or any other superimposed data.

Applying algorithms to sets of tierboxes (or other region boxes) There are numerous scenarios where subdividing images into smaller regions can produce more efficient algorithms. This includes pixel-based computations (e.g., global color histograms) where pixels' contexts vis-à-vis surrounding pixels are not considered, so images could be subdivided simply for purposes of parallelization. Other analyses—such as those involving *local* color histograms—explicitly employ image subdivisions as a unit of computation. Finally, some forms of image processing are based on a lattice of "seed" points, which in a "subdivision indexing" scenario can readily be provided as the centers of tierboxes, or 3×3s/9×9s.

All of these are operations that would be relevant for images encoded via XCSD, potentially used with a diversity of front-end applications. Software components recognizing XCSD would therefore need a protocol to invoke the relevant procedures, exposed by XCSD code (yielding cross-component integration requirements along the lines analyzed in Chap. 20). The situation is analogous with many other image formats: most image-processing applications will accept images in multiple formats, and many formats have specific operations or information unique to that format, rather than generic to any image data. This raises the question of how software may be extended to support new formats that may come along.

Supporting formats entails accessing data proper for computer vision algorithms, but also incorporating format-specific GUI capabilities. For example, XCSD includes code to model individual tierboxes via distinct graphics-scene objects, so each tierbox receives user events independently, with its own mouseover and context-menu effects. The earlier discussion about GUI integration mentioned basic-level interactions such as viewing pipeline stages via OpenCV **imshow**, but some image formats will support a much more extensive suite of GUI tooling (e.g., formats that have a color-map option for channel reduction—mapping all pixels onto one of a small group of colors—should allow users to examine the specific tones in each color map). Further cases of GUI functionality will be outlined at the end of next section.

22.4 An Example Image-Processing Pipeline

Having defined several operations that could be exposed by an XCSD-based image-processing component, the following discussion will outline how various functions could be pieced together into a larger workflow. This summary will point to steps in the pipeline wherein data sharing between the XCSD component and a host application might be warranted, thereby serving as concrete examples of how each side could expose feature for interoperation.

One premise of XCSD is that conventional formulae for converting images into grayscale are often flawed. The rationale for analyzing "grayed" versions of images—rather than the full-color originals—is that most statistical analyses rely on color discontinuities. Analyses look for points in an image where there is an above-average gap between two adjacent pixels. Discontinuity, however, is an intrinsically "scalar" or one-valued concept. For example, given a sequence of numbers, a "spike" in value would correspond to a point where two sequential numbers have an unusually large difference, compared to typical adjacent number pairs. Generalized to two dimensions, the same notion of discontinuity works across different directions (up, right, down-left, etc.), giving rise to a matrix of difference measures, because any pixel is "adjacent" to several others (eight others, if we include both orthogonal and diagonal directions). However, quantifying discontinuity by value-difference only makes sense, intrinsically, when there are just two values to compare. Full-color spectrums distort such measures. In the typical RGB format, colors can be compared along three axes (red, green, and blue). If adjacent pixels have fairly large red differentials but small differences in green and blue, say, should the measure of their mutual difference be weighted more toward the larger gap (the red) or the smaller gaps (blue/green)?

To head off these sorts of questions, computer vision algorithms usually perform a "channel reduction," wherein the three channels of a full-color image are reduced to one single channel (which can be visualized as an image "in gray," although in principle one-channel graphics can be rendered with any one single color, such as red or blue, fading to white and/or darkening to black at points of greater or lesser magnitude). A simple way to reduce channels is, indeed, merely to take a grayscale version of the original image, as if it were an old black-and-white Polaroid. The single channel thereby measures color intensity—how dark or light it is—ignoring color-hue entirely. This tactic might discard potentially useful color information, because very different colors can have similar grayscale values. For instance, medium-red and medium-blue would, if juxtaposed, strongly suggest some ground-truth discontinuity, but they could both map to almost indistinguishable grayscale values if they are similarly mid-range in intensity (neither dark nor light). For these reasons, some image-processing frameworks prefer alternative channel-reduction techniques.[10]

[10] We admit that this outline is a little oversimplistic. Some gray-scaling algorithms are more sophisticated than simply averaging out red/green/blue measures; there are ways to incorpo-

The XCSD format is designed to support several alternative reductions to one channel. One tactic employed by XCSD code involves "toroidal" color space, which is related to HSV. A toroid is a solid torus; topologically, a torus is the result of one circle rotating around a central point, tracing its own circular path in the process. A point on a torus surface can therefore be identified by two angles, one measuring how far the "rotating" circle has gone around its anchor point, and one fixing a spot on the rotating circle itself (of course, we initially have to choose an angle to represent "zero" both on the surface and at the center—the latter axis would point out in space away from the center of the torus "hole," like a rod threaded through some bagels on which they hang by their centers). In HSV, the *hue* dimension is typically understood as circle-like (embodying the "color wheel" whose opposing colors are green/magenta and blue/yellow). For any given hue, the colors generated by varying "value" and "saturation" in the HSV model naturally form a triangle, with the pure color fading to white along one side, and decaying to black on another, while the white-to-black side spans progressively darker shades of gray. Near the midpoint of the triangle would be a "muddied" version of the pure color, appearing as the blend of that color with grayish tones.

More broadly—setting aside computational image processing for a moment and considering human vision—people appear to experience color "thematically," rather than purely chromatically, in that we are receptive to the optical effects through which color shades become visible as well as to actual color tones. The experience of color is not primarily oriented to monochrome color spreads, but rather (we might say) colors as "themes," wherein one thematic color carries expectation of some textural and lighting-effect variation. A straightforward example of such a theme might be a canonical foreground color; or, we may have an image foreground composed of multiple objects with their own predominant color, each thereby becoming distinct themes for the overall image (but still encompassing textural patters across objects' surfaces). In nature, it is rare to encounter either solid-hue patches or quality pigmentation; more common are surfaces that reflect light with some scatter and inconsistency, altering the primary color value and modulating apparent shades as the eye scans laterally. At the same time, this is an intrinsic feature of visible appearance—it is actually oversimplistic to describe surface appearance *only* as a matter of color in terms of chromatic channels (red/green/blue,

---

rate information about how our eyes see color, so that some hues "feel" brighter or darker [2, 21, 26, 31, 32]. Equal saturation among the three red/green/blue channels does not yield equal psychological experience of dark or light, and these differences can be incorporated into "realistic" techniques for color-to-gray conversions. Also, even a relatively small gap in gray measure will be detected by most computer vision algorithms, e.g., to identify boundaries between different image segments. So in some sense even grayscale formats incorporate *some* hue data, indirectly. Nonetheless, differently hued segments that have similar psychological brightness valence will still be experienced as distinct color patches. Likewise, small grayscale difference may complicate algorithms because it could be ambiguous whether such differences result from textural or light/shadow variation within one relatively homogeneous color patch or actual segment partitioning. In other words, even a more-refined gray-scaling algorithm is subject to processing errors for similar reasons to simpler gray-conversion formats.

etc.). Vision is about obtaining information optically, and all facets of light waves' interaction with objects are candidate data sources.

If a surface's primary color is red, say, we would not expect that object to appear uniformly and smoothly red, like a half-Polish/half-Indonesian flag. Instead, "redness" would be phenomenologically experienced as a distribution of relatively reddish tones, shading to black/white or variants (browns, grays) suggesting shadow, illumination, and the muddying effects of rough (material) texture. Collectively, these are sometimes called "textural effects" and analyzed through computer vision through texture analysis [3, 11, 12, 16, 22, 25, 27, 30, 33, 39, 40, 43, 44, 46, 50, 53]. Because the material coarseness (or, inversely, shininess) is an intrinsic physical property of objects—akin to red/green/blue color—effects such as texture and glare/shadow may even be modeled via distinct channels, alongside chromatic tones themselves (other potential supplemental channels may be glossiness, also called "specularity," and transparency). Image analysis cannot examine physical materials directly, of course, so texture-related qualities need to be inferred computationally, and color models might be evaluated in terms of how efficiently they feed algorithms that address these optical dimensions of color appearance.

In practice—returning to variations on the HSV model—most real-life colors do not appear with pure hues; instead, because of glare/shadow effects and/or material texture (which causes light to scatter off images, so that the visible color at given points has a grayer appearance than the underlying pigmentation would suggest) a patch of color (representing one evenly colored surface) would typically span several HSV triples (with varying amounts of white, black, or gray mixing with the primary color). From this observation, we can envision an HSV triangle being centered at a midpoint between pure gray and the pure primary hue, a point from which all other parts of the triangle can be defined in polar coordinates (i.e., as a rotation away from a line between the pure hue and pure gray). This rotation is *additional* to the hue's own rotation in the color wheel, so the overall color space is defined by two angular coordinates, which is why the space could be described as a toroid.[11]

Given this construction, we can define a color-comparison metric in terms of two rotation angles and one radius length (the "rho" measuring distance from a neutral grayish manifestation of the pure color, with larger rho values being pure color, pure black, or pure white). By weighting these parameters, the HSV triangle could be distorted (into something like a disk, more in keeping with the toroid shape), with the goal being to derive a color-distance metric that accommodates how empirical texture and glare/shadow effects can alter *apparent* color distance (see also [49]). Whereas a visual illustration of the global HSV space would appear like a cylinder,

[11] Unlike a toroid proper, such a space has a "pure" white-to-black line that yields identical colors for every hue, which distorts color distances in ways that do not match a toroid shape; however, actual color-distance formulae within the toroid should be weighted anyhow, so the internal geometry does not match a toroid in Euclidean space to begin with.

the resulting variation on HSV results in something like a toroid (hence, a "toroidal" color model).

An assumption behind this model is that because even "solid" colors (vis-à-vis an object's surface pigmentation) will appear sometimes lighter, darker, or grayer, these textural/lighting effects should be compensated for when computing color distance. Exactly how to weight toroidal parameters is a matter for future research; the existing XCSD implementation employs a straightforward balancing mechanism that tries to compress distances for near-black and near-white shades and expand distances for purer (high-value, high-saturation) tones.

Code for XCSD applies a toroidal model to derive several channel-reduction strategies. One computation involves estimating a good reference color typical of an image's *background*, and another canonical color (also called a "pole" in XCSD's code) for *foreground*. Via distance from these colors, XCSD computes a probability that any given color belongs to a foreground object or background area. This probability is a scalar quantity, so it serves as a reduction to one channel, one that is typically noticeably different from conventional grayscale (Fig. 22.2).[12]

For the case-study workflow discussed here, XCSD estimates foreground/background poles by assuming that foreground colors are likely to be featured near the center of an image, and background colors closer to the margins (particularly on top). Such heuristic may not comport with all images, but it is a good place to start. The workflow then leverages XCSD's "subdivision" layout to grab dominant colors from histograms local to tierboxes both near the image center and top periphery, averaging out these two cases to obtain working foreground/background poles.[13]

Having assigned foreground/background probability to all pixels, this workflow then reifies the probability scalar as a channel and can invoke different feature detectors. The illustrations here are based on a BRISK (binary robust-invariant scalable keypoints) detector [23], which yields a fairly extensive set of keypoints almost all of which are positioned within regions of the image that, on visual inspection, clearly belong to the foreground.

[12] The probability metric is actually a combination of two other values, difference *from* foreground and from background, with the assumption that pixels *closer* to the foreground color and *further* from the background pole are the most likely actually to be in the foreground—but unless the two poles are actually at a maximum inter-hue distance, which is unlikely for real-world pictures, the two distance-from-pole parameters can be varying independently, which is why they must be averaged together (arithmetically inverting one of the pairs) to yield a one-channel reduction.

[13] A variation of this pipeline would allow human users to fine-tune this selection manually, by selecting specific tierboxes and examining their histograms. It is helpful to support interactive image processing along these lines, even if only to fine-tune workflows eventually designed to run without human input (e.g., users might engage interactive methods to finalize parameters for an image series in the context of one or several sample images, and then fix those parameters for non-guided analyses extended to the series as a whole). Here, though, the discussion will stay focused on fully automated workflows; interactive options will be touched on at the end of the section.

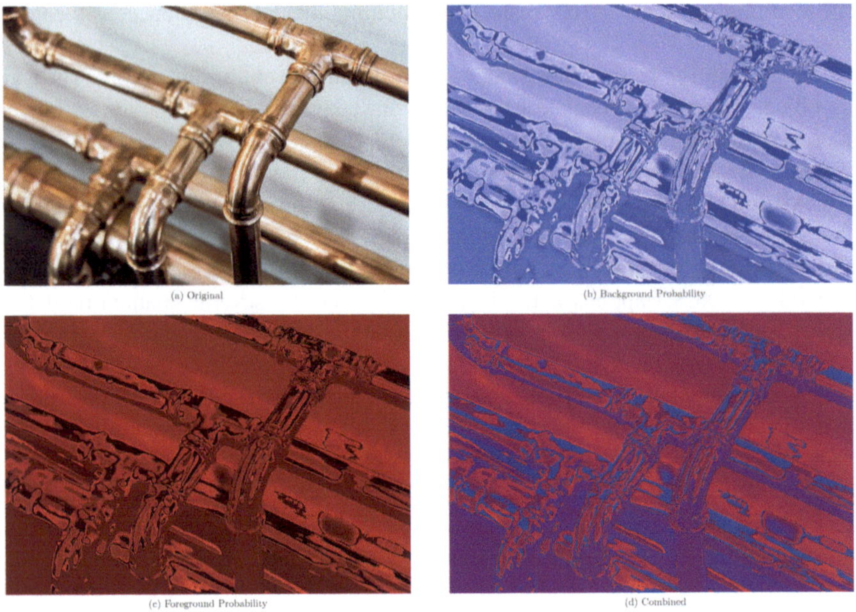

Fig. 22.2 Background and foreground probability scalars for channel reduction. (Foreground probability is shown red to black, background is shown blue to white, and the two scales are merged in the combined graphic). (**a**) Original. (**b**) Background probability. (**c**) Foreground probability. (**d**) Combined

22.4.1 From Keypoints to Superpixels

Once useful keypoints have been identified, most pipelines will then generalize keypoints to image areas, either enclosing or bounded by relevant keypoints. In the current workflow, a superpixel segmentation (see, e.g., [1, 17, 19, 24, 36, 47]) of the image is performed, and superpixels that encompass some of the BRISK keypoints are identified; because the keypoints are localized near the foreground, the sum of keypoint-containing superpixels then serves as a reasonable approximation of the foreground as a whole (Fig. 22.3). This is not the only way to use zero-dimensional keypoints; a more sophisticated analysis might consider keypoint attributes so as to more rigorously screen for those in the foreground proper.[14] For the sake of discussion, however, consider the more immediate calculation where the overall set of keypoints is treated as a point "cloud" roughly distributed through the foreground. Intuitively, keypoints at region-of-interest boundaries would seem to be the most valuable, because they would directly yield useful segmentations;

[14] That is, attempt to find statistical signatures strongly suggesting that a given keypoint is indeed part of the region's foreground (or near the foreground/background boundary, in which case such boundary keypoints could serve as a scaffolding with which to isolate the foreground).

Fig. 22.3 Localizing keypoints to the image foreground. Two keypoint detectors yielding good results for this workflow, BRISK (Binary Robust-Invariant Scalable Keypoints) and SURF (Speeded Up Robust Features) (**a**), (**b**). Combined with a superpixel segmentation, the foreground could be detected by extending the feature point cloud to the set of superpixels containing one keypoint (or some count of keypoints, beyond a chosen threshold) (**c**). This superpixel segmentation uses a different channel reduction via mapping the full set of image pixels onto a small collection of the most frequent colors, according to toroidal distance (**d**)

however, complex/occluded images may require more circuitous techniques because feature detectors will generate *too many* keypoints. To compensate, one option is to actively seek out algorithms that identify large numbers of features and then apply heuristics to separate those belonging to the background and foreground (this also applies to two-dimensional features, e.g., edge detectors: the graphics in Fig. 22.6c intimate how detected edges can be more concentrated in the foreground than the background, which provides options for leveraging kernels that detect seemingly superfluous edges, rather than trying to restrict algorithms to contour boundaries).

Returning to the example, visual inspection suggests that the BRISK results do indeed match the foreground well, but the workflow will attempt to calculate this result systematically. To do so, the code loads a data structure (defined in an XCSD-specific language) that describes a series of image transforms that rotate and slightly skew the image, then blur the image to solid-color 27 × 27 blocks or boxes, and cancel (white out) all boxes that are not sufficiently close to the foreground color. These operations manipulate the image so that the solid blocks may act as classifying bins, where the same transforms are then applied to the keypoint set, such that each keypoint gets mapped to a specific box, which could be labeled as either foreground or background. The goal is to test, first, that the keypoints are indeed mostly found in the foreground (i.e., most keypoints map to a foreground bin) and also that the keypoints are distributed relatively thoroughly across the foreground (i.e., most foreground bins include at least one keypoint).

In combination, the two statistics just outlined confirm that the aggregate of keypoints—particularly, when extended to the containing superpixels—approximates the foreground, both by *including* most of the foreground and by *excluding* most of the background. Such assessments could be explicitly quantified by measuring the percentage of keypoints that get mapped to a foreground (not background) box, and also the percentage of foreground boxes with one (or more, above some desired threshold) keypoint—the higher both percentages, the more effective the workflow.

As a final detail, a different XCSD distance metric is employed to calculate "Hough" lines—one form of line detector [7, 8, 29, 34, 42, 45]—so as to derive a good angle with which to rotate the initial image (Fig. 22.4). Here XCSD dispenses with one-channel reduction and instead considers "local" color distance, considering color variation within a relatively small (e.g., 7×7) area. Unlike converting a full image into grayscale (or any other one-channel reduction), in which a useful color-distance metric should apply across the image, when measuring color distance locally, we can focus on how far apart two nearby pixel's coloration appears in the local context, without considering the presence of other pixels elsewhere in the image.

Multi-channel color distance is complicated by the fact that distance magnitudes may take on different meanings in different contexts; we might measure two colors as relatively far apart and then find a third color that is equally distant from the original two, leading to questions of how the respective distances should be ordered. However, consider the border between two different image segments: there will in general be relatively little color variation *inside* each segment, but a notable differences between sample colors in one segment and in the other. Near the border, then, color variation on either side should be reduced, but cross-border comparisons will be much larger. So long as we only consider local neighborhoods where the latter cross-border distance clearly outpaces intra-segment color distances, we can set aside the question of how local distances compare to "global" color distance vis-à-vis the image as a whole. Applying this metric, consequently, we may isolate pixels that appear to lie near segment boundaries.

In the context of line detection, this latter strategy for producing a single image channel—measuring local-neighborhood inter-pixel color difference—can potentially enable inter-segment lines to stand out clearly. For the case study described here, all Hough-detected lines have almost identical angles of inclination (they appear parallel when overlaid on a base image), which is a good indicator that the average of the Hough lines gives an angular orientation for the image as a whole. If so, noteworthy boundaries or other one-dimensional features tend to be aligned along a specific angle (or at right angles to that) so that rotating the image by that angle allows those prominent features to be oriented parallel to the image sides—which, in turn, allows the image to be partitioned into orthogonal boxes for

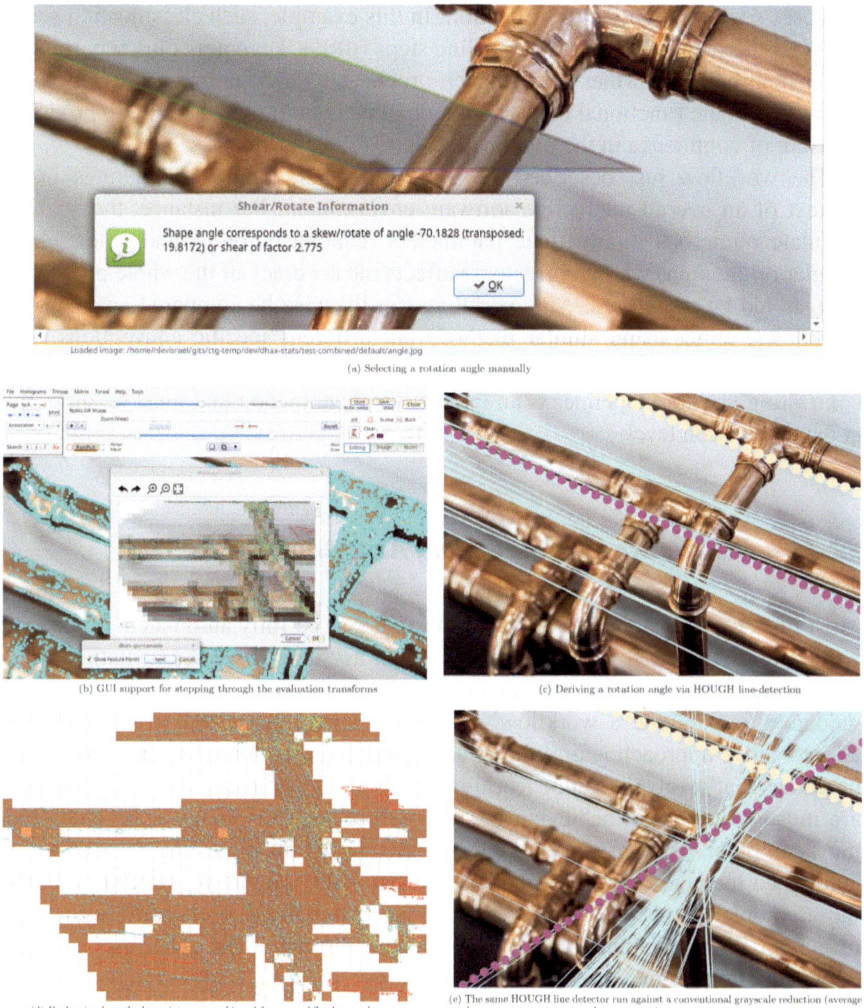

(a) Selecting a rotation angle manually

(b) GUI support for stepping through the evaluation transforms

(c) Deriving a rotation angle via HOUGH line-detection

(d) Evaluating how the keypoints map to binned foreground/background areas

(e) The same HOUGH line detector run against a conventional grayscale reduction (average angles are magenta, target angle tan-orange)

Fig. 22.4 Deriving a rotation angle for evaluating the keypoint detector. To calculate how well the keypoints match the foreground, we transform the original image to yield orthogonal boxes, where it is easy to calculate how many keypoints (if any) lie on each box. Visually, note that an automated HOUGH line detector operating on an XCSD toroidal color space yields results comparable to manually selecting a global angular orientation for the image (**a**), (**c**). Mapping the image to a skew/rotated and discretized box set to quantify how closely the keypoints map to the foreground (**b**), (**d**). Watershed methods (based on color average inside each box and their distance to the highest-probability foreground color) classify each box as approximating a foreground or background region. For comparison, (**e**) shows a duplicate of [c]'s line detector only using a conventional (realistic) grayscale conversion in lieu of XCSD channel reduction; the latter analysis shows an entirely different (and mutually inconsistent) set of HOUGH lines

purposes such as keypoint classification. In this example, such classification serves to assess the quality of the prior pipeline steps (BRISK keypoints plus superpixels), rather than being a further analytic step proper, but such summary evaluation can play a legitimate functional role within a pipeline overall (e.g., helping to produce a measure of confidence in the pipeline results).

The workflow just described is fully automated, but it still belongs in the context of an overall interactive software environment. For instance, the analyses calculate numerous intermediate parameters (canonical foreground color, global angular orientation) whose correctness affects the accuracy of the whole process, so they should be monitored. Any workflow can likewise be compared against other workflows, either using similar methods but different specific conventions (e.g., different channel-reduction formulae) or very different methods. Initiating and then contrasting *multiple* pipelines is an interactive phenomenon that must be supported at the application level.

22.4.2 Interactive Workflows and Assessments

Most image-processing algorithms are designed to be fully automated, proceeding without user input. However, techniques such as "interactive segmentation" (plus comparing *different* techniques, as just cited) are an exception, relying on human guidance to fine-tune a workflow's behavior. There are several scenarios where an interactive approach is useful, including training data for subsequent automated processes that involve machine learning, establishing parameters for workflows that are mostly automated, achieving greater performance for image series that elude most computer vision methods, and developing protocols for testing the accuracy of automated methods. Whatever the specific rationale, interactive workflows require more complex application-level support than automated ones because users need to act on image graphics and workflow data through GUIs. This user-interface dimension implies that interactive image analysis presents a representative use case for general themes in software engineering (including VMs).

To illustrate, consider a fairly simple interactive workflow that could be juxtaposed with the automated workflow outlined above (Fig. 22.5). Our goal is to isolate the copper-colored foreground from a gray background and measure the relative size of the foreground. Unlike the automated process, we can proceed without special mathematics (matrices or Gaussian kernels) or attempting to isolate the foreground with some level of detail—we only want to get an approximate cover for the foreground (what is sometimes called an "object proposal" [4, 6, 9, 10, 13, 18, 20, 28, 35, 41, 48, 51, 52]). If there is sufficient and consistent background/foreground color difference, we could take color alone as a good indicator of foreground probability. The automated workflow tried to estimate this probability without human intervention, but in an interactive session users could manually identify representative foreground and background color, perhaps via a GUI color picker (which reads the color under the cursor when clicking via a mouse, after the user

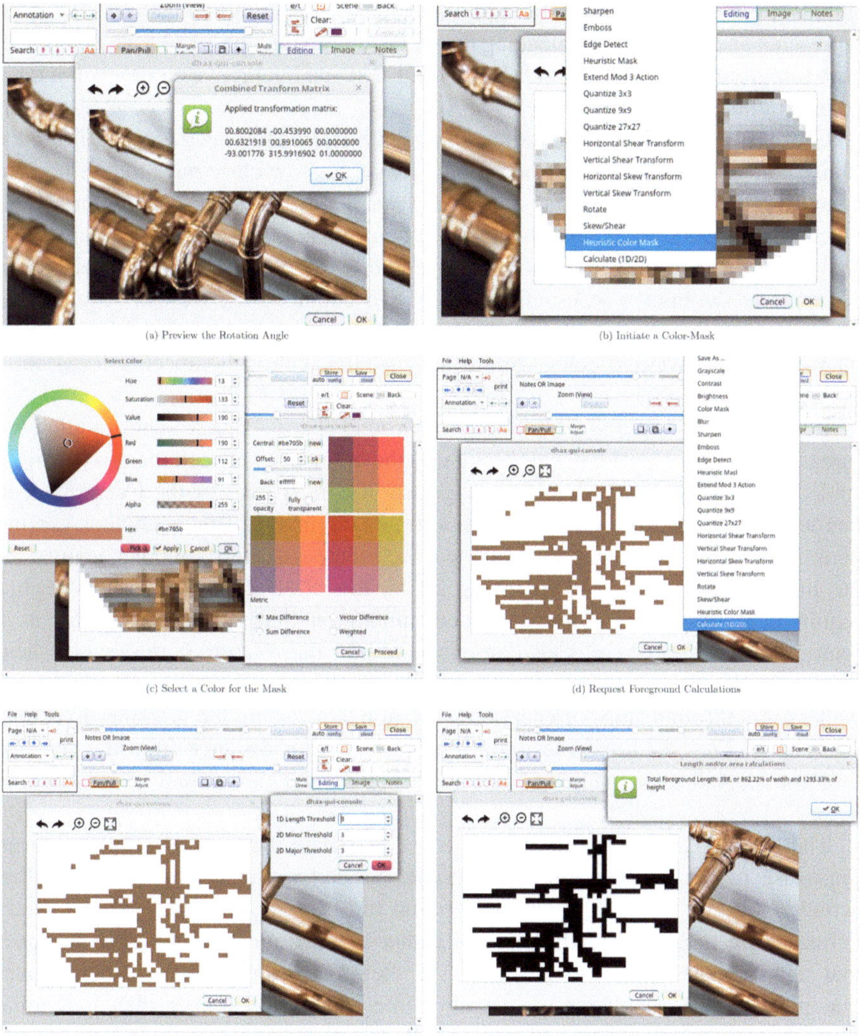

(a) Preview the Rotation Angle

(b) Initiate a Color-Mask

(c) Select a Color for the Mask

(d) Request Foreground Calculations

(e) Selecting a Connectedness Threshold

(f) Requesting Measurements

Fig. 22.5 An interactive workflow with XCSD. Defining an image-transform workflow interactively: first applying a selected skew/rotate angle (**a**); selecting a color, difference threshold, and difference formula for foreground isolation (**b**), (**c**); choosing a threshold to group connected foreground boxes (**d**), (**e**); measuring the size of the foreground approximation against the total image dimensions (**f**). As shown in the previous figure, a similar workflow can be defined to test the automated results of a keypoint-plus-superpixel workflow that tries to isolate keypoints within the image foreground

signals an intent to temporarily apply the mouse press in this manner). This step represents one parameter influencing the workflow's behavior (or two, if users select both foreground and background shades) and also one interactive modality that must be programmed into a GUI.

Given a selected foreground color, all pixels within some specific color difference (this is another detail that may be set interactively—how close must that difference be and via what formula to calculate it) are mapped to the foreground color and all others to the background, resulting in a two-color image. This may not actually yield an isolated foreground, however, because the foreground-classified pixels could be scattered without a well-defined shape. To counter this, we can segment the image into boxes, treating each entire box as foreground or background depending on whether the average color within that box fits within the specific foreground difference.

In order for boxes to align with the foreground shape—rather than straddle both foreground and background, cut through by the boundary between them—we might start the workflow by rotating the image, so that most lines of the foreground, to the degree possible, run parallel to the sides of the image. This is another varying parameter—how much to rotate any specific image—and also another feature for the interactive GUI. As shown in Fig. 22.4a, the XCSD code implements such a requirement through a rhombus-shaped annotation that users can register as a rotation instruction (there is also an option to pair the rotation with skew or shear actions, further aligning foreground boundaries to horizontal/vertical axes). The GUI walks users through steps of enacting a rotation, converting the granular pixel data into coarse 27×27 boxes, testing each box for color distance from the selected foreground, and then mapping each box to either the foreground or background color.

The last steps in this demo workflow are to isolate the foreground proper from imprecise pieces, perhaps image anomalies or perhaps indeterminate boxes that actually straddle the foreground/background line. Again, the goal is not to neatly delineate the foreground but simply to gather a collection of medium-scale boxes that approximate its size and shape. We therefore want to include only boxes that piece together coherently—forming lines and areas, rather than individual foreground boxes dispersed within the background. This implies a requirement that boxes be filtered out unless they form one-dimensional chains or two-dimensional rectangles where, for example, there should be at least three consecutive boxes in a straight line for them to be finally classified as foreground. The specific thresholds for this filter are a further workflow parameter, and a further detail that users might set interactively. With this final piece of information, the program can filter extraneous pseudo-foreground boxes and sum up the areas of the boxes that remain, yielding an approximate measure of the foreground's extent relative to the overall image.

This is a fairly coarse set of algorithms, which may or may not yield usable results when deployed against real-world images. However, even if more sophisticated methods (such as line or keypoint detectors) are ultimately employed instead, the interactive workflow just outlined could still be useful for preliminary experimentation with an image series. Similarly, the *evaluation* of the automated workflow described earlier can proceed via a series of image transformations constructed via the interactive workflow. Whereas the latter workflow is designed to *calculate* the foreground area by rotating and reducing the image to a two-color, coarse-boxed simplification, the evaluation performs similar steps so as to yield a binning classification (the foreground/background boxes become bins into which keypoints are sorted). As this example suggests, an interactive workflow can provide *assessment* of automated workflows even if the goal is for computer vision operations to run without human input. Such evaluations could help ensure the accuracy of automated pipelines and enable engineers to identify which image-processing tactics and which workflow parameters yield the best results for a given image series.

For an image-processing environment to support interactive workflows, it must implement multiple GUI and event-handling requirements, typically ones that are not addressed by computer vision libraries themselves (such as OpenCV). Even the mathematically simplistic workflow reviewed here necessitates a number of special-purpose GUI controls, such as a color picker, a tool for image annotations (rhombus-shaped, in particular, to get images' angular orientation), dialogs for setting thresholds and parameters on image difference and criteria for filtering morphologically non-cohering foreground boxes, and controls for reporting pipeline/calculation results, graphically and/or numerically. Other GUI controls come into play when interactive workflows are adopted for evaluating automated pipelines (see Fig. 22.4, especially 22.4a and b).

22.5 Conclusion

Interactive workflows are a prime example of how image-processing applications rely on GUI-level programming, but there are numerous other use cases for which similar comments apply. For example, users might want to visually compare the results (or intermediate stages) of diverse workflows, spanning multiple techniques, to determine which approaches work best for a particular image series. We can assume that in many cases parameters and algorithms that yield promising results on sample images within a series may be automatically applied to the image series, particularly if such images are thematically related (such that the series elements have similar color patterns, textural complications, perspectival orientation—e.g., whether canonical background colors tend to arise from the top center, as in outdoor/landscape pictures, or the left/right sides—etc.). Figure 22.6 shows a variety of GUI components associated with XCSD that demonstrate how applications can support users' experimenting with different algorithms (e.g., edge-detecting methods), visualizing color schema, examining interactive color histograms, and otherwise reviewing and fine-tuning data that will ultimately be incorporated into engineered workflows.

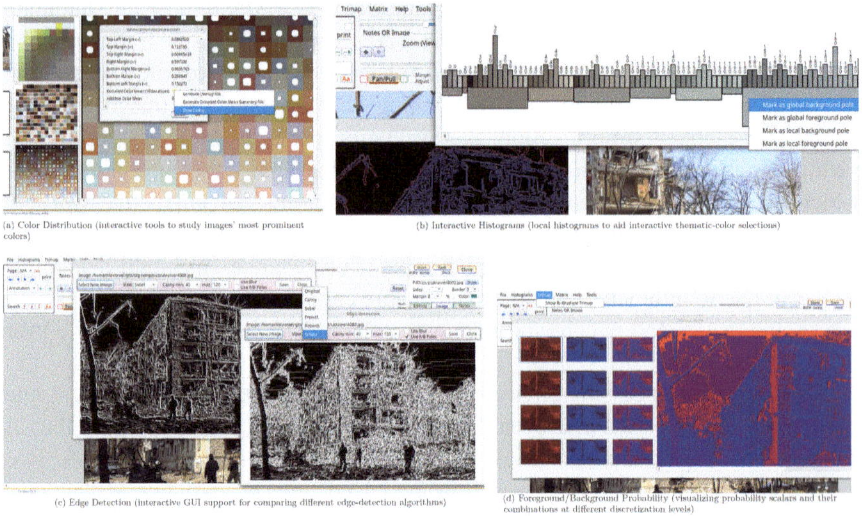

(a) Color Distribution (interactive tools to study images' most prominent colors)

(b) Interactive Histograms (local histograms to aid interactive thematic-color selections)

(c) Edge Detection (interactive GUI support for comparing different edge-detection algorithms)

(d) Foreground/Background Probability (visualizing probability scalars and their combinations at different discretization levels)

Fig. 22.6 GUI tools for interactive image processing. Screenshots of GUI components for interactive design or evaluation of computer vision workflows. Viewing color distributions and histograms (**a**), (**b**). Comparing edge-detection algorithms (**c**). Visualizing foreground/background probabilities (**d**)

The more that applications seek to provide GUI support for image-processing operations, the more complex and expansive becomes the integration between computer vision libraries themselves and the GUI components that supply data and parameters to computer vision workflows. In this scenario, one approach to engineering and maintaining the proper cross-component interoperability would be to support GUI controls, event handlers, and data exchange among the relevant software libraries via a virtual machine. This could entail VM code for designing GUI classes, during testing and development; VM intermediaries for handling events/signals generated by user actions through these GUI elements; VM-based protocols for connecting the GUI data to image-processing workflows implemented via isolated computer vision modules; or any combination of these.

Section 22.2 outlined specific image-processing workflows, enacted through the XCSD image format. That outline points to several details within an XCSD pipeline (arguably indicative of common concerns within many different computer vision workflows), which could involve cross-component interop, insofar as image analysis is provided via dedicated plugins, extensions, or third-party libraries adopted by a host application. As mentioned above, a pipeline would often defer to the host at least for graphics displays and initial image acquisition, but the XCSD workflow points to other moments at which interop may be necessary.

First, we mentioned the idea of consulting local histograms interactively so that human users might define foreground and background poles for sample/reference images in a series. In this case, the host application would need to take responsibility for loading histogram displays and processing user actions. For "local" histograms, the color counts are drawn from a small area—via XCSD the image is predivided into tierboxes anyhow, but in an interactive case users need to select one tierbox for which to view histogram data, which would be an action handled by a front-end component. Assuming XCSD code calculates the actual histogram and then sends this data back to the front end, which in turn presents it visually, we see a multi-step exchange of data between the two components. Likewise, histogram displays may themselves be interactive—given that these visuals typically rank colors by showing bars filled with a particular color with a height documenting its occurrence in the histogram area, the bars themselves can serve as color selectors, so for example users could use histograms to select colors assigned specific thematic roles (such as foreground/background). Such actions would then generate further data routed back to the processing component (e.g., selected foreground/background poles applied to form a new one-channel reduction).

In the exchange just described, we assume that a front end exposes capabilities to display histograms when provided the requisite data, while image analysis components leverage this functionality by assembling histogram data. Generalizing from this example, image-processing front ends could support a variety of features relevant to initiating and examining computer vision pipelines, such as displaying images with annotations and overlays—this would include showing intermediate graphics wherein an image being analyzed is marked with icons, lines, or contours representing features extracted by detectors for Hough lines, scale/rotation-invariant keypoints, boundary contours, and similar algorithms. Host applications would thereby anticipate the general kinds of interactive/presentational features that are important to provide for an image-processing context, but they would remain open-ended with respect to the specific analytic modules that may target their front-end capabilities. In particular, host applications could then support a flexible spectrum of analytic methods, insofar as new sorts of algorithms may be introduced with their own mathematical constructions; each component can provide distinct analytic code so long as it accepts a common protocol for basic functionality involving acquiring, displaying, and invoking front-end features for analyzed images and image series.

In short, the interop between image-processing host applications and analytic modules is a good example of bidirectional data sharing between semi-autonomous components, and fall under the rubric of interop situations where protocols could benefit from formalization or (at least partial) implementation via virtual machines.

As a concluding observation, readers may be interested in checking the chapter's demo code, which includes scripts (mostly build tools) linked to over 100 cross-platform image analysis libraries, offering a useful overview of the common practices and design patterns currently favored by computer vision developers. One noteworthy insight, reviewing the code for this collection of libraries, is that very few image-processing resources provide GUI tools or attempt to interoperate with GUI front ends. Most of these repositories are focused on supplying processing algorithms (for segmentation, feature detectors, image tagging, and so forth) and implement workflows with little connections to outside components, after acquiring an initial image to work on. Considering both the vast range of research that has been done on computer vision methods (evinced by the number of distinct open-source tools, many of which codify methods presented through published research), and (by contrast) the paucity of application-integration tools, we might conclude that there is room for common standards and shared technologies for interoperating GUIs with image-processing modules to fill in a noticeable gap in the current computer vision ecosystem.

References

1. Radhakrishna Achanta, *et. al.*, "SLIC Superpixels Compared to State-of-the-Art Superpixel Methods". https://www.cs.jhu.edu/~ayuille/JHUcourses/VisionAsBayesianInference2022/4/Achanta_SLIC_PAMI2012.pdf
2. Martin Čadík, "Perceptual Evaluation of Color-to-Grayscale Image Conversions". http://cadik.posvete.cz/color_to_gray_evaluation/cadik08perceptualEvaluation.pdf
3. A.C. Chadwick and R.W. Kentridge, "The perception of gloss: A review". https://pubmed.ncbi.nlm.nih.gov/25448119
4. Darren M. Chan and Laurel D. Riek, "Object Proposal Algorithms in the Wild: Are they Generalizable to Robot Perception?". https://cseweb.ucsd.edu/~lriek/papers/chan-riek-IROS2019.pdf
5. Jim X. Chen and Xusheng Wang, "Approximate Line Scan-Conversion and its Hardware Design". https://cs.gmu.edu/~jchen/report/hardwareLine.pdf
6. Mingliang Chen, *et. al.*, "Interactive Hierarchical Object Proposals". http://www.shengfenghe.com/qfy-content/uploads/2019/12/458b6d07423a8844d3129412b717a0b3.pdf
7. Ming-Ming Cheng, *et. al.*, "Deep Hough Transform for Semantic Line Detection". https://arxiv.org/pdf/2003.04676.pdf
8. Gideon Kanji Damaryam, "A Method to Determine End-Points of Straight Lines Detected Using the Hough Transform". https://www.ijera.com/papers/Vol6_issue1/Part%20-%201/H601016775.pdf
9. Matt Duckham, *et. al.*, "Efficient Generation of Simple Polygons for Characterizing the Shape of a Set of Points in the Plane". http://geosensor.net/papers/duckham08.PR.pdf
10. Mark Ewald, "Content-Based Image Indexing and Retrieval in an Image Database for Technical Domains". http://www.ibai-publishing.org/journal/issue_mldm/2009_july/mldm_2_1_3-22.pdf
11. Davit Gigilashvili, *et. al.*, "Perceived Glossiness: Beyond Surface Properties". https://jbthomas.org/Conferences/2019aCIC.pdf
12. Xiaoying Guo, *et. al.*, "Analysis of Texture Characteristics Associated with Visual Complexity Perception". https://link.springer.com/article/10.1007/s10043-012-0047-1
13. Bumsub Ham, *et. al.*, "Proposal Flow: Semantic Correspondences from Object Proposals". https://www.di.ens.fr/willow/pdfscurrent/ham2017.pdf
14. Kunimatsu Hashimoto, *et. al.*, "KOSNet: A Unified Keypoint, Orientation and Scale Network for Probabilistic 6D Pose Estimation". http://groups.csail.mit.edu/robotics-center/public_papers/Hashimoto20.pdf
15. Zheyuan Hu, *et. al.*, "Object Pose Estimation for Robotic Grasping based on Multi-view Keypoint Detection". http://www.cloud-conf.net/ispa2021/proc/pdfs/ISPA-BDCloud-SocialCom-SustainCom2021-3mkuIWCJVSdKJpBYM7KEKW/264600b295/264600b295.pdf
16. Csaba István Kiss, *et. al.*, "On Feature Extraction for Texture Analysis". https://core.ac.uk/download/pdf/236628682.pdf
17. Sifan Ji, *et. al.*, "Image Clustering Algorithm using Superpixel Segmentation and Non-Symmetric Gaussian-Cauchy Mixture Model". https://ietresearch.onlinelibrary.wiley.com/doi/pdf/10.1049/iet-ipr.2020.0402

18. Zequn Jie, *et. al.*, "Scale-Aware Pixelwise Object Proposal Networks". https://www.sysu-hcp.net/userfiles/files/2021/02/28/3a62f36bc1fe7fb9.pdf
19. V. R. Jyothish, *et. al.*, "An Efficient Image Segmentation Approach using Superpixels with Colorization". https://www.sciencedirect.com/science/article/pii/S1877050920310620
20. Philipp Krähenbühl and Vladlen Koltun, "Geodesic Object Proposals". http://vladlen.info/papers/gop.pdf
21. Giovane R. Kuhn, *et. al.*, "An Improved Contrast Enhancing Approach for Color-to-Grayscale Mappings". https://www.inf.ufrgs.br/~oliveira/pubs_files/MS_C2G/Kuhn_Oliveira_Color_to_Gray_TVCJ_Pre-print.pdf
22. Rushi Lan, *et. al.*, "Quaternionic Local Ranking Binary Pattern: A Local Descriptor of Color Images". https://ieeexplore.ieee.org/document/7350143
23. Stefan Leutenegger, *et. al.*, "BRISK: Binary Robust Invariant Scalable Keypoints". https://www.research-collection.ethz.ch/bitstream/handle/20.500.11850/43288/eth-7684-01.pdf
24. Nannan Liao, *et. al.*, "BACA: Superpixel Segmentation with Boundary Awareness and Content Adaptation". https://www.mdpi.com/2072-4292/14/18/4572
25. Maiko M. I. Lie, *et. al.*, "Joint Upsampling of Random Color Distance Maps for Fast Salient Region Detection". https://www.sciencedirect.com/science/article/abs/pii/S0167865517303215
26. Wei Hong Lim and Nor Ashidi Mat Isa, "Color to Grayscale Conversion Based on Neighborhood Pixels Effect Approach for Digital Image". https://www.emo.org.tr/ekler/14281fad9e3d0cc_ek.pdf
27. Xiu Liu and Chris Aldrich, "Deep Learning Approaches to Image Texture Analysis in Material Processing". https://mdpi-res.com/d_attachment/metals/metals-12-00355/article_deploy/metals-12-00355-v3.pdf?version=1645497882
28. Yao Lu and Linda Shapiro, "Closing the Loop for Edge Detection and Object Proposals". https://homes.cs.washington.edu/~shapiro/yao-aaai.pdf
29. Subhransu Maji and Jitendra Malik, "Object Detection using a Max-Margin Hough Transform". https://www2.eecs.berkeley.edu/Research/Projects/CS/vision/shape/smaji-cvpr09.pdf
30. Fereshteh Mirjalili and Jon Yngve Hardeberg, "On the Quantification of Visual Texture Complexity". https://www.mdpi.com/2313-433X/8/9/248
31. László Neumann, *et. al.*, "An Efficient Perception-based Adaptive Color to Gray Transformation". https://diglib.eg.org/bitstream/handle/10.2312/COMPAESTH.COMPAESTH07.073-080/073-080.pdf
32. Chuong T. Nguyen and Joseph P. Havlicek "Color to Grayscale Image Conversion using Modulation Domain Quadratic Programming". https://ieeexplore.ieee.org/document/7351674
33. Shin'ya Nishida "Image Statistics for Material Perception". https://www.sciencedirect.com/science/article/pii/S235215461930035X
34. Ivan Papusha and Matthew Ho, "Hough Transform for Directional Orientation". https://stacks.stanford.edu/file/druid:rz261ds9725/Ho_Papusha_RealTimeHoughTransform.pdf
35. Jordi Pont-Tuset, *et. al.*, "Multiscale Combinatorial Grouping for Image Segmentation and Object Proposal Generation". https://upcommons.upc.edu/bitstream/handle/2117/105287/1503.00848.pdf
36. Antonio Rubio, *et. al.*, "BASS: Boundary-Aware Superpixel Segmentation". https://digital.csic.es/bitstream/10261/166227/1/bassbound.pdf
37. Giorgos Sfikas, *et. al.*, "Quaternion Harris For Multispectral Keypoint Detection". https://www.cs.uoi.gr/~sfikas/icip-quatharris.pdf
38. Hadi Sutopo, "Bresenham's Lines Algorithm Visualization Using Flash". https://ggn.dronacharya.info/ITDept/Downloads/QuestionBank/Odd/V%20sem/Bresenham_Lines_Algorithm_Visualization_Flash_IIISem.pdf
39. Mihran Tuceryan and Anil K. Jain, "Texture Analysis". https://cs.iupui.edu/~tuceryan/research/ComputerVision/texture-review.pdf
40. Takeshi Uejima, *et. al.*, "Proto-Object Based Saliency Model With Texture Detection Channel". https://www.frontiersin.org/articles/10.3389/fncom.2020.541581/pdf

41. Esteban Uriza, *et. al.*, “Efficient Large-scale Image Search with a Vocabulary Tree”. https://www.ipol.im/pub/art/2018/199/article.pdf
42. Xinming Wang, “Real-time Classified Hough Transform Line Detection Based on FPGA”. https://www.atlantis-press.com/article/25848213.pdf
43. Zhi-Zhong Wang and Jun-Hai Yong, “Texture Analysis and Classification With Linear Regression Model Based on Wavelet Transform”. https://hal.inria.fr/inria-00517305/document
44. Gunnar Wendt and Franz Faul, “Factors Influencing the Detection of Spatially-Varying Surface Gloss”. https://journals.sagepub.com/doi/pdf/10.1177/2041669519866843
45. Jianping Wu, *et. al.*, “Real-time Vanishing Point Detector Integrating Under-parameterized RANSAC and Hough Transform”. https://openaccess.thecvf.com/content/ICCV2021/papers/Wu_Real-Time_Vanishing_Point_Detector_Integrating_Under-Parameterized_RANSAC_and_Hough_Transform_ICCV_2021_paper.pdf
46. Minglong Xue, “Deep invariant texture features for water image classification”. https://link.springer.com/content/pdf/10.1007/s42452-020-03882-w.pdf
47. Yi-Xuan Zhan, *et. al.*, “Superpixel Segmentation Using Improved Lazy Random Walk Framework Based on Texture Complexities”. http://www.csroc.org.tw/journal/JOC30_5/JOC-3005-20.pdf
48. Ziming Zhang and Philip H.S. Torr, “Object Proposal Generation using Two-Stage Cascade SVMs”. https://zimingzhang.files.wordpress.com/2014/10/ziming_tpami_obj_proposal_final.pdf
49. Hanli Zhao, *et. al.*, “Image Recoloring using Geodesic Distance Based Color Harmonization”. https://core.ac.uk/download/pdf/81904494.pdf
50. Shan Zhao and Jing Liu, “Image Descriptor Based on Local Color Directional Quaternionic Pattern”. https://www.spiedigitallibrary.org/journalArticle/Download?urlId=10.1117%2F1.JEI.28.4.043003
51. Yu Zhou, *et. al.*, “Object-level Proposals”. https://openaccess.thecvf.com/content_ICCV_2017/papers/Ma_Object-Level_Proposals_ICCV_2017_paper.pdf
52. C. Lawrence Zitnick and Piotr Dollár “Edge Boxes: Locating Object Proposals from Edges”. https://link.springer.com/content/pdf/10.1007/978-3-319-10602-1_26.pdf
53. Jana Zujovic, “Structural Texture Similarity Metrics for Image Analysis and Retrieval”. http://users.eecs.northwestern.edu/~pappas/papers/zujovic_tip13.pdf

Chapter 23
The Missing Links Between Computer and Human Languages: Animal Cognition and Robotics

Nathaniel Christen and Amy Neustein

23.1 Introduction

Chapter 20 began with a review of the differences between natural and programming languages, and this chapter, as well the next, will explore such differences (and a few similarities) in greater depth. These chapters will also consider human/robot communication as a case study in forms of language that may be intermediary between human language (in all its complexity and nuance) and the mechanistic, artificial limits of computer code. Once we step beyond communication between *people*, what should we understand by "language" specifically; what significance should be attached to calling a sign-system *language* and not something more general (some less specific form of iconification or symbolic representation)?

To explore such questions, it is interesting to consider other examples of language/communication—apart from both programming and natural/human languages. Hence, robotics, looking ahead to Chap. 24. Our idea in *this* chapter is to approach these topics from the perspective of "animal cognition": that is, to consider whether communication between non-human animals, or (especially) between people and animals, can offer insights into forms of discourse that are simpler than human speech but also, unlike computer code, organically evolving (not artificially manufactured) and interpersonal, emerging in the context of animals' social lives and interactions.

Examining "non-human" languages—from computers to animals—has several different practical and theoretical payoffs. On the one hand, analyzing simpler modes of communication can shed light on human language, which is arguably the most complex discursive framework we know of: its origins, formal mechanisms,

N. Christen · A. Neustein (✉)
Linguistic Technology Systems, Fort Lee, NJ, USA
e-mail: amy.neustein@verizon.net

A. Neustein et al. (eds.), *AI, IoT, Big Data and Cloud Computing for Industry 4.0*, Signals and Communication Technology, https://doi.org/10.1007/978-3-031-29713-7_23

and cognitive foundations. On the other hand, simpler sign systems might serve as models for constructed computer/robot languages more so than natural human speech proper—unless we intend to equip robots with full-fledged NLP engines (which could run up against limits in how thoroughly natural language processing may ever fully comprehend human dialog) [11, 14, 21, 41]. If we envision people "talking to" robots, for instance, accuracy depends both on correct speech recognition (e.g., speech-to-text conversion) and then on correct parsing of generated text. Even theoretically strong results, such as an engine achieving accuracy over 90%, could yield an interface that people still experience as hard to use—it would be frustrating to conduct a conversation with a partner who misunderstands one out of every 10 words.

Moreover, even if engineers overcome or just accept such barriers, it is hard to envision a scenario where an NLP interface should be the *canonical* means of communicating with robots or computers. Presumably natural language inputs would be parsed into intermediate structures reflecting a more computationally tractable language, and for testing or development—plus as an alternative interface that users might find more convenient—presumably robots and software should be equipped to accept inputs in "languages" structured around computers' innate instructions and capabilities.

The prior chapters in this section have focused on virtual machines, in contexts such as cross-component interop protocols and image processing. Virtual machines are intrinsic to the implementation of "computer" languages because they may serve as an execution engine for bytecode derived from parsing programming language expressions into machine-readable intermediate representations. Looking ahead, Chap. 24 will address more practical issues in deploying languages within robotics, for example. This chapter therefore serves as a kind of bridge or interregnum. If we intend to design languages for robots, and other *en situ* installed computing environments—and moreover we feel a need to find communication models other than human language as guides or prototypes—it is worth examining whether language use by non-human animals carries potential lessons for discourse engineered as a computational capability.

Most animal communication research has focused on how animals express or coordinate thoughts with one another (in the same species), and scientists have developed ingenious methods to measure the sophistication and infer the meaning of animals' vocalizations (see, for instance, [44, e.g., pages 24 and 30]). While this is important, equally valid insights can be gleaned from watching animals such as dogs and gorillas learn to speak *our* language, whether through signs/gesture or through programming devices that produce words as if one were "talking." Because they are already pets, studying dog cognition is logistically simpler than with most other animals. This century has seen a noteworthy uptick in studies focusing on dogs' reasoning and neural architecture, such as Gregory Berns's "Dog Project"—obtaining the first Functional MRI imaging of dogs in a fully alert and engaged state, allowing their patterns of brain activation while playing certain games and interacting with their caregivers to be imaged and analyzed [3]. More recently, several dog owners have conducted what are in effect "citizen science" projects where they train their pets to "talk" via Augmentative and Alternative Communication (AAC) buttons. Their published videos suggest that dogs, after adapting to AAC devices, are capable of linguistically interacting with people on a nontrivial level, forming compound sentences and grasping the epistemic and interpersonal dimensions of human speech (e.g., using language to convey new information to someone, or comment on present situations).

Such observations paint a preliminary but compelling picture of the level of linguistic understanding that dogs can attain. We will argue that language at this level offers a plausible model of "mid-level" language systems that are richer than computer code but still more rudimentary and tractable than human language itself (as practiced between fully competent speakers, in the context of collective human activities or situations).

The primary sources we will analyze here are reports published by Christina Hunger, a speech therapist who works professionally with young children, about her experiences when she encouraged her dog Stella to communicate using AAC buttons (allowing Stella to produce word sounds by stepping on the buttons with her paws). Inspired by Stella's progress, "talking dogs" has become something of an Internet meme over the last few years.[1] Following Stella, a Sheepadoodle named Bunny, and her caregiver, Alexis Devine, have also developed a wide social media following. A data set accompanying this chapter (github.com/scignscape/dog-cognition-research) includes links to many videos featuring these two dogs, along with summarial annotations and computer code (such as procedures implemented to locate their video links). There is growing data supporting the notion that dogs have cognitive faculties that with some learning can be manifest in a surprisingly sophisticated level of language-like communication, albeit through the use of mechanical speech aids as intermediaries.

The point is not that dogs conduct conversations on even a child's level, but the level of understanding they do possess is noteworthy—especially when we consider that they are not evolved to vocalize words and have to adjust to mechanical artifacts to communicate with humans on our own terms (adding a layer of complexity to the whole process that would not be evident in, let us say, bird songs or Chimpanzee calls). How dogs nonetheless adapt to human language, we will argue, presents some interesting insights on the nature of human language itself—and, by extension, on how some properties of human languages can be simulated in more primitive contexts, such as robots' communication.

[1] Needless to say, this has nothing to do with cute videos of dogs (especially Huskies) barking or howling in ways that with some imagination we might perceive as trying to produce human words. Analysis of innate canine calls is a legitimate scientific topic, but at best only tangential to the new AAC-inspired research.

23.1.1 Comments on Methodology

Animal cognition, robotics, and even computer science lie at an interesting intermediary position within the spectrum of scientific or academic disciplines. The emergence of computers, "digital humanities," artificial intelligence (and related research areas), computational linguistics, and so forth have all added to the scope of hybrid disciplines that in various ways straddle humanistic and physical sciences.[2] Methodologies here can run the gamut from quantitative models similar to those of chemistry or physics—consider brain scans or artificial neural networks—to interpretive projects descended from philosophical schools such as phenomenology, or the kind of multidisciplinary "humanities theory" associated with (for example) gender studies or critical race theory.

Instead of ranking methodological paradigms as "more" or "less" scientific (or rigorous, unbiased, etc.), hybrid disciplines suggest how different research protocols address similar topics from different directions. For example, macroeconomic analyses (often based on quantitative models) work with empirical data, but they average over large aggregates of individual economic choices. On the other hand, psychological and/or neurological research into economic decision-making (e.g., Gregory Berns's work prior to his branching into dog cognition) centers on each individual's thought processes. The two perspectives are complementary. From a different angle, *socio*economic research can address how economic systems overlap with other social structures. Each approach has its own sense of "empirical" data, which might be economic data sets, brain scans, or sociological case studies. In contrast to natural sciences, where there are certain core paradigms—typically, mathematical models justified by experiments and/or observations—to the extent that research projects are neither purely "scientific" nor "humanistic," it becomes harder to privilege one evidentiary protocol or theory of argumentation over others. Warrants for researchers' claims have to be evaluated on a case-by-case basis.

Linguistics, in particular, illustrates the expanse of science/humanities hybrids in microcosm. Human languages have both fairly rigorous rule structures and nondeterministic interpretive nuance. Analyzing the former structural realm permits quasimathematical formal models, while delving into the latter calls on forms of theory grounded in conceptual precision rather than empirical or computational observations (consider how cognitive linguistics implicitly justifies treatment of sentences' epistemic status or parse structures based on collective/introspective assent—authors and their readers mutually agreeing that certain reconstructions

[2] In "academia"—e.g., university departments, journal foci, book classification—one implicitly recognizes a baseline distinction between humanities and physical sciences. A lot of research, however, does not fit neatly into either category. Traditionally of course, we have a notion of "social" sciences, which tend to overlap with humanities disciplines in terms of subject matter but with "natural" sciences in terms of methodology. This has been, perhaps, a tenuous admixture for generations. This is all the more true the more that humanistic and social sciences become influenced by computers and technology.

of cognitive processing appear to be plausible given our innate sense of how *we* understand language). Meanwhile, although programming languages are not "interpretive"—compiler theory and runtime engineering are governed by the mechanistic limitations of CPU (or other processing) architectures—some facets of software engineering, such as optimal design patterns and human–computer interaction, are more subjective.

Animal cognition and ethology span multiple forms of empirical data, including reasoning experiments (involving manufactured problem-solving scenarios) and fMRI scans, as well as (less carefully staged) observations of animal behavior. With respect to animal *language*, even in the nascent "talking dog" community, there is already a nontrivial amount of empirical data, and projects such as TheyCanTalk at the University of California at San Diego promise much more. These data immediately raise questions of where to fit analytic methodology on the science/humanities continuum, and how "dog language" should be situated in its hybrid intellectual space.

Dog cognition research and interpretations thereof may lead in multiple directions: there is a psychological/experimental stratum of controlled studies exploring canine reasoning; a linguistic interest in dogs' behaviors in contexts related to human language; the ethology of canine adaptation to human environments through domestication; and even an anthropological perspective drawing insights from dogs' "pro-social" comportments to speculate about human history and society (encapsulated by the title and phrase coined by Brian Hare and Vanessa Woods, "Survival of the Friendliest" [18]). Given this thematic range, we can expect that a comprehensive picture of dog cognition—and what it means for humans—can only be pieced together with bricks that reveal a spectrum of topical foci and academic styles. This is not a discipline that should rely on facile argumentational templates as landmarks for methodological gravitas. Or, in other words, dog cognition reveals multiple metatheoretic personalities no less than any other human/social science, from paper to paper (or even paragraph to paragraph). By analogy, queer theorists' subaltern dialectology, phenomenological introspections into cognitive-schematic acceptability, and category-theoretic proofs of the Chomsky–Schützenberger representation theorem are all sibling projects in the linguist's ecosystem.[3]

As a subset of dog cognition investigations, if we consider communication via AAC devices—as opposed to dogs such as Chaser or Rico (Border Collies famous for large vocabularies) *responding to* rather than *generating* speech acts, as per John Pilley's or Juliane Kaminski's (their respective owners) methods [25, 26, 38] (and likewise, influentially, Stanley Coren [9, Chapter 6])—this is a very recent trend, and little academic work addresses this subject directly (we have yet to see the peer-review dialectics and methodological back and forth that might guide research in how best to formulate theories and hypotheses). Nonetheless, Internet discussions reveal a lively exchange of claims and counterclaims about "talking" dogs, such that the parameters of future academic protocols are perhaps already visible.

[3] We will cite [34], repeated from Chap. 21.

Perhaps the most substantial debates have centered on the question of "over-interpretation," an implication that observers "read too much" into dogs' use of talking buttons. This is not, indeed, a subject area that is conducive to neutral observation—ethology (the study of animal behavior) probably tends to attract animal lovers, and documenting animal intelligence and emotions is doubtless an important step toward promoting animal welfare and humane treatment. Such a scenario risks confirmation bias for people who *want* to find evidence of (for example) canine intelligence and the dog/human emotional bond.

The proper response to issues of over-interpretation is to adopt protocols that can mitigate its deleterious effects, without eliding how *some level* of interpretation is patently appropriate. Dogs, obviously, are not "fluent" in human language, and their attempts to communicate with us via our own mechanisms—(arguably) innate for us, not them—are inevitably provisional and trial-and-error-based, comparable to human toddlers (not to imply that dogs have a similar level of linguistic proficiency, but just that their use of language can be more confident in one situation and less in another, similar to that of young children). Also, dogs' communication via AAC boards is restricted to the set of buttons (typically less than one hundred) that boards provide. Sometimes dogs have to guess at the meaning of words—i.e., at what *humans* will think upon hearing them. This is a natural part of *learning* language.

In and of itself, then, *interpreting* dogs' communications is not unwarranted. When Bunny, for example, sounds out "ouch stranger paw" because she has a thorn embedded between her toes, we must indeed "interpret" this phrase, but it is situation-specific and reasonable to the point where we can assert with some confidence that she chose those specific words intentionally (i.e., *linguistically*). Other interpretations might be more tenuous. Bunny appears curious about time: "soon afternoon," "night yesterday," "yesterday went"[4]—and about, say, the concepts of "human" and "animals." Her asking broad (even "philosophical") questions via AAC buttons is an intriguing behavior, but also one we should hesitate to extrapolate equally to more situationally grounded episodes.

We would argue that an optimal methodology would accept interpretations but also introduce a series of metrics framing the degree of guesswork or confidence such interpretations reveal. For example, any given video segment showing dogs' button use might be annotated with parameters such as their "deliberateness"—whether the dogs seem to press buttons in a premeditated fashion, anticipating a specific set of buttons, and reacting only *after* a total phrase is enunciated before seeking a response. We could also measure the degree of confidence we as human observers have in our interpretations—are we guessing at what the dog might be saying—and at the communiqué's situational appropriateness. Also—consider Hunger's distinction between "core" and "fringe" vocabulary programmed

[4] See https://www.youtube.com/watch?v=echpdOTH78w—this video certainly seems to show Alexis Devine "teaching" her time words and concepts, and Bunny "learning" interrelationships between these concepts. Many of her videos show a similar pattern; Bunny appears to be genuinely interested in the proper way to use many of the buttons.

into AAC boards [22, page 97]—seemingly well-motivated phrases involving less-common words would serve as stronger data for something like bonafide linguistic understanding than capably employing baseline words such as "outside" or "play."

Moreover, we can check how the dogs' *other* (non-AAC) behaviors coincide or resonate with interpretations of their "talking." Saying something about going outside and then walking to the door is a strong confirmation that sounding "outside" is proficient and deliberate (the dog is not just guessing about what *that* word means). Likewise, Bunny jumping on a sofa and showing her paw after the "ouch stranger" phrase—or, in another video, her expressing "want cat" and then looking in the direction of her feline housemate; or Stella saying "friend" before kissing her (human) baby brother—are behaviors that would be puzzling if the dogs were *not* using buttons situationally.

Actually watching Stella and Bunny videos, it does appear that many times their choice of buttons is deliberate and premeditated. Sometimes the results are enigmatic, and it is reasonable on such occasions for the scientifically minded to question how well they actually understand words they produce. But even if dogs fail to grok the nuances of humans' construals and semantic encodings of abstract concepts, the fact that they show interest in words for these concepts—and *attempt* to incorporate them—reveals a basic understanding of the underlying linguistic principle that words (sound gestures) designate "objects," sometimes including abstract concepts.

Moreover, often it seems their discourse is haphazard and uncertain—probably dogs do not rise to a level of linguistic competence we associate with young humans, at the time when children begin to internalize principles of grammar and produce complete sentences ([39, page 269], for instance). But so what? These points, even if eventually borne out by empirical data, would only refute claims that dogs progress through language learning as rapidly and extensively as human beings. There is still a large terrain of linguistic understanding and prelinguistic awareness that lies anterior to formal grammar, but building off of generic "gestural" communication toward bonafide language use, via words as convention-driven icons for generalized concepts and via linguistic expressions as media for exchange of ideas.

If we find *cognitive* linguistics, in particular, an especially compelling approach to structure and processing, it is actually very interesting to witness a stratum of cognition that lies within the threshold of language proper but is also organized by intuitive cross-conceptual patterns rather than under the influence of syntactic regularities, gradually internalized. Treating grammar as a formal chisel that molds provisional linguistic awareness into structured human language: to glimpse that awareness in its "raw" or primordial mentality, prior to the dictates of syntactic form, is a valuable opportunity. This is especially true for projects oriented around "languages" outside human contexts proper, such as robots. Equally, we should acknowledge that linguistic understanding *without* syntactic entrenchment still reveals high-level cognitive acuity, perhaps reflecting dogs' highly evolved ability to "read" humans.

Methodologically, even within the recent "dognition" research trend, studies of dogs' reasoning have been oriented to artificially manufactured experiments that are not necessarily proper assessments of canine cognition in their everyday lives. Most of this research apparently follows protocols of experimental psychology typically applied to humans (including children) or to animals in controlled habitats, not household pets. Obviously, this situation reflects logistical problems in obtaining empirical data in non-controlled settings. Against this backdrop, however, dogs' use of AAC buttons is an intriguing alternative. Such buttons may be deployed in their familiar domestic environments and with some training (which many dogs appear to enjoy) can become part of their daily lives. At the same time, what dogs choose to enunciate through AAC boards, and their behaviors while doing so, indirectly suggest properties of their cognitive states.

This may be indirect evidence, requiring inferential reasoning to fill out; but no less so, we would argue, than extrapolating from psychological metrics in artificial setups. The evidentiary frameworks for dog language assessments belong more to cognitive linguistics than cognitive psychology, and therefore bring on board a suite of theories from an alternative and complementary perspective. For these sorts of reasons, extant "talking dog" data should be approached as an important scientific resource. We believe this is how we should read testimonies such as *How Stella Learned to Talk.*

23.2 Animal Cognition and Talking Dogs

Anyone who has cared for a dog understands that they are highly communicative and can understand a lot of human language by hearing, even if they do not natively "talk" themselves. It is not surprising, with this in mind, that dogs can extrapolate from *hearing* human speech, as well as from their own ways of communicating with people (because dogs are certainly capable of letting caregivers know what they want)—learning to use "talking buttons" as an extra medium on top of all their existing communicative toolkit. Reading "talking dog" accounts one might be inclined to expect dogs' language to essentially grow out of their gestural/expressive repertoire in its familiar form. Case studies such as Christina Hunger's dog Stella, however, suggest a more complex language development process. We contend that for this reason these case studies deserve particular attention in the general area of linguistics and philosophy of language.

When assessing dogs' communicative instincts vis-à-vis their human companions, we should also bear in mind how there is a "dialogic" element even to purely nonlinguistic interactions. Consider two people intending to lift a heavy piece of furniture: it is not unusual in these cases for us to exaggerate gestures to some extent, using eye contact and nonverbal cues to coordinate our actions. The goal of course is for both people to lift at the same time, and we instinctively try to ensure such coordination by enacting the relevant physical routine (bending our knees, positioning our arms) conscious of the other person watching and timing their similar actions to match ours. These gestures therefore migrate from being elementary motions to achieve some physical task to being expressive signs at some level, analogous to how a person standing up from a chair often signals their intention to leave the table, or how a host extending a bottle of wine to a guest serves as an invitation for the latter to extend their glass in turn for a reup. Dogs of course signify through gestures in similar ways; standing by the door in a posture that makes it convenient to attach their leash is an unambiguous way for a dog to prompt their caregiver for a walk.

Given that certain physical movements are recurring tropes when carrying out some repeated task—and that people (plus animals too) see one another performing such movements and can often infer their goals—movements in and of themselves are sometimes a way to make another person aware of one's own desire or intention. Acting out gestures typical of *starting* some enactive sequence is, in particular, a prompt to solicit another's attention to complete the task collaboratively; holding out a bottle is the first in a multi-step process that ends (presumably) in a glass being filled. We cannot really say that the holding-out is a physical gesture with no expressive intent, but nor is it a pure sign; in a sense, it hovers between being a nonverbal signifier and an actual attempt to initiate a pragmatic sequence. The very fact that people coexist in co-enactive and co-visible worlds means that almost any enactive gesture, insofar as it is witnessed by others, could potentially become also a dialogic sign; especially if there are ways for the initiator's actions to be continued with physical participation from another person.

It is reasonable to speculate that language itself emerges from these prelinguistic cues. Such gestures as holding out a bottle, or deliberately practicing how one should grasp a piece of furniture, can be performed without a conscious awareness of producing a signifier of one's intentions; but as soon as we realize that another person observing our actions could potentially respond to them interactively, we inevitably will become disposed to construe our gestures as communicative devices as well as pragmatic mini-steps. It is not hard to see how such dialogic awareness can progress to the point where we perform gestures *primarily* for their communicative effects. Consider a person holding an elevator door for someone else a few steps away—this is often done in such a way as to show the fellow rider how the doors are being deliberately obstructed, perhaps inviting them to clarify whether they are in fact trying to catch up to the elevator. People will perform such a gesture even if putting one's arms over the door does not actually work to keep them open (we might use our free hand to press a button instead)—the movements to prevent the doors from closing then serve as a communicative signal more than a pragmatic act, although we may not consciously register the difference.

The point here is that we do not necessarily choose between carrying out gestures for their dialogic impact or their pragmatic effects, or even explicitly deliberate on how much an act may be communicative versus enactive—instead, we instinctively perform gestures that have some mixture of practical and discursive content and allow the admixture between the two to evolve organically. In the "elevator" example, if we realize that we need to press a button and cannot force the doors open just with our arms, we would still maintain the door-blocking gesture to signify that others can board the elevator if desired, so in that scenario the gesture migrates to being mostly dialogic and not pragmatic; but conversely the elevator might have a sensor that reopens the doors on resistance, so the same gesture in that case would be pragmatic as much as significatory. We may not know how the episode will play out *a priori* and, in effect, have learned just by recurring interactions with other people how to act collaboratively and allow for nonverbal gestures to mutate back and forth between communication and enaction, recalibrating our intentions accordingly.

Given recent "talking dog" evidence, it seems that dogs essentially do the same thing and are able to transition from performing gestures instinctively to self-consciously soliciting people to observe what they are doing so as to convey some idea. A dog will walk to the door when they want to go outside, look you in the eye when they want to share your food, or position themselves by a water dish when they are thirsty. In *How Stella Learned to Talk*, Hunger describes how Stella as a puppy, a day or two after being adopted, pawed at her water dish requesting that Christina fill it up (we will use first names when warranted because those names eventually became part of Stella's vocabulary, per Hunger's account). Of course, a dog might use her paws to move their water dish so it is easier to drink, but Stella obviously was aware that having Christina see her doing this might alert the person of her need for water *in* the dish.

In the context of "gestural" communication, moreover, we should also bear in mind how dogs (and, presumably, most or all other animals) have evolved in the context of a natural world in which almost all actions involve physical *inter*actions with other objects. To lift a tree branch that is partly entwined on roots, you need to tug at it until the force you import disentangles the branch from its surroundings; to unearth something underground, you need to dig a hole by displacing soil with your paws. Force and causality in these situations are clearly localized and experienced directly through your physical movements—pulling a branch with your mouth, carving the ground with your paws. This force-dynamic reality is a natural conduit for gestures being eventually metastasized to signs, once we become aware that someone else can *see* us pulling or pawing and maybe infer what we are trying to do.[5] As such, dogs can instinctively translate their enactive cognition to gestural dialog, as with pawing at a water dish or walking to the door to go outside.

At the same time, however, dogs in a (in many ways unnatural) human environment also adapt to a world that can be counterintuitive against force-dynamic instincts. As humans, we may forget how we have become habituated to environments where our enactive patterns have apparent causal anomalies: we hit a switch on a wall and a light comes on across the room. We pull a sink faucet handle to the right and water comes out of a pipe some 10–12 inches away from where our hand gesture took place. For an animal evolved for a world without such engineered artifacts, making sense of when such "leaps" of causation across space happen—and when they do not—understandably takes some time and watchfulness. Dogs have to figure out that (say) people can get water by pressing a button but have to manually open the door of a refrigerator or cupboard to get food.

[5] Indeed, recent dog cognition research strongly suggests that dogs learn by observing other dogs, perhaps even to a degree more often associated with higher primates.

Nevertheless, once dogs start to realize that cause and effect in the human world does not always follow localized force-dynamic patterns, this possibility seems to open up more expansive linguistic potential. Consider Hunger's account of her initial experiments involving Stella's "talk" buttons. As she describes it, she placed an "outside" button near the door, a "water" button near Stella's water dish, and "play" by her toys. If we imagine ourselves in her (Stella's) shoes, it is reasonable to guess that at first (she being just a few-month-old puppy) the buttons were not experienced as *communicative* devices per se. Indeed, evidently it took a little over two weeks for her to pay them much attention. But recall our earlier comments about water coming from a faucet upon turning a handle—a dog would surely realize that humans have, in some sense, rigged up systems where water becomes available somewhat mysteriously, compared to out in nature where one would, let us say, walk over to a stream to drink. So eventually Stella evidently figured out that the button that produced the sound of the word "water" had something to do with her people putting water in her dish (Christina and Jake, her other caregiver, also said the word "water" often so that she would have been familiar with its sound).

Given that Hunger describes Stella's acclimation to her buttons as slow at first, we can certainly hypothesize that Stella only gradually came to understand their purpose as being communicative—after all, if people could press a switch for light, why not press a button for water? True, she also adapted to the "outside" button to signal when she wanted to take a walk, and there is no obvious corollary to the relation of a faucet handle to running water that could serve as an analogy to a water button in the context of going outside. But humans have their own routines when going outside—they grab keys, put on jackets, and so forth—so we can imagine Stella guessing that maybe the button served some functional purpose, analogous to how someone might remotely unlock a car door. If we grab our keys, it signals to a dog we are probably going out; they will then associate the keys with outside, even if they may not understand why the keys have an operational purpose. For all Stella knows (at that early stage), the "outside" button might have been functional in the same way—but, still, on the theory that initiating pragmatic sequences is a way of conveying wishes, functional devices can be pressed into service for dialogic ends.

In effect, we can guess at her reasoning: if humans have this thing where they press a black button to go outside, maybe I can prompt them to go outside by doing the same, hoping this initiates the rest of the process? As Hunger put it, describing one of the first occasions when Stella seemed to notice her buttons,

> There was no way for me to know Stella's intention as she stood in front of the buzzer and looking down at it. She might have wanted to get my attention to go outside, or she might have simply been curious about the object on the floor. Either way, I did not care. Providing positive praise showed Stella that I enjoyed what she was doing, which would hopefully lead her to continue exploring the buttons in the future. (*How Stella Learned to Talk*, page 59).

By now she evidently realized that the button had something to do with going outside, so she may have reasoned that Christina seeing her paying attention to the button would be a way to suggest her interest in doing just that. Or she may have been trying to figure out *why* or *how* the button factored in to the going-outside process. Huger writes about Jake's noticing Stella "barking" at the outside button and wagging her tail when Jake pressed it, behaviors suggesting that Stella at first interpreted the button as operationally part of going outside, not realizing that *she* could press the button too, which we can imagine would take a further paradigm shift—after all, she did not grab the keys, open the door, or any of the other details of going outside that the humans performed. "Stella looked down at the buzzer, then up at me, and back down to her 'outside' button. This time, there was no questioning her eye gaze. She was gesturing for me to push it" [page 61].[6] Hunger reports that over time Stella's awareness of the buttons was initially focused on her watching *people* press them, in the midst of all their other goings-on; she did not at first grasp upon the buttons as something for Stella *herself* to use [page 67].

Retracing Stella's thoughts is purely speculative on our part, of course, but our point is that we can see the origins of linguistic communication in nonlinguistic circumstances. Even if we accept the notion that language is an "innate" human ability, and that some preconfigured intellectual faculty is a precondition for possessing language on a full human level, observing cases such as Stella's acclimation to her "water" and "outside" buttons gives us intuitive feedback that animals *without* innate linguistic faculties (in the sense of actual human language) can nevertheless acquire the foundations of language by branching off of other cognitive modalities that *are* innate (to them).

In short, at some point Stella must have realized that the point of the buttons was to mimic the words that Jake and Christina say to each other (and to Stella herself) rather than some other functional role like the faucet handle vis-à-vis the faucet or the keys to the door. That is, the buttons are not causally efficacious (even in the indirect engineered way that things sometimes happen in the human world, at least the modern human world of electricity and running water); their sole effect is to trigger other people's understanding. Stella learned to say "water" to achieve the effect of Jake or Christina becoming aware that she was thirsty (and maybe needed water in the dish). Obviously, this understanding of communicative effects piggybacks on a dog's acute awareness of their people's thoughts and feelings. Stella presumably sought to express her need for water to Christina because she believed that: (1) Christina did not know she needed water and (2) Christina would make sure to give Stella water if she sees the empty dish.

Even in this still-rudimentary use of "language"—or, at least, single words—we can see some of the foundations of language as a human phenomenon. Not surprisingly, dogs' communication often reflects their wants and desires, and (as we alluded to at the start of this section) we might guess that dogs use "talking buttons" as basically an extension of all their other communicative gestures. If a dog instinctively (and surely self-consciously) acts theatrically cute-sweet and charming when she wants a treat, it is not hard to imagine similar instincts guiding them to solicit our attention by stepping on a button when they want water. But reviewing the evolution of Stella's linguistic abilities suggests that more was involved, as her use

[6] Future page references without further citations will refer to *How Stella Learned to Talk*.

of talk buttons expanded, than the kind of emotional-solicitous gestures that dogs often perform when they want something.

In particular, Stella's use of the buttons was clearly grounded on a seemingly advanced capacity to infer what the people around her knew and intended. Obviously, some level of mind reading is also implicit in, let us say, "solicitous" performances—a dog might realize how acting affectionate to a person will cause the person to mirror that affection toward the dog in turn, hopefully scoring a treat or an ear scratch. That kind of mind reading is emotionally charged and has a certain lap dog tactility. What is interesting about dogs like Stella (and, say, Bunny) is that their ability to coordinate with human's epistemic states is equally evident even when taken outside the tactile contexts that presumably evolved as dogs became domesticated—in effect, the evidence is that dogs read people's minds beyond their presumably instinctive awareness of how affection and tactile interactions can promote dog–human bonding.

By way of example, one interesting anecdote Hunger describes, moving forward a few months in Stella's development, concerned an occasion when Stella and Christina had returned from a walk. Once Stella realized that her talk buttons were not simply gadgets for humans to use—Jake and Christina encouraged *her* to press them—Stella picked up on their communicative purpose quickly. (Hunger suggests that only a day or two after she first used the "outside" button herself to go out in the yard, she also pressed her "play" button—after Christina had done so too when Stella went over to her toy bin—to signal that now she wanted to play inside. In effect, she immediately generalized from the "outside" button to the "play" button.) Hunger gradually expanded the number of buttons with programmed words, and eventually decided to group the buttons together on a board (rather than positioning the buttons thematically: the "water" button by the water dish, and so forth). Having the buttons in one place made it easier for Stella to use two or more buttons to create phrases. Hunger's account of Stella gradually forming increasingly complex and detailed expressions includes numerous examples that are interesting from a linguistic point of view. We will mention in particular several examples that touch on Stella's (so to speak) "epistemic" awareness.

The incident we started to mention last paragraph (the family had moved to San Diego in the interim) happened on a day when Christina and Stella went to a dog park; Jake was not at home when they left, but had come back by the time they returned. Seeing him, Stella used her buttons to say the two-word phrase "park play," and a likely interpretation is that Stella wanted to clarify for his sake where the other two members of the family had gone [page 198] (contrast with [24]). The day before, they had all gone to another favorite spot, by the ocean (programmed as the "beach" on another button)—beforehand Stella was apparently impatient waiting for the humans to finish eating before they could leave; upon returning, Stella "said" (via the buttons) "bye Stella bye good outside": "She was using her words to comment on what she just did, not what was happening now or what she wanted." [page 198] (She had figured out to use the "bye" button as a proxy for "leave"). Similarly, somewhat before then, "Jake woke up earlier than usual [and] took Stella on a much longer walk than they normally took before work" [page 179] and upon their return she said with her buttons "good Jake" (perhaps expressing to him that she liked the walk or to Christina about their morning, or both).

Two-word pairs such as "park play" and "good Jake" were actually simpler than a lot of Stella's word use by this time, but these three occasions are noteworthy, we believe, because they all involve her commenting on activities that were completed, and so they could not be read as her merely translating her instincts for goal-seeking communication to the talking button arrangement. Stella seemed to be pursuing dialog for its own sake, not as a means to some immediate end. Moreover, at least part of her intent appears to be describing things that have recently transpired to the third party—telling Jake that she and Christina had played in the park, or telling Christina that she was happy with Jake for a long walk. Such conversations indirectly reveal Stella's implicit understanding of "belief formation"—an awareness, for example, that Jake returning from work after the rest of the family had left would not know where they went. Stella was (quite possibly) using language to fill in gaps in (what she accurately assessed as) belief states of other people, at her own initiation, a level of linguistic maturity that would be impressive coming from a human toddler, let alone a one-year-old puppy.

Similar "epistemic" cognition is evidenced on other sorts of occasions. Hunger describes, for instance, how Stella once knocked something over unintentionally, then said "no" to explain that it was an accident—obviously she was worried Christina might *think* that her actions were deliberate [page 205]. Once when she wanted to go out early in the morning, her people still a little bleary-eyed, she heard Jake ask Christina to take her, to which Stella explicitly said "Jake"—that is, Stella inferred from their conversation their jointly forming an intention for her to do the honors and not him, inferring humans' planning and intentions [page 182]. A similar occasion involved Stella hearing the two debating whether to go out yet one more time, and—when they seemed to be inclined to retire for the night instead ("I think we can stay in ... She's probably good")—Stella proceeded to "hop off her bed" and say "mad outside come come outside," a reaction probably triggered by hearing their conversation (since she was resting at the time, it seems that she only explicitly requested to go out once she realized that they were inclined not to).

Another time, when their

> neighbors were dog sitting [and] Stella saw them walk past the window with an unfamiliar dog, she barked, said "help no help" and ran to the window to bark again. Stella recognized that our neighbors were bringing a stranger into our complex. She did not bark when they walked with their own dog into their unit. I wondered if Stella was using her words to tell us why she was barking [page 197].

At the very least, she was clearly trying to alert her family about something suspicious. What is noteworthy is how she deliberately incorporated her talk buttons, instead of just barking—she actually ran over to the buttons and used them, before returning to the window. Also, she acted as if aware that Christina, elsewhere in the house, could *not* see the dog. As with Jake coming home while they were at the park, Stella was logically aware of epistemic *absence*, of what someone would *not* know, and how language can fill in such gaps. Against this backdrop, it is worth noting that philosophers (even in recent times) have openly debated the question of how extensively animals possess a "theory of other minds"—an awareness of the structural patterns of belief formation and the perspectival limits of knowledge. Even after it has become generally accepted that an animal's world can be richly social and experiential, philosophers and behavioral scientists have perhaps been conservative in attributing to them sophisticated awareness of what *other* animals believe—perhaps projecting from human development, since it takes some growth for children to show signs of grasping epistemic patterns intuitively.

Reading about multiple occasions where Stella—still a puppy and with only a few months' exposure to "talk buttons"—employs her buttons to communicative effect where understanding "other minds" is likely an explicit factor, Stella's incorporating this awareness into explicit conversations through talk buttons, without human prompting, makes the breadth of her "theory of mind" especially concrete. The buttons add a new layer of specificity to how we might observe animal behavior. In particular, Stella's communications with her two caregivers are often implicitly three-way conversations where Stella distinguishes Jake's beliefs from Christina's. An operational theory of mind at this level, we can surely posit, reflects a qualitatively more sophisticated form of linguistic capability than a dog's use of communicative gestures to achieve specific goals, prompted by somatic (more than "cognitive") issues such as thirst or hunger.

Another dimension to Stella's emerging language use is how she evinces high-level conceptual understanding. In particular, she appears to generalize concepts to different contexts and situations. There are multiple examples in Hunger's account that could be cited in this context—for instance, very early at the stage where Stella had just started to use her buttons (not yet grouped onto a board), she was watching Christina water some plants and used her button to *say* "water." Hunger reports thinking first that Stella needed water in her dish, but then realized the dish was full—instead, Stella was *commenting* on how Christina was spraying water on the plants. It is interesting to begin with that Stella was using her button even outside the context of asking for something—again, this happened while the buttons were still new for her[7]—and also that she unhesitantly perceived the concept "water" as applying to the water Christina gave to the plants, poured from a can, just as it applied to her water dish (moreover, she made this inference visually; she did not drink or touch the water for the plants).

After their move to San Diego, Hunger also describes when Stella's "beach" button broke, and they had to wait for a replacement. Wanting to go to the beach, she decided to use the combination "water outside" as a (de facto) "lexical" substitute for "beach" [page 213]. She actually said "mad" (after the button stopped working) and then "help water outside," having learned to sound "help" when something was

[7] "This was the first time Stella said a word without requesting an object or an action from one of us. She walked out of the sunroom, down the hall, and into the dining room all simply to tell me what she observed in the world around her." [page 84]

wrong with her buttons.[8] Stella obviously understood that the beach was adjacent to the ocean and that the ocean had water. In short, she generalized the concept "water" from what was in her dish to what was watering the plants and to what was in the ocean.

Hunger documents similar generalizations with other words. For Stella, "outside" could function as a signifier for the "immediate" outside (e.g., their yard) or something further (e.g., the beach); she said "help" when she needed the people to fix or resolve something (like a toy inaccessible under the sofa) but also to add oomph to her requests. After a pair of rainy days, for instance, Hunger reports how Stella said "help beach love you" [page 168], evidently thinking she needed more entreating to get her humans' attention ("help" implying a need for attention beyond the typical level of interest she might have when just saying "beach" by itself).[9]

Stella also showed an appreciation for conceptual nuances that were not limited to physical things ("nouns," lexically speaking). She used "bye" like a verb, saying "Christina bye" one morning after she had left for work [page 180], or "bye" when they had dropped her off for a stay at a dog sitter [page 191] (a few days later, when they got back to San Diego, Stella saw the sitter packing up her belongings and realized they would be coming soon, saying "Jake Christina" with the buttons). Likewise, a video shows Stella frustrated because she wants to go out, saying "want want love you bye bye Stella" then walking to the door.[10] The pattern underlying *bye* in these cases is going from inside to outside, but the situational details are different—sometimes Stella talks about *her* being inside and then going out (or wanting to), other times about her *staying* inside while someone else goes out. Also, Stella presses "bye" in contexts where she's thinking specifically about going *from* inside *to* outside, as opposed to saying "outside" or (e.g.) "walk" when she is focused instead on what she will do *after* leaving.

Also noteworthy was that Stella seems comfortable with "no," which does not have a concrete correlation to objects or actions but rather plays a more indirect role of altering the rest of the conversation. Hunger describes trying to give her a homemade dinner one night, which evidently she found unappetizing, using the buttons to say "eat no"—next day Hunger restocked on Stella's usual food, and after her meal Stella made a point of saying "happy eat" [page 214]. She also said "eat no" when her sitter did not feed her on time; and also, when Christina did not feed her when expected because of the Daylight Saving time adjustment, said "help eat" and then—Christina was still trying to get her to adjust to the new clock—"love you no" [page 150]. Once "when Jake went out of town for a few days, Stella said, 'Jake no

[8] Hunger describes how—after having moved the buttons to a board, so that Stella needed to learn the placement of different words—she said "help" repeatedly [page 161].

[9] In another example of the ocean-water generalization, she said "love you water"—probably pleading to go to the beach, with a solicitous "love you" for emphasis.

[10] Apparently, Stella also started to use "bye" to indirectly designate peanut butter because her people left her peanut butter as a treat when they went out: "bye eat" became her homespun phrase signifying the treats they left her before going "bye."

bed', before she hopped up on his side of the bed and slept there for the night" [page 196]—which shows her generalizing the concept of "bed," but also commenting on Jake's absence via "no"; if he is not here, I might as well sleep on his spot. She used "no" to model both logical negation (what has not happened, or—recall her conveying that she had knocked something over accidentally—what was not intentional) and emotional states, alongside words/phrases such as "happy," "mad," and "love you." In short, her conceptual repertoire, as manifest in her expressions, covered the gamut from nouns to verbs to more abstract ideas and was sensitive to concepts' adaptability across distinct situations.

When seen through the lens of our scientific interest in the extent of animals' linguistic capabilities, Stella's communication is noteworthy less for the size of her vocabulary (in contrast to Chaser, say) then for how realistically she fits words into situational contexts. The actual words are clearly just one thread in an acute awareness Stella has of the routines and thoughts of the people around her, so that for example after eating Christina (on a video) can say "all done eat, what now?" and Stella confidently respond "play outside," an exchange that would not be out of place if it occurred between a parent and a (human) child, at least if we morphosyntactically decorate "eat" to "eating." Stella did not just use words statically, like the name of a toy; she packaged words in different combinations given the situation at hand.

Hunger gives many examples that reveal Stella's highly situational use of language. Once she found a "ratty scrap of some other dog's toy" and brought it home [page 212]; Christina was reluctant to actually bring the toy in the house. Stella was not happy. She stood by the door and then said "toy inside" (probably a novel combination of words, because all of her toys usually *are* inside).

One evening when friends were visiting, Stella got impatient, while they were all talking by the door preparing to leave, finally hit the "bye" button before looking up at the guests [page 109]. Another visitor anecdote has her playing with guests when Christina came over to pet her, and Stella reacted by saying "Christina later"—again, a combination that would not fit except in specific kinds of circumstances (here she wanted to play with her new friends without being disturbed, presumably). Stella got into the habit of saying "all done bed"[11] in the morning to announce when she was awake (and ready for food and walk) or "help bed" at night when she wanted to go to sleep and wanted someone to open the bedroom door [page 196]. She said "love you" when Jake and Christina were hugging each other, squeezing between them [page 137]. She said "all done" when she wanted Christina to stop vacuuming [page 202] and "all done walk ... want" and then "Stella bye love you," standing by the door, when they were setting up furniture after a move, indicating that her annoyance with all the commotion and that she wanted to go for a walk instead [page 210]. She said "love you" to guests that she especially liked [page 178], and likewise when Christina was sick in bed [page 154], both times getting up from the bed or sofa on her own volition to do so then returning to her prior spot. She used the

[11] Video, https://www.youtube.com/watch?v=g5xtDyOju8E.

buttons for "park" and "beach" to indicate on a given day which way she wanted to walk, after the family moved to an apartment a few minutes from both her favorite spots.

All of these situational, conceptual, and interpersonal details suggest that Stella is bringing to language intellectual dispositions that actually mirror those of human speech. The breadth of Stella's understanding, in short, is evidenced most strongly in this cognitive infrastructure, more so than in lexical or syntactic scope. The semantics of Stella's "language" is relatively informal and ad hoc—she developed her own routines for using words in specific ways, not necessarily those Hunger anticipated when recording the words, or their normal English meanings. Thus, "bye" for "leave"; "no" for "not"; "help" or "love you" for "please" [page 168]. Stella's "grammar" could also be haphazard. She would use word sequences to suggest patterns in time—"bed eat outside," say, to communicate how, in the morning (after she is out of bed) she wants to eat and then go for a walk [page 197]—but she probably was not in these contexts following what she perceived as linguistic rules, but rather pawing-out words in combinations that seemed to make sense in context. The point is not that Stella's use of language lacked syntax or semantics, but that she does not seem to believe that rigid semantic and syntactic rules are prerequisite for successful communication. Instead, she seems prepared to stretch words' meanings and combinatorial possibilities informally, almost experimentally, trying to modulate the language available to her in response to the specifics of each situation in itself.

Stella also clearly has self-awareness. She used the button recorded with her own name in ways that almost prove her being aware that "Stella" named herself in the same way that Jake and Christina named her people. She stood looking at herself in the mirror when she had carried a toy or a tree branch [photo after page 150]. Stella clearly understands herself to be a participant in situations alongside other people, each with their own (perhaps slightly different) beliefs and perspectives, and realizes that she can use language to alter how situations play out. In short, Stella's communication shows that she understands the world in situational terms and sees situations through an agent/participant lens, cognitively structuring states of affairs in terms of the beliefs and intentions of cognitive agents (including herself, and the people whose lives she shares).

23.2.1 Lessons for Natural Language

To what degree can we generalize from Stella's story to broader insights about animal cognition, and even human language itself? At one level, Stella's speech acts are primitive, when compared to human language but perhaps even when compared to animal sign systems that have strict patterns and, through those patterns, permit complex variations. Animal communication research has shown how many primates and cetaceans—or even apparently more primitive mammals, such as prairie dogs (who are not canines, but rather related to groundhogs and squirrels)—can modulate their vocalizations to express several different levels of detail about, for example, lurking predators. These sorts of communications (likewise for many species' birdsongs as well) reveal a kind of modularity, where parts of coded messages may take on various forms, with the messages in their entirely therefore able to span a variety of uses and circumstances. This would seem to be analogous to phenomena in human language such as verb tenses, noun declension, and other lexeme-level markings that adjust sentences' sense of time and place.

Because AAC buttons do not allow that kind of modular fine-tuning, we cannot directly compare dog's use of speech-assisted human words to other species' native vocalizations, which could have evolved for subtle modulations alongside the evolution of the vocal apparatus itself. Still, it is entirely possible that some species' communications reveal a syntactic and/or morphological complexity that outstrips Stella and Bunny on AAC boards. And yet, it certainly feels as if "talking dogs" are *communicating* with *us* on a level that would be hard to conceive with almost all other animals.[12] We would suggest that this reinforces one's impressions wherein lexical and syntactic norms are not the predominant criteria for (human) language's sophistication.[13]

From the perspective of cognitive linguistics, the structural rigor of human language derives less from semantic and syntactic formality and more from structural correlations between conversants' situational understanding. There is a prelinguistic co-awareness that lies behind our intersubjective interactions, and we are disposed to treat units of language—both words and syntactic paradigms—as tools that can and should be adapted for different scenarios. It is easy to imagine or recall contexts where people stretch word meanings in novel ways; sometimes clever turns of phrase become popular and end up being just another word meaning (consider "primaried" as a verb or "Watergate" as an adjective: Donald Trump's "Watergate moment").

In essence, we cross the threshold of language not by acquiring sufficient competence in static lexical-grammatic systems but rather in engaging with other people with sufficient prelinguistic understanding that we are able to extend our interactions via linguistic constructions, with some degree of legibility. On this theory, the building blocks of language are those which undergird collaborative situational awareness, more so than those that are structural units of morphosyntax, dictionaries, or syntactic templates. Animal cognition observations such as Stella's speech acts can help us understand the cognitive and experiential foundations of *these* prelinguistic faculties, which appear to bring us to the core of language itself.

[12] Perhaps except higher primates, as evidenced by Koko and other gorillas and bonobos who learned (something like) sign language.

[13] Indeed, spoken language tends to be more ad hoc than the norms of written dialects dictate—during conversation and in the midst of concrete activity, people regularly speak in sentence fragments, overlapping conversational turns, fast-paced enunciation that elides morphological markings, and in general produce speech acts that can trip up computational systems engineered to recognize relatively formulaic and well-structured discourse.

As implied above, there are three dimensions to Stella's discourse that—relative to what we might expect from non-human animals—seem surprisingly advanced. Specifically, she has a highly developed situational awareness, comfortable in the midst of people and participating in human activities; she has a detailed and logically coherent "theory of other minds"; and she understands concepts in flexible ways, extending and reusing concepts (like "water") in variegated forms across situations. A linguist or philosopher grounded in cognitive-grammatic paradigms might be inclined to focus on these specific strata—conceptual plasticity, situational awareness, and "theory of mind"—as language's inner nexus anyhow, but it is intriguing to witness this paradigm be embodied in a one-year-old puppy.

Experimental dog cognition research certainly confirms (or at least strongly intimates) that dogs do have an acute understanding of "human" thoughts and situations. Talk buttons can add a further dimension to this line of investigation, not only because there is a new kind of empirical resource (videos or accounts of dogs' speech acts via the buttons) but also because they add an extra cognitive layer that we need to understand in the context of their thought processes. It is one thing to innately or implicitly appraise a situation, in the sense of overlaying structures of categorization and action planning over raw experience (which we could assume is a natural instinct: the world can be classified into things that move, things that provide sustenance, and so forth), with their own action possibilities (things that move may be chased, food and water can be consumed). It is another to deliberately destructure this awareness and intentionally present it to someone else in the form of (button-assisted) spoken words.

Incorporating speech acts into her pragmatic repertoire requires, on Stella's part, something akin to (what is sometimes called) meta-cognition. A lot of our situational awareness is subconscious—we plan and react instinctively, playing out innate (or entrenched) enactive scripts, calibrating our actions to our (not entirely predictable) surroundings but without sustained deliberation. Presumably, it takes a level of cognitive distance from such fully "immersive" functioning to communicate *about* our surroundings, especially when our communicative mechanism is a device that takes time and planning to use.

Situations, conceptualizations, and construals about other people's "epistemic attitudes" are constituent strata of prelinguistic understanding. How do these layers coalesce into language proper? Certainly, the strata overlap, and they are threaded together: concepts are adaptable, for example, because one concept has multiple variations that fit divergent situations (water in a dish and the ocean, say). We form cognitive schemata for situations by appraising how relevant concepts are narrowed and modulated via their manifestation in *this particular* state of affairs. Other people likewise, we realize, perceive the same situations from their own perspectives and are engaged in analogous conceptualization; our awareness of that process—as something generic to context actors, not just in our own mind—and its concordances and deviations from our own conceptualizations gestate our "theory of other minds." So concepts build to situation schemata, and multi-person situations situate us in a shared epistemic space.

At the same time, intersubjective awareness supplies a structuring principle regulating situations. Stella is, of course, quite self-aware; again, she uses her name to refer to herself when "talking" and studies herself in the mirror.[14] Self-awareness then implicitly extrapolates to situational schema; if she wants to go outside, it is *her* location that orients the inside/outside contract. *She* is inside now and will (or would) be outside afterward. We conceptualize inside/outside *qua* schema by projecting ourselves as a "trajector" sited first here then there. Of course, schemas exist where someone or something *else* participates in the inside/outside duality—consider "help outside" on Stella's seeing a dog stranger—but this very conceptualization depends on consciousness of "what it is like" to be one place or the other. The designation "outside" is meaningful because of prior (retained) and, at some level, anticipated (protained) consciousness of phenomena there.[15]

It is easy to see how drive-based self-awareness propels schemas like inside/outside to become phenomenologically entrenched. A dog wants to go outside to pee, play, and so forth; there is an explicitness to such desires and a vivid contrast between inside and outside experience such that the particular inside/outside schema, we envision, could become cognitively registered merely via subjective intensity.

As a structuring principle, however, schemata take full form as we project outward from *our* intentionality to other people's, drawing in "other minds." We can take schemata nurtured through experiences where *we* entertain actions or observations and restage their structures with others in our place: another person going outside, "having" left, "eating," and so on (here citing schemas reflected in various Stella speech occasions). Stella for example said "bye" when a house guest had departed after a several-days stay—Stella was resting on the spot where the guest had slept (Stella had also seen her leave at the airport); she generalized "bye" from a temporally localized reference to the *act* of leaving to the more generic state of *having* left [page 139]. Stella might say "eat" when seeing Jake and Christina doing so, without intending to ask for food [page 134]. Cueing the others to look outside ("help outside") implicitly translates the inside/outside trajector to something more abstract: instead of *going* outside, we just *look* outside.[16] Apparently, she generalizes "outside" from the physical dynamics of going/having-gone to the more epistemic concept of "outside" as a surrounding place to be observed/monitored.

---

[14] In particular, as we cited earlier, Hunger reports that she goes over to the mirror *when carrying* a toy or tree branch, an occasion-specific and deliberate act that strongly suggests that she identifies herself, rather than merely happening upon the appearance of a dog that she might ponder without self-recognition.

[15] Phenomenologically, we cannot think of an "inside" without conceiving that there is around there some "outside."

[16] That kind of generalization applies to the unfamiliar dog outside, but also one case Hunger mentions, early on, where Stella typically used her "outside" button to prompt going out to their back yard—except on one occasion, hearing Jake using a lawn mower at the *front*, asked for "outside" and ran that direction instead, apparently both seeking to investigate the unfamiliar noise and alerting her people about the (for her) confounding sound and appearance of a lawn mower [page 83].

Ideas of look/gaze, of who-sees-what, invite other-mind issues because belief formation is driven primarily by seeing things. As such, intrinsic to many cognitive schemata are notions of how we visually navigate through surroundings; in practice this includes tracking other peoples' gaze, as well as cooperative phenomena such as "joint attention" that Hunger discusses in the context of her preliminary realization about how Stella's actions as a young puppy resembled babies during the early stages of language development [page 41]. Recognizing where other people are looking and what they are seeing is presumably an important component of our emergent understanding of other people's minds, which in turn generalizes to schemata having one structural scaffolding comprising of people's physical orientation within situations and how this permits certain lines-of-sight (or blocks others), part of a more general sense of belief formation.

Here, then, we can see a kind of schematic expansion outside of somatic/subjective schemas driven by our own experience to multi-person collaboration, which involves both generalizing from *our* positioning vis-à-vis situations (going outside in the sense of *our* movement from here to there) to the schematic form with our own presence elided (something *else* going outside, or, something outside when we see it from inside, or, via counterfactuality, something outside which *could be* inside, like Stella's toy find)—and then generalizing to schematic configurations centered on someone else; some *other* person going/looking/being outside, say. Schemata may be tuned along these axes, like knobs being dialed, with configurations being more or less grounded on specific people at spatiotemporal schematic positions, and with such agent positions being understood more subjectively (one's own experience) or more via extrapolation (either imagining other minds immersed in situations or inferring the epistemic invariants investing situations with coherent belief-forming potentialities that can be abstracted from any one subject's perceptions).

Awareness of other minds, on this account, is a central element in how we generalize subject-relative schemata to an understanding of more neutral/impersonal belief formation; this is what a *typical* epistemic agent would believe, or what we should *expect* someone to believe, experiencing such-and-such a situation from their specific perspective. Such argumentation alludes to a phenomenological tradition where metaphysical notions of the world's intrinsic facticity are grounded in collective action and intersubjective consciousness. Traditionally, of course, phenomenologists have understood the truth-disclosing quintessence of the world's givenness, its ability to be known and availability for knowledge, as a cooperative resource of the collective *human* community; we, speaking globally, through communication and collaboration, invest reality with epistemic potential. In light of animal cognition research, perhaps we now need to expand this intersubjective vision to encompass animals as well.

Understanding "other minds" has two overall dimensions—we have awareness of *particular* individuals' beliefs and also the *general* beliefs that would be warranted circumstantially. In the context of experiments where, for example, animals or children watch scenes playing out some form of hidden-in-box trick, the experiment's subjects often take the place of a "neutral observer" who knows a treat has been, let us say, surreptitiously moved from one box to another, whereas someone who wants to find the treat is misled. At a sufficiently advanced stage, children will anticipate *false* beliefs and expect someone to look in the box where they *think* the treat is to be found, whereas the experiment subjects know the treat is elsewhere. As such, they distinguish between beliefs that are *warranted* by all the available facts, from beliefs that are coherent for a party with limited facts, or an occluded perspective, but are actually erroneous. To conceive of a "neutral" perspective, one with maximal access to relevant information, however, is essentially to imagine an epistemic collective, the sum total of everyone's perspectives to the degree that we can aggregate all facts available to all people (at least all people witnessing a given scene). The fact that the treat is in its actual box, for instance, is witnessed by how someone watching the scene neutrally can *see* the treat being placed there.

In the context of Stella's communication, it seems clear that she is implicitly able to reason about other minds both from general and from particular perspectives. Her communiqués sometimes reflect assessments of one specific person's thoughts (like telling Jake that she and Christina had been to the park), and sometimes comment on, or perhaps seek confirmation about, general facts (like her saying that their weekend guest had gone "bye").

Considering these details about Stella's "epistemic" awareness, alongside her situational understanding and agility in extending and modulating concepts, there should be little doubt that Stella's communication reveals much of the cognitive architecture underlying human language. Her level of "prelinguistic" competence, in short, is more apparent if we look at the circumstances surrounding particular speech acts than if we consider what she says in isolation, just as strings of words. Insofar as Stella observations are so far mostly anecdotal, drawing comparisons between this research and prior analyses conducted in a more rigorous, laboratory setting cannot be more than provisional; but, with that said, we do think there are subtle clues that Stella's thought processes deviate from prior models of canine intelligence, even those carried out with a strong belief in dogs' innate intelligence.

23.3 Joint Attention and the Foundations of Language

Videos of Stella and Bunny show their dogs in familiar, day-to-day environments. In that milieu, dogs appear to have a cognitive repertoire more expansive than previously documented, even by researchers largely sympathetic to claims of canine intelligence. One theory is that within more conventional scientific experiments there is a higher threshold for empirical rigor, so Bunny and Stella observations remain tentative at best. Alternatively, perhaps cognitive studies are *more* accurate when sited in normal surroundings, rather than experimental mockups.

For example, backed by neurophysiological research—including fMRI brain scans while dogs played simple games with their owners—Gregory Berns and his colleagues have suggested that dogs' reasoning is oriented (at least, more than people) toward holistic activities rather than abstract/generic concepts. He suspects, say, that his dog Callie perceives the word "hedgehog" as a phrase—essentially meaning *get hedgehog*—rather than an isolated noun: her concepts are episodically pragmatic, more than invested in specific objects or people [4, page 168]. In terms of language, limited grasp of words as individual signifying units (such as verbs or nouns) implies limited ability to join words into variegated phrases.

This model would also have to question dogs' capacity for conceptual generalization. Logically, the more that concepts are grounded in *objects*, the more that we can perceive a conceptual continuity when we experience that object in different contexts. Arguing for an activity-centered conceptual predilection, conversely, implies that dogs do not (at least as readily) adapt concepts to multiple different activities.

The same would go for types and kinds, such as "water": this concept re-appears in settings that we, from an activity-based standpoint, will experience in quite different ways—drinking water; watering a plant; playing in the ocean. Berns's models imply that dogs experience concepts with less acuity and granularity in their cross-contextual guise, being more attuned to specific, familiar situations, experienced holistically, with less attention to their conceptual building blocks.

By extension, Berns questions to what degree dogs recognize their name as a signifier for themselves, in the sense that (while certainly knowing to pay attention when they hear their name called) dogs may equally well interpret that sound as meaning something like "come here" or "look at me." If dogs experience the world activity-wise, their conceptual focus on themselves as a participant in activities would be less intense than thoughts about the activity as a whole, and so there would be less conceptual specificity required for attaching their name to themselves in particular, as opposed to a more diffuse acknowledgement that the sound of their name tends to presage activities of which they are a part.

Here we are emphasizing only one facet of Berns's analyses, but these are the issues that seem most relevant to assessments of dog's "language." If dogs understand linguistic expressions bound up in specific activities, they are less aware of individual concepts in a more abstract vein—perhaps Callie fails to distinguish "hedgehog" (referring to a toy) qua concept from a command or suggestion ("get/bring [me] hedgehog"). Again, similarly, he expresses skepticism about dogs' grasping their names as "direct" symbols for themselves, rather than solicitations for attention (a caregiver's request for the dog to come to them or look at them).

The implication of these positions is that the "semantics" of dogs' innate language would mostly quantify over holistic activities, like *fetch a toy* and *come to me*, rather than singling out objects in the guise of discrete concepts. Analogously, he questions whether dogs cognitively distinguish nouns from verbs, claiming that only Chaser (the Border Collie who knew over a thousand toy names) has been documented as clearly making such a distinction, and voices doubts that dogs (including Chaser) could understand significatory aggregates of more than one or two words.

It is these sorts of analyses that seem counter-indicated by Stella and Bunny observations. There is little reason to question that they understand *objects* as distinct *concepts* which are both individuated and perdure across situations; or that they refer to themselves by name, and because this is self-referential cannot be understood as responses to their own names mostly in reaction to people asking to do something, like "come here." Instead, they use their names to describe feelings to other people, via phrases such as "Stella mad" and "bye bye Stella." Stella also uses her name to describe what has already happened, as in "Stella Christina park." Hunger's accounts also strongly suggest that Stella understands concepts in generic/counterfactual contexts, not just bound up operationally à la "fetch hedgehog." Her entreaties like "toy inside" are proposing a state that the toy should be in, not describing an activity such as "play with the toy."

And, of course, Stella and Bunny routinely create multi-word expressions, perhaps five or six words long (apart from repetitions for emphasis), like "love you come outside bye Stella."[17] Looking only at the surface form of their language —the actual words they create via buttons—misses the larger complexity of the ideas they clearly have in mind. This is because their communications are highly context-sensitive. When Stella wants to "eat" and gets impatient because of the clock change, her subsequent "love you no" is a two-part phrase in isolation, but it actually continues the prior exchange and refers to the current situation, where she is exasperated that her people seemingly have forgotten to give her supper. Moreover, she presumably believes that saying "love you no" will *cause* Jake and Christina to realize that she is unhappy (she did not just keep repeating "eat" the way a dog might scratch at a closed door, hoping someone would eventually realize they want to come inside). In other words, she was not acting out of instinct but commenting on the situation, for the benefit of people around her. The larger meaning of her statement is that she is upset *because* Jake and Christina have not yet fed her. Similarly, "eat no" after returning from the park when she did not eat earlier is just a two-word phrase, but its overall content is a broader idea, along the lines of "I haven't eaten yet (this evening)."

Once we factor in context, we realize that Stella's language can encompass many distinct components in a single phrase, referring backward and forward in time, spanning multiple people and perspectives, multiple facets to one activity (park play outside), and so on; we cannot measure the complexity of her thoughts just by counting actual words. Simple phrases are no less vehicles of language than elaborate ones. Or, to put it differently, ideas signified linguistically may span multiple time frames and situational perspectives, not all of which are marked by their own linguistic units; a simpler surface expression has to refer to such complexities obliquely. But there is still sophisticated *communicative* intentionality involved because the enunciator needs to anticipate *how* addresses will absorb such expressions (even simple ones), filling in the gaps.

Given how her talking relates to situations, we can infer the ideas Stella intends to communicate—and, indeed, perceive that she anticipates how her people will

[17] Compared to actual human speech, of course, Stella's discourse is fragmentary, but that must be at least in part a product of the relatively primitive communication tools at her disposal (human's own conversations on, e.g., social media, where are more logistically cumbersome than talking in person, are similarly typified by incomplete sentences and abbreviations, or emojis).

understand her communication on the basis of her words *in context*. She probably would not say "love you no" unless she guessed that Jake and Christina would recognize this as describing her impatience and, in the context of that moment, understand *why* she was angry.

It is true that Stella's and Bunny's language does not have particularly complex syntactic structure, and one might reasonably doubt whether dogs could learn a sort of language that rivals humans in the sense of using grammatic rules and morphological markings to form complex linguistic constructions. Another way to say this is that dogs might not be able to *encode* complex ideas into equally complex language. However, this limitation is a different issue than dogs' possessing complex thoughts to begin with. A lot of cognitive detail may get filtered out by the primitive nature of AAC buttons; so "I have not eaten (tonight)" becomes just "eat no." But that does not really affect the important cognitive point, which is the complexity of thoughts held in the mind. Even Stella's simple language expresses her thoughts, in context: the main point of language is communication. She uses simple phrases to convey complex ideas, and we can assume much of that dialog is deliberate and self-conscious given her assessment of how people hearing what she "says" with the buttons will interpret that—again, *in context*.

Moreover, almost surely Stella's self-awareness is more individuated than Berns's model implies. Consider how she coherently and pointedly uses her name when talking, often for emphasis, or to solicit her people's attention. That is, she might say "come outside" when she has a rather neutral opinion about going out [page 143, 147, 174, 181, 197], but "bye bye Stella" when she was especially frustrated (video cited earlier). She obviously understands that the people around her use this name to designate her specifically and would pay attention to her through the name the way they attend to other objects referenced by name, and to one another. Her self-awareness, in short, is individuated and singular in the sense that she knows how people have a concept for her in particular, a sense probably reinforced by mirror recognition. She is Stella in the same way that Jake is Jake and Christina is Christina. By extension, Stella uses her people's names to refer to them as she would use her own name to refer to herself. Such self- and other awareness is surely a factor in how Stella employs simple expressions to communicate complex thoughts. Stella has the patience to produce compound speech acts because she believes they cause her feelings and opinions to be known to Jake and Christina (and sometimes others).

To clarify, Berns's neuroscientific research in general yielded strong evidence of dogs' intelligence, and he personally has forcefully advocated that dogs get proper respect and ethical credit, proposing, for example, "personhood" legislation that would prohibit unpleasant experimentation, dog racing, and other potentially inhumane treatment. We raise no larger-scale dispute with any of this work, which would be out of context here anyhow; we merely want to consider specifically the ideas Berns puts forth about dogs' intelligence in the context of cognitive schemata. In that vein, Stella and Bunny appear to refute almost every claim Berns makes about dogs' cognitive limitations. More subtly, perhaps, they appear to counter-indicate generalizations made by other prominent dog cognition researchers, such as Brian Hare.

In one revealing passage, Hare (a few years before Stella was born) was discussing Alexandre Rossi, a Brazilian researcher who programmed speech buttons somewhat similar to Hunger's; Hare then wondered whether Rossi's dogs could form simple phrases like two-word combinations [20, page 138], apparently not even imagining the loquaciousness of later "talking" dogs. Like Berns, Hare also voices skepticism about the extent of dogs' self-awareness and theory of (other) minds.

In this context, it is interesting to note that—although outside an "academic" or scientific context proper—Hare was interviewed about Stella on slate.com and raised questions that deserve response from anyone taking "talking dog" accounts mostly at face value, as we admit to doing here.[18] The tenor of Hare's comments implies that he accepts the premise wherein dogs may indeed learn to associate individual words with certain actions or outcomes, and perhaps short phrases, but he apparently believes more work is needed to demonstrate that dogs can formulate compound phrases characteristic of language proper. Granted, in isolation (or even, say, paired with one other word), a single word is not really a linguistic artifact, but rather a sound correlated with some idea at the same level as other gestures might be, such as touching a paw to a bowl to connote "water"; spoken words without linguistic structuration are in effect just vocal gestures. Hare also appears to base his assessments on published videos of Stella and does not refer to Hunger's book-length treatment.[19]

However, multiple published videos reveal Stella communicating via multi-part phrases, such as "want, love you, Stella bye bye" (mentioned earlier) or "help Christina inside blanket" when Stella was looking for her blanket. These videos are consistent with Hunger's descriptions of similar episodes, so it is reasonable to consider the existing videos to be relatively strong indicators that Hunger's textual descriptions are accurate enough to serve as (at least preliminary) data points. Hare also implies that Hunger is projecting interpretations *onto* Stella's words. A cautious reading of Stella data might suggest that her communication has some legitimate features of language—e.g., she specifically uses words to convey ideas or desires, with some understanding of what the words mean (or at least a correspondence between the word sound and the correlated concept), and that the people she is addressing understand her communicative intent. Nonetheless—or this is an explanation we can consider—her word combinations might be rather haphazard: she could guess that since "beach" has something to do with water that is outside, sounding out these two words will get people's attention focused in this direction; her strategy of stringing words together may not have sufficient structure in her own mind to qualify as properly linguistic phrase formation. Language proper has an intentionality that cannot be directly equated with experimentally producing

[18] https://slate.com/technology/2019/11/stella-christina-hunger-instagram-dog-speech.html.

[19] It is true that the book's descriptions of Stella's communication sometimes cite incidents that have not been recorded, or at least the book does not cross-reference text to video. But even if we take these accounts as preliminary work that would need to be generalized to reach optimal scientific status, perhaps statistical comparing videos from multiple dogs, Hunger's and Devine's pioneering work with Stella and Bunny are a valuable starting point.

words for several related concepts in the hopes that, on aggregate, the totality of the concepts expressed will communicate some underlying idea.

While it is true that "experimental" attempts to approximate ideas by sounding out multiple related concepts should be distinguished from using words in linguistic ways, it is not unreasonable to speculate that such experimentation is a *stage* on the way to acquiring language proper. In that case, we can consider whether Stella has *only* reached that stage or whether she has gone on to mix words with *deliberate* and *premeditated* intent and structure. In the latter case, we should argue that Stella has progressed to cognizing and using words in a properly linguistic vein. Even when dogs' language-use modulates or cycles between properly linguistic episodes and something more hesitant, like exploring buttons and word combinations to assess humans' responses, would not this actually suggest that dogs are *learning* some language?

In the interview mentioned above, Hare cites Occam's razor to look for the "simplest" explanation of Stella's behavior—maybe she associates the buttons with specific concepts and specific positive outcomes, and so she uses them *intelligently* but not necessarily *linguistically*? Given the totality of evidence, however (combining video and descriptive observations), we think the simplest explanation is actually that Stella (and also Bunny, often) forms multi-word phrases deliberately and with attention to aggregate conceptual structures. On video, for example, their sequence of button presses appears to be premeditated and to construe a group of produced words as a unit: notice how Stella looks down at the button set, while she is walking through a sequence, and only after the last word looks up to see Christina's response. This implies that Stella knows in advance what multi-step word group she wanted to produce and was not only adding on new words by "trial and error." The fact that Stella often "talked" of her own initiative, including when people around her had their attention directed elsewhere, implies that Stella was not just getting cues from humans on which button to press (contra suggestions that her apparent language understanding was mostly mimetic, analogous to Clever Hans's arithmetic), again pointing toward the *simpler* theory being how Stella *does* use words linguistically (albeit at a relatively primitive, child-like level) rather than that she *does not*.

Likewise, implications that Hunger misreads or even deliberately misrepresents Stella's actions are plausible but *more* far-fetched than reading the contents of *How Stella Learned to Talk* as largely reliable. Finally, Stella's use of words in context-specific ways—commenting on people arriving or departing, sounding out "friend" to express affection for her little brother (that is, her caregivers' human child), and so forth—invite us to consider Stella's grasp of conceptual aggregates not only in terms of actual word combinations (which might be subject to interpretation vis-à-vis how deliberately they are intended), because even a one- or two-word phrase reveals multi-faceted underlying conceptualizations when it only makes sense in a detailed and situation-specific context.[20]

[20] Stella's relationship with the Hungers' baby boy (born after *How Stella Learned to Talk* was written) is not profiled in published writing, but she discusses Stella's affection for the child on social media.

If we also consider Stella's apparent conceptual extrapolations, with "water" for instance recurring in many contexts, we should be inclined to conclude that Stella experiences situations as aggregates of conceptual building blocks, rather than concepts as localized to situations—arguing against suggestions that dogs think almost entirely through gestalts such as "hedgehog" meaning "get hedgehog" rather than through concepts referring to individual objects or object kinds. Stella's deployment of concepts in situation schemas, in short, would tend to be "analytic" in human-like ways, where concepts (modulated by context) chain up to form situational representations [40, page 150], as opposed to "holistic," where we experience situations as gestalts and have a vaguer sense of concepts extending across situations.

Berns's research suggesting otherwise is perhaps an artifact of his experimental methodology: he did not scan dogs in their normal lives living with people (which would be infeasible if special tools are needed for brain scans); instead, he could only scan subjects doing a narrow range of activities that are possible when laying still for an protracted time, for the scan to work (e.g., 30 seconds). Or, in the case of fetch-toy style games, these are supposed to be fun activities for dogs who might get bored before long, so their attention wanders elsewhere. If the kinds of activities Berns's team set up for intelligence tests or fMRI scans are different from the norms of their day-to-day lives, it is not surprising that they would register different cognitive patterns, being more conceptually isolated and self-contained as compared to situations that are more familiar—the very divergence of such situations from dog's habits would incline concepts to be more narrowly rendered as parts of the situation than would be the case for experiences that are less atypical.

Similar comments could be made about Hare's experiments, even if they are carried out in a more natural setting than the limited tasks dogs can perform in an MRI scanner. Strange scenarios fabricated to quantify animals' reasoning—such as knowing where to look for a treat hidden beneath one of two cups—are an imperfect stand-in for the routines of a dog's life shared with humans. Dogs reason intelligently about the totality of their life-world in conjunction with their care givers, where they can make inferences informed by habits and routines, observing humans' actions and emotions, even by the current place and time of day (e.g., dogs usually get food in the kitchen or treats when they are walking or playing outside).

Most of these internalized contexts and cues are absent from the settings where cognition tests would typically be performed. Of course, problem solving in unfamiliar contexts is one dimension of intelligence, but measuring *only* this aspect of cognition does not properly observe thought processes in the plurality of subjects' lives (humans, too, are less intelligent when navigating unfamiliar territory—consider trying to find an address in a foreign city, or a book in a new library).

We think a hypothesis worth pursuing is that recent dog cognition research has accepted an overly simplistic *emotional* portrait of our canine companions, which end up distorting their *cognitive* theories. At one point in *The Genius of Dogs*, for example, Hare cites experiments about dogs "disobeying" people, which would run sort of as follows: owners place food on the floor but also instruct dogs to "lie down," implying that the dogs may not go to take the food [20, page 245]. The owners then engage in various activities differentiated by how clearly they appear to be distracted and/or are positioned so that they can or cannot observe the dog's behavior. Experimenters then test when and whether the subjects break from the "laying down" position to get the food. Hare discusses these experiments in terms of the dogs being tempted to counter owners' instructions, but hoping they don't notice.

Issues of obedience are also endemic to "trainability," which is often treated as a central factor (perhaps the most quantifiable metric) for dogs' intelligence. It is hard to measure whether a dog who fails to perform a certain trick or action does not understand what a person is requesting or simply isn't motivated to play along. However, it is generally assumed that dogs want to please their humans, and if that is not sufficient incentive, their interest can be sparked by treats. The overall picture is that dogs have motivation for praise and treats and—when behaving intelligently—should regard humans' requests as means to obtain one or both; which may include solving some sort of puzzle or reasoning task. At the same time, dogs' motivation for rewards like treats could override their interest in human praise.

This account might be too superficial because it breaks down dogs' motivations into simple units—treats, praise—and expects those to act as motivating factors in their using intelligence to think through hindrances. Such paradigms simplify the role of intersubjectivity, treating dogs either as pliantly calculating how to get rewards or as obstinate rebels willing to counter owners' wishes. A more nuanced picture is that dogs perceive their relationship with caregivers as cooperative: after all, humans feed dogs multiple times every day, and in other ways care for them, so (at least properly loved) dogs implicitly understand that humans care about their well-being. Dog lovers are happy when their dogs are happy, an emotion which the latter surely perceive, and reciprocate; dogs are protective of people in part because of a certain sympathetic mirroring—dogs are contented when their people are secure and protected. Dogs and caregivers share bonds of mutual empathy, where positive and negative emotions for one party become instinctively shared by the other.

In this context, dogs presumably follow along with human requests because they perceive that doing so will make us happy. Probably, this often takes the form of dogs experiencing activities as games: they may bring a ball or toy to humans to initiate play, and they probably believe we are doing the same with tasks and routines such as "sit," "lie down," or the (sometimes rather convoluted) actions or puzzles formulated by cognitive scientists. While researchers may be interested in problem solving or fMRI scans, the subjects probably think that these requests are a kind of game and that we want to play because doing so makes us happy. So dogs play along and expect the activity to progress the way playtime usually does. Putting ourselves in their shoes: if we then follow up with gestures that seem out of place in a game scenario—like putting down food and saying "lie down" but then turning away to read a book, or engaging with the dog by pointing in different directions but then turning our back (supposedly to test for "directional preference") —we can easily see dogs struggling to understand what we are actually trying to accomplish.

By analogy, imagine if your best (human) friend suddenly starting carrying out unusual and disjointed behaviors, like putting food underneath a cup (instead of a normal plate, for people, or bowl, for dogs) or randomly stretching their arms in strange directions. You would probably find their behavior weird and be unsure how to react. If dogs perceive many of their interactions with us as akin to games, then they may perceive our specific requests and instructions as attempts to follow a certain "play" script—in the same way that humans playing games together accept certain rules (without which games would not be much fun). That would explain why we care about whether a dog is sitting or lying down, for example, or whether she moves when we make one sort of hand gesture but not another. But game playing involves expected behaviors and comportments; we anticipate participants to show signs of enjoying themselves (smiling, laughing, and so forth) and to act physically as if they are having fun. When we fail to act as if the interactions we are having with dogs are play-like for us, then dogs plausibly get confused and start trying to figure out what we actually want them to do. Probably most dogs would never imagine that activities we rig up to test their intelligence are some sort of scientific experiment rather than a form of play time (after all, you probably would not guess that your bizarrely acting friend was actually carrying out cognitive research on you).

In short, dogs' cognition naturally takes place in a cooperative mode with humans where both sides mirror each other's fun/happiness, a resonance that inspires cognitive intersubjectivity. Humans acting in ways that seem nonsensical relative to this normal relationship can understandably leave dogs confused and uncertain about what their humans want, a lack of clarity that potentially manifests itself as difficulty reasoning through situations intelligently. After all, it is not as though dogs realize they are being tested. For example, if a person places food on the floor—especially a person who has always fed them—why would they not allow dogs to go get the food?

To put it formulaically, experiments may be poorly designed if their implementation can easily yield "emotional" anomalies that distort dogs' normal thought patterns. Reading emotional cues (for both dogs and people) is an essential aspect of situational understanding. Experiments that force people into unusual behaviors also prevent us from engaging with dogs through our instinctive emotional patterns, which in turn makes it harder for them to adapt to the situation at hand—but humans can also get befuddled if other peoples' emotional reactions do not seem to match current situations in an expected or rational manner.

The point here is not that artificial experiment situations never yield worthwhile data, but that we should be skeptical about making generalizations about canine intelligence (especially negative ones) on such a basis. The optimal milieu to assess canine cognition is through the lens of normal daily lives among people; in particular, here emotional cues evolve alongside situations organically, insofar as we instinctively project emotions in familiar ways, so that dogs can recognize these emotional patterns alongside observing our behaviors. Emotion and reason go hand-in-hand in multiple ways, and one manifestation of this synergy in particular is that observing other's apparent emotional state helps us reason through intersubjective situations.

While not equating dog and human intelligence (any more than we would find it meaningful to compare adult reasoning to toddlers'), in their familiar contexts we should, we contend, be inclined to postulate that dogs experience situations much like ourselves, in the fundamental patterns of cognitive structures and attentional dispositions. In particular, we should hypothesize that they experience discrete objects (and persons/animals) as self-contained particulars that over time will re-appear in different kinds of situations, each one structurally different. Such recurrence of specific objects grounds significatory potential, because it endows things with a continuity and individuation that make it sensible for them to the "named." Of course, people have names like objects do—and so do dogs, who in turn have a selfhood, a capacity to be part of situations (with some autonomy) alongside humans. Stella is clearly aware that "Stella" is *her* name.

It is also reasonable to speculate that patterns of conceptualization—with objects' perdurance and individuation foregrounded—are an important part of Stella's self-awareness. Insofar as situations are logically formed from conceptual building blocks, Stella can reason and even communicate about her surroundings, acquiring greater agency by, at least in part, influencing them. When saying "toy inside," for example, she is not signaling desire for an *activity*, or something relatively non-individuated, like "play outside." Rather, she mentally and linguistically singles out one object in particular (the toy she found) and envisions a specific state, of affairs structured around the toy (now it is in the yard; she wants it in the house). Intellectually, these communications involve frames of reference and counterfactuality that are built up atomistically; they have structural networks that gestate compositionally, constructions that may be embodied in linguistic form (in the sense of, say, Construction Grammar).

Situations are also sometimes represented temporally with analogous structural assemblage, such as sounding "bed eat outside" to assert that, having gotten out of bed, Stella intends to have breakfast and then go for a walk [page 197]. It is reasonable to guess that Stella's self-awareness includes a notion of *agency* and that this is bound up with a certain "analytic" tendency to understand situations as concept aggregates, insofar as concepts serve as the points of contact between multiple parties' perspectives on (their shared) situations. Analytic schemas allow co-operating agents to share ideas, plans, and opinions.

We think one should therefore question Brian Hare's own skepticism about whether dogs entertain to any substantial degree plans or anticipations about the future, or Berns's skepticism about dogs recognizing their own name or image. Hare *has* emphasized that dogs' intelligence draws heavily on "social" communion and intersubjectivity, and he has documented pro-social, cooperative behavior among other animals (notably bonobos).[21] Hare and Vanessa Woods have even generalized

[21] See, for instance, *Bonobo Handshake* [51] (a clever title because it refers to a gesture where what is being shaked is, um, not the ape's *hand*), written by Hare's sometimes co-author, Vanessa Woods (they are also married) but describing research they conducted in consort (for this context particularly Chapters 9 and 35; experiments reported as [19]).

ideas about pro-sociality and the cognitive benefits of intersubjective awareness to a kind of ethicopolitical thesis whereby humans have "self-domesticated," relying on a baseline instinct for collaboration and empathy to survive. This "Survival of the Friendliest" philosophy is explicitly inspired at least in part by the evolutionary success of dogs—who gained protection and care from humans precisely by *being* friendly to us. So dogs' behavior is thematically linked to pro-sociality in bonobos and people, even though Hare hesitates to endorse their exceptionalism in some facets of cognition and intelligence. Even if we accept that dogs do not have neurological resources akin to people and higher primates, however, it is possible that they have a more human-like sense of agency and selfhood than Hare's arguments imply.

Along these lines, it is worth noting that Stella was possessive of her buttons and acted protectively toward the part of the house where her buttons were set up. She would often nap on the floor next to the buttons; when Hunger first moved the buttons to a larger board, Stella asked for the old board back (gesturally) but, it turns out, wanted to use it as a mat (she lay down on it) [page 195]. This behavior implies that she had positive connotations with the buttons in general and with their overall setup. When they took the button board on vacation and placed it, for instance, in motel rooms, she employed the buttons as usual, so that her experience of talking was not specifically connected to the location of buttons in her own house. It does not seem as if Stella experienced the buttons as toys, per se, but she did clearly consider them part of her territory and her personality. Perhaps she feels empowered because the buttons allow her to express herself and solicit attention from her people. Her behavior, at least, certainly points to her using the buttons on her own volition, not to impress or appease her humans, as if they were some kind of trick. She talks not only to get what *she* wants, but also to send encouragement and positive emotions toward people (rather than the other way around), like saying "I love you" on a day Christina felt sick. In short, Stella appears to deem the buttons an extension or magnification of her personal agency.

Perhaps, in short, one reason that Stella and Bunny were motivated to pay attention toward AAC-like buttons—and to experiment with different phrases, observing humans' responses—is that the dogs wanted to be part of the human world. Talking to each other is one of our most essential human traits, and dogs joining in this activity is a way for them to experience people including them as "peers" in our collective behavior.

23.3.1 Learning from Humans

Perhaps Stella enjoys the buttons because she perceives them as tools people use, and they allow her to participate in people's activities. Certainly, there are many examples of dogs mimicking humans so as to be more fully included—one can find online videos of a dog who climbs up and sits down on a chair by the dinner table (in the partly upright "sit" position so her head can be above the table top); Porter, a dog who rides a bicycle and a scooter; several dogs trained by a group in New Zealand to drive a car—that is, literally start a car and use the gas pedal and steering wheel to maneuver around a circular track (there are no people in the cars with them); Eclipse, the Black Lab/Bullmastiff who became emblematic in Seattle for habitually riding the bus (alone) back and forth to his favorite dog park; and of course Stella and Bunny are joined by other dogs who have learned to talk with speech-assist buttons (videos show a yellow lab named Copper, for instance, communicating in context-specific ways such as sounding "outside mad pool" after a wading pool that Copper loved was disassembled, or "mad Copper no" after not being allowed to eat Halloween candy as treats).[22]

In mimicking human beings, we should actually *expect* dogs to process cognitively in human-like ways, even if only because that is what the activities they are trying to master call for. Dogs "in the wild" may deploy, in most situations, cognitive schemata that are less structurally resonant with human thought (at least modern house/apartment-living, urban-dwelling humans). Dogs who have since the start of their lives spent almost all their time in the same rooms and sharing activities with people, however, would naturally adapt neurologically to the patterns of human life. In particular, schemata that foreground conceptual aggregative architectures and synergy between multiple persons' epistemic attitudes predominate in human experience. It is nonetheless fascinating that dogs have enough neuroplasticity for their cognitive systems to adapt to environments where such schemata present themselves.

In fact, many "human traits"—e.g., language, self-recognition, altruism, advanced planning (cf. Stella's saying before leaving the house whether, that day, she wants to visit the beach or the dog park [page 211])—are, we have increasingly realized, shared at least on some level by many animals. That may be hard to reconcile with people's supposed skepticism, not too long ago, that animals were even conscious, or at least whether they possessed conscious experiences defined by intelligent thinking rather than rote instinctive drives.[23]

---

[22] See https://www.facebook.com/VTRND/videos/copper-the-talking-dog/1663327744059171/.

[23] Simplistically mechanistic or behaviorist models of animal minds are long out of favor, and moreover, it is not clear whether anyone literally believed that dogs, say, were actually unfeeling machines. Philosophers will sometimes put down in writing arguments that are stronger than they believe in practice, probably with the goal of advancing a certain skepticism toward commonsensical platitudes that are hard to prove logically—by analogy, a modern scholar who defends claims wherein it is "possible" we are "brains in a vat" (like in the Matrix movies) probably does not believe such claims are actually "possible" in the normal sense of the word, but rather pointing out that they cannot be facilely disproved the way that we can reject a mathematical theorem (e.g., by finding a counterexample). Since (as evidenced by portraiture, say) many people in (e.g.) Descartes' time loved dogs, and the practice of keeping dogs as pets was a feature of that society, we should not naively credit Descartes himself as actually believing the contrarian views that dogs are mere machines, whatever he may have written technically. Still, while extreme behaviorism (completely rejecting any notion of understanding animals, or other people for that matter, from their own perspective as opposed to simply noting their overt behaviors) may be discredited, it is possible that older paradigms still influence our concepts of animal intelligence and may still compel us to exaggerate the degree to which humans are intellectually and morally superior.

While "behaviorism" has been largely abandoned, many scientists seem to be cautious about anthropomorphising animals—arguably behaviorism's opposite extreme—as if we should not attempt, from our own human perspective, to understand animals' experiences. To which perhaps we should respond, why not? Granted anthropomorphising taken to an extreme could hold the risk that we become insensitive to animals' differences from us, potentially yielding a *lack* of empathy and compassion. Obviously, it is unfortunate when people punish dogs for doing things that are natural for dogs but not for people, due to our failure to see the difference. But it is not clear that the anthropomorphising is what causes such reactions; after all, people can be similarly prejudiced against other people who are different—in terms of race, religion, culture, and so forth.

Rather, we would be inclined to argue that concerns about anthropomorphising actually evince a kind of species essentialism that is itself unscientific. No one should claim that talking, sitting on a chair at dinner, riding a bus, or driving a car or bicycle are "natural" canine behaviors; doubtless, these are idiosyncratic habits that individual dogs acquired in a world where (among humans) they are commonplace. But by the same argument they are not "natural" for humans either. There is no reason to posit that eating at a table or getting around via motor vehicles are "innate" human behaviors. People adapt to the reality around them, which often means acclimating to an engineered and constructed (e.g., urban) environment. Projecting that animals can potentially adapt to the same environment, via neuroplasticity, is only in a benign sense "anthropomorphising" them—suggesting otherwise would mask how much for humans too it is neuroplasticity and adaptation which enables us to experience constructed environments as "second nature." There is no reason to believe that our cognitive dispositions, shaped by our environments, evolved to mimic the world around us any less than that of animals'.

Indeed, ethnolinguistic research suggests that cognitive schemata implicit in language will vary among communities with noticeably different lifestyles (nomads versus city dwellers, say), differences that can be detected in linguistic patterns (spatiotemporal markers in some indigenous peoples' dialects, for example, where the relevant language community is closely tied to a specific geographic area—whose natural features form a day-to-day backdrop encoded in language—function notably different than modern English, say) [40, page 141]. Neither genre of cognitive schema should be considered "true" to human nature; instead, we possess adaptable intellects that evolve to prioritize certain forms of mental representations over others, depending on what we need to function in our present environments. Guessing that the same holds for animals is not "anthropomorphising" them; it is instead making a credible generalization or extrapolation—from reasonable theories of how *our* cognition works—to other animals, or at least other mammals (who after all are relatively close to us in terms of evolutionary/genetic distance).

Data about individual dogs' abilities perhaps remain somewhat sporadic at this point because different researchers employ different methods. Gregory Berns is focused on fMRI scans and in general derives cognitive models from apparent brain activity, while Brian Hare assesses reasoning through problem-solving activities. John Pilley communicated with Chaser by talking to her, whereas Stella and her people communicate back and forth through talk buttons. The contexts and protocols in each case are different. Pilley demonstrated that Chaser had (comparatively) advanced language skills (or perhaps more appropriately cognitive abilities that lay at the foundation of language)—not just knowing many names for toys—because, for one, Chaser could react to noun categories and not just single names. She could, for instance, "get the ball"—properly locating one toy matching the category—similar to but at a more abstract level than fetching a toy referenced by its unique name. Chaser also understood different verbs; she would at Pilley's request "nose," "paw," or "get" objects, each word corresponding to a different action on her part. Obviously, Chaser therefore understood two-part phrases.

Evidence of Stella's understanding is different, because Hunger taught her a variety of words spanning different categories and uses, rather than a very large number of words with similar semantic profiles (like proper names). One question for dog cognition researchers is presumably how to synthesize multiple experimental protocols, accepting that it can take a long time to acclimate dogs to specific kinds of communication and activity, so they probably cannot just be layered together haphazardly.

Still, it seems plausible to generalize from examples such as Chaser and Stella so as to form a broader picture of dogs' potential language understanding, construing observations from individual dogs like patches in a collage. If we consider, in particular, Berns's neurological studies, any dog's linguistic capabilities are presumably rooted in brain states that are not significantly different between different individuals—it would be implausible to argue that Chaser, Stella and Bunny are black swans whose "experiments" present no ideas that could be generalized further. Any cognitive capability of one dog has to arise from a neurological architecture presumably shared, modulo subtle details, by the whole species. Any dog's nature is an intrinsic embodiment of what is biologically possible.[24] Only recently have dogs cohabitated with people to the degree seen in (at least urban, first world) households, suggesting that this arrangement also incubates latent canine intelligence, particularly that derived from observing and interacting with people.

Talk of animals adapting to "urban" environments usually refers to non-domesticated birds or mammals that learn to build nests or burrows and find food in urban landscapes, but of course here we are referring to how dogs living with people acclimate themselves to the accoutrements of modern human lifestyles, like riding in cars (or buses), living in apartments, visiting stores, and so forth. Over time dogs learn the routines necessary for them to accompany people about their day, even if this involves behaviors (like jumping into the back seat of a car) that would be bizarre in the "natural" world. Much of this learning presumably happens by

[24] As such, we can see the linguistic faculties apparently evident even in just a few dogs as latent possibilities that all dogs have to varying degrees, whether they are properly loved or laying in a crate by a conveyor belt soon to be euthanized (Benoit Denizet-Lewis, *Travels with Casey*, page 171 [12]).

watching people. If this is true, though, it means that intersubjective consciousness is a crucial part of dogs' neuroplasticity.

On this theory, dogs' cognitive schemata are highly subject-centric, in the sense that schemata are structurally organized around individual people—their kinesthetic gestures, their thoughts and feelings, their belief formation. Dogs, we might say—much like people—live their lives as a sequence of intersubjective episodes, recurring "stories" involving themselves and the people in their life: someone comes home, someone leaves, people cook and eat dinner, go to sleep and wake up, get sick and recover. If Stella observations can be generalized, dogs perceive the world through representations that are oriented around individuals (including themselves), projecting themselves into others' perspectives as a kind of epistemic expansion that transitions from inner experience to objective facticity. In phenomenological terms, dogs have an *epistemological* intentionality, triangulating with other minds to metastasize transient consciousness into (implicitly) propositional attitudes. Apparently stepping over Berns' estimation of its limits, we seem to have fair grounds for perceiving within dogs' intelligence the kind of interpersonal–propositional faculties Husserl theorized in works such as *Experience and Judgment* and *Cartesian Meditations*: we can download the Husserlian system of collaborative intentional facticity into our appraisal of dogs' cognition.

Analysis of animals' grasp of (for example) "object permanence" or "physical causation," or of intersubjective phenomena such as joint attention and perception of others' belief states, skews toward experimental psychology. This may be true for humans too, or at least especially young children (where experiments could take forms such as playing with toys where games are rigged to yield puzzle-solving or behavioral observations, or showing infants scenes which do or do not appear to violate physical norms, as adults understand them, and noting that babies stare more attently at counterintuitive goings-on; or, for older children, telling stories, perhaps with the aid of toys/props, and then asking questions) [39, pages 68–9, 156], [28, page 100], [7, 13, 17, 27, 29, 35, 47–50, 53]. While doubtless informative, such experiments present the relevant "folk-physics" notions as intellectual faculties that one merely has to varying degrees, perhaps (or as if) hard-wired in the mind. From this perspective, it is easy to rank different species based on innate smarts.

But such a psychology-based paradigm perhaps does not tell the whole story. Consider object permanence: consider moreover how cognition occurs *in the context of* empathic, nurturing, collaborative co-existence among people and/or animals who interact, communicate, and participate together in unfolding situations. It is easy to speculate that shared situational understanding carries over to how we experience objects themselves (in the sense of ordinary, inanimate, medium-sized physical things around us): the more we interact in the context of a chair, a ball, a toy, and so on, the more we instinctively factor in how others perceive that object, which in turn probably "triangulates" that thing in our mind. It is not just a sense impression, or a somatic intensity that only I experience. Awareness of other minds compels me to transition from experiencing objects as raw sensory aggregates to grasping their coherent behavior (and behavior potentials), their perdurance as rational objects in others' minds, and in the world that we share overall.

Our point here is not that we can reconstruct animals' cognition by some sort of metaphysical deduction, but that these observations suggest *instinctive* processes by which "theories" of other minds translate to an awareness of the world that we, under the influence of philosophical reflection, would deem rational (and appropriately prelinguistic, *poised for* something like language). There is no reason to suggest that dogs contemplate ontological questions; but it *is* at least a plausible hypothesis that other-mind cognition compels dogs toward a form of world awareness with structuration similar to what philosophers identify self-consciously as a groundwork for metaphysical analysis. In other words, experience of or reasoning through object permanence (say) perhaps emerges not so much as an intellectual faculty but as a byproduct of other-mind awareness. This thesis seems entirely in sync with the idea of "distributed" or "extended" cognition [1, 2, 8, 15, 16, 30, 36, 42, 45, 46, 52]; that perception of an object is not a private sensory dance between an isolated mind and a static external thing, but an emergent reality within a total complex wherein other minds also are often present: it is *we* who cognize, not *I*.[25]

This is, we would argue, at least one route to begin formulating a version of phenomenology that recognizes "mind" more broadly than just *human* intellect. The idea of a non-anthropocentric phenomenology has impressive precedents, notably the volume *Phenomenology and the Non-Human Animal: at the Limits of Experience* [37]. Charles Brown's chapter here is noteworthy in tieing intentionality to ethics, which would be an organic continuation of an other-mind/extended cognition theory of intentionality such as we, implicitly, just sketched:

> Consequently, while the fundamentals of our moral psychology may start off as gut instinct, these basic proto moral sentiments are not just reactions to outside stimuli but have the quality of being directed to something or being about something. Empathy, for instance, is always empathy about something. Altruism, similarly, is aimed at the other. Moral sentiments are experienced by humans not as "raw feels," or unstructured qualia, but as referring to the other in an attitude of empathy. These are psychological–somatological moments that exhibit a pre-discursive intelligibility and a prima facie rationality. These social instincts, as Darwin refers to them, i.e., compassion, sympathy, and feeling, are not simply Humean impressions (as Darwin seems to think) but display an intentional structure that provides or denies justification of the lived sense of that experience. [5, page 88]

Even scholars whose methodology leans psychological rather than phenomenological seem to agree that dogs' interactive dispositions—closely observing humans, sharing human space to a degree that we can without hyperbole surely describe as unique among all non-human animals—lie at the heart of canine intelligence. It is not a radical further step to suggest that intersubjective situational awareness, as it evolves and gets refined in animals' minds (and probably humans too), can in turn mold cognitive attitudes and dispositions (see also [6, 10, 23, 31–33, 43]).

[25] Not to suggest (to be precise) that the functionalist orientation of "extended mind" hypotheses as formulated notably by David Chalmers and Andy Clark is identical to perspectives of "distributed" cognition, as if "extended" and "distributed" were synonyms. However, further development of the Chalmers arguments does seem to have expanded upon the former notion so that it overlaps more and more with the latter.

Such a paradigm, if pursued, would involve in particular approaching canine experience from the perspective of (propositional) *intentionality*, not just conscious experience in general, which would also influence how we appraise dogs' feelings and desires. No one questions, for example, that dogs appreciate human affection and value their caregivers.[26] Subjectively, then, we can certainly imagine how a dog feels reassured and joyful in the presence of an affectionate person (after all, we ourselves have a comparable experience in the presence of the dog). Such positive emotion, however, is relatively undifferentiated. Properly respecting dogs' intelligence implies that we should model dogs' comportment to their humans as more detailed than *merely* experiencing people as a zone of positive energy. Dogs do not just settle into our presence and feel good; they observe our behaviors and often learn to mimic them.

Watching people and figuring out how to insinuate themselves into human activity positions dogs in a context of treating human behavior as a kind of puzzle to be structurally resolved, not just the habits of protectors to be experienced emotionally. Dogs, that is, have to observe people as individual subjects operating in situations whose pragmatic and epistemic formations can be organized around them, potentially and perspectivally.

Doubtless the emotional bonds between dogs and their people add weight to their interest in the latter, so that the underlying emotional experience plays a motivational role in focusing attention. It is because Stella cares about Jake and Christina, for example (we could speculate) that she watches them with special emphasis, which in turn allows her to perceive the situations *she* observers from *their* perspectives as well. But if the onset of these motivations is emotional, the end result is cognitive schemata that are structurally organized around human beliefs and behaviors, rather than the less "propositional" experience of love, hunger/thirst and satiation, play/boredom, and other basically somatic modes of consciousness that we might instinctively assume form the majority of canine experience.

Moreover, it is hard to image that Stella would learn to communicate at the level that she did without acute observation of the people she ended up talking to. Her linguistic capabilities piggybacked on her ability to form mental models of situations from different perspectives (her own plus others). Language is not only structured around cognitive schemata; it is structured around schemas whose organizing principle is the configuration of situations around each individual person, and our empathic/interpersonal ability to migrate orientation from person to person so as to understand situations collectively. We would argue that on these terms intersubjectivity grounds the two other foundational structures of language we have endorsed here (following the paradigms of cognitive grammar as we read them), namely situational understanding and conceptual plasticity.

[26] Berns's research, in particular, showed that reward centers in dog's brains responded as or more intensely to their "owner's" kind words as to treats [4, page 150].

23.4 Conclusion

Confirmation bias manifests itself in different ways. While scientific consensus should not be casually flouted, this maxim does not legitimate bias *in favor* of a status quo. Skepticism should be balanced, prepared to question counter-paradigmatic arguments but also to flag when hypotheses may be perpetuated more by force of habit than substantial supporting evidence. Gregory Berns, Brian Hare, and Juliane Kaminski are leading dog cognition researchers, but even among their writings we believe we can find conclusions that are counter-indicated by plausible evidence from talking dog videos, some of which we have tried to briefly mark in this chapter (more detailed analysis/counter-argumentation remains a project for future research).

Even dog-loving experts such as Berns, for example, have proposed (in the recent past) that dogs do not associate their names with themselves. Stella and Bunny videos dramatically suggest the opposite: they both on occasion sound out their name in consort with follow-up behaviors that would be hard to explain *without* assuming that they know what "Stella" and "Bunny" means. At some level, we reach a threshold where perpetuating older assessments of dogs' limitations carries *less* evidential warrant than revising them. This applies to self-awareness, and also, quite probably, to issues such as whether dogs understand noun/verb distinctions or multi-word phrases.[27] Ranking video observations via measures such as deliberateness, interpretive confidence, situational appropriateness, and follow-up behaviors is an example of methodological protocols that would guard against biases both *pro* and *contra* the status quo.

A final methodological point is that "dog communication" research does not need to show dogs' formulating phrases according to the rules of *proper* language as such (with multiple parts of speech—not just verbs and nouns—and morphosyntactic patterns). Probably, no one expects dogs to reproduce the complex system of syntactic interconnectors in natural language, and this would be infeasible via AAC buttons anyhow. For example, Stella's "beach inside"—referring to an inflatable boat and toys they take by the ocean being in the living room—implies a perceptive and situation-appropriate blend of concept in her mind. It is not especially important

[27] Granted, we might speculate that some dogs are more inclined to "learn" human language. It is often said that *canis familiaris* shows a wider morphological range—not least size variation—than any other species on the planet (among adult individuals), and some dogs are faster than others, stronger, more or less interested in different activities (fetching, chasing, herding, guarding), so it seems to reason that there could be similar variation in dogs' ability or motivation to mimic human speech. It is still an important scientific insight, however, if we accept that only a small percentage of dogs can achieve a level of linguistic understanding comparable to Bunny or Stella. Dogs who may not evince linguistic intelligence presumably focus their attention elsewhere (as with people, less proficiency in one area may be balanced by more proficiency somewhere else). A therapy or seeing-eye dog probably reveals equal intelligence to Stella and Bunny, but in their own line of work, no less than a Hadza hunter with spears and arrows is equally expert in his metier than urban professionals in theirs.

that she uses "beach" metonymically for things encountered *on* the beach, rather than devising a more intricate phrasal complex to orient her discourse toward beach-related objects rather than the beach itself.

Those questioning whether dogs are literally "talking" perhaps infer that some people are attributing to dogs linguistic competence at that quasi-human (adult) level. The reality is perhaps more interesting: such research gives us a window into how situational understanding and conceptual aggregation (*beach inside*, say, or *stranger paw*) yield proto-linguistic potentialities that can be analytically separated from the complexity of language's surface-level structures—wherein discursive articulation explicitly presents concept-connections that may be merely implicit in shared mental models. Thus "toy inside" depends on mutual understanding to convey *which* toy and *where* inside, and why, compared with how a person might express it ("I want to take the toy I just found inside the house").

Apart from the animal cognition dimension, then, research into talking dogs deserves attention from linguists proper because it helps separate prelinguistic cognition from lexico-syntactic processing. This applies to parts of AI as well, such as natural language processing. To the degree that dogs communicate via AAC buttons, their abilities lie in joint attention and habitual, situational collaboration with people, rather than by commanding large-scale vocabularies or inferring complex parse graphs. Compared to computational NLP systems, dogs' linguistic processing appears to emerge from their shared *umwelt* with people rather than from innate abilities to master discursive, lexical, and morphosyntactic conventions.

If this analysis has merit, it should reinforce our skepticism that robots or AI could be engineered to understand human language, at anything less than a partial and superficial (albeit perhaps still quite useful) level. Robots, for example, cannot have (we may reasonably claim) interpersonal experience analogous to the shared affection between people and dogs, and therefore cannot have that underlying communal awareness that becomes leveraged in understanding "other minds." In this sense, robots appear to lack an essential ingredient—perhaps *the* essential ingredient—in the emergence of human language. This does not foreclose some level of human/robot communication, because we already intimated through talk of "mid-level" language that a robot's discourse probably would not be human language in any case. But it does mean that the architecture of meaning and practical action that serves as structuring principles for robots' "language" needs to be conceived against a foundation notably divergent from the foundations of human language.

23.4.1 Robotics and Environment Models

If human language is not a reliable basis for human/robot communication, what should be the source for potential theories of "languages" appropriate for robotics? Here, research into dogs' communication with people can perhaps provide one crucial insight: our ability to interact at some linguistic/gestural level with dogs depends on a significant overlap between dogs' and humans' models of their shared environment, which in turn get leveraged for communicative intent. If this analogy may carry over, any human/robot language would have to be grounded in robots' "perceptual" models of their environment, which in turn humans have to implicitly understand so as to orient robots' actions in context.

A significant part of most robots' understanding of their environment, of course, comes from visual processing, which would involve algorithms such as those touched on in Chap. 22 (e.g., for feature detection and image segmentation). Robots' models of their surroundings would need to start with preliminary object detection—recognizing where there are discrete physical objects that can be studied and potentially moved or manipulated, perceptually distinguishing such potential foci of attention from the background environment—but apart from just *spatially* individuating objects, robots have to estimate their physical properties (which objects, or parts of objects, can be moved, for example, and in what directions).

Robots' environment models need not be completely open-ended; typically, robots are built for specific tasks in specific kinds of environments, so it is possible to anticipate what *sorts* of objects robots are likely to encounter. In this sense, they are not making inferences about external objects' behavior based *only* on visual input, but can be presupplied with data structures summarizing environing objects' properties and affordances. The interplay of robots' visual perception and their leveraging such *a priori* data then yields a correlation between spatial and function-organizational analysis similar to the cases examined in Chap. 20. That is, by combining spatial perception with preliminary data structures modeling categories of external objects, robots can develop hybrid environmental models that combine spatial perception—which situates objects and movement affordances in surrounding space—with profiles of objects' functional connections, which in turn would guide a robot's behavior and tasks to complete.

The examples in Chap. 20 considered analogous spatial/functional hybrid models such as Industry Foundation Classes (IFC) in the Architecture, Engineering, and Construction (AEC) context. IFC data (at least for many kinds of objects) combines CAD-oriented assets that model AEC artifacts visually and spatially, so that IFC object collections underlie blueprints and/or 3D models of buildings, building interiors, and other architected sites. At the same time, IFC objects can be aggregated to provide functional models of buildings as organized systems, so as to assess their usability (from the point of view of their residents/occupants), risk mitigation (in terms of, say, fires, floods, contamination, and other environmental hazards), or energy efficiency/carbon footprint. Such analyses are outlined more completely in Chap. 20; in the present context, we simply want to point out that system-functional models are in this context tied together with spatial representations, such as CAD-based 3D scenes or 2D floor plans. A similar integration can inform systems where spatial representation is not a matter of *designed* space, intended to visualize architectural/engineering plans, but actual space where robots are situated.

Consider a robot equipped with data structures conformant to IFC or similar (modeling) protocols. Upon perceiving an object (which it sees and/or physically contacts), we can imagine the robot matching its shape, color, and related apparent qualities to a database of object profiles, presumably matching the perceptual inputs to a class or category with an IFC-like description. Such a description, in turn, would offer both a 3D model of the object's geometry/kinetic properties (how the object moves holistically and its parts move relative to each other) and schematic data aggregation specifications related to the objects' materials, manufacturing, physical attributes (such as weight, density, fragility), and other information that would be relevant to robots' instructions. If an object is a piece of industrial equipment, for example, the IFC-like data could include part/model numbers and industrial specs; a robot might be programmed to test (and notify human monitors) if a part appears to be damaged.

The functional/spatial convergence evinced by formats such as IFC can be taken as an intuitive example of the environment models through which robots could operate. Insofar as functional/spatial hybrids may likewise be represented through computer applications and GUIs, we might assume that users guiding robots could similarly access views of robots' environments with analogous functional/spatial duality—e.g., a user through a robot's visual feed would see a 3D scene (embodying spatial data) and then select visible objects to examine through data-presentation windows (transitioning to functional-organization style data). In this sense, users and robots can synchronize their respective views on robots' environment, which, inspired by animal (or at least dog) cognition research, we should take to be the crucial first step toward communication.

On these considerations, we believe that a prerequisite for effective human/robot communication would be well-organized GUI systems representing robots' environments *for* human users, where GUIs encompass components for viewing both spatially articulated data (video/camera feeds, 3D scene models, etc.) and functionally oriented summaries of objects' industrial, physical/kinetic, environmental, or project-specific attributes ("project-specific" in the sense that robots typically interact with objects manufactured to collectively form a physical system engineered to specific ends; analogous to machines, but on the scale of a complex whole, such as a factory, building site, environmental cleanup operation, etc.).

Such observations are just a starting point, of course; further analysis of these proposals would need to address how human users should view and interact with robots' environments. We will return to this issue in Chap. 24.

From this chapter's perspective, the key foundation of linguistic communication is shared situational and "environment" models. Looking beyond human language proper, we can see clear evidence of human communication with animals, so long as both parties' mental models are in sufficient synergy. To say that Stella *talks to* Jake and Christina (and vice versa) is not a metaphor or exaggeration; albeit with the aid of a mechanical device based on AAC tools (that some people also use), Stella and Bunny are *literally* "talking dogs." Their use of buttons is too sophisticated to be equated with performing tricks or enacting gestures (as if the sound produced by talk buttons were proxies for nonlinguistic gestural communication), given how they string words together, initiate talk in contexts other than requesting something specific like a treat or a walk (e.g., when they are expressing affection, or commenting on what is happening around her), and almost certainly regard their ability to employ human language as a source of agency and connection to their human friends.

None of these intersubjective phenomena belong in models of human/robot interactions, but we may still talk of humans and robots sharing environment/situational awareness, which could potentially lay a foundation for (at least something like) a common human/robot language.

Humans and robots can share mental models of an "environment" because robots are specifically engineered to simulate the intelligent and semi-autonomous behavior of animals (in that robots typically process inputs visually and/or tactilely, and are able to move around or reposition their physical configuration, often without explicit guidance). Most computational systems are not equipped with analogous capabilities, so the picture of humans communicating with computers through *language* becomes more strained or analogic when we go from robots to computers in general. Certainly when interacting with familiar desktop computers it is only in a highly metaphorical sense that humans and computers "share" common mental models. Nevertheless, although joint awareness is a foundational prerequisite for language proper, there are other syntactic and semantic structures that emerge on that foundation, and there is some merit in treating computer programming languages as instantiating at least some of these linguistic patterns, in some (not entirely metaphorical) sense. The specific details of how computer languages might borrow structures from language proper (even if, as we claimed at the start of Chap. 20, we should not take literally the convention of calling the dialects of computer code "languages," even artificial ones) is a theme that will be taken up in greater detail in Chap. 24.

References

1. Miranda Anderson, *et. al.*, "Distributed Cognition and the Humanities". https://marksprevak.com/pdf/paper/AndersonWheelerSprevak---Distributed%20cognition%20and%20the%20humanities.pdf.
2. Eric Arnau, *et. al.*, "The Extended Cognition Thesis: Its significance for the philosophy of (cognitive) science". https://www.tandfonline.com/doi/full/10.1080/09515089.2013.836081.
3. Gregory Berns, *How Dogs Love Us*. Houghton Mifflin, 2013.
4. _____, *What It's Like to be a Dog: And other adventures in animal neuroscience*. Basic Books, 2017.
5. Charles S. Brown "The Intentionality and Animal Heritage of Moral Experience: What We can Learn from Dogs About Moral Theory". https://link.springer.com/chapter/10.1007/978-1-4020-6307-7_7
6. Cameron Buckner, "The Semantic Problem(s) with Research on Animal Mind-Reading". http://cameronbuckner.net/professional/semanticproblems.pdf
7. Susan Carey, "Do Constraints on Word Meanings Reflect Prelinguistic Cognitive Architecture". https://www.harvardlds.org/wp-content/uploads/2018/05/Carey.-1997.-Do-constraints-on-word-meanings-reflect-prelinguistic-cognitive-architecture.pdf
8. David Chalmers, "Extended Cognition and Extended Consciousness". https://philarchive.org/archive/CHAECA-9
9. Stanley Coren, *The Intelligence of Dogs*. Free Press, 1994.

10. Laura Danón, "Modest Propositional Contents in Non-Human Animals". https://www.mdpi.com/2409-9287/7/5/93/pdf
11. Edoardo Datteri, *et. al.*, "Going Beyond the "Synthetic Method": New Paradigms Cross-Fertilizing Robotics and Cognitive Neuroscience". https://www.frontiersin.org/articles/10.3389/fpsyg.2022.819042/full.
12. Benoit Denizet-Lewis, *Travels with Casey*. Simon & Schuster, 2014.
13. Sebastian Dörrenberg, *et. al.*, "How (not) to Measure Infant Theory of Mind: Testing the replicability and validity of four non-verbal measures". https://www.psych.uni-goettingen.de/de/development/pdfs/doerrenberg_rakoczy_liszkowski_2018.
14. Anna Frohnwiese, *et. al.*, "Using Robots to Understand Animal Cognition". https://pubmed.ncbi.nlm.nih.gov/26781049/.
15. Riccardo Fusaroli, *et. al.*, "The Dialogically Extended Mind: Language as skillful intersubjective engagement". https://pure.au.dk/ws/files/55171506/2013_cognitive_systems_research_Dialogically_Extended_Mind.pdf.
16. Hajo Greif, "What is the Extension of the Extended Mind?". https://www.ncbi.nlm.nih.gov/pmc/articles/PMC5686289/.
17. J. Kiley Hamlin and Karen Wynn, "Young Infants Prefer Prosocial to Antisocial Others". https://www.ncbi.nlm.nih.gov/pmc/articles/PMC3076932/pdf/nihms248773.pdf.
18. Brian Hare and Vanessa Woods, *Survival of the Friendliest: Understanding Our Origins and Rediscovering Our Common Humanity*. Random House, 2020.
19. Brian Hare, *et. al.*, "Tolerance allows bonobos to outperform chimpanzees on a cooperative task." 2007 https://pubmed.ncbi.nlm.nih.gov/17346970/
20. _____, *The Genius of Dogs*. Penguin, 2013.
21. Matej Hoffmann and Rolf Pfeife, *et. al.*, "Robots as Powerful Allies for the Study of Embodied Cognition from the Bottom Up".'https://philarchive.org/archive/HOFRAP-2.
22. Christina Hunger, *How Stella Learned to Talk: The Groundbreaking Story of the World's First Talking Dog*. Harper Collins, 2021.
23. Simon P. James, "Phenomenology and the Problem of Animal Minds". https://www.environmentandsociety.org/mml/phenomenology-and-problem-animal-minds
24. Juliane Kaminski and Patrizia Piotti, "Current Trends in Dog-Human Communication: Do dogs inform?". https://journals.sagepub.com/doi/abs/10.1177/0963721416661318
25. Juliane Kaminski, *et. al.*, "Domestic Dogs Comprehend Human Communication with Iconic Signs".https://asset-pdf.scinapse.io/prod/2148275342/2148275342.pdf
26. Juliane Kaminski, *et. al.*, "Word Learning in a Domestic Dog: Evidence for 'Fast Mapping'".https://quote.ucsd.edu/cogs156/files/2013/03/Kaminski05Rico.pdf
27. Melissa M. Kibbe and Alan M. Leslie, "Conceptually Rich, Perceptually Sparse: Object Representations in 6-Month-Old Infants' Working Memory". https://www.bu.edu/cdl/files/2019/01/2019-KibbeLeslie-PsychScience.pdf
28. Christine Kenneally, *The First Word: The Search for the Origins of Language*. Viking Press, 2007.
29. Jonathan F. Kominsky, *et. al.*, "Simplicity and Validity in Infant Research". https://www.jfkominsky.com/PDFs/Kominsky_simplicity_validity.pdf
30. Ronald Langacker, "Interactive Cognition: Toward a Unified Account of Structure, Processing, and Discourse". https://www.proquest.com/openview/8f77fd4dc8af81470c9cc73483c0c6e7
31. Dominique Lestel, *et. al.*, "The Phenomenology of Animal Life". https://pdfs.semanticscholar.org/6a7d/9b86779a783ed92ebf42f70170911667f5a7.pdf
32. Dieter Lohmar, "How do Primates Think? Phenomenological Analyses of Non-language Systems of Representation in Higher Primates and Humans". https://link.springer.com/chapter/10.1007/978-1-4020-6307-7_5
33. _____, "Language as Means and as an Obstacle of Communication: Phenomenological grounds for intercultural understanding". https://revistes.uab.cat/enrahonar/article/view/v57-lohmar/1023-pdf-en
34. Paul-André Melliès and Noam Zeilberger, "Parsing as a Lifting Problem and the Chomsky-Schützenberger Representation Theorem". https://hal.archives-ouvertes.fr/hal-03702762/document

35. Henrike Moll and Michael Tomasello, "How 14- and 18-Month-Olds Know What Others Have Experienced". https://dornsife.usc.edu/assets/sites/311/docs/hmoll_publications/Moll-Tomasello2007DevPsy.pdf
36. Zachariah A. Neemeh and Luis H. Favela, "Beyond Distributed Cognition: Towards a Taxonomy of Nonreductive Social Cognition". https://cogsci.mindmodeling.org/2017/papers/0528/paper0528.pdf
37. Corinne Painter and Christian Lotz, *Phenomenology and the Non-Human Animal: At the Limits of Experience*. Springer, 2007.
38. John W Pilley and Hilary Hinzmann, *Chaser: Unlocking the genius of the dog who knows 1000 words*. Houghton Mifflin, 2013.
39. Steven Pinker, *The Language Instinct*. William Morrow, 1994.
40. _____, *The Stuff of Thought: Language as a window into human nature.* Viking, 2007.
41. Katie A. Riddoch, *et. al.*, "Exploring Behaviours Perceived as Important for Human-Dog Bonding and their Translation to a Robotic Platform". https://journals.plos.org/plosone/article?id=10.1371/journal.pone.0274353
42. Mark Rowlands, *et. al.*, "Extended Cognition and the Mark of the Cognitive". https://www.ida.liu.se/~729A10/mtrl/Rowlands.pdf
43. Kenneth J. Shapiro, "What it is to be a Dog: A Qualitative Method for the Study of Animals Other than Humans". https://www.wellbeingintlstudiesrepository.org/cgi/viewcontent.cgi?article=1081&context=acwp_arte
44. Con Slobodchikoff, *Chasing Doctor Dolittle: Learning the Language of Animals*. St. Martin's, 2012.
45. Paul R. Smart, "Toward a Mechanistic Account of Extended Cognition". https://eprints.soton.ac.uk/454082/2/Toward_a_Mechanistic_Account_of_Extended_Cognition.pdf
46. Michael Tomasello, *The Cultural Origins of Human Cognition*. Harvard University Press, 1999. http://ektr.uni-eger.hu/wp-content/uploads/2015/11/tomasello-the-cultural-origins-of-human-cognition.pdf
47. Michael Tomasello and Nameera Akhtar, "Two-Year-Olds Use Pragmatic Cues to Differentiate Reference to Objects and Actions". http://people.uncw.edu/hungerforda/Graduate%20Developmental/PDF/tomaselloetal1995.pdf
48. Michael Tomasello and Katharina Haberl, "Understanding Attention: 12- and 18-Month-Olds Know What Is New for Other Persons". https://www.eva.mpg.de/documents/AmericanPsychologicalAss/Tomasello_Understanding_DevPsych_2003_1555821.pdf
49. James L. Yu, "A Phenomenological-Hermeneutic Analysis of Experimental Infant Research". Dissertation, Duquesne University, 2009. https://dsc.duq.edu/cgi/viewcontent.cgi?article=2410&context=etd
50. Amanda L. Woodward, "Infants Selectively Encode the Goal Object of an Actor's Reach". https://yzhu.io/courses/core/reading/07.woodward.pdf
51. Vanessa Woods, *Bonobo Handshake*. Gothan Books, 2010.
52. Jiajie Zhang and Vimla L. Patel, "Distributed Cognition, Representation, and Affordance". https://psycnet.apa.org/record/2006-21234-010
53. Thomas R. Zentall and Kristina F. Pattison, "Now You See It, Now You Don't: Object Permanence in Dogs". https://www.jstor.org/stable/44318955

Chapter 24
GUIs, Robots, and Language: Toward a Neo-Davidsonian Procedural Semantics

Nathaniel Christen and Amy Neustein

24.1 Introduction

With respect to the foundations of human language, previous chapters have pressed two claims in particular: first, that this form of communication depends on shared (pre-linguistic) situational awareness; and, second, that full-fledged interpretations of discursive meaning derive from *extra*-linguistic cognitive processes. In other words, we have discussed both the *pre*- and the *extra*- linguistic, implying that language proper bridges the two. In *that* guise, there may be room to identify overlaps between natural and other languages (those of animals, computers, or other mechanical/artificial engineered devices). Without a true *intersubjective* basis for shared situationality, or without human-like cognition, certain structures shaping the language-proper "bridge" may still be similar (enough for some formal cross-analysis), even between very different sorts of dialogic actors. This chapter will expand on that argument in both natural language and computational contexts.

For reasons such as discussed in Chap. 20, computer programming languages offer only a flawed and limited simulacrum of "natural" language (i.e., human, spoken language). This does not rule out how structural features of programming languages could potentially shed light on human linguistics (shortly, we will present in greater detail where we think the most promising or prominent structural correlations exist), but the tenuous connections between human and computer languages writ large invite questions related to the nature of languages that might be intermediate between them—more complex and/or context-dependent than computer code, but also more computationally tractable than human language itself.

N. Christen · A. Neustein (✉)
Linguistic Technology Systems, Fort Lee, NJ, USA
e-mail: amy.neustein@verizon.net

A. Neustein et al. (eds.), *AI, IoT, Big Data and Cloud Computing for Industry 4.0*, Signals and Communication Technology, https://doi.org/10.1007/978-3-031-29713-7_24

This chapter is motivated in part by how these sorts of questions could be applied to robotics, which is an area of concern for several of this book's chapters. One foundational question when engineering robots is how humans should communicate with them. If scientists' goals are to fabricate "life-like" robots, perhaps taking a vaguely human-like physical form, then it would be natural to envision robots equipped with sophisticated natural language processing capabilities. Such an aspiration, however, could eventually run up against the outer limits of computational NLP, in the sense that any inanimate machine—even ones with sophisticated AI capabilities—may simply be incapable of parsing and responding appropriately to human speech on a truly realistic level. Of course, even today NLP is sufficiently advanced to be a very useful tool, but not to the point where people in most cases would experience actual communication with AI agents as believably life-like, or a credible substitute for the emotional, gestural, and contextual nuances of linguistic interactions between people (which are almost never "purely" linguistic).

We think these issues are still open-ended. Philosophically, it is a reasonable hypothesis to assume that most of the "human" qualities of conversation are driven by cognitive activity that might be simulated at least in theory, so a truly advanced robotic or AI agent could—again, at least in theory—asymptotically approach a state of being considered almost human-like in how it communicates. A counter-argument to this kind of analysis might be that language-processing sophistication is necessary but not sufficient for being an effective discursive partner: witness how people from different cultures can have trouble understanding one another, even if baseline semantic and syntactic impediments are removed. Sports fans from London and New York, say, somehow eavesdropping on each other's conversations, could very well fail to comprehend their respective dialogs despite the two groups speaking essentially the same language. Likewise for comparing discussions about US versus UK politics, or navigating Brooklyn or Camden Town like a local—inevitably, in short, there must be some shared cultural and lifestyle familiarity between conversants in order for reasonable human dialog to take place, because so much of the semantic and referential details of our speech remain unstated—an implicit part of the discursive background rather than explicitly invoked lexically or descriptively. No one in a typical conversational context would need clarification on how "Spurs" in Texas probably refers to a basketball team but a football club in the UK, or "Soho" to a part of Westminster in London but a part of Lower Manhattan in New York. These are matters of shared life experience, not linguistic competence.

How, in this sense, could a robot "share" life experiences with people? On the face of it, such questions should lead us to question whether realistic NLP is ever truly feasible, because even in a hypothetical world where a machine with human-level abilities of raw linguistic processing could be built, that machine would still not be "living" in a human world and absorbing local culture and dialect. Could all of our enculturated background knowledge be adequately simulated on a computer? What about the emotive and empathic dimensions of language, without which it can be difficult to distinguish, for instance, sarcasm from sincerity?

Short of some empirical disproof of mind–body reductionism—i.e., the belief that all of our feelings and experiences ultimately derive from neurological phenomena, which are not known to have any physical properties that are truly beyond the scope of computational approximation—we certainly cannot rule out the possibility that truly human-like AI is theoretically possible. But the absence of impossibility is a weak argument for something being realistic and feasible in practice, and cultural/interpersonal issues demonstrate that the hurdles for NLP are not only linguistic in nature.

In short, even if we cannot be certain that realistically human-like AI (at least in terms of linguistic competence) is *not* possible, there seems to be no reasonable argument to guarantee that it *is* possible, so any technological project that depends on (something approximating) human-like NLP is currently on shaky grounds. At the same time, we *should* aspire to robot/human communications that are more user-friendly than artificial constructs such as programming languages. Note that this is an issue pertaining to robots' "users"—people who at some level guide a robot's behavior in the course of its assigned task, as compared to developers who engineer their capabilities in the first place. Robots' onboard operating systems can be programmed in a variety of languages, as is true for other embedded systems, and low-level computer code provides the minimal functionality allowing robots to operate (any physical movements or audio/visual processing of their surrounding environment). When a robot is performing a task on behalf of a person, however, we can assume that in most contexts the user would not be communicating with the robot through such low-level code but would instead give instructions more conceptually. For a robot to move a box, say, the person should be able to designate the box in question and its destination, allowing the robot's low-level machinery to calculate the object's pose and location, and its own physical configuration and movements appropriate for that goal.

Here, then, is a good example of the "intermediate" language question we alluded to before: what forms of communication exist that are higher level than artificial languages such as computer code, but simpler and more computationally deterministic than natural language? This actually encompasses two topics: how such "mid-level" languages should be *designed* (their semantic and syntactic properties—whether their vocabulary and/or grammar is based on actual human languages, albeit at a rudimentary level, or something more endemic to a robot's assigned tasks), first, and, second, how to translate expressions in such mid-level languages to low-level operations that can be programmed as robots' baseline capabilities.

We will touch on these issues later in the chapter, but here we will also be focusing on the theme of language in that gap between humans and computers—what resources can help us think through the nature of language outside the prototypes of mechanical computer languages and holistic human language? We argued at the start of Chap. 20 that calling implemented coding systems such as C++ or JavaScript "languages" is misleading, and in some ways the science of building and using programming languages (sometimes called software language engineering) has very little connection to linguistics proper. Note, however, that some sophisticated lines of linguistic research are driven or inspired at least part by concepts that emerge, or are formalized, in the programming language context

([3, 6], or [4], for instance). So there are at least some technical formations in software language implementation that may arguably shed light on how people understand our own human languages (some of these we reviewed in Chap. 20 and will revisit later in this chapter as well). With that said, though, the kinds of artificial languages employed to program computers and machines have at best only sporadic and limited structures that shed light on natural language, and as such are only an imperfect model for the kind of intermediate communication system that we might envision possessed by robots and other "Industry 4.0" machines.

We argued in Chap. 23 that intersubjective awareness serves as a central motivation that underlies other foundational dimensions of language—specifically, situational understanding and conceptual plasticity. Interpersonal understanding allows us to structure situational schemas around people (or, more abstractly, around epistemic agents and situational actors) so as to form coherent logical models of situations, and these models in turn are built up from concepts. Concepts, for their part, exist in more generic and more specific guises, and the manner in which broad concepts become particularly applicable to narrower situations provides one facet of situations' internal logic. Via this sort of architecture intersubjectivity evolves, through a kind of dialectic, into a roughly effective logical and co-operative picture of our surroundings, which ensures the pre-linguistic synergy of our communication.

Controversies over how to interpret results in artificial intelligence perhaps lend support to this point: enacting computer vision or processing natural language via statistical inferences (even if augmented by convolutional neural networks or machine learning) may or may not qualify as legitimate "understanding." Computers' sometimes human-like performance in translation, or (for instance) image tagging, could give the impression that AI modules are progressively approximating human reasoning. However, the so-called adversarial examination of AI programs has led some researchers to question such assumptions. These analyses are able to produce images, for instance, that "trick" computer vision technology into radically incorrect or nonsensical results—tagging two almost identical images into completely different categories, one accurate and one not (e.g., a school bus and an ostrich), or applying the same tag to a realistic photograph and nonsensical picture looking like random noise [63, pages 110–113]. Similar effects have been produced against NLP engines [63, pages 211, 227–231].

The issue here is not merely occasional inaccuracies, but that AI tools can be *so* wrong on select kinds of input that we have to question whether the "intelligent behavior" they usually exhibit truly approximates what we would think of as (human-like) intelligence. After all, people too can mislabel images and misread sentences, but we would never confidently assert that incoherent mumblings or "nonrepresentational" graphics (with appearance akin to TV screen "snow") were normal sentences or pictures of Milla Jovovich. One way to express these concerns is a doubt that AI agents—even if statistically accurate most of the time—truly share *situational understanding* with us.

Removing intersubjectivity (and subjective awareness) from language and communication does not necessarily disassemble the situational and conceptual formations that emerge on its basis, but they leave open-ended the process of how linguistic competence can take hold *without* interpersonal, collaborative foundations. Our intent last chapter was not to argue that the triad of intersubjectivity, situations, and conceptualizations are constitutive of language per se, but rather of the pre-linguistic firmament in which language can take root. Linguistic constructions encode situational details, but how they relate to contextual schemata remains a further question we did not address in any detail.

So, what exactly do we accomplish via speech acts insofar as linguistic activity adds a further layer to our preliminary cognitions and co-operation? We propose a "foundations of language" story which says in effect that our common situations and "joint attention" are *pre*-linguistic, while language proper concerns *changes* to our joint environments (whether past, present/ongoing, or futurally planned/contemplated). This perspective fits nicely with "verb-centric" theories of semantics and grammar, according to which sentences' parse graphs are organized around verbs, with nouns (and noun phrases) slotting in verb details and thematic roles expressed by "theta" relations (such as verbs' subject and object components) and by case/declension (where nominals in the locative, instrumentive, benefactive, and so forth add specificity—how, where or to where, by whom or for whom—to their verb's profiled event, process, or action) [55, e.g., page 39], [9, 11–13, 40, 53, 81]. It is true that almost any theory of language will focus on verbs because verb-plus-noun combinations (VP+NP) are considered to be the minimal requirements of a complete phrase. But some paradigms reinforce the verb-centric perspective still further, with the idea that verb meanings form the central organizing motif through which sentences are cognized—and that verbs dominate syntax in that all grammar elements (including categories such as nouns, adjectives, prepositions, or conjunctions, plus morphosyntactic marking rules/conventions) play roles ultimately determined by how they modify and add information to phrase- or sentence-level verbs.

Let us assume that in order to exchange speech acts, people must have situation/environment models that are sufficiently in sync. Because this agreement *precedes* anyone saying anything in particular, we can assume that language itself is not usually *about* this shared awareness, albeit we sometimes verbally correlate our mutual understanding (along the lines of "do you see ...?" or "did you know that ...?"). Normally, the role of language is to address what *deviates* from common understanding (or could cause respective models to become misaligned). Someone might propose an action that would alter the current situation (e.g., "Let's go to the park"), or inform others of a change they may not know about.

As a rule of thumb, then, we should look at a sentence and consider what details are expressed there that appear to propose, observe, or describe some *deviation* in shared understanding to which the speaker wants to call attention (or to request/recommended). Sentences' meaning would then be structurally marked by the contrast or gap between existing shared situation/environment models and some belief idea held by the speaker and (potentially) no others. Insofar as verbs profile the difference between one environment state and another, it makes sense that verbs would then be the central organizing anchors for syntax and semantics.[1]

This overall approach to sentence meaning highlights common understanding as a linguistic precondition, because it construes meaning as an effect of difference where that prior understanding is one differend—we measure significance *against* the common model, with the meaning proper consolidated in how a proposal or observation departs *from* that model. Linguistic competence would thereby be a two-part phenomenon that includes both sharing with others pre-linguistic joint understanding *and* being able to communicatively propose or effectuate changes to shared situation/environment models (including by changing the situation itself). In accord with our discussion in Chap. 23, we contend that we should credit dogs such as Stella with credible human-like language so long as we feel confident that they have cognitive resources to participate in our language at both of these levels. Data we summarized there certainly suggest both that Stella shares mental models with her caregivers, with some understanding of their context-specific beliefs, *and also* that she understands (even if at a rudimentary level) how she can *alter* others' beliefs or *suggest* changes to their shared situation by utilizing language (in Stella's case, by sounding out phrases via her talk buttons).

Because language proper expresses *deviations* from shared models—not withstanding how common understanding is a prerequisite—there must be other significatory facets to language that we can study via syntax and semantics. This, we would argue, qualifies as the origin of structures that can be applied to phenomena such as computer languages, which have some overlap with syntax or semantics in natural language proper but (for reasons we have outlined) do not, we believe, reflect the preliminary phenomena of joint attention and intersubjectivity. This chapter will consider some of these structures from a linguistic perspective and then pivot to discussing notions of "semantics" that may actually be appropriate for artificial settings such as robotics and software programming.

24.2 Semantics and Situational Change

Most of the time, language reflects our pragmatic concerns. Situations provision a backdrop for goal-directed activity, and the circumstances obtaining at any moment serve as both a parameter on and, to the degree that we act deliberately, a byproduct of our actions. If we ride a bus to the dog park, we are on the bus *because* we have decided on that course of action for the afternoon, but the bus is also a constraint; it

[1] Although "verb-centric" paradigms are tangentially related to Davidsonian "event semantics," we are thinking of verbs in the most general sense of *profiling* states, relations, or processes as well as events, following Langacker in particular [46, Chapter 4], [50, page 24], [49, e.g., pages 21–25], [48]. Also, for reasons sketched below, we are wary of semantic structures presented as *logical models* of sentences themselves, rather than as patterns organizing cognitive *responses* that may be *triggered* by linguistic constructions but are not prefigured or reciprocated within them.

creates situations for us that we occupy rather than explicitly architect. If the bus is stuck in traffic, we have no way to get there faster. Rain keeps us indoors; cars cannot drive except on roads; locked doors must be opened with keys. The affordances of constructed objects are also djinns producing states of affairs that we must adapt to. In this sense, our situational awareness, while partly present tense, is always too projecting forward, protaining the immediate future, subtly modulating plans in response to the not-fully-anticipated turns that situations take.

Ultimately, we would argue, it is this enactive-protaining register that forms the predominant mode of linguistic semantics. What we *mean* is usually reflective of how we anticipate our actions in the environing context of our situations to proceed, moving forward, in the immediate future. Linguistic utterances are, in short, canonically, formulaic expressions of temporal intentions in this sense, i.e., "intentions" in the sense of intending-to-act or intending-to-do. Of course, we do not spend all our linguistic lives simply asserting plans, but such pragmatic intentionality forms the substratum of language proper, we would argue. Such intentionality is still an underlying semantic paradigm when we switch from raw expository statements to questions, conversation turns, prompts for others' opinions, and in general the techniques we use to form collaborative intentions via dialog.

In sum, then, language departs from underlying cognitive schemata largely through the gap between merely experiencing present situations and co-ordinating near-future actions that follow from present-moment situational episodes. Stella (referring back to Chap. 23) is accordingly working in a linguistic register when she uses words to project forward in time, via articulations that incorporate the present moment but are not wholly contained in it ("eat park outside," "want park/beach," "come bye outside," "play outside" after Christina's "what now," and so forth). The complexity of linguistic constructions, of course, will mirror the degree of detail and breadth of planning spanned by the current discourse. Stella's relatively simple phrases stand at one extreme; something like a political party's platform paper, sketching out years' worth of legislation and programs (however realistic or not) at the other. But it is reasonable to argue that the crucial dimension of language proper is not syntactic complexity itself; instead, complexity of form is a metaphenomenon reflecting the intricacy of the anticipations which language communicates. Simple plans are no less vehicles of language than elaborate ones.

On this account, it is impossible to theorize linguistic meaning without describing how hearers receive speakers' statements as asserting or formulating near-future plans. Addressees must understand how word meanings, in the context of their proper grammatic forms, translate to anticipated actions. Such meaning depends heavily on pre-linguistic knowledge and on all parties to a conversation having succinct correlations in their respective beliefs to co-operate productively. If two people talk about taking a bus to the park, let us say, they need a prior understanding of what that would entail, such as the pragmatics of boarding a bus, where to get off, and so forth. This embodied knowledge is not a predicate that could be built via syntax and semantics alone, like a journalistic reportage ("today the new Prime Minister visited the King") stating a (relatively self-contained) fact. We might indeed say that the meaning of that Prime Minister sentence is wholly expressed through its propositional content, but a more quotidian speech act such as "let's take the bus to the park" is less illocutionary, its meaning deriving less from any proposition mimicked by its construction and more from how it solicits pre-linguistic knowledge—pre-linguistic "scripts," one might say—that would enable the actions proposed.

From an enactive/pragmatist perspective, we (as language users) have a substantial repertoire of "scripts" that represent learned know-how, enabling our competent achievement of concrete goals (riding to the park, and so forth). Such scripts are not propositions, nor are they facts that language can convey in propositional form (like "The Prime Minister visited the King"). Rather, they are competences which we have at some level abstracted and represented such that they could be referenced indirectly through language. There is nothing in the phrase "ride the bus" that conveys the propriocentric patterns of climbing onboard (and the geospatial acuity to know where to get off). Nonetheless, expressions such as "take the bus" serve as proxies to these pragmatic packages. Linguistic meaning, in short, often takes the form of nominating various enactive-prelinguistic scripts with which an addressee (by assumption) is acquainted.

Not all expressions work by "proxy" to this degree, of course (as compared for instance to describing facts via logical constructions), but insofar as language such as "ride the bus" depends for its semantics on already-familiar enactive "scripts," we cannot treat meaning as something *internal* to language, something structurally replicated in linguistic form. We have attempted more detailed analyses to motivate what we see as an essential deviation between language and logic (see footnote 2), a line of argumentation which we will not pursue here for long; but a thumbnail version is that most language does not work the way "logical positivist" (among others) paradigms have claimed, where language serves in effect as an informal version of the kind of predicate compositionality modeled more rigorously by formal logic. Sometimes it does: "(Rishi) Sunak visited (King) Charles" individuates two actors in a logical relation and names this relation directly, all through straightforward lexical and referential means. But most linguistic constructions do not work by reciprocating propositional constructions to this extent.

A more typical speech act is something like "let's take the bus to the park," whose apparent meaning does not have a direct propositional content in the first place (the intended idea is more involved than just, say, the expected fact of them soon *being at the park*); moreover, to the degree that predicates are implicit in the thought conveyed, the sentence does not work by structurally embodying them. Instead, the sentence's effect is, more, to inspire in the listeners imaginings of, and (to the degree that they endorse the plan) anticipations of the practical and experiential episodes that would comprise, in sequence, "going (by bus) to the park." The sentence works by invoking a script which the hearer (presumably) knows reasonably well.

Arguments to the effect that language "invokes scripts" more than "models propositions" are consistent, we think, with various theories of language that purport to offer an alternative to "analytic" philosophy of language arguably over-invested in language being a kind of symbolic (or symbolized) logic. These (counter-Analytic-paradigm) approaches would include cognitive grammar and construction grammar, probably alongside "embodied" notions in cognitive linguistics (George Lakoff and Mark Johnson, for example), or Jean Petitot/René Doursat "morphodynamics," speech act or conversation-oriented models in the tradition of Austin, Grice, and Searle, or ethnolinguistic (and "eco-"linguistic) approaches to language as an anthropological phenomenon more than a "system" to be structurally analyzed [10, 14, 23, 29, 34, 35, 37, 41, 44, 47, 62, 65, 72–75, 88, 100, 101]. To this list, we would add Orlin Vakarelov's "Interface Theory of Meaning," [91, 92] which we contend may be applied to artificial (programming) languages as well as natural language. In the formal language context, the value of such "interface theories," as we intimate them, displaces the formal semantic paradigm (manifest to some degree in computer science, for example in the design of the "Semantic Web") according to which programming language semantics should be evaluated in logical (and set-theoretic) terms. That is to say, expressions in programming languages are (within the paradigm being disputed) logical elements such as predicates/relations, sets, and set members, and by sufficiently intricate aggregates of such logical building blocks we can form computational simulacra of empirical objects. Logically constructing propositions about such objects (or their states and properties) enables objects to form the "semantics" of programming and database/data modeling languages, so that such languages' semantics is constituted by the logical approximations relating computational models to their presumptive referents (again, according to the paradigm being disputed).

Against such a "logical" paradigm, we argued in Chap. 20 for a "procedural" perspective, in which the semantics of computational languages are constituted by the collection of procedures that may be invoked through (any given) language (in any given context). This yields what has been called a "procedural semantics" and applied both to linguistics and robotics [19, 28, 85, 87]. Computational (or, respectively, by analogy, cognitive) *procedures* semantically implicate empirical objects (by presenting information about them, showing them pictorially, simulating then virtually, and so forth), but we should not reason as if the structures of computer code somehow "describe" the objects in question. Instead, computer code engineers a space of procedural responses to user actions in which object representation, to the degree that it is perceived as such by users, is an "emergent property" of the overall computational system. Outside the indeterminacy of such "emergence," the actual "semantics" of computer languages should be sought in the procedures that create representations of objects for a user's benefit.

Of course, computer code exists on multiple levels, and while we might agree that higher-level code depends for its meaning on procedures that are bound to code symbols, those procedures are implemented via lower-level code, and so on down the compiler stack. Eventually, we reach a machine language whose semantics is logic gates and memory addresses rather than procedure calls as such. However, these lowest-level layers—where computations are actually manifest on a physical level—have less resemblance to real-world objects than computer code at higher levels. The layered nature of code implementation motivates a model wherein any "natural" semantics for computational systems is an emergent property, not a logical construction. We defer to Vakarelov's "interface theory" in both contexts: the overt meaning of a linguistic expression (a speech act in natural language, or a code statement in programming) is an "interface" to some underlying script/procedure that resides not at the level of the explicit language, but on a deeper or hidden level (the repertoire of pragmatic scrips held by a person using language; the space of compiled routines known to a programming language runtime and available to be called from application code). Natural language invokes "cognitive procedures" (e.g., "imagine riding the bus") much as high-level code notates instructions to call low-level functions.

We contend, therefore, that something like an "interface theory" yields a useful paradigm with payoffs in both natural language and software language engineering. Granted, philosophical paradigms are not especially consequential in the latter domain unless they imply some specific tactic or norm for the design and/or implementation of coding languages (their compilers, static analyzers, and/or runtimes). It may not be obvious how argumentation that the semantics of computers is basically one of "procedures" and not "logic" practically affects any of these latter topics. In earlier chapters, we presented analyses hoping to show that there is at least some practical consequence that might compel new language engineering ideas, for example in more rigorously formalizing non-set-theoretic type theories and potentially new strategies for code analysis (alluded to via "constructor channels" and our overall suggestions to the effect that we should formulate compiler design and code verification via structures that can be translated to query evaluations over hypergraphs). This is, however, we concede, a fairly narrow domain of practical application. We have not clarified how detecting a concordance between "interface theories" in natural and computer languages yields any substantive insights in either context. The following, as such, are brief comments addressing these questions.

24.2.1 The (Provisional) Semantics of Syntactic Dis-ambiguation

The central idea of "interface" theories of meaning, as we see them, is that linguistic structures do not (at least fully) encode meanings in complete detail, but rather defer to extra-linguistic cognitive processes—language proper defines parameters and specifications on such extra-linguistic faculties, but does not logically articulate, at least in full detail, whatever mental operations substantiate cognitive processes themselves. To be rigorous—given that *reciprocate* usually connotes something coming *after the fact*—we can imagine a variation on the reciprocation theme wherein the idea of a structural alignment, a systematized lining-up of structural elements, is lifted from the "reciprocate" notion but switched to that which is temporally prior, like an anticipation or foreshadowing. According to an implicit linguistic paradigm—probably shared by many branches of "Analytic" philosophy as well —language produces logically organized meanings because elements in language are patterned, during the course of enunciation, to model (in the sense of *illustrate* or *prototype*) the logical form of ideas that then get decoded by addressees. The prior logical structure then guides addressees to assemble a notional understanding of expressions by reconstructing their predicate outlines. Despite this temporal ordering, though, it seems best to say that the expression structure "reciprocates" the later reconstruction—on this account—because addressees are grasping the idea that is anterior to and contextualizes the expression as a communicative artifact. The goal is to achieve a cognitive synergy between multiple parties' thought processes; the linguistic form is merely a device to achieve that telos.

This is, at least, one way to gloss the "logistic" paradigm which some writers have critiqued ([45, Chapter 25], say). Cognitive grammar and (at least as presented by Vakarelov) interface theories present an alternative paradigm; our gloss on this latter perspective is that whatever actually reciprocates logical form emerges from cognitive processing, not from linguistic structure (whether shallow or "deep") itself. The form of communicative expression has *some* correlation with logical structures, but this is not a mimicry in the sense of reciprocation or organizational alignment; instead, linguistic elements are patterned so that their structural relations, along with semantic and lexical cues themselves, serve as "prompts" or instigators molding processes of cognitive reception in different directions, so that the telic logical form emerges at the end of a cognitive arc. The resemblance of linguistic form to propositional structure is more like an embryo to a newborn than a closed term in symbolic logic to the proposition it encapsulates.

For lack of a better word, we might fall back on something like "temporally inverted reciprocation" to capture the notion that sentences signify propositional content by modeling its logical form (more or less directly—rather than prompting addressees to infer that logic via extra-linguistic cognition). There is no room here to go into detail as to what sorts of cognitive processes we have in mind, or linguistic evidence that we think argues for this perspective. A more expansive empirical and speculative defense of such arguments forms the thesis of a series of articles deposited as part of, and an analysis of, datasets published alongside [54] (we will reference the paper series in a footnote rather than citing them because this supplemental material was developed as part of the dataset, not submitted as journal articles)[2]—with a mashup of computer code, empirical data, and cognitive-grammar-style analyses, trying to achieve methodological plurality like we sketched in Chap. 23. The overarching theme here is a critique of "logicism" or symbolic-logic encoding of linguistic expression/signification, which of course rests in the shadow of similar counter-paradigm projects within the cognitive linguistic canon, represented on the philosophical side by Lakoff and Johnson in particular, and

[2] Here are the respective links (the dataset to which these belong also includes a collection of language samples, analyzed in the papers):

From "Naturalizing Phenomenology" to Formalizing Cognitive Linguistics (I): Cognitive Transform Grammar.

From "Naturalizing Phenomenology" to Formalizing Cognitive Linguistics (II): Grounding and Center/Peripheral Relations.

From "Naturalizing Phenomenology" to Formalizing Cognitive Linguistics (III): Externalism and the Interface Theory of Meaning.

from the linguistic side by, e.g., Cognitive Construction Grammar (Langacker's Cognitive Grammar arguably being intermediary, part formal linguistic and part-philosophical).

We claim however that these extant analyses intending to supplant "logistic" philosophy of language are still incomplete, and there is room to solicit more empirical data, more correspondences with formal and computational linguistic models, and more extensive investigation of cognitive processes *outside of* logicomathematic propositional models. That is to say, the cognitive-grammatic counter-arguments should be more thoroughly integrated with formalizable counter-theories of linguistic structure, perhaps from the perspective of "interface theories" and/or generalized Neo-Davidsonian event semantics (e.g., [15, 16, 21, 22, 32, 59, 79]). Again, though, alas, the analyses and supporting data we have in mind are outside the scope of the current discussion. Here, we will focus on the narrow band within this territory that fits the points of contact between natural human discourse and artificial (e.g., programming) languages.

Having argued above that shared understanding is *pre*-linguistic, we would add that cognitive processing is *also* outside language, so that language proper is mostly a link or interconnection from one to the other: syntax and semantics *start with* joint situation/environment models and compel each conversant to supply new cognitions, somehow altering or re-envisioning these surrounding, with language being the conduit between joint attention and a subsequent re-staging of the environment, both of which are predominantly *extra* linguistic. Linguistic utterances then become figured as intermediate constructions that bridge but largely remain separate from these cognitive subsystems.

One potential counter-argument to this formulation is that fully constructing a linguistic expression (e.g., a sentence), so that it may serve such intermediary role, would therefore appear to be a mental operation outside of actual meaning comprehension because on this theory *extra*-linguistic cognitive processes flesh out meaning proper. This suggests that *as* linguistic, or during the "processing stage" where expressions are *only* linguistic artifacts (not yet subject to extra-linguistic cognitive scrutiny), expressions do not "have" meaning per se, and so are in effect meaningless. But, if so, how can we have coherent linguistic expressions in the first place? After all, what differentiates valid utterances from (logical or grammatical) nonsense is (apparently) that proper expressions are meaningful in some sense. For example, properly parsing a sentence would appear to be a necessary step toward aggregating linguistic content to serve as material for extra-linguistic cognitions. If there is no preliminary meaning, how can we actually distinguish correct parses than ones that yield gibberish, even if they satisfy certain criteria (such as pairing up words based on part of speech)? How can we rule out grammatically plausible but logically nonsensical parses, or filter interpretations that are syntactically reasonable but depend on mapping words to lexical options that are contextually unlikely?

In short, a further complication with regard to "interface" semantics, as we see it, concerns the role of meaning anticipation in syntactic parsing. An idealized view of language might hold that we perceive parse trees first and fill in semantic content later; i.e., that parsing and lexical comprehension are two distinct processing stages.

This might be how one would design an artificial language, but in practice there seems to be at least a circling between the two processes, where appeals to semantic coherence help resolve syntactic ambiguities. Assuming, with link and dependency grammars, that parse structures can be modeled as graphs, most sentences have many candidate graphs that are not ruled out by syntactic desiderata alone (such as parts of speech). Barring outright "eats shoots and leaves" style ambiguity, only one parse structure will typically emerge as the most sensible reading *given* the resulting meaning, but assuming that a full semantic interpretation functions as a litmus test for rejecting less-likely parses would violate the model according to which we mentally form a parse graph *first* and *then* deduce its intended meaning, by filling in the words' lexical (and, as needed, referential/anaphoric) content.

Of course, a simplistic way to address this problem is to envision our minds subconsciously testing many different parse graphs, potentially "reading" thousands of different interpretations of a sentence—giving each a fully specific meaning if possible—and only after all this work scoring one reading as the most likely. Such an explanation, which by stipulation we could never consciously verify anyhow, is not very satisfying (barring strong neurolinguistic evidence that processing really does branch out in parallel to this extent), because it basically just papers over an apparent theoretical gap by vague appeals to unknowable cognitive machinery. A more compelling analysis would provide some model of how the parse-graph space could be pruned *without* full semantic reconstruction, perhaps allowing *partial*, but noticeably limited, appeals to sentence-constituents' lexical (and semantic/referential/anaphoric) interpretations. Dependency grammar, for example, provides empirical data indicating how metrics of word distance and "edge crossings" factor in to parses being more or less likely [20, 24, 25, 36, 61, 78, 97], but these are still only provisional analyses—many sentences' actual parse graphs have configurations that would be deemed improbable based on such metrics in isolation, and even when they prune the parse space correctly, there can still be quite a few reasonable readings that must be further disambiguated.

We will not present empirical data supporting these proposals, but in light of our arguments in Chap. 20, we would like to summarize at least the rudiments of a theory to further refine our models of parse space. In general, we seek to find criteria that would eliminate candidate graphs by considerations *other than* explicit semantic meaning, because with such restrictions in effect one can "quickly" narrow the scope of parse structures that are legitimate contenders to be "correct" readings of each sentence. Part-of-speech data, for example, might play such an eliminative role, since only in specific contexts can a pair of nouns, say, or a pair of verbs, enter into head/dependent relationships. In practice, however, part of speech is fluid; quite a few constructions implicitly treat nouns as adjectives alongside other nouns (like "pet" in "pet store") or treat nouns as verbs (e.g., "to primary" in politics), nouns-as-verbs-as-adjectives ("iced tea"), etc. Graph pruning has to ensure that valid part-of-speech migrations are not misclassified, which again begs the question of how to know what interpretations to reject *without* fully realizing the potential meanings that could be attributed to such interpretations.

A minimal criterion for parse structures being acceptable is that they present sentences which build up to some (relatively) complete idea, although this still should account for partial expressions that addressees mentally complete (imagine someone holding out a glass and saying “could you please?”, implying without stating a proposition related to someone filling the glass). This appears to be a fairly rigid system within language, which can without oversimplification be modeled via mathematical formalisms, such as lambda calculus. Actually, the role of lambda calculus in this context would lean toward type theory—even in untyped calculi, there is a provisional distinction between functions and their arguments, and type checking can be progressively refined as our type models become more granular. So the essential detail, represented in lambda calculus via “beta” reduction, would be how type tuples under the influence of functional types collapse to a single type, and iteratively such reduction may occur over multiple stages (for sub-expressions), yielding, in the linguistic context, the stipulation that (complete) sentences have “propositional” types.

In Chap. 21, we argued that lambda calculi and their variations (sigma calculus, calculi with exceptions, and so forth) can be represented via “channels” and encoded through hypergraphs; this represents a small extension to the concept of type reduction because reductions are type-checked across multiple channels. Aside from that, we can further analyze to what extent semantic information may be encoded in type systems; in the aforementioned [54] dataset analysis for example, we distinguished between different levels of type granularity (“macro-types,” “meso-types,” and “micro-types”), with the idea “intermediate” granularity could provide a parse-pruning mechanism that rejects semantically implausible graphs on bases that are still heuristic and simpler than full-scale semantic interpretation.

Elaboration of our attempted extensions along these lines—via channels and meso types—is outside the scope of the current discussion; we merely want to point out that a type-theoretic approach may offer rigorous models of parse pruning and that the effectiveness of pruning “algorithms” increases to the degree that type systems become finer-grained. This is true both for natural and programming languages, as evidenced by how strongly typed languages support more rigorous modes of static analysis. Type systems that encompass “dependent types,” type states, and/or detailed “type-class” extensions progress further into the realm of encoding semantic information via type attributions; these seem like good candidates for emulating how humans create provisional semantic models of sentences without fully completing all lexical/semantic/referential/anaphoric details. However fine- or coarse-grained our recognized type system, the key structural claim is that type reductions cascade from terminal nodes to root nodes in tree-like (or directed-acyclic) subgraphs and that this represents one universal constraint through which candidate parses can be eliminated.

Alongside such type-theoretic models, we also believe that interword relations along the lines investigated by Link Grammar supply a different and orthogonal class of constraints because sentences appear to have predictable regularities in terms of contexts where specific forms of interword links are permissible. In [67, Chapters 6 and 9], we argued (inspired by work such as that of Lucas Champollion, as well as more generically from Cognitive Grammar and Link Grammar [64, 80, 82, 93] overall) that the most important framework for modeling pair relations as parse criteria is that of "thematic relations," focusing on the theta-roles and grammar cases that link verbs to their profiled constituents (subjects and objects, and then via further detailing the various locative, benefactive, instrumentive, and etc. case forms). Each pairing between a verb and its thematically related associated words or phrases adds a level of detail to the state or event profiled by the verb, as this is postulated by the speaker. The total collection of a verb's thematic relations represents the verb's state/event within epistemic packages, mediated by the speaker's propositional and perspectival attitudes. Parse graphs only make sense if thematic relations sum together in ways that are epistemically coherent. This, then, presents an additional layer of criteria that can be imposed on parse spaces to distinguish credible from dubious parse candidates. We defer to work such as Langacker [51, 52], [46, Chapter 9] for actual descriptions of "epistemic coherence" (or see [76, page 59], [77, page 135], [5, 7, 17, 42, 66, 69–71, 84, 86], with respect to speaker-relative grounding and epistemic contingencies);[3] here, we will simply say that certain lineups of thematic relations are more coherent than others, and this can be grounds for accepting or rejecting candidate parse graphs.

In sum, then, we propose that both type reduction and thematic relations are systems that introduce parse-acceptability criteria, and both systems can operate at once over possible parse graphs to eliminate spurious candidates (while still remaining uncommitted to full semantic realizations). Moreover, we believe these two systems are quite apparent in both natural and programming languages. Deciding amongst competing parse graphs is not necessarily an issue for compilers in the latter case, because most computer dialects do not accept any sort of syntactic ambiguity comparable to natural language; the parse structures of code expressions should be fully determinate from rules that can be enforced mechanistically with compile-time information. However, parse graphs for computer code only indicate how arguments are connected to procedures; there can still be ambiguity in the relation of procedure symbols to the specific implementations that are intended to be called at the relevant code points at runtime. In this context, compilers and/or runtimes have to apply heuristic tests for symbol/procedural mapping, often involving "topological sort" algorithms and metrics grading type coercions. Here a type-theoretic infrastructure

[3] With the caveat that we believe notions such as grounding and epistemics should be read more broadly than *merely*, say, semantics of modal verbs, or figure/ground relations—that indeed the entirety of a linguistic expression is patterned around speakers signaling through semantic and syntactic means their particular epistemic comportment vis-à-vis shared situations. Even if the purpose of a discourse fragment is not to thematize speaker-relative epistemics directly (*I think that.... Surely/Probably..., It looks to me like...*), we cannot help but reconstruct the particularities of our perspectives through linguistic form and communicative pragmatics because the overlap and possible contrasts between our situational beliefs and other people's are intrinsic to their cognitive reception of our speech acts. Such cross-epistemic elaborating, we claim, factors directly into syntax/semantics and serves as the overarching architecture of linguistic form to the degree that it does not just logistically simulate propositional structure; again the dataset article-trio linked in footnote 2 develops these claims in more detail.

plays a pruning role analogous to parse-graph multiplicities in natural language. Also, procedures' pre- and post-conditions serve as potential criteria on graph edges structurally equivalent to thematic relations in natural language: consider the idea of overloading a procedure operating on two lists (or vectors, arrays, stacks, queues, etc.) by implementation bodies differentiated via type states: two empty lists, one empty list, two same-sized lists, and so forth.

With respect to "interface theories of meaning," the type-reduction and thematic-relation systems offer a plausible model of how parse graphs can be selected even while parsing is understood to be functionally prior to cognitive "scripts" that yield the full-scale semantics for linguistic expressions. Although it would take more work to make this proposal rigorous, we believe that the simultaneous domains of type reduction and thematic relations serve as structural criteria in both natural and programming languages, yielding constraints on syntactic and/or semantic processing that allow some decisions to be made—parse-graph pruning in the former case, function-symbol resolution in the latter—that foreshadow but do not depend on complete semantic information.

24.2.2 From Natural to Computer Languages

While similarities between programming and natural languages are in general pretty superficial—to reason otherwise would minimize the subtlety and indeterminacy of human speech, which is completely different than the mechanistic engineering of software code—we think these systems in particular (type reduction and thematic relations) are structurally complex, relatively speaking, and operationally non-trivial, revealing interesting overlaps between human and computer languages notwithstanding their ontological contrasts.

If these are shared structures across both forms of language, one might reason that similar formations would likewise apply to "robot" languages, to the degree that—as we hypothesized above—an optimal robot language would lie at some intermediate distance between human speech context-sensitivity/indeterminism and computer code's mechanism.

If it is true that the proper semantics for computers and analogous computationally engineered artifacts (such as robots) is *procedural*, so that the "meaning" of an expression in "language" (however rigid or artificial) is the procedures which it designates or which supply its implementation, then analyzing this semantics requires explaining the environment within which procedures operate. This does not mean only one single procedure—the context that obtains prior to its starting and subsequent to its ending—but from a broader vantage point the settings in which we can speak of groups of procedures executed in sequence, forming an aggregate unit of computational activity. As we pointed out at the start of Chap. 20, most software applications are designed to enter passive states awaiting user input. It is only after a user-driven *event*—due to *gestures* such as clicking a mouse button or tapping a keyboard letter/symbol—that applications launch a series of procedures responding to the user's (presumed) intent. Usually, the series concludes with some visible change to application state (typically something the user sees on-screen) and then the initial passive state is re-entered. As such, the canonical context for software procedures on aggregate is bounded by the *initial* user action (setting parameters on how the application reacts) and the eventual (usually on-screen) content presented to the user. To the degree that computers have "semantics," we should analyze such semantics in terms of whatever intermediate processing connects the start of these sequences (user gestures) to the end (user-visible content).

In the case of robotics, the details are similar, except that most robots do not primarily function merely by presenting information on-screen (unlike personal computers or even *en situ* computer-like devices with touch screens, e.g., interactive transit displays). A GUI console can be one part of a robot's equipment, but more broadly we expect robots to fulfill user intentions by moving themselves and/or external objects. Although robot actions might be preprogrammed (and they can be semi-autonomous, deciding some actions on their own), to sustain the analogy with other computing systems, let us consider the case where a robot launches a series of actions in response to an explicit request by a human user ("lift the yellow box" or something similar). In that case, the robot should respond by moving and/or entering a configuration where it may physically comply.[4] As with GUI front ends to traditional applications, the robot's procedures propagate in response to user input and have a well-defined end point, but here the terminus of a procedure sequence would (primarily) not be something visible on-screen, but rather a robot changing its location and/or configuration in physical space (possibly affecting some external object in the process).

The units of GUI presentations are "application windows" (visible screen areas), inside of which are "controls" such as buttons, labels, scroll bars, magnitude indicators (consider how we drag an icon to partially fill or empty a bar, to zoom in or out on a text or graphic display) and dividing lines; within those controls, in turn, can be text (including numbers) or other displays of textual and/or quantitative information. Therefore, the "end state" of a series of procedures, at the application level, would be a change in visible GUI display through some of these constituent elements. Likewise, in the case of robots, the components of a robot's "configuration" would be angles of gears, extenders, wheels, and whatever mechanical devices robots utilize to move and reposition themselves. At the culmination of a procedure set enacted in response to user input, then, the robot will in general exhibit a different state with respect to their location and the orientation of their arms, graspers, mounted cameras, or whatever other mechanical equipment they have onboard. We can see these physical gadgets as analogous to windows, controls, and text/number displays on a GUI.

[4] In some cases, the user input might be provided in advance, with the expectation that the robot only responds to such information actively under certain circumstances; for example, robots could be instructed to perform some task only after they sense a particular kind of change in their environment.

Based on that analogy, deciphering the semantics of reactive GUI programming can be an indirect pathway to formulating semantic models appropriate for robots, because the constituent parts of GUIs provide structural analogs (with respect to the origination and telos of procedure sequences reacting to user input) to robots' movements and configuration. In both contexts, the "semantics" of the computational system is determined by how procedures react to user actions so as to yield state changes (to GUIs' visible contents and robots' orientations, respectively) that are "correct," or consistent with their initial intent, from the user's point of view ("correctness" in this sense taking the place of being "meaningful," for the purpose of semantic formalization).

Given the analogy of GUI state to robots' orientation, an analysis of GUI structures (intended to be included in an overall procedural semantics for computer code), which will be the focus of the following section, can perhaps serve as a starting point for further work in robotics and other areas that generalize and extend the familiar desktop-computing environments.

24.3 GUIs, Robots, and Environments

Current paradigms in desktop-style GUI engineering have stayed relatively consistent since roughly the start of the century. The heyday of GUI applications (often called "rich client" or "thick client" interfaces, see for instance [98] or [68], to distinguish them from Web applications—"thin" clients—which need to run in an Internet browser) largely coincided with a period in programming methods where object-oriented techniques were predominant. Neither object-orientation (OO) nor desktop-style software is canonical to the same degree as two decades ago, given the emergence of popular programming languages that de-emphasize OO structures on the one hand, and Web or smart phone apps as an alternative to desktop environments on the other. However, for many categories of software—for instance, scientific computing, or for that matter the IDEs (Integrated Development Environments) that programmers use to write other programs (of all varieties)—desktop technology remains the only feasible option. It is impractical to engineer applications that work in Web browsers or within the limited screen size of a smart phone, once applications get past a certain threshold of functional and presentational complexity. For this discussion, then, we will refer to GUI applications as, canonically, native-compiled software components that run on a desktop computer or a similar environment (as opposed, for example, to Web applications that are viewed in a browser and run by code interpreted through the browser rather than compiled to native machine instructions).

Object-oriented techniques remain the predominant method for GUI development in this sense, largely withstanding the familiar criticisms leveled against the OO paradigm in other contexts (such as database or server-side programming). It is reasonable to conclude that OO is an effective modeling tool for GUI functionality, one that cannot readily be replicated by functional programming (FP) languages, for example (to prove the point, consider that FP bindings to GUI software libraries, particularly cross-platform ones—i.e., intermediate libraries that allow FP code to construct and manipulate GUIs—are notoriously buggy, poorly maintained, and/or difficult to use; for example, compiling and merging these libraries with an application's overall code base tend to be more difficult with FP bindings than using OO libraries written directly in languages such as C++). So, again narrowing the focus of the current discussion, we may consider general-purpose models of GUI elements and structures to be preeminently object-oriented.

Accordingly, GUI elements are canonically objects with state (FP languages often are designed around "stateless" values, but in the GUI context recognizing mutable state appears to be the most natural digital encoding of the phenomenon being programmed, namely visible windows and parts thereof wherein applications present information to users). There are several different facets of GUI object state, which may be grouped into three overall categories:

Visual Style Effective GUI design often depends on conscientious choices for colors and other visual details. This is particularly true because the visual appearance of GUI elements includes not only how they appear in their "normal" state, but also how their appearance is subtlely modified as a cue to users trying to orient themselves vis-à-vis applications' interactive features. For example, GUI buttons and clickable spans of text almost always provide some visual indication that these are not only static elements but can respond to user events, such as clicking the mouse while the cursor hovers over a button. To help guide users, buttons (and text, when appropriate, plus other clickable GUI controls) take on a different style, usually with a variant color scheme, when the cursor "hovers" over the element (sometimes called a "mouseover" or "rollover" state)—this visual cue shows that the control is indeed clickable and that the mouse is currently positioned on the item, so that clicking the mouse (via a physical mouse button) is understood to be an action for "clicking" the (virtual) GUI button/control. Once a button *is* clicked and while the response is progressing, a button may appear in a "down" state, which has a style distinct from both normal and rollover.

Alternatively, some buttons (or other controls, such as checkboxes) are not primarily designed for requesting a specific action, but rather to set some flag or indicator (e.g., a checkbox in a GUI area related to text searching would be set when users want to do a case-sensitive match—in which a capitalized word for instance would not be considered a match of a lower-case search phrase—whereas the default setting, indicated by an unchecked box, would be case-insensitive). In these contexts, there would be a visible contrast between "up" (or "unchecked") and "down"/"checked" states.

Moreover, buttons and other GUI elements might sometimes be "disabled," meaning that one cannot interact with them according to normal patterns because (given present application state) their associated actions would be impossible or illogical. For instance, a button which saves a file (i.e., signals the application to enact such functionality upon being clicked) would not make sense if there were no file open. In that case, a disabled state and its style would indicate to the user that the button is not presently able to accept interactions (it is common to use gray colors to cue such states), but the styling of these states should not obscure the control's primary purpose when it *is* enabled. A disabled "save" button, say, should be given a style where users can still clearly discern the word "save" (or whatever icon replaces it—it is a widely recognized convention to employ graphics of a floppy disk in this case) even while indicating that there is no possible save action available at the moment.

As illustrated by button states, most GUI controls have multiple different visual styles, and their changes are intended to cue users on how the controls may be utilized interactively. For this reason, GUI styles have to be designed attentively because each control state should be visually distinguished from others, but also the styles should not interfere with the control's primary functions. For example, a button's descriptive label/text (and/or its icon) must remain visible even while its background and/or outline color (or other style-state effects, such as a color gradient) changes. Style rules for GUI elements in general need to adopt color schemes in which coloring is relatively nonobtrusive (colors should not be overly dark or high contrast) but still allows components to show a variety of different states.

Many applications do not use a single style scheme, but rather allow their styles to be adapted for different users and/or different platforms. Most operating systems have preferred stylistic concepts, so that applications on a Mac tend to follow different conventions than those on Windows or Linux. Consistent visual patterns from one application to another help users apply their familiarity with existing programs to learn new software more quickly, so applications generally try to conform to operating-system standards. As such, cross-platform applications often employ multiple style schemas, depending on the kind of computer where the application happens to be running. In addition to these operating-system variations, applications may also recognize flexible styles that can be adapted for different users, in case someone wants to view larger text, say, or wants to switch between different (human) languages for GUI labels.

Information Content Most GUI elements exist to show data, so their state is obviously influenced by the nature of information they seek to present at any given time. In the simplest case, these data are in textual form, or is numeric data presented textually. For example, consider a label that shows the name and folder path of a file currently being viewed; when a new file is opened, the text shown on that label will be updated. Here the relevant data are presented just through a line of text. Or, a digital map might have labels to represent the current latitude and longitude coordinates under the cursor. These are numeric values but can be represented for users merely by showing the numbers in character form.

Other numeric displays, however, might represent magnitudes more pictorially. Temperature, for example, may be expressed by coloring a bar at a height proportionate to degrees Fahrenheit or Celsius; the bar filled higher to the top would indicate higher temperatures, and lower bars show colder temperature. Color could also reinforce these visuals, with warmer temperatures represented via reds and yellows and near-freezing conditions cued by blue or purple. For some GUI controls, numeric values might be conveyed by visuals mimicking real-life physical gadgets, so the volume on an audio player could be shown via circular indicators mimicking the knobs on a stereo.

One relevant contrast between numeric and textual data is that the latter can be long and open-ended; GUI controls designed to show paragraph-like text could represent content spanning many lines. On the other hand, numerical indicators typically show only one or two quantities, particularly if they are gadget-like displays (relying on bar/line heights or angles to connote magnitude, rather than printed numerals). Multi-part numeric data would then normally be represented via collections of different controls. Where appropriate, quantitative datasets could also be shown via charts or diagrams, either static images or aggregate components with multiple independent controls.

Aside from textual and numeric data, some GUI controls are engineered to show 2D images, or even interactive 3D graphics (in the latter case, users can modify 3D scenes—moving or rotating the display to show different angles onto the scene, which creates the illusion of three dimensions even though the graphic at any moment in time is confined to a 2D computer screen). Although we can speak of graphics displays as presenting "information," this is a more informal usage if a control shows, for example, a photograph. However, images can also be generated on-the-fly to visualize data structures, in which case we could say that image graphics (at least when showing tools such as charts, graphs, or diagrams) represent one strategy (separate from text and indicators) for data visualization. Image data, then, occupy a gray area where we can sometimes speak of controls simply showing users the contents of an image file and sometimes instead treat image displays as vehicles for presenting structured data analogous to other kinds of GUI controls.

To some degree, this can also be true of text displays, when for example a control is intended to show contents of an HTML or PDF file. The PDF format, in particular, is engineered so that each individual document page becomes treated by applications as if it were an image or photograph (PDF text is not a data structure; pages instead are rendered and provided to GUI front ends via mechanisms similar to photo-displays). On the other hand, PDF files also often include word data so that applications can perform searches inside PDF text (although such data tend to be imperfect, with situations such as hyphenated words, ligatures, and foreign accents causing transcription errors; in short, PDF does not directly provide an accurate machine-readable transcript [8], [94, page 8]). Text in HTML is similarly largely static, but users can select character ranges and activate hyperlinks in both formats (this is true for any HTML document and most PDFs). Controls showing full text files, then, analogous to image displays, are in some sense intermediate between organized information displays that directly map data structures to GUI aggregates, and raw file viewers that only in an indirect sense represent "information," instead primarily rendering file contents into a static visual form.

The more canonical case of GUI controls showing structured data is one where GUI screen areas are subdivided into units that represent individual data fields on a fine-grained scale. For example, tabular data or CSV (comma-separated value) files may be presented via spreadsheet-like controls, which are subdivided (by rows and columns) into cells that typically include one single numeric (or short textual) data point. Information shown in these sorts of displays might originate from files—insofar as such GUI controls access files they need for loading data into computational structures first, and then map individual parts of these structures onto parts of the GUI display—but the correlation between files and front-end controls is therefore granular and systematic, whereas image viewers simply calculate how to render pictural data via on-screen pixel areas and display the resulting graphs visually, without constructing distinct areas or sub-windows to show specific parts of a loaded file. Text displays can be between these two alternatives. They are largely unstructured (like pictures) but portions of text may be isolated as if they were interactive windows (such as hypertext links) and users are able to select character ranges (e.g., for copy and paste) in which case the screen area spanning selected text needs to be treated as an interactive window in some sense.

GUI Controls as Containers for Other Controls Because controls whose explicit purpose is to render individual data units tend to handle relatively small pieces of information (e.g., single numeric values), multi-part data structures have to be displayed through higher-scale controls that group together such more narrowly focused controls. Partly, containers along these lines serve to aggregate smaller controls in an orderly manner; for example, a display of several numeric fields could be arranged with a single key–value pair on each line of a multi-line list (tables and spreadsheets, grouped by both rows *and* columns, are a more complex organizing motif). GUI containers do not need a fully symmetrical layout; often smaller controls are arranged in table-like patterns but with varying numbers of elements on each line/row, depending on the amount of space each element needs. The important detail is that users can visually isolate and comprehend specific points of information, which is easier if GUIs are visually balanced and logically arranged. Grouping related controls—and leaving spaces (or drawing lines) between different such groups—is also a useful layout convention.

Apart from just organizing controls, GUI containers can also compensate for limited screen real estate. GUI windows often have more data available than can be shown on-screen all at once. To compensate, containers could use various tactics to show some contents and hide others, such as scroll areas (where users manipulate scroll bars to alter which controls are visible and which are hidden); tab/notebook windows where entire (sub)windows are "stacked" or lined up so that only one window is visible at a time; dialog boxes (which are typically temporary displays that become visible to show the user specific information and then are closed); "wizards" (in effect, multi-page dialog boxes); and "collapsible" displays, whose parts can be toggled between visible or hidden states. A canonical example of the latter is lists of files in folders, where individual folders may be seen in an "open" state (such that folder contents, including other folders, are listed) or "closed," and likewise for subfolders inside (open) folders, and so on. These kinds of controls are sometimes compared to an accordion because as folders and subfolders open or close their screen area expands and contracts, respectively, creating a visual effect vaguely reminiscent of accordion leaves' folding patterns.

There are several notions of GUI controls' *state* covered by the above breakdown. Controls can have *functional* states, reflected in style changes, such as a button in "normal" or "hover" modes (or also, say, "down," "checked," and "disabled"). Most controls' state is also determined by the data they present (e.g., for single-value numeric indicators, the specific quantity expressed by the control at a given point in time). And, for containers (GUI controls that group together multiple smaller controls), state is also characterized in many cases by the interplay of which elements are currently visible, and which hidden.

The forms of GUI state just enumerated refer specifically to controls' features that are visible to users, so that state changes convey information. This includes specific data points which a control is designed to display, such as a numeric value; it also includes hints guiding the user to interoperate with a GUI most effectively. Mouseover effects, for instance, communicate which controls can be "clicked" and which not. There are other, more subtle forms of cues as well. For example, when the mouse hovers over the corner of a resizable control (so that "dragging"—moving the mouse with the left button held down—adjusts the control's size by pulling the corner inward or outward) typically the cursor, which marks the current mouse position, will switch to a different icon (one drawn to imply something getting larger or smaller); there might be extra coloration or thickened lines at that corner too. Similar patterns apply to other "drag" scenarios, such as moving a divider line between two parts of a "split" container, where different controls lie on either side of the line; moving the line resizes both parts of the container (one gets larger, the other smaller) potentially hiding or unhiding different contents. Another pattern is "drag and drop," typically involving icons being moved (figuratively "dragged" by the mouse) from one container to another: in some contexts, the receiving container takes on a colored background to convey that it is possible to drop an item on its interior.

Some visual hints are explicitly designed to inform users about applications' features, or the status of current operations. For example, many front ends feature "tool tips" where hovering the mouse over a control causes a box of explanatory text to temporarily appear, explaining the purpose of that GUI element. Tooltips serve both as didactic guides and as further visual markers that the relevant control accepts interactions. Other examples of indicative GUI features include progress bars (which are usually visible only while an operation is ongoing, such as downloading a file); message boxes (presenting users with brief informative text, such as declaring that a download has completed); and status bars (typically single-line controls with text related to recent activity, e.g., "download complete": unlike a message box, which might also inform the users of something like a download completing, status bars are usually permanently visible in the front end). For container controls that alternately show and hide inner contents, visual cues sometimes clarify what material is currently visible: e.g., a "notebook" container, which has multiple tabs each showing a different window, will color and outline the tab corresponding to the currently open window with a different (more emphasized) style than the other tabs (whose windows are invisible at present). Details might also be conveyed by discreet changes to textual data: for example, if a file has been modified and not yet saved, a label or status bar showing the file name will often append a small marker, such as an asterisk, to the file name, reminding users that they may need to save the file.

Style conventions also inform how GUIs help users follow the current mouse position. In general, the mouse is tracked with a cursor, which travels across the screen when users move the mouse (or run their fingers against a touchpad); a mouse click (or tap against a touchpad) is understood by the application to signal the user's intent to execute a "click" gesture at the screen position where the cursor currently hovers. The cursor will then take on different icons depending on how the click would be processed, given the control under the mouse position. An example mentioned above would be a "resize" icon when the mouse hovers over a corner (this works also for some controls' edges); cursors tend to be given an arrow-like icon in these settings. When hovering over text, on the other hand, cursors usually take on shapes reminiscent of a capital letter "I."

In some cases, the on-screen mouse position can be correlated with some physical or empirical coordinate system—a good example is digital maps, where positions inside the map correspond to latitude and longitude numbers. Here it is common to print the geographic coordinates next to a corner of the map, so that the numbers change as users migrate the mouse around the map view, which both aids users if they are trying to find a specific latitude/longitude pair and helps visually reinforce the idea that mouse positions map directly to geospatial points. A similar pattern in one dimension would be using a mouse to advance or rewind an audio or video player, wherein a numeric control shows the time elapsed at different points relative to the file's start (the convention being to pair audio/video controls with a line bar and position indicator such that pulling the indicator right/left implies forward/backward in time, respectively). Finally, some controls display graphics or data charts/diagrams against a background of orthogonal gridlines (similar to "graph paper" for drawings), or sometimes just the lattice points at gridline intersections. In these contexts, mouse position may be indicated by highlighting the gridlines which a mouse passes over as it moves.

One takeaway from this overview is that well-engineered GUIs are highly interactive. There are many conventions that govern how front ends should present subtle visual cues and state changes to assist users when interacting with the application. Users may not appreciate the full scope of these effects, becoming habituated to "responsive" features and perhaps absorbing them at only a subconscious level. However, front ends that fail to implement expected interactive patterns—where GUIs are not sufficiently responsive in terms of giving users frequent feedback on current state (such as mouse position) and interactive possibilities—would almost certainly be experienced by users as erratic and confusing.

Most of the front-end patterns mentioned here are probably familiar to people who work on desktop-style computers, but it is nonetheless useful to outline them explicitly, precisely because well-engineered responsiveness effects will be subtle and non-distracting. When we appreciate the full scope of interactive patterns, we can perceive the complexity of standard GUI programming: a well-featured front-end application will typically have a large collection of associations and interconnections between GUI controls, and mappings between control state and visual styles/interactive cues. Systematically engineered applications will carefully document these interconnections and implement the correlated front-end functionality via a consistent set of coding patterns.

As discussed in this chapter (and also touched on at the start of Chap. 20), application functionality may be analyzed in terms of operations and procedure sequences that are, in effect, bookended by user actions (which invite responses from the application) and by visible GUI changes (changes in the state of one or more controls, communicating to users the results of the actions they requested). Reviewing forms of GUI state serves to explicate the second half of this equation. We can also examine user gestures, constituting the first half of the "equation," insofar as gestures signal user intent, which leads to application responses, and finally to GUI changes. Gestures are, indeed, closely connected to GUI state because a user's actual (literal and physical) action—such as pressing a mouse button—needs to be interpreted in light of current GUI state, to be seen as an "action" in the "virtual" context of a front-end display. We talk, for example, about "clicking" a GUI control: in reality, the "click" happens on the physical mouse, but the presence of cursors and mouse tracking permits this physical movement to be read as an interactive gesture within the scope of a software component. In this sense, the "semantics" of user actions builds off of GUI state analogously to the semantics of GUI visuals conveying information, so that the two may be analyzed together.

24.3.1 The Semantics of GUI Control State

To briefly clarify terminology: a user *gesture* refers to some physical action that is *interpreted* as a specific kind of request, input, or instruction. The gesture is interpreted as an *action*, in the sense that gestures signal actions according to patterns that reappear in the logistics of how people interact with software (patterns that users recognize). As a canonical example, the physical gesture of tapping a mouse button is interpreted as the contextual action of "clicking" a GUI button or other GUI control. The relation of gestures to actions can perhaps be compared to that between audible words and their intended lexemes.

Mapping gestures to actions depends on GUI state, such as the current mouse position. Accordingly, GUI state overall provides the structures against which individual actions are defined: an action of *clicking a (GUI) button*, for example, is meaningful in the context of front-end environments where that specific button is among many controls that could be "clicked," and given that it is enabled and visible.

Some GUI coding contexts will refer to "actions" as akin to "requests" that may be cued by more than one user gesture. For example, instructions to save a file could derive from clicking a "save" button, but also selecting a save option from a context menu, or hitting a combination on the keyboard (usually "control"-plus-"s"). For more rigorous discussion, we suggest referring to actions in this multi-indication context as "requests," so that the various gestures signaling one common request (like "save file") each have distinct actions. In any case, all actions then get routed to interpretations of users' intentions. We may adopt the term requests for these interpretations in general, so that some requests are cued by one single action and some are targets of more than one possible action. Current GUI state determines both how *gestures* are interpreted as *actions*, and how *actions* are interpreted as *requests*.

By design, applications are engineered to *respond* to such requests, and GUI state (alongside application state overall) contributes to how software models the request to be handled. Consider a notebook control that contains multiple windows, each displaying an editable text document: when the user requests a save, the application would infer *which* file should be saved based on which document's window is currently visible. This is an example of GUI-specific state. Whether or not the document has been modified (affecting whether a save is necessary)—and the file path that would be overwritten—would be an example of application-state proper (because these details are not determined by the visible appearance of GUI elements; although the GUI can convey to users such information as whether the currently viewed document is modified and unsaved).

Some user requests may be handled in different ways depending on specific properties of application state or the surrounding computer environment. Consider a request to save a file: if the file already exists, saving the updated version involves replacing the contents on disk; in some cases, the current user is not authorized to make such a change, so the save action (with the current file name) cannot be completed. On the other hand, if this is a *new* file, the user needs to assign it a name. There are, then, at least three possibilities: the file can be overwritten with its current name; the file cannot be saved with its current name but may be saved with a new name (maybe in a different folder), so that there will be two different versions on disk; and the file cannot be saved until a name is provided, which will then be one single version of the file. How the application fulfills the user's request will be different in each of these scenarios. Some of these differences might be noticed by users: if the file needs a new name, users will generally be shown a "Save As" dialog configured for choosing file/folder names, whereas if a file is overwriting a prior version users might see a message box confirming that they intend to do so (maybe with an option to make a backup of the older file).

The point of this example is that responding to user requests demands a rigorous calibration of GUI and application state. This is one reason why object-oriented methods predominate in front-end programming; the logic of such state correlations is typically simpler to reason through via OO paradigms than alternatives. Indeed, GUI controls are best encapsulated as objects-with-state, and because GUI controls are functionally interconnected with many other application elements, similar OO concepts tend to be most effective on the application side as well. For example, if a notebook window shows one document, the window itself corresponds to one GUI object, from the front-end point of view; on the application side, the actual document visible *in* the window might be a different object, encompassing data fields such as file path, file type, length, edit history, and permissions (e.g., the identity of the user who created the file).[5] User requests relative to the *GUI* object would be carried out, by the application, in the context of the corresponding application object. These functional interconnections are best managed by maintaining association between application and front-end objects (which implies an overall OO system unifying these two contexts).

24.3.1.1 Extending Object Orientation

One might argue, however, that there are details in application/GUI object systems that are inadequately addressed even by conventional OO models and call for extensions to the OO framework (which is a different matter than replacing such a framework with something like FP methods oriented to stateless values).

Here is one example: in some contexts, a front-end display will include two (or more) GUI controls that are styled and positioned so that the user views them as an integral unit. A simple case is a button that has both an icon and a descriptive text label; this arrangement could be implemented with a label control placed next to a button control, but configured so that the label appears to be a logical continuation of the button itself. In this setup, some facets of GUI state should be shared between the two controls as if they were one single element. For example, if the button is disabled (causing it to adopt a new style), the label should be restyled as well. Or, consider tooltips, as mentioned earlier—explanatory text that temporarily appears near controls when the mouse hovers over them. If users are to experience a button and a label as functionally integrated, this pattern is reinforced by them sharing a tooltip, which means that the same tooltip content is applied to both controls.

Consider a procedure to define this tooltip text. Ideally, the same procedure would act upon both objects because we want to reinforce in code the idea that two controls share some state details, including their tooltips (we do not want this correlation to be dependent on two different procedure calls, partly because programmers may forget one of them and partly to remind those maintaining the code that the two objects are interdependent). In OO systems, the basic pattern is that procedures that affect the state of a specific object are implemented via "methods," in which a specific object is singled out as a distinct form of input value, sometimes called the method "receiver" (in languages such as C++, C#, and JAVA, "receiver" values are designated with the special keyword **this**; elsewhere the most popular name is **self**). The problem with this convention in a context where procedures are intended to synchronize state between *two* objects is that it seems artificial to choose one or

[5] Sometimes deciding whether to qualify state as GUI or application level can be a gray area: consider undo/redo history. Should user actions that can be reversed be tracked at the GUI level, because these actions are signaled by GUI actions, or within the application logic, because requests are managed by application code? This question determines whether undo/redo history should be categorized as part of GUI or application state.

another as "the" receiver—a more fluent paradigm would be to allow *two* receivers, in the same way that procedures can have two or more normal input arguments.

In Chap. 21, we mentioned the idea of grouping procedural parameters into "channels," including a "sigma" channel that can generalize receiver objects to object lists (derived from the "sigma calculus," which attempts to model OO methods by extending lambda calculi [1, 2, 18, 26, 39]). This would be one way to tweak OO systems—permitting multiple receivers. There is no technical reason why programming languages could not support more than one **this** value; by analogy, most languages only recognize a single procedure *return* value, but there are outliers (e.g., Common Lisp), and generalizing returns to multi-value lists does not present special syntactic or implementational difficulties in those contexts.[6] By way of reference, the code accompanying this chapter includes some functionality that emulates "sigma channels" in ordinary (albeit unconventional) C++.[7]

Apart from multiple method receivers, a more substantial potential departure from conventional OO suggested by GUI state involves user actions and corresponding "signals." A common pattern in GUI programming is to generalize the idea of procedures calling one another to a model wherein procedures can post and/or respond to signals; such a system is usually referred to as a *signal/slot* mechanisms, where signals can be "connected" to slots. Rather than one procedure calling another directly, in this architecture, the first procedure "emits" a signal, which the second then "receives," having the effect that the second procedure gets called, with parameters (if any) provided by the "sender." Unlike direct inter-procedure calls, these connections are routed through an intermediary component, and they might be dynamically reconfigured: signal/slot connections may be newly created or dropped, while an application is running. When a procedure emits a signal, there is no guarantee that any other procedure will implement a "slot" that receives it (but also *multiple* procedures can potentially connect to the original slot). Compared to direct inter-procedure calls, the two (or more) procedures involved are thereby less tightly connected.

Programming environments that employ signal/slot connections often use this technique to handle responses to user gestures. After a user clicks a (GUI) button, for example, an object representing that button broadcasts a "click" signal, which in turn is handled by procedures to initiate the application response. If, say, the button in question represents requests to save a file, there will be some procedure whose purpose is to wait for "save" signals and then initiate the application-level process of actually saving the relevant file. The slot thereby connected to the "save" signal only is responsible for initiating save operations; it does not consider where

[6] Disambiguating methods called on "multiple receivers" may prove more complicated when two **this** objects are of different types—which class should be searched for corresponding methods?—but compilers could adopt similar rules to those used when matching procedure names to functions wherein argument lists have multiple types.

[7] See https://github.com/scignscape/dog-cognition-research. This repository was constructed specifically for Chap. 23, but we are using it to host content for other chapters belonging to the current book section (Part 5) as well.

the save button is located on screen, or its GUI state and state changes (mouseovers, tooltips, etc.). On the other hand, the button object would track these GUI-specific details but not get involved with actual save operations. Signal/slot mechanisms are as such (as this case study suggests) valuable tools for enforcing "separation of concerns," where each procedure and each object have a relatively narrow set of responsibilities. Signals and slots prevent objects from becoming tightly coupled because signaling objects can be designed in isolation from procedures that receive (and act on) their signals.

Although a powerful pattern that fits organically within OO paradigms, signals and slots are not directly supported by the OO model, and few programming languages have intrinsic signal/slot features. In C++, several libraries emulate the signal/slot patten. The QT framework, which is both a GUI library and a general-purpose technology for application development, supports a coding style wherein signals and slots function essentially as native language features—however, this is only possible because QT implements an extended compiler workflow involving pre-compilation parsers and code generators, in effect implementing an extension to the C++ language itself.[8] Because most OO languages do not intrinsically support signal/slot connections—that is, they cannot be expressed directly through the language's syntax and semantics, but rather have to be approximated by code libraries (potentially including language extensions through code generators)—this pattern is for all intents and purposes an extension to OO that is not part of the basic OO model.

One further detail concerning signal/slot systems involves properly aligning multiple parameters of GUI state that should adhere to certain design regularities—for example, different lengths or magnitudes in the front display preserved in a constant ratio. Consider a scroll area showing an image: if the image's size (at its current zoom level) exceeds the dimensions of the scroll area, then the visible part of the image is "clipped" and users could scroll left/right and top/down to alter this visible area. On the other hand, if the image size is *less* than the scroll area, it is conventional to center the image so that left and right margins, and likewise the top and bottom, are balanced. If the image is resized (by zooming in or out), its top-right coordinate therefore should be recalculated. Similarly, if the scroll area's viewport is resized because the user expands or contracts its surrounding window, the image itself would either have to be resized proportionately, or else moved to a

[8] More recently, parts of QT's "meta-object" system have been re-implemented to work with inline code annotations (attributes), which are more of a C++ standard than the code generators which QT has traditionally used. However, attributes themselves are not consistently handled by C++ compilers. The QT example actually shows how features initially implemented orthogonally to a language's own formulations can help inspire evolution in the language itself. That does not, however, change the fact that even a hypothetical cross-compiler and standards-compliant signal/slot framework would then just be utilizing attribute techniques to implement reactive programming idioms within (in effect) a code library, not as a fundamental extension to C++ per se. How QT allows programmers to employ signal/slots and other meta-object capabilities via de facto special keywords and syntax suggests that signal/slot concepts actually belong to the realm of full-fledged enhancements to the underlying language, which is the point of the current example.

different top-left position, to preserve the margin's desired left/right and top/bottom symmetry. In this example, there are several different numbers that must be tracked in consort: the overall window's size; the width and height of the scroll area's viewport (its visible screen area); and the image's position, dimensions, and zoom factor. Changes to any one of these measures may need to trigger changes in some or all of the others, to ensure that the GUI continues to feel visually balanced.

Because multiple quantities are involved, orchestrating the proper symmetry in this example may require co-ordinating multiple signal/slot connections. For example, an object representing the main window might emit a signal when it expands or contracts. Most GUI systems employ "layout managers" that calculate how controls should be oriented relative to application windows, so that front ends can have a roughly tabular, grid-like, or top-down organized layout, with controls that need a fixed amount of space (such as labels whose text does not change) remaining constant, while others gain or lose space in sync with available screen area (scrollable containers, for example, could be downsized because users expect to scroll around their interior anyhow). Layout managers will therefore tend to automatically resize certain controls, which can potentially trigger new signals: a screen area resizes because its surrounding window does, and both end up producing signals broadcasting their new state. Code responsible for the image display accordingly should keep track of either or both signals and modify the image's own size and/or on-screen placement accordingly.

Quantitative changes such as window size may also trigger other kinds of GUI state changes. For example, if an image is *smaller* than its scrollable viewport, then scrolling per se is not possible; this situation is typically cued via scrollbars being visible but grayed out (users perceiving the scrollbars, even if they are disabled, reinforces that they *will* become active when necessary). Then, given either the image zooming in or the viewport shrinking, once internal scrolling becomes necessary, the scrollbars should be enabled, taking on a new visible appearance. Resizing and zooming may therefore both affect the scrollbars' styling.

For a related example: when an image is *first* opened many picture viewers resize the main window so that the image can be seen, in its entirety, in its native size. At this point, the scroll area and main window dimensions would be calculated on the basis of the image; the user can then resize the window if desired. However, *after* the image is loaded, ordinarily the user is free to expand or contract the window irrespective of the image size (scrolling being available to compensate for when the window gets too small to show the whole image). There are, accordingly, two different scenarios for how the relation between window size (and by extension a scroll area's viewport size) and image dimensions should be calculated. Application code needs to keep track of GUI "history" to properly separate these different contexts. For example, if the main window is resized immediately after (and because of) a new image file being opened, the window's resize signal should be processed differently than a similar signal emitted due to users manually adjusting the window (typically by pulling at one corner or by clicking maximize/unmaximize buttons).

The point of these examples is that signal/slot connections often do not occur in isolation, but must be coordinated to maintain desired style and layout conventions in the front end. Often these involve groups of interrelated parameters reflecting states from multiple different GUI controls. GUI state can be guided by sets of "rules," such as that an image top-left corner must be positioned and repositioned to ensure balanced margins, or that scrollbars should be visible but disabled when image size is *less* than the available viewport area, but automatically enabled when the former expands to beyond the viewport dimensions.

In the image-view case, there are at least four quantities or coordinate pairs that are numerically interdependent: the image width/height; its top-left corner position; the viewport dimensions; and the image zoom. Most programming languages do not have an intrinsic feature for sustaining predefined mathematical relations among two or more independent variables along these lines—there is no way in code to declare interrelated variables such that changes to any one (or more) value will automatically propagate to the others. The closest approximation might be to group all these values into one data structure and then make any changes to specific parameters via procedures specific to that structure, which can be engineered to update the other fields as needed. In that case, attributes such as an image's size would not be magnitudes (or pairs thereof) directly, but would instead be handles onto specific parameters held within the coordinated data structure (one added complication is that while that structure's fields are interrelated in the sense of demanding specific mathematical invariants, they are also properties of different controls' states, so they need to be properly classified vis-à-vis the control they are associated with).

Orchestrating these sorts of interrelationships among multiple objects and data fields is a common problem but without a canonical design pattern, in particular in the OO context (this is one area where functional programming has produced somewhat more concise formulations, that OO could perhaps borrow). Therefore, another kind of potential OO extension relevant to GUI state would be supporting groups of interrelated parameters, distributed across multiple objects but bound by stipulated invariance requirements. Such features would overlap with signal/slot systems because signal/slot connections are sometimes defined specifically to ensure proper inter-parameter relationships (by receiving signals notifying change in one interlinked value and then updating the others proportionately).

From a practical point of view, these examples point to OO extensions that could potentially be incorporated into new language standards. More theoretically, we can also study how such extensions add representational features to OO models, enhancing the OO repertoire for capturing the state and interrelationships of complex systems. The "theory" in this case could provide a further perspective on OO "semantics" in the sense that object-oriented principles can inform data models, "knowledge systems," and other technologies that are designed to have some explicit connection with real-world objects and phenomena. More direct applications of an expanded OO "theory" might also lie with tools such as code analyzers, software development tools, and compiler extensions, collectively serving to accelerate the implementation and quality of GUI front ends.

Front ends discussed here thus far have implicitly been focused on two-dimensional text and graphics, either viewers of 2D images or layouts presenting structured information, via text, printed numbers, or magnitude indicators. In contexts such as robotics, GUIs also often work with 3D graphics, videos, and other multimedia assets that have properties and present complications different in kind from the more 2D contexts and issue raised to this point. Partly so as to circle back to the theme of robot environments, these multimedia concerns will be the topic of the rest of this section.

24.3.2 3D Graphics and Robotics Front Ends

When displaying 3D graphics, applications need to support a further group of user actions because there are a larger set of possible view transformations, compared with 2D resources. Such graphics can be "panned" in three different directions (moving the virtual camera along **x**, **y**, and **z** axes) and could also be rotated (which is implemented as an effect akin to the virtual camera being spun along the outside of a sphere centered at a point along the camera's line of sight). Also, the scene may be zoomed in and out (preserving the camera's line of sight but moving its point of focus closer to or further from the camera). With a three-button mouse, pan and zoom in *two* dimensions are typically signaled by activating the central mouse wheel (for zoom) and pulling scroll bars (to pan, moving left/right or up/down) around the screen display. Sometimes panning can also be achieved by "dragging" inside the scene, repositioning the mouse while keeping its left-button down, with the cursor positioned inside the relevant view window (similar to drag gestures for dropping and resizing controls). When working in *three* dimensions, these kinds of drag actions are most commonly interpreted as rotations. Most 3D viewers support panning and zooming by recognizing "keyboard modifiers," wherein users hold down a key (often **shift**, **control**, or **alt**) while also moving the mouse.

In general, 3D displays are "responsive" in the context of users modifying the visible scene, but many of the interactive protocols that are commonplace in 2D structured front ends do not apply to 3D graphics (nor, for that matter, to 2D images such as photographs, rather than organized data visualizations). Some 3D environments, tentatively, have tried to get beyond these limitations. The premise here is that users may have an interest in getting information about a location or object which is one *part* of a 3D scene. With 2D structured displays, it is easy to allocate a small screen area specifically to represent its own data structure, where the application may respond by presenting users with a more complete review of the relevant information (perhaps in a separate window). For example, users clicking a "print" button will in most programs see a "print dialog" that contains multiple points of information about the current page and printer setup. In general, when a front end does not have enough space to show a complete data structure, a GUI control can show an icon or a short description, giving users a chance to click at that point for a more complete view of the associated information.

Similar effects are sometimes warranted in 3D. In a virtual "street" tour, for example, users will digitally "walk" past different streets and buildings, and platforms such as Google StreetView insert icons onto the scene allowing users to click for location-specific information—with a restaurant, for example, a page showing menus and reviews; or, with a transportation center, a link to transit routes and maps. Via similar methods, virtual tours *inside* buildings are sometimes annotated with icons that lead to information about individual rooms or objects, or else with short textual descriptions (potentially including hypertext links that can lead to more detailed views). Interactive icons in these contexts are sometimes called "tags" or "place-marks" and allow 3D scenes to have a select group of interactive features.

The above comments apply mostly to graphics rendered via 360° photography, which creates the illusion of three dimensions by piecing together many individual "panoramic" pictures (often taken from special-purpose cameras that capture images from all directions, rather than having a single view angle). Also, in the context of engineered 3D views, such as those rendered via mesh geometry, computer-aided design (CAD) programs will sometimes isolate specific object parts and surfaces and make these interactive. For example, a CAD file modeling an industrial part may permit users to select one piece of the object and examine information such as its dimensions or thickness—in the CAD setting, someone could work within the application to actually create and modify industrial designs, so selecting specific segments of objects may lead to operations that actually edit their properties within the overall design (e.g., making industrial subparts larger or smaller relative to their containing physical object).

Despite these familiar use cases for interactive 3D, there are multiple technical limitations that prevent 3D graphics from being responsive to the same degree and with the same granularity as structured 2D front ends. Computationally, the geometry of 3D scenes—most often derived from triangular mesh complexes (point clouds where triangles are formed by connecting nearby vertices) or 360° images—is fundamentally different than the layout managers powering 2D front ends. Layout managers ensure that GUI containers remain well-organized and usable even when a window is resized or scrolled. In general, GUI *controls* are not placed *within* 3D scenes the way that they are situated in GUI layouts. As one progresses through a 3D view, different parts and objects come into view or disappear—it is not clear how one would implement 3D analogs to layout managers, which would need to continuously update layouts while scenes are rotated, panned, and zoomed.

Apart from the mathematical computations involved in translating scene-transform to visible geometry (each rotation or camera movement alters all angles and causes different parts of the scene to be hidden, or unhidden)—which would apply to GUI controls immersed in 3D scenes no less than mesh triangles or other units of 3D geometry, and which are exponentially more complex than calculations related to 2D front ends—interacting with controls visible *inside* scenes could easily become cumbersome for users. For example, in order to click on controls, they would have to rotate or migrate around within the scene, and the graphics engine would have to determine how to represent icons, text, or areas for controls at different icons or zoom levels. Taking commercial tools such as Matterport Tours or Google StreetView as examples, immersing even simple "tag" icons in Virtual Tours yields inconsistent and uncomfortable GUIs—it is common in these environments for tags to appear and disappear for no apparent reason, and after having seen a tag (or a clickable hyperlink), it can be frustrating to maneuver the on-screen view so that the tag/link could be activated.

Such issues with 3D GUIs are directly applicable to robots' environments, because in most cases robots will be receiving visual inputs via mechanisms similar to panoramic photography. Human users wishing to remotely monitor and/or guide robots by watching video feeds based on robots' own inputs—adapting this paradigm in "Human-Robot Interaction" (HRI) [38, 43, 89, 95, 99]—will therefore be presented with displays similar to 360° tours. Unlike virtual tours, however, robots are not primarily designed simply to create visual feeds for human viewers (although that can indeed be one of their uses: consider robots deployed to gather real-time images from locations that would be dangerous to inspect in person [30, 57, 58]). Because robots typically process and structurally analyze their environments, inferences they make concerning object boundaries and affordances could be incorporated into graphics accompanying their visual feed. For example, robots may be programmed to categorize objects in their visual field and/or identify ones which they can move or manipulate. Within their video feed, then, certain objects will be singled out through computer vision, making it possible either for views of such objects to be visually altered/highlighted or for a separate front end display to be presented alongside the video feed proper—one which schematically diagrams the robot's environment as it is modeled through image analysis.

Perhaps the most effective paradigm for robotics GUIs, along these lines, would be a pairing between 3D capture of their raw video feeds and 2D structured windows presenting their environments schematically, with an emphasis on diagramming inferred relationships between objects rather than spatial accuracy. If a robot travels through a room with multiple objects that it could move, for example, a schematic GUI container might represent each of these physical objects with icons or screen areas arranged so that users could interact with the 2D diagram as a proxy for controlling the robot through its 3D video feed. For example, GUI controls for each movable object may be programmed with interactive functionality analogous to controls tied to application features—users might, for instance, click on or select context-menu items for specific objects as a way of guiding robots to move, inspect, or otherwise manipulate them. Here the combined system merging the robot with the user-facing front end would react to user actions similar to a desktop application, but instead of the reactions being restricted to digital operations localized in the computer itself (like moving a file from one folder to another), users could employ such a setup to mold robot's actions on external, physical things (e.g., moving a box from one room to another).

In such a scenario, humans would be relying on robots' visual processing to convert a raw video feed into a structured diagram (one that could be rendered in two dimensions, making interactive gestures easier). For this reason, there is a sense that humans would be interacting with robots by virtue of shared environment models: people would find it difficult to guide robots in this kind of system unless they analyzed their visual inputs similarly to how humans would if they were seeing the same objects around them. In short, robots' mapping of immersive visual inputs to schematic environment summaries would have to feel natural and intuitive for humans. This kind of phenomenological correlation may be much less vivid and intersubjective than joint attention between people themselves (or between a person and an animal, in the right circumstances), but it at least lays the foundation for something resembling human/robot communication—if one accepts the premise, which we emphasized in Chap. 23, that shared situational understanding is the origin point for language.

Note that in this kind of setup humans would be talking to robots indirectly, mediated by diagrams of robots' environments. For example, if a display shows (as a summarial view onto the robot's visual input) outlines representing boxes that could be moved, the human user would request such an action by virtually moving the object within the diagram, or by some analogous gesture (like selecting a "move" option from a context menu). The front end, in short, would be managed by a software component that translates user intentions as expressed through a GUI into commands that the robot understands. There would, in effect, be two different languages or sign systems linked together: the idioms of a GUI front end with which humans would interact, and then a distinct robot language engineered to convey instructions to robots at a tractable level, i.e., where there are deterministic algorithms to "compile" expressions in this language to low-level commands affecting robots' physical movements and orientation.

The idea of people "communicating with" robots probably calls to mind a more direct exchange where humans perhaps "talk directly" to them, their words parsed via natural language processing (this is the kind of scenario discussed elsewhere in this book, see Chapter 11; or [27, 31, 33, 56, 60, 90, 96], etc.). However, within the relevant NLP pipeline, there would presumably be a distinction between components that consume the spoken language directly and those that model specific robot instruction suitable for low-level responses. In short, even a medium of exchange which users perceive as communicating "directly" to a robot would most likely involve multiple stages, organized analogously to the GUI-based scenario where people interact with a GUI display in lieu of "talking" to the robot explicitly. Because of these functional parallels, designing systems based on GUIs is a strategy for fine-tuning robotics communication to develop intermediate algorithms that might then also be applicable to interfaces based on natural language. With a working pipeline to map GUI operations onto robot-action code, engineers could swap out the GUI front end for an NLP interface, perhaps preserving much of the lower-level protocols on the robot's side specifically. Working with GUI front ends first permits the two different dimensions to the latter system to be engineered independently—the NLP parsers would have their own set of requirements and complications (effective speech-to-text translation, for example) that could be worked through independently of how the robot's target languages (to which either NLP input or GUI actions would both have to be translated eventually) operate.

None of this discussion in the robotics context represents new theories or proposals—the techniques we have summarized here appear to be relatively mainstream approaches to robotic engineering. However, it is worth emphasizing that building GUIs to serve as intermediate communication devices may provide both practical benefits (in well-engineered robotics systems) and serve as one stage toward direct human/robot communication (for reasons just sketched out). While GUI front ends similar to those we have summarized are sometimes used for robot interactions, there is relatively little literature on implementing robots' GUIs as a distinct subject area within both software engineering and robotics (in the intersection of these two disciplines). This is a topic that deserves further research, partly because effective GUI programming is itself not given due attention as far as a systemic approach to GUI engineering, to make front ends more user-friendly and simultaneously to improve development tools. Many details of robotics-specific front ends, such as how to pair 3D graphics with 2D schematic diagrams, and in general how to make 3D displays more interactive, are themes applicable to other software domains apart from robotics proper.

In particular, it is an open question how best to associate parts of 3D views with independent GUI controls: should controls be *immersed* in 3D scenes, so that they are subject to the same variation in view angle, rotation, zoom, and other 3D effects along with all other facets of 3D geometry; or should they "float" above 3D views when their associate object/scene locations are visible; or should they be arranged diagrammatically in a separate window? Of course, these options could also be combined. To our knowledge, rigorous answers to these questions have yet to be determined, let alone implemented—nothing in current WebGL, OpenGL, or panoramic photography software appears to provide the level of interactive granularity that would be implied by direct associations between 3D scene locations and 2D-style GUI controls, whether these are visible within the scene (immersed or floating) or separated out to a separate display.

24.4 Conclusion

This chapter has considered the semantics of programming languages and GUI displays alongside semantic and syntactic theses related to natural language. Combining investigations into human language with computational environments produces an admittedly idiosyncratic style of exposition, cycling from a philosophical dialect (as much philosophy of language as linguistics proper) to the tone of a Stack Overflow discussion board (as much programmer's workshop as mathematically inspired computer science). Despite the fact that these two subject areas lend themselves to different discursive conventions, we contend that there are some structural parallels between human and computer languages, which such a juxtaposition may highlight. In particular, something like a "procedural semantics" can be theorized in both domains. Also, type theory offers a strategy to work through notions of preliminary comprehension: in both human and computer languages, a "procedural semantics" paradigm needs to explain how parse graphs (for sentences and computer code statements, respectively) can be isolated without the full feedback loop of grammar cross-referenced against semantic interpretation (which, for reasons sketched in Part 1, we believe is deferred from the language-processing stage to some extra-linguistic cognitive or computational faculty). One way to approach "partial" interpretation (grasping enough semantic detail to disambiguate parse structures) is via type systems, modeling provisional but only approximate semantic data (e.g., detailed enough to recognize parts of speech but not specific word meanings).

As a final comment: referring back to the discussion about disciplines straddling science and the humanities, such as cognitive science and linguistics, methodological integration may proceed from multiple directions. In linguistics, for instance, we can start with a mostly humanistic paradigm for studying communication and signification and then gradually incorporate formal or computational models. Alternatively, we can develop systems in computational or mathematical linguistics and then consider how well they perform or apply to actual human discourse. The "flow of ideas" could travel in multiple directions. Concepts emerging in a more speculative, philosophical vein—Davidsonian event semantics, Chomskyan tiers of grammar classification, Barwise/Perry situation theory, or Kripke frames for modal logic come to mind—propagate into computer science, artificial intelligence, or even pure mathematics. Crossing back the other way, Hindley–Milner type systems, Montague grammar, Complexity/self-organization theory, or convolutional neural networks have appealed to researchers as plausible models or explanations for different aspects of human language, thought, and society.

In the philosophy of science, it is commonplace to invoke an "explanatory gap" between what we might physically or biologically learn about human (and animal) minds and what we can actually experience, or how we talk about and represent experience through dialog and culture. Such a gap, perhaps, points to a metaphysical incompleteness on both sides of the science/humanities divide (to the degree that we can even talk as if knowledge has that neat cleavage). We know that human life and civilization belong to the physical world and therefore is explained by physical laws, but enigmas of (say) the neuro-anatomical origins of consciousness perpetuate a sense that our theories about society and culture, however compelling their structure and presentation, lack the kind of causative closure we expect of biology and other natural sciences. Conversely, theories in the physical sciences are models in search of explananda. Mathematical abstractions are necessarily idealizations, and we need human insight to connect the formality saturating the model with the empirical realities—be they at the scale of molecules, cells, bodies, continents, or galaxies—it seeks to explain.

When models are formulated as explanantia for *conscious* or *experiential* empiricity—when they are addressing thought, or language, that we actually witness ourselves enacting—the externality of model to ground is highlighted, because we recognize that such theories are not approximating something inert like a rock or sheet of metal, or something distant like seen in a microscope or telescope, but ourselves: our awareness, feelings, sense of community, lifeworld. It is reasonable that we should hold high standards for the accuracy of formal models entertaining ambitions to unmask the science of *this* reality. A natural language processor that is 90% accurate is interesting and useful, but that one-in-ten error still leaves a certain *frisson*, like seeing a very believable wax figure that is still not a person. Just like the wax figure is so far from being human, we get the uncanny sense that however close to just-right an NLP engine might be, there is something fundamental missing there. What is perhaps most thought provoking or disquieting is that the gap seems wider than a number such as 90% suggests: humanness cannot be quantified; an error rate of just one-in-ten, even if it got progressively better, would still not actually be advancing toward human quintessence. We are left wondering if this humanness that seems so real in our experience cannot be measured at all, and as such stands in the context of science as essentially a metaphysical fiction.

And yet, perhaps the explanatory gap is actually just the unavoidable separation between theories and phenomena: models do not contain the reality they simulate. It would seem reasonable but also unremarkable to say that models of galaxy formation do not fully capture galaxyness, or theories of biological systems have some ontological separation from the notion of life itself. Perhaps quintessential humanness is no more a speculative phantom than galaxyness or planetness or metalness or organic-compound-ness.

Ultimately, formal systems purporting to describe thought and language—be their origins in mathematics, computers, artificial intelligence, or physical science—should be taken as approximating idealizations, like the Newtonian construction of our solar system. Natural science tests theories against empirical data, often collected in laboratories or specially constructed observational equipment, like the Grand Canarias Telescope or CERN's Large Hadron Collider. But when the relevant explananda belong to human reality—thought, language, problem solving, motivations—at some level, the empirical data that would advance or disconfirm theories belong to our own experience, and so *we* are the lab tests, the sites where theory is compared to reality. Our *sense* that models of language (for example) comport with the dialects we speak serves as one vein of evidence. So, much of linguistic reasoning is driven by the "sense" of sentences or expressions being acceptable or not. This is not an empirical or analytic result; in philosophical terms, it would be considered speculative, rooted in internal judgment. We might generalize this to assessments about theories of reasoning originating in artificial intelligence, for example. Speculative thought thereby has an evidentiary purpose in the ecosystem of formal models of thought and language, by analogy to empirical phenomenology in the natural sciences.

So, epigrammatically, formalizations targeted at language and reasoning seek completion in several directions, among them "empirical" grounding that in this case actually takes the form of human intuition and introspection, which to the degree that it might be systematized would be a matter of philosophy rather than science per se. And at the same time—again, the bi-directional flow of ideas—philosophy produces speculative frameworks that we can collectively dispute (disputation not in the sense of disagreement, necessarily, but a thought-community testing theories' coherence) but for which, for a further dimension of evidentiary status, we might seek scientific backing; to the effect that a theory of language, say, appears formally well-motivated, or a theory of human behavior is supported by neuroscience. Even if we argue that NLP or AI engines are not physically relevant simulations of human (or animal) minds, couching linguistic models in formal presentations that may be investigated in light of how NLP systems process language can be one way to give them a scientific substratum: theories that jive with generalized models of "language processing" that might be realized in multiple *physical* systems—computers, brains, robots—gain traction within the reasonable paradigm that language is a system with its own rules, capable of being isolated to some level of abstraction, independent from the electrical or biological machinery through which it is implemented.

However they may spell it out, philosophers have probably sensed the mutual co-extensibility of humanities and natural science that we have attempted to capture via "bi-directionality": our experience tests science when its investigations turn to realms that overlap with human reality, and science tests humanistic theories when we seek causative (and not just speculative) assent. In some cases, we may appeal to biology or neuroscience to explain human behavior—evolutionary forces shaping the emergence of language, say, or the neurochemistry of our unconscious judgments and responses—but more prevalent in philosophy and related disciplines is to turn, not so much to physical sciences, but to formal logicomathematical systems (which bring the same kind of explanatory rigor as mathematical models inject into natural phenomena, reconstructed quantitatively). The history of "analytic" philosophy is in part a sequence of borrowings stepping through generations alongside the evolution of twentieth century logic and mathematics: first-order symbolic logic, intuitionist logic, models of computation at the dawn of the digital age, modal logic, category theory, type theory (and homotopy type theory and the Curry–Howard isomorphism), formal proof assistants, symbolic AI, connectionist AI, neural networks, machine learning, and Artificial General Intelligence.

The linguists and philosophers working within the humanities-oriented bases of their disciplines have largely looked toward logic and mathematics to find that scientific scaffolding, that supplement to the closed loop of speculation, which scholars in other social sciences might find in psychological studies or socioeconomic data. These humanistic fields at some level thereby coexist with computational models of (e.g.) NLP, software language engineering, and computer vision, insofar as both appeal to logicomathematical frameworks as supporting structures, but with some indirection. In the humanities, logical systems are only approximate guides because thought and language have nuances and embodiment that seem outside the scope of formal abstraction. On the other hand, computer systems stand apart from logicomathematical foundations because they are not abstract theories; software has to be *implemented*, with trial and error and refinement via continuous feedback, leveraging design patterns that get formulated often with the wisdom of concrete programming experience rather than formal theories. In short, logical models can reveal substrata or optimizations that get incorporated into a computational system, but the final form such systems take on often emerges through the process of building them through computer code itself, its important structures emerging from that pragmatic context rather than the original mathematical abstractions.

It is interesting that, despite their being co-tethered to a shared set of logico-mathematical canons, philosophers have not often appealed to computer science or programming directly as a source for formalization conventionally sought, in the analytic tradition, within logic and mathematics. In other words, analytic philosophers turn to branches of mathematics or symbolic logic to model phenomena such as syntax and semantics, or classification, deductive reasoning, and so forth—but they have not similarly attended to the experience of working software programs; or sought to integrate philosophy with software platforms, to anywhere near the same level. Arguably, approaching logicomathematical formalisms via their manifestation in the real-world pragmatic contexts of software implementation—rather than the abstract presentation of such formalisms as, in effect, a kind of mathematics—would be more consistent with their proper status vis-à-vis philosophy and linguistics, as idealizations that have to be reconnected with the nuances and practicalities of human experience.

Along similar lines, it is interesting that we evaluate computational models of (e.g.) language and image processing mostly in terms of their payoffs in building accurate technology—even if part of our interest in such systems lies with how they reveal some of the substrata underlying human reasoning. We are interested in how NLP or computer vision engines work because they may explain some of what goes on in the brain when we understand language or experience visual perception. A derivation of sentence parse graphs via link grammar connector-mating/expectation edge-annotations, or a superpixel segmentation via morphological operators and watershed thresholds, is of interest partly because we can extract data from the parsed sentence or the segmented image; but partly also because the workings of such technology may help reveal how our own minds process syntax and semantics (for language) or color and geometry (for vision). But *if* we are interested in computational models that *could potentially* replicate some of what goes on mentally, there may be explanatory merit to such computational simulations that transcends their practical accuracy.

In this chapter, and also Chap. 22, we have outlined potential computational approaches to language and vision that are motivated by appeal to, potentially, one's own experience: the idea of "thematic" colors and how color experience is invested not with one single hue but rather an expectation of how coloration varies across textured and variably illuminated (and occluded) surfaces; and, in language, by the idea of provisional type-theoretic semantic coherence and parse graph evaluation in the context of merely provisional anticipations of sentence meaning, consistent with a "procedural" semantics or "Interface Theory" according to which meaning itself tends to emerge from *extra-linguistic* cognitions. In the reigning paradigm, we take these kinds of experiential intuitions and, to the degree that we are working in a computational environment, try to encode them in working software, which may then yield slightly better NLP or computer vision engines. So, for example, a philosophical understanding of neural connectivity helped inspire connectionist models that revolutionized AI (and displaced "good-old-fashioned" symbol-processing paradigms) at the start of the twenty-first century. But the inspiration from philosophy was important because the *new* AI was (in this case *much*) better, more accurate by orders of magnitude, with potentially society-changing consequences.

However, we would argue that whether computational models "work" when emulating human *reasoning* is a different question than how well they shed light on *how* we reason. We might say that a technologically successful approach invites comparisons to the human mind: if convolutional neural networks achieve excellent results in human-like reasoning tasks, does this not count as some evidence that our own minds work in similar ways? But, on the other hand, the cellular realization of consciousness and cultural embodiment of the human person make us very different from *in silico* software. There is no guarantee that computational systems optimized for AI would be equally successful in the brain and vice versa. Suppose we have formal approaches to sentence parsing or image segmentation that are compelling because they appeal to our intuitions about how we *experience* vision and language but are not necessarily better as technological artifacts than other paradigms (maybe ones that seem artificial, such as "big data" statistics). Should we conclude that the more life-like models are inaccurate? Or does this reveal perhaps that humans process language and visual perception in ways that would be difficult for computers to emulate?

We believe these are interesting questions, and in an ideal world, if reflections such as these (wherever they originate) provide food for thought, they could help reorient our perception of the proper relationship between technology and the humanities. Perhaps disciplines like philosophy and linguistics should seek integration with technological platforms such as those for language processing and computer vision, partly so as to explore how computer systems can emulate facets of human reason and perception—perhaps reciprocally explaining some of their causal origins—but keeping an open mind with respect to the explanatory value of computational simulations. The AI modules giving the best results are not necessarily the most life-like, or the most valuable for revealing how the mind actually works, because perhaps our brains are optimized differently than computers. Once we decouple our appreciation for successful AI as a useful tool from computational models of actual human reasoning, we lose an orientation that could oversimplify our construal of the role of technology in human sciences. We cannot just let AI paradigms compete from a distance, waiting for the preeminent solutions to appear and regard them as prototypes for simulated intelligence. We need to be more actively engaged. Different computational platforms will bring different sorts of merit to hybrid methodologies: some may be valuable because they embrace hypotheses that have technological payoffs, which is a data point that such models are *successful* emulators of human reason. Other models may be valuable because they appear to formalize processes that we might believe on speculative grounds to be consistent with how the mind actually works.

Different approaches to (say) NLP and computer vision could offer benefits in either direction, and it may not be obvious which is which. These issues have to be worked out on a case-by-case basis. Projects in these computational disciplines can overlap with humanistic concerns either because of behavioral correctness or because of intuitive lifelikeness, but the proportion of each virtue may be uncertain as the project is ongoing. Given that uncertainty, the methodological cross-pollination between humanities and computational science should be actively pursued in the context of individual projects. This could take multiple dimensions: being more proactive in incorporating technological resources (such as datasets and/or code libraries) into human-science publications; scientists on the computational end being more willing to embrace the priorities and intellectual commitments of the humanities as one dialect within their writing and their bird's-eye view assessments of what technology has (or has not) accomplished, and vice versa; a denser, more widely dispersed paradigm of interdisciplinary collaboration at the borderlands of natural (or "formal") and human sciences.

If we accept that the notion of a rigid bifurcation between human and natural sciences is outdated—that many intellectual disciplines straddle such an imaginary line—then a natural correlation is that the "two cultures" [83] no longer serve as an organizing motif. Under the sway of a presumptive physical/social divide, it seems like we instinctively line up disciplines as if on a spectrum from one end to another—not only assigning some forms of research to one side or the other, but implicitly ordering humanistic fields according to their imaged degree of similarity with physical science, and vice versa. Thus anthropology, political science, or sociology, which work with empirical data, are more science-like than philosophy or literary theory; and psychology—which works with empirical data obtained *empirically*, according to conventionalized protocols—is more scientific still. Meanwhile, across the divide, fields such as zoology and geology often resemble the more empiricist social sciences, based on observation and on approximate models (such as population dynamics) rather than the heavy mathematical machinery of cosmology or physics.

A corollary to this instinctive intellectual roadmap is that disciplines such as philosophy—and those humanities oriented to culture, politics, or ethics—would find less relevance in empirical, mathematical, or computational models than, say, economics, sociology, or psychology. Yet such impressions are not borne out by actual scholarship. In digital humanities, for example, or projects such as "cognitive Phenomenology," there is a profound and intense engagement with technology and with computational resources, even if humanities practitioners are more likely to be skeptical of facile AI boosterism or naively optimistic techno-ideology.

If the line between hard-core humanities and hard-core physical science is an arc that bends rather than a diagonal marked with disciplinary notches, we should not be surprised if computational modes of investigation can find synergy with the more theoretically inclined corners of human science, those at home with abstract argumentation and its dialectics more than, say, anthropology in the field or psychology in the lab. Even if just as a matter of intellectual style: empirical data gives researchers a grounding that can anchor raw speculation, but it also dials down the rigor involved when nuances of arguments' linguistic exposition or conceptual hair splitting form the entire weight of theoretical dialog and progress, as they do in most manifestations of "Continental" philosophy (phenomenology, structuralism and its post-, critical theory, hermeneutics). Such abstract rigor is not so far removed from creating software, which demands the programmer to become immersed in opaque worlds with their own special word-meanings and conceptual tics.

To take just one example: whether or not computer vision modules work *similarly* to human perception or not—and articulating *differences* may be just as revealing as finding correlations—it is valuable to explain vision from both sides and compare the results. There is no reason that research rooted in Husserl, for example (see Jean Petitot for a case in point), could not spearhead this comparison no less effectively than, say, experimental psychology. But trying to understand computer vision is not only a matter of learning some mathematical theory; it involves tangible, feedback-driven, process-oriented experience and experimentation with code and software. Precisely because software is imperfect, because it can create useful or realistic tools only with great mental effort—experientially, something very different from the ambient objectivity of (psychological) experiments or (socio-)economic "data"—arguably the programming milieu serves as a better formalizing corollary to "hard core" humanities than logic or mathematics, with (when all goes well) their lack of *frisson* between models and realizations, their lack of frustration that what seems straightforwardly truthful cannot be easily engineered, which should in turn deconstruct presumptions that logical models or empirical data have unproblematic objective status (see: "crisis of replication").

In sum, we can hope that the project-oriented, trial-and-error pragmatism of computational engineering—the ecosystem of code libraries to be maintained, applications to be architected, procedures to be implemented (because other procedures depending on them are to be implemented thereafter), online communities where developers work their way through work-in-process complications—becomes (more a) part of the arsenal of humanities theory, even an inspiration; but also that the humanities continue to press the importance of justice and values as a reminder to technologists. As a reminder of why technology matters.

References

1. Martín Abadi and Luca Cardelli, "A Semantics of Object Types". Proceedings of the IEEE Symposium on Logic in Computer Science, Paris, 1994. http://lucacardelli.name/Papers/PrimObjSemLICS.A4.pdf
2. _____ "A Theory of Primitive Objects: Untyped and First-Order Systems". http://lucacardelli.name/Papers/PrimObj1stOrder.pdf
3. Nicholas Asher and Zhaohui Luo, "Formalization of Coercions in Lexical Semantics". https://semanticsarchive.net/sub2012/AsherZhaoHui.pdf
4. Ash Asudeh and Gianluca Giorgolo, *Enriched Meanings: Natural Language Semantics with Category Theory*. Oxford, 2020.
5. Jerry T. Ball, "Double R Grammar: The Grammatical Encoding of Referential and Relational Meaning in English". https://citeseerx.ist.psu.edu/document?repid=rep1&type=pdf&doi=34c258564f72b6dc6455aeaac6e6be1a4a07d262
6. Chris Barker and Chung-chieh Shan, *Continuations and Natural Language*. Oxford, 2014.
7. Lawrence W. Barsalou, "Grounded Cognition". http://matt.colorado.edu/teaching/highcog/spr10/readings/b8.pdf

8. Hannah Bast and Claudius Korzen, "A Benchmark and Evaluation for Text Extraction from PDF". https://ad-publications.cs.uni-freiburg.de/benchmark.pdf
9. Marcus Berger, "The Internal Structure of Nouns, Nominals, and Gerunds: An Analysis of Their Relationship and Representation". https://journals.flvc.org/floridalinguisticspapers/article/view/93063/89240
10. Hans C. Boas, "Cognitive Construction Grammar". https://sites.la.utexas.edu/hcb/files/2011/02/CCxG-October22-2010.pdf
11. Greg Carlson, "Thematic Roles and the Individuation of Events". https://www.sas.rochester.edu/lin/people/faculty/carlson_greg/assets/pdf/them-roles-events.pdf
12. Lucas Champollion, "Covert Distributivity in Algebraic Event Semantics". *Semantics & Pragmatics*, Volume 9 (2016), pages 1–65. https://semprag.org/article/view/sp.9.15
13. _____, "Parts of a Whole: Distributivity as a Bridge Between Aspect and Measurement". Dissertation, University of Pennsylvania, 2010. https://repository.upenn.edu/cgi/viewcontent.cgi?article=2117&context=edissertations
14. Nancy Chih-lin Chang, "Constructing grammar: A computational model of the emergence of early constructions". Dissertation, University of California, Berkeley, 2008. https://digitalassets.lib.berkeley.edu/techreports/ucb/text/EECS-2009-24.pdf
15. Robin Cooper, "Perception, Types and Frames". https://link.springer.com/chapter/10.1007/978-3-030-50200-3_8
16. Bridget Copley and Heidi Harley, "A Force-Theoretic Framework for Event Structure". http://heidiharley.com/heidiharley/wp-content/uploads/2016/09/CopleyAndHarley2015Published.pdf
17. Astrid De Wit and Frank Brisard, "A Cognitive Grammar Account of the Semantics of the English Present Progressive". https://www.cambridge.org/core/journals/journal-of-linguistics/article/abs/cognitive-grammar-account-of-the-semantics-of-the-english-present-progressive1/DBF16D3E3A05293E90CDDE9C16BA5D34
18. Matt Duckham, "Object Calculus and the Object-Oriented Analysis and Design of an Error-Sensitive GIS". http://www.geosensor.net/papers/oo.pdf
19. Marie Duží and Michal Fait, "Integrating Special Rules Rooted in Natural Language Semantics into the System of Natural Deduction". https://www.scitepress.org/Papers/2020/93696/93696.pdf
20. Jason Eisner and Noah A. Smith, "Favor Short Dependencies: Parsing with Soft and Hard Constraints on Dependency Length". https://www.cs.jhu.edu/~jason/papers/eisner+smith.iwptbook10.pdf
21. Jean Mark Gawron, "Circumstances and Perspective: The Logic of Argument Structure". https://escholarship.org/content/qt7sd5987t/qt7sd5987t_noSplash_14d78590a8657a71ed9ad46dc6e69064.pdf
22. _____, "Paths and the Language of Change". https://gawron.sdsu.edu/paths_change.pdf
23. Chris Genovesi, "Metaphor and What is Meant: Metaphorical content, what is said, and contextualism". https://www.tcd.ie/slscs/assets/documents/Clinical-speech/genovesi_2019_metaphor.pdf
24. Daniel Gildea and David Temperley, "Optimizing Grammars for Minimum Dependency Length". https://www.cs.rochester.edu/~gildea/pubs/gildea-temperley-acl07.pdf
25. Carlos Gómez-Rodríguez and Ramon Ferrer-i-Cancho "The Scarcity of Crossing Dependencies: a direct outcome of a specific constraint?" https://arxiv.org/pdf/1601.03210.pdf
26. Andrew D. Gordon, *et. al.*, "Compilation and Equivalence of Imperative Objects". https://www.microsoft.com/en-us/research/wp-content/uploads/1997/01/maya.pdf
27. Sergio Guadarrama, *et. al.*, "Grounding Spatial Relations for Human-Robot Interaction". https://citeseerx.ist.psu.edu/document?repid=rep1&type=pdf&doi=e64fd4eb33199fc2cc731aec1c25d4e88a4d4f40
28. Robert F. Hadley, "A Default-Oriented Theory of Procedural Semantics". https://onlinelibrary.wiley.com/doi/pdf/10.1207/s15516709cog1301_4
29. Daniel W. Harris, "Speech Act Theoretic Semantics". Dissertation, CUNY, 2014. https://academicworks.cuny.edu/cgi/viewcontent.cgi?article=1221&context=gc_etds

30. Manjula Udayanga Hemapala, "Robots for Humanitarian Demining". https://cdn.intechopen.com/pdfs/56656.pdf
31. Padraig Higgins, *et. al.*, "Towards Making Virtual Human-Robot Interaction a Reality". https://iral.cs.umbc.edu/Pubs/VAMHRI2021-Higgins.pdf
32. Darrin Louis Hindsill, "Its a Process *and* an Event: Perspectives in event semantics". Dissertation, University of Amsterdam, 2007. https://eprints.illc.uva.nl/id/eprint/2058/1/DS-2007-03.text.pdf
33. Seng-Beng Ho and Zhaoxia Wang, "Language and Robotics: Complex Sentence Understanding". https://ink.library.smu.edu.sg/cgi/viewcontent.cgi?article=6486&context=sis_research
34. Kenneth Holmqvist, "Conceptual Engineering: Implementing cognitive semantics". In Jens Allwood and Peter Gärdenfors, eds., *Cognitive Semantics*, pp 153–171, Amsterdam, Philadelphia: John Benjamins, 1999.
35. Kenneth Holmqvist, "Implementing Cognitive Semantics: Image schemata, valence accommodation, and valence suggestion for AI and computational linguistics". Dissertation, Lund University, 1993. https://www.degruyter.com/database/COGBIB/entry/cogbib.5625/html?lang=en
36. Juneki Hong and Jason Eisner, "Deriving Multi-Headed Planar Dependency Parses from Link Grammar Parses". https://www.cs.jhu.edu/~jason/papers/hong+eisner.tlt14.paper.pdf
37. Laurence R. Horn and Istvan Keckes Peirce, "Pragmatics, Discourse, and Cognition". http://www.albany.edu/faculty/ikecskes/files/Horn%20and%20Kecskes%2065-45-R1026-Horn.pdf
38. Yue Hu, *et. al.*, "Towards Active Physical Human-Robot Interaction: quantifying the human state during interactions". https://hal.archives-ouvertes.fr/hal-03567276/document
39. Radha Jagadeesan, *et. al.*, "Monotonic Typestates for the Object Calculus". https://fpl.cs.depaul.edu/jriely/papers/2009-monotone.pdf
40. Kyle Johnson, "What Makes a Theta-role". https://www.qmul.ac.uk/sllf/media/sllf-new/department-of-linguistics/hagit-borer-celebration/Johnson.pdf
41. Mark Johnson, "The Body in the Mind: The Bodily Basis of Meaning, Imagination, and Reason". University of Chicago Press, 1990
42. Werner Kallmeyer, "Verbal Practices of Perspective Grounding". https://d-nb.info/120819738X/34
43. Andrey Kurenkov, *et. al.*, "An Evaluation of GUI and Kinesthetic Teaching Methods for Constrained-Keyframe Skills". https://www.sim.ece.utexas.edu/static/papers/kurenkov_iros2015.pdf
44. George Lakoff and Mark Johnson, *Philosophy in the Flesh: the Embodied Mind and its Challenge to Western Thought*. New York: Basic Books, 1999.
45. _____, *Metaphors We Live By*. University of Chicago, 1980.
46. Ronald Langacker, *Cognitive Grammar: A Basic Introduction*. Oxford, 2008.
47. _____, "Cognitive (Construction) Grammar". https://www.degruyter.com/document/doi/10.1515/COGL.2009.010/html?lang=en
48. _____, "Constructions in Cognitive Grammar". https://www.jstage.jst.go.jp/article/elsj1984/20/1/20_1_41/_pdf/-char/en
49. _____, *Grammar and Conceptualization*. de Gruyter, 1999.
50. _____, "How to Build an English Clause". https://oaji.net/articles/2016/3124-1458506162.pdf
51. _____, "Evidentiality in Cognitive Grammar". https://benjamins.com/catalog/pbns.271.02lan
52. _____, "The English Present: Temporal coincidence vs. epistemic immediacy". https://benjamins.com/catalog/hcp.29.06lan
53. Gianluca E. Lebani and Alessandro Lenci, "A Distributional Model of Verb-Specific Semantic Roles Inferences". https://colinglab.humnet.unipi.it/wp-content/uploads/2012/12/Lebani-Lenci-2018.pdf
54. Joseph Lehmann, *et. al.*, "Age-Related Hearing Loss, Speech Understanding and Cognitive Technologies". https://link.springer.com/article/10.1007/s10772-021-09817-z

55. Sander Lestrade, "The Space of Case". https://repository.ubn.ru.nl/bitstream/handle/2066/82611/82611.pdf
56. Rui Liu and Xiaoli Zhang, "A Review of Methodologies for Natural-Language-Facilitated Human-Robot Cooperation". https://arxiv.org/pdf/1701.08756.pdf
57. Giacomo Lunghi, *et. al.*, "An Advanced, Adaptive and Multimodal Graphical User Interface for Human-robot Teleoperation in Radioactive Scenarios". https://www.scitepress.org/Papers/2016/59714/59714.pdf
58. Giacomo Lunghi, *et. al.*, "Multimodal Human-Robot Interface for Accessible Remote Robotic Interventions in Hazardous Environments". https://cds.cern.ch/record/2699858/files/untitled.pdf
59. Claudia Maienborn, "Event Semantics". https://www.researchgate.net/publication/236898346_Event_semantics
60. Cynthia Matuszek, *et. al.*, "Learning to Parse Natural Language Commands to a Robot Control System". https://homes.cs.washington.edu/~lsz/papers/mhzf-iser12.pdf
61. Ryan McDonald, *et. al.*, "Non-projective Dependency Parsing using Spanning Tree Algorithms". https://ryanmcd.github.io/papers/nonprojectiveHLT-EMNLP2005.pdf
62. Chukwualuka Michael Uyanne, *et. al.*, "Ecolinguistic Perspective: Dialectics of Language and Environment". https://ezenwaohaetorc.org/journals/index.php/ajells/article/viewFile/5-1-2014-010/61
63. Melanie Mitchell, *Artificial Intelligence: A Guide for Thinking Humans*. Macmillan, 2019.
64. Erwan Moreau, "From Link Grammars to Categorial Grammars". https://hal.archives-ouvertes.fr/hal-00487053/document
65. Stefan Müller and Stephen Wechsler, "Lexical Approaches to Argument Structure". https://hpsg.hu-berlin.de/~stefan/Pub/arg-st.pdf
66. Kobeen Raouf Mustafa and Azad Hasan Fatah, "Nominal Grounding Elements in English: A Domain-based Account". https://www.researchgate.net/publication/326960775_Nominal_Grounding_Elements_in_English_A_Domain-based_Account
67. Amy Neustein and Nathaniel Christen, *Innovative Data Integration and Conceptual Space Modeling for COVID, Cancer, and Cardiac Care*. Elsevier, 2022.
68. Ana Cristina Ramada Paiva, "Automated Specification-Based Testing of Graphical User Interfaces". Dissertation, University of Porto, 2006. https://web.fe.up.pt/~apaiva/PhD/PhDGUITesting.pdf
69. Péter Pelyvás, "Epistemic Modality: A choice between alternative cognitive models". https://litere.uvt.ro/publicatii/BAS/pdf/no/bas_2011_articles/22%20247-259.pdf
70. _____, "On Epistemic and Deontic Grounding". https://argumentum.unideb.hu/2019-anyagok/special_issue_I/pelyvasp.pdf
71. _____, "The Development of the Grounding Predication: Epistemic Modals and Cognitive Predicates". https://brill.com/display/book/9780585474267/B9780585474267_s010.xml
72. Jean Petitot, "Landmarks for Neurogeometry". http://jeanpetitot.com/ArticlesPDF/Petitot_Landmarks.pdf
73. _____, "Morphological Eidetics for a Phenomenology of Vision". http://topologicalmedialab.net/xinwei/classes/readings/Petitot/MorphEideticsPhenPerc.pdf
74. _____, "The Morphodynamical Turn of Cognitive Linguistics". https://journals.openedition.org/signata/549
75. Jean Petitot and René Doursat, *Cognitive Morphodynamics: Dynamical Morphological Models of Constituency in Perception and Syntax*. Peter Lang, 2011
76. Steven Pinker, *The Language Instinct*. William Morrow, 1994.
77. _____, *The Stuff of Thought: Language as a window into human nature*. Viking, 2007.
78. Emily Pitler, *Models for Improved Tractability and Accuracy in Dependency Parsing*. Dissertation, University of Pennsylvania, 2013. https://repository.upenn.edu/cgi/viewcontent.cgi?article=2070&context=edissertations.
79. James Pustejovsky, "The Syntax of Event Structure". https://www.cs.rochester.edu/u/james/Papers/Pustejovsky-event-structure.pdf.

80. Gerold Schneider, "A Linguistic Comparison of Constituency, Dependency and Link Grammar". Diploma, Zurich University, 2008. https://files.ifi.uzh.ch/cl/gschneid/papers/FINALSgeroldschneider-latl.pdf
81. Jae-Young Shim and Samuel David Epstein, "Two Notes on Possible Approaches to the Unification of Theta Relations". https://jaeyoungshim.weebly.com/uploads/1/2/3/1/12312623/linguistic_analysis_two_notes_jae-young_shim.pdf
82. Daniel D. Sleator and Davy Tamperley, "Parsing English with a Link Grammar". https://www.link.cs.cmu.edu/link/ftp-site/link-grammar/LG-IWPT93.pdf
83. C.P. Snow, "The Two Cultures and The Scientific Revolution". Cambridge University Press, 1961. https://sciencepolicy.colorado.edu/students/envs_5110/snow_1959.pdf.
84. Michael Spranger, "Incremental Grounded Language Learning in Robot-Robot Interactions". https://ieeexplore.ieee.org/document/7346140
85. _____, "Procedural Semantics for Autonomous Robots – A Case Study in Locative Spatial Language". https://www.researchgate.net/publication/280830547_Procedural_Semantics_for_Autonomous_Robots_-_A_Case_Study_in_Locative_Spatial_Language
86. _____, "The Evolution of Grounded Spatial Language". https://langsci-press.org/catalog/book/53
87. Michael Spranger, *et. al,*, "Open-ended Procedural Semantics". https://martin-loetzsch.de/publications/spranger12openended.pdf
88. Sune Vork Steffensen and Alwin Fill, "Ecolinguistics: The state of the art and future horizons". https://findresearcher.sdu.dk/ws/files/87218550/Steffensen_2014_Language_Sciences.pdf
89. Daniel Szafir and Danielle Albers Szafir, "Connecting Human-Robot Interaction and Data Visualization". https://cu-visualab.org/papers/vis_hri_final_acc.pdf.
90. Stefanie Tellex, *et. al.*, "Robots That Use Language: A Survey". https://h2r.cs.brown.edu/wp-content/uploads/tellex20.pdf
91. Orlin Vakarelov, "Pre-cognitive Semantic Information". https://link.springer.com/article/10.1007/s12130-010-9109-5
92. _____, "The Cognitive Agent: Overcoming informational limits". https://philarchive.org/archive/VAKTCAv1
93. Linas Vepstas and Ben Goertzel, "Learning Language from a Large (Unannotated) Corpus". https://arxiv.org/pdf/1401.3372.pdf
94. Lucy Lu Wang, *et al.*, "CORD-19: The Covid-19 Open Research Dataset". *ArXiv*, 2020. https://www.ncbi.nlm.nih.gov/pmc/articles/PMC7251955
95. Alexander Wilkinson, *et al.*, "Design Guidelines for Human-Robot interaction with Assistive Robot Manipulation Systems". https://www.degruyter.com/document/doi/10.1515/pjbr-2021-0023/html?lang=en
96. Peratham Wiriyathammabhum, "Computer Vision and Natural Language Processing: Recent Approaches in Multimedia and Robotics". https://www-hlb.cs.umd.edu/sites/default/files/scholarly_papers/PerathamW.pdf
97. Himanshu Yadav, *et. al.*, "Are Formal Restrictions on Crossing Dependencies Epiphenomenal?" http://socsci.uci.edu/~rfutrell/papers/yadav2019formal.pdf
98. Poonam Yadav and John Darlington, "Conceptual Frameworks for Building Online Citizen Science Projects". https://hcjournal.org/index.php/jhc/article/view/72
99. Holly A. Yanco and Jill Drury, "Classifying Human-Robot Interaction: An Updated Taxonomy". https://www.sci.brooklyn.cuny.edu/~sklar/teaching/f06/air/papers/yanco-drury-taxsmc04.pdf
100. Jordan Zlatev, "Embodiment, Language, and Mimesis". https://lucris.lub.lu.se/ws/portalfiles/portal/4514085/1044802.pdf
101. _____, "Phenomenology and Cognitive Linguistics". https://citeseerx.ist.psu.edu/document?repid=rep1&type=pdf&doi=ebc99d9442497d36be7477f393dcc38d2d53ab5b

Index

A. Neustein et al. (eds.), *AI, IoT, Big Data and Cloud Computing for Industry 4.0*, Signals and Communication Technology, https://doi.org/10.1007/978-3-031-29713-7

MIX
Papier aus verantwortungsvollen Quellen
Paper from responsible sources
FSC® C105338

If you have any concerns about our products,
you can contact us on
ProductSafety@springernature.com

In case Publisher is established outside the EU,
the EU authorized representative is:
Springer Nature Customer Service Center GmbH
Europaplatz 3, 69115 Heidelberg, Germany

Printed by Libri Plureos GmbH
in Hamburg, Germany